QUICK REFERENCE LIST: SOCIETAL TOPICS

QUICK REFERENCE LIST: PHILOSOPHICAL TOPICS

QUICK REFERENCE LIST: GREAT SCIENTISTS

Physics

PHYSICS
Concepts
and Connections

Second Edition

ART HOBSON

University of Arkansas

Prentice Hall, Upper Saddle River, New Jersey 07458

Library of Congress Cataloging-in-Publication Data

HOBSON, ART.

    Physics : concepts and connections / Art Hobson. — 2nd ed.
      p.   cm.

    Includes bibliographical references and index.
    ISBN 0-13-095381-4 (pbk.)
    1. Physics. I. Title.
QC23.H739   1998
530—dc21   98-14593
                                  CIP

Executive Editor: *Alison Reeves*
Production Editor: *Debra A. Wechsler*
Editor in Chief: *Paul F. Corey*
Editorial Director: *Tim Bozik*
Assistant Vice President of Production & Manufacturing: *David W. Riccardi*
Executive Managing Editor: *Kathleen Schiaparelli*
Assistant Managing Editor: *Lisa Kinne*
Director of Marketing: *Kelly MacDonald*
Marketing Manager: *Jennifer Welchans*
Director of Creative Services: *Paula Maylahn*
Associate Creative Director: *Amy Rosen*
Assistant to Art Director: *John Christiana*
Art Manager: *Gus Vibal*
Interior/Cover Designer: *Rosemarie Votta*
Cover Art: Hubble deep field by Robert Williams and the Hubble Deep Field Team (STScI)/NASA; *The Thinker* by Auguste
   Rodin/Explorer, Paris/Superstuck
Cover Research: *Karen Sanatar*
Copy Editor: *Jane Loftus*
Manufacturing Manager: *Trudy Pisciotti*
Assistant Editor: *Wendy Rivers*
Photo Research: *Beth Boyd*
Photo Editors: *Lorinda Morris-Nantz/Melinda Reo*
Illustrator: *Charles Pelletreau*

 © 1999, 1995 by Prentice-Hall, Inc.
Simon & Schuster/A Viacom Company
Upper Saddle River, New Jersey 07458

PRINTED IN THE UNITED STATES OF AMERICA

10 9 8 7 6 5 4 3 2 1

ISBN 0-13-095381-4

Prentice-Hall International (UK) Limited, *London*
Prentice-Hall of Australia Pty. Limited, *Sydney*
Prentice-Hall Canada Inc., *Toronto*
Prentice-Hall Hispanoamericana, S.A., *Mexico*
Prentice-Hall of India Private Limited, *New Delhi*
Prentice-Hall of Japan, Inc., *Tokyo*
Simon & Schuster Asia Pte. Ltd., *Singapore*
Editora Prentice-Hall do Brasil, Ltda., *Rio de Janeiro*

For my daughter Ziva

Contents

*Chapters and sections marked by asterisks can be skipped without destroying the continuity of the
remaining material.

CONTENTS   xi

CHAPTER 16

Fusion and Fission: —And A New Energy   416

16.1   Fusion: the fire in the sun   417
16.2   The nuclear energy curve   420
16.3   Creation of the universe and the atoms: we are star dust   422
16.4   *The discovery of fission: passage to a new age   426
16.5   The chain reaction: unlocking the strong force   429
16.6   *The Manhattan Project and fission weapons   432
16.7   *Fusion weapons: star fire on Earth   438
Summary, review, project, discussion, exercises, problems   441

CHAPTER 17

*The Energy Future   445

17.1   A brief history of energy   446
17.2   Energy use today and future possibilities   448
17.3   Nuclear power: how it works   452
17.4   Technology assessment: assessing the nonrenewables   454
17.5   Issues for nuclear power   456
17.6   Future energy options: renewables   459
17.7   Future energy options: conservation   465
Summary, review, projects, discussion, exercises, problems   467

CHAPTER 18

*Quantum Fields: Relativity Meets the Quantum   472

18.1   Quantized fields: the reason there are particles   472
18.2   Quantum electrodynamics: the strange theory of electrons and light   474
18.3   Antimatter   477
18.4   Electroweak unification: neutrinos   482
18.5   Grand unification: quarks   486
18.6   Grander unification: gravity and the inflationary universe   491
Summary, review, discussion, exercises, problems   496

Epilogue: Summing Up   499

ANSWERS TO ODD-NUMBERED EXERCISES AND PROBLEMS   502
GLOSSARY   513
PHOTO CREDITS   526
INDEX   529

Preface

This is a liberal-arts physics textbook for nonscience college students. Its central premise is that nonscientists deserve a true *liberal-arts* physics course. Thus, this is not a watered-down version of the standard technical introductory physics textbooks for scientists. Far from being a simplified version of anything, this book is designed for a course that has a cultural sophistication not found in more technical courses. It is a cultural, rather than a technical, physics textbook. It presents physics as a human endeavor in its full philosophical and social context.

Many organizations have recommended new approaches to science education and science literacy.* The features in this book that reflect these recommendations include:

Scientific literacy. This book addresses the values, philosophical meaning, and social impact of science, and stresses the methods of science.

Modern physics. Fully half of this book is devoted to relativity, quantum theory, nuclear physics, high-energy physics, and other post-Newtonian topics.

Societal connections. This book applies the ideas of physics to such socially relevant topics as ozone depletion, global warming, technological risk, energy resources, nuclear power, and nuclear weapons.

Appropriate quantitative skills. Nonscientists should become *numerate* as well as literate in science. Quantitative skills such as interpreting graphs, thinking probabilistically, making estimates, and understanding powers of 10 are important for everyone. On the other hand, traditional algebra-based physics problems are less important for nonscientists. The mathematical details are important to professional physicists who must confirm, apply, and extend the principles, but not essential for nonscientists who seek to understand science and its relation to their lives. Physics is, first and fore-

---

*American Association for the Advancement of Science *Project 2061,* reported in *Benchmarks for Science Literacy* (New York, Oxford University Press, 1993); National Science Board *Undergraduate Science, Mathematics and Engineering Education* (National Science Foundation, Washington, 1986); American Institute of Physics *Introductory University Physics Project* (IUPP).

most, about ideas, and ideas are expressible in words. For example, the magic of Newton's law of gravity resides in the simple but profound fact that all material objects attract each other. By comparison, the rest is details. Although this book presents some of the quantitative detail, it focuses on the qualitative concepts, without algebra.

Less is more. While presenting most of the great ideas of physics, this book omits many narrower topics and applications normally included in introductory courses. Many studies have found that it is a mistake to try to present all of the traditional physics topics in a single course. Rather than skimming a lot of topics, it is preferable to dwell at greater length on fewer topics. It is especially important to delete nonessential topics if one wants to present much of the modern physics that is included here.

Unifying themes. Four unifying recurrent story lines, discussed below, represent the book's main pedagogical goals.

The first and foremost of the book's four story lines is *how we know what we know in science.* Science isn't simply a body of knowledge about the universe. It is instead a way of knowing—a process for proposing, disposing, testing, and refining explanations about nature. The notion that knowledge comes from experience and is subject to testing by experience and by rational thought is science's most basic value, and probably its most important benefit. The methods of science dominate Chapter 1 and form a constant refrain throughout the book.

The second theme is *the significance of post-Newtonian physics.* "Modern physics" (i.e., physics since 1900) reveals a universe of fields and energy structured according to relativity and quantum theory, a view that differs considerably from the older mechanical picture of particles and forces. Because the Newtonian mechanical universe runs so deeply into the cultural roots of industrial civilization, and also because science has accepted the relativistic and quantum view of nature for about a century now, it is incumbent on any general liberal-arts physics course to treat post-Newtonian physics in depth, including its philosophical and cultural significance. In the book, this theme appears in connection with many Newtonian and modern topics, beginning with the atomic idea in Chapter 2.

Energy is the book's basic physical concept, and it forms the third recurring theme. From the fall of a pebble, to the workings of the atomic nucleus, to the evolution of the universe, the principles of energy give us a wonderfully unified view of all natural processes. Furthermore, many of our science-related social problems are related to uses and misuses of various forms of energy. I hope that you, the student, will come away from this book with a habit of viewing all of the changes around you as transformations of energy and that you will perceive the great laws of energy manifested in these changes. Although energy is not introduced until Chapter 6, it is the basis for nearly all of the ensuing physical analysis.

The final theme is a practical and pressing topic in today's world: *the social context of physics.* It has been said that, in this scientific age, the itch is to accelerate, but the metaphysical direction is unknown. The great power conferred on us by science and technology demands great responsibility from each of us to see that that power moves us in humane social directions. Thus science is far too important to be left to the scientists. This is, in fact, why I have written this book. It is written for you, the teachers, poets, politicians, historians, business people, journalists, and others who must help us find a rational and humane path through the scientific age. If we are to prosper, and indeed if we are to survive, we need your perspective and your informed leadership.

I have tried to develop the scientific ideas slowly, with attention to the learning process.* You will find the following learning aids in the book:

Quotations in the margins lend different perspectives, and depth, to the philosophical and societal topics. Quotations, especially those on controversial topics, represent a wide range of views. Please understand that the inclusion of a particular quotation does not mean that I necessarily agree with that viewpoint, or that the scientific community necessarily agrees!

Footnotes provide additional details for students who want them. It is a perennial problem for writers of introductory textbooks, especially books for nonspecialists, to be accurate while not burdening students with too many details. Footnotes seemed to me to be the best way to handle such situations.

Dialogue questions in the main body of each chapter are my way of conducting a dialogue with the reader, to check and reinforce the reader's understanding. Readers should formulate their own answers to each dialogue question as they come to it, before reading the answer that appears in a footnote.

How do we know? is a frequently asked question in this book, and subsections bearing this title appear regularly. It cannot be emphasized too strongly that scientists have specific observation-based reasons for their theories and principles.

Making estimates is one of the skills that this book seeks to develop. Examples and exercises bearing this title appear frequently.

Summaries of ideas and terms, following each chapter, summarize and clarify the main concepts. These should be helpful when studying for exams.

Review questions, about 30, follow each chapter. Organized by sections within the chapter, they go over the main points. Answers can be found by looking in the appropriate section.

Home projects, a few of them, follow most chapters. There is nothing like "hands-on" experience to bring out the essential experimental nature of science. These projects could be done either at home using common household items, or in physics laboratories, or as demonstrations or desktop experiments in lectures.

Discussion questions, a few of them, follow most chapters. These are meant to stimulate thinking about open-ended questions, such as values questions, that have no single correct answer. They can be used for class discussion, essays, or individual thought.

Exercises, about 30, follow each chapter. They are qualitative, or conceptual. These are designed to exercise the mind, the way that jogging exercises the body. They are based on the text, but they require original thought. Answers to all odd-numbered exercises and problems are in the back of the book.

Problems, usually about a dozen, follow each chapter. These are quantitative. Many are purely numerical, and many are based on formulas. Some of the problems require algebraic manipulation; these could be used in courses in which algebra is appropriate.

As you study or teach from this book, you will undoubtedly have comments, suggestions, criticisms, or corrections. Please send them to me![†] A textbook, like

---

*I have benefited, especially in my presentations of Newtonian concepts, from Arnold B. Arons' book *A Guide to Introductory Physics Teaching* (New York: John Wiley &Sons, 1990).

[†]Art Hobson, Department of Physics, University of Arkansas, Fayetteville, AR 72701; phone 501-575-5918; fax 501-575-4580; internet: ahobson@comp.uark.edu.

learning and like life itself, is never really finished. I welcome your contributions to this book's continuing evolution.

WHAT'S NEW IN THIS EDITION?

Because this book emphasizes modern physics and contemporary societal topics, updating is needed to keep abreast of new developments. Although the book's general format is unchanged, every page contains many updates and revisions, and many sections are entirely rewritten. Besides updating everything and improving some of the explanations, I have tried to tighten the prose, deleting superfluous or repetitive material and making explanations more concise. This has shortened several sections and chapters.

The end-of-chapter material is expanded by the addition of about 15 somewhat mathematical problems, some requiring algebra, for those courses in which such problems are appropriate. However, the text itself remains nonalgebraic while still presenting the needed background for the Problems. I have added, at the back of the book, a comprehensive glossary that includes all of the book's main points and all of the terms that are highlighted by boldface type in the text.

Here are the more significant changes: Sections 2.4 and 2.5, on atoms and the structure of matter, are updated and more concise. The explanations of speed, velocity, and acceleration in Chapter 3, and the connections between force, mass, and acceleration in Chapter 4, are rewritten for clarity and brevity. Section 6.7, on energy efficiency and new power plant construction, is updated and expanded. Chapter 7, on the second law of thermodynamics, is thoroughly revised: The section on entropy is rewritten, expanded, and given a more prominent place early in the chapter; the sections on the automobile and the steam-electric generating plant are updated and revised. In Chapter 9, ozone depletion and global warming needed a lot of updating, and some expansion, to reflect recent events and scientific findings.

Many adopters have commented favorably on the book's modern physics emphasis and have suggested new or expanded topics. Because these comments fit nicely with the book's overall goals, I have incorporated several of them. Section 11.7, on general relativity and cosmology, now includes the large-scale geometry, density, and fate of the universe, along with the inflationary theory's prediction of a flat geometry and the related question of the search for the (possibly) missing mass in the universe. Chapter 12, on the search for extraterrestrial intelligence, is updated to include the new possible evidence for life on Mars, new evidence for planets orbiting other stars, and new evidence for extraterrestrial amino acids. In the chapters on quantum theory, the electron double-slit experiment (Section 13.4) and the evidence for Bell's interconnectedness principle (Section 14.5) are rewritten for brevity and clarity. In Chapter 16, nuclear energy is now introduced by putting nuclear fusion first as a concrete example, followed by a new general discussion of nuclear energy that is now more concrete and compact. Later in the chapter, there is less emphasis on the technology of nuclear weapons and more emphasis on the physics behind these weapons. Many of the energy-and-society topics of Chapter 17 have been updated and rewritten, resulting in a livelier and more compact chapter. In the final chapter, on quantum fields and high-energy physics, Section 18.6 has been expanded to include the cosmic inflationary theory of the big bang, Higgs fields, and the predicted Higgs particles.

A WORD TO THE INSTRUCTOR

This book takes a conceptual, nonalgebraic approach to physics for nonscientists. There is a significant amount of educational research showing that all students, including science students, grasp the essential ideas of physics better if they conceptualize before they calculate. Furthermore, because the average nonscience student is deterred by a mathematical presentation, a conceptual, nonalgebraic approach is especially appropriate for these students. See the *Instructor's Manual* for further discussion of the conceptual approach to physics teaching.

One of this book's goals is to connect physics with its full cultural context. For this, it is essential to present the science and the context together, in an integrated fashion. Thus, I have placed the societal issues side by side with the pure physics topics rather than relegating them to separate chapters or to the end of the book. This leads to some juxtapositions within chapters that will be unfamiliar to many instructors. My choice of placement for societal topics was guided by two principles: I wanted to present societal topics as early as possible, because student interest in these topics deserves to be satisfied sooner rather than later. And, on the other hand, I didn't want to present any particular societal topic before presenting all of the relevant physics. Thus, for example, ozone depletion appears within the chapter on electromagnetic radiation, because electromagnetic radiation is the last of the book's physics topics needed to discuss ozone depletion. Similarly, the chapter on the search for extraterrestrial intelligence (SETI) follows two chapters on relativity but precedes two chapters on quantum theory, because relativity is needed to discuss SETI but quantum theory is not.

To allow you flexibility in your choice of topics, the book is purposely too long for one semester. I usually omit the equivalent of about three chapters. Chapters and sections that may be omitted without seriously affecting the remaining material are indicated with asterisks in the Contents. Many other omissions are possible, provided you are careful in choosing the remaining material. The flow chart following this Preface shows connections between topics and suggests alternative course structures. The first chapter, about scientific methodology, is one of the deletable chapters, because there is material on this topic throughout the book. However, if Chapter 1 is skipped, students should read its first section anyway because it introduces the book. Some topics could be assigned for reading only.

Perhaps you want your course to focus on modern topics, or on societal topics, or perhaps you simply want to cover less material. For a modern focus, omit Chapter 1, the starred (with asterisks in the Contents) sections in Chapters 3–7, most of the starred sections devoted to societal topics, Chapter 17, and perhaps Chapter 12. Omitting the starred sections in Chapters 3–7 reduces the Newtonian material to an amount that can be covered comfortably in 8–10 lectures. For a societal focus, omit Chapters 1, 13, 14, 18, and the starred portions of Chapter 11. There are many options for a course covering less material: You could omit Chapters 15–18, and/or Chapters 1, 12, 13, and 14. Please consult the *Instructor's Manual* for further suggestions about alternative course designs.

Do get a free copy of the *Instructor's Manual.* It contains, in addition to the standard material that you expect in an instructor's manual, general tips for teaching liberal-arts physics, class activities, a student handout entitled "How to Succeed in Physics," alternate course designs, an annotated bibliography on physics teaching, a list of objectives for each chapter, general notes and specific classroom teaching suggestions for each chapter, and an annotated bibliography for each chapter that can be helpful for instructors and students alike.

ACKNOWLEDGMENTS

As I ponder my acknowledgments, I realize how vast is the network of people and institutions that contributed to the origin and evolution of this book. A six-month residence at the Stockholm International Peace Research Institute inspired me to further develop the interdisciplinary and social interests that led to this book. The University of Arkansas has provided a great place to work, as well as two sabbatical leaves that have been crucial to my professional development. The students in my general physics courses have always been a source of energy and ideas, and have provided feedback during class testing of the book. The following instructors class-tested the book: Charles Bottoms, Laurel Busse, Bernard Hoop, Mardi John, Carl Kocher, Joan Kowalski, Bill Lankford, Tom Meeks, Diane Claire Shichtman, and Robert Socolow. They, as well as their students, provided feedback. My colleagues in the Department of Physics provided help and encouragement, and have put up admirably with my questions; this is particularly true of my friends Julio Gea-Banacloche, Raj Gupta, Bill Harter, Claude Lacy, Mike Lieber, Charles Richardson, Suren Singh, and Min Xiao. Across campus, I had inspiring discussions with Lowell Nissen and Ben Anderson of the Department of Philosophy. For communications about various aspects of the book, I thank Ralph Baierlein, Al Bartlett, Timothy Burt, Malcolm Cleaveland, Martin Gardner, J. Richard Gott III, Leonard Mandel, N. David Mermin, Rita Massey, David Renfro, Robert Resnick, F. Sherwood Rowland, John Shultz, Cynthia Sexton, Abner Shimony, and Richard Wilson. Beatrice Shube gave me advice about writing, as did Michael Hennelly and Michael Alley, especially through his fine book *The Art of Scientific Writing.* The American Physical Society's Forum on Physics and Society has played a crucial role in providing emotional and professional support; I especially thank my Forum friends Dave Hafemeister, Ruth Howes, Barbara Goss Levi, Richard Scribner, and Peter Zimmerman. The American Institute of Physics' Introductory University Physics Project has given me helpful ideas, human interactions, and professional encouragement. David Bodansky, Bruce Daniel, Paul J. Dolan Jr., Robert Ehrlich, Norman Hackerman, Carl Kocher, William F. Lankford, Charles Leming, Theodore Lopushinsky, Mark Modera, Dwight Neuenschwander, Carl T. Rutledge, Steven Schneider, Dietrich Schroeer, Diane Shichtman, Clifford Swartz, Shelia Tobias, Carl T. Tomizuka, and James Watson have reviewed and commented on the entire manuscript. I would also like to acknowledge the following physicists and authors for their inspiration and their influence on this book: Arnold Arons, John Bell, Niels Bohr, E. A. Burtt, Albert Einstein, Richard Feynman, Werner Heisenberg, Paul Hewitt, Leon Lederman, Robert Romer, and Carl Sagan.

I could not have seen this project through to completion without the support of my family and friends: My wife Marie Riley, son David, daughter Ziva, and friends Justin Riley and Dick Bennett. In addition to their emotional support, they have given me valued advice.

In preparing this second edition, I especially thank the supportive and cheerful staff at Prentice Hall: Alison Reeves, editor; Debra Wechsler, production editor; Jane Loftus, manuscript editor; and Milt McGowan, local sales representative, advisor, and friend. I thank the reviewers for this edition: Neal Cason, University of Notre Dame; Paul D'Alessandris, Monroe Community College; Susan Mini, Northern Illinois University; and John R. Thompson, DePaul University. In addition to the people who helped me with the first edition, I also received assistance for this edition from Hans Christian von Baeyer, D. J. Bate, Stephen Brush, Brian Cressman, Leonard Finegold, Nathanial Fortune, Ulrich Harms, Hubert Klar, Harvey Leff, Matthew J. Moelter, Frank Munley, John Roeder, Al Saperstein, and Clifford Swartz.

I hope that you have as much fun using this book as I had writing it and teaching the related course! Enjoy!

ART HOBSON

Flow chart of topics

For a course emphasizing modern and philosophical topics, many societal topics could be omitted. For a course emphasizing societal topics, many modern and philosophical topics could be omitted. For chapters and sections that can be omitted without affecting the continuity of the remaining material, see the table of contents. For further details on alternative course structures, see the *Instructor's Manual*.

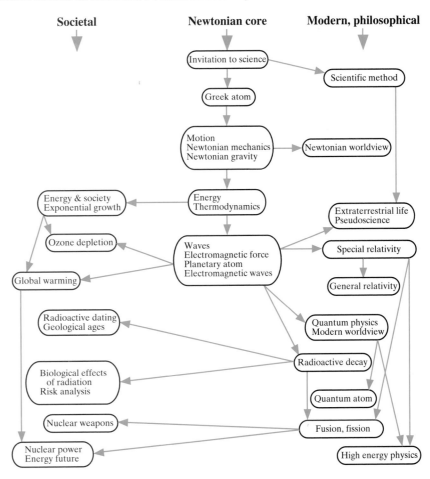

1

The Way of Science

experience and ideas

1.1 STARDUST: AN INVITATION TO SCIENCE

We came from the stars. We are, literally, made of atoms created and blown into space by ancient stars. This fact is only one strand in a network that connects us with the rest of the universe. **Science**—observing and understanding the natural world—is a path toward embracing that network.

An expanding awareness of nature is discernible in the long history of life on Earth. There is good reason to believe that our planet formed about 5 billion years ago and that the earliest simple living organisms formed over 3 billion years ago (see Chapter 15). Since then, organisms have evolved biologically to interact with their environment in increasingly complex ways. Looked at from the human perspective (an amoeba would no doubt view things differently), humankind is the latest in a sequence of increasingly aware biological organisms.

We could even say that through biological evolution, the universe has become more aware of itself. Education and science can be viewed as an extension of this process. And you, as you learn about the universe, are part of that process of expanding awareness.

Albert Einstein spoke of this widening circle of awareness when he wrote:

A human being is a part of the whole, called by us "Universe," a part limited in time and space. He experiences himself, his thoughts and feelings, as something separated from the rest—a kind of optical delusion of his consciousness. This delusion is a kind of prison for us, restricting us to our personal desires and to affection for a few persons nearest to us. Our task must be to free ourselves from this prison by widening our circle of compassion to embrace all living creatures and the whole nature in its beauty. Nobody is able to achieve this completely, but the striving for such achievement is in itself a part of the liberation and a foundation for inner security.

I hope that this book will help you discover many links between you and the universe. In writing this book, my constant criterion has been, Is this material really relevant to readers who want to participate fully in our science-related culture but who do not necessarily use science in their professional lives? I

The essence of education is the conquest of parochialism, that is, expanding one's horizons. You've got to look beyond the limits of your own narrow horizons.

WALTER MANGER, GEOLOGIST,
UNIVERSITY OF ARKANSAS

have tried to use language that is meaningful to literate nonscientists. There are no extraneous technical terms and no extraneous mathematics—in particular, no algebra. The text does, however, make wide use of numbers, proportionalities, graphs, and numerical estimates, because quantitative ideas are essential to meaningful communication today. Literate people must also be *numerate*.

We have discussed one reason for learning science: expanded awareness. A second and related reason is to help develop cultural and social values appropriate to the scientific age. Take a moment to list a dozen problems that are important to the nation or the world. A typical list might include population growth, poverty, crime, species destruction, illiteracy, global warming, urban decay, drugs, wars, air pollution, AIDS, and famine. Most or all of these problems have a science component.

Now try listing solutions to these problems. A typical list of solutions might include birth control, economic growth, education, sustainable farming, democracy, international law, environmental protection, disease control, better government, rational use of energy, more understanding among people, and concern for the environment. Most or all of these solutions have a science component.

The problems and the solutions of our times are bound up with science and its close relative, technology.* That is why we call this the scientific age.

The world needs your help. We dare not simply turn over these critical issues to "experts" or to governments. If we are to govern ourselves, we must make decisions about the issues that determine our lives, and these issues are related to science.

In his book *Of a Fire on the Moon*, about humankind's first venture to the moon, novelist and journalist Norman Mailer wrote pessimistically:

> The [twentieth] century would create death, devastation and pollution as never before. Yet the century was now attached to the idea that man must take his conception of life out to the stars. . . . A century devoted to the rationality of technique was also a century so irrational as to open in every mind the real possibility of global destruction. . . . So it was a century which moved with the most magnificent display of power into directions it could not comprehend. The itch was to accelerate . . . the metaphysical direction unknown.

Other eras have had their own problems. But if we are to resolve today's problems, we must find our metaphysical direction in this scientific age.

Science confers great power. You use this power daily when you switch on a light, a television set, an automobile, or a computer. Every such device has powerful effects on the world, both good (light to read by) and bad (pollution from electric generating plants). The classic moral dilemma of the scientific age—a dilemma symbolized, for example, in Mary Shelley's nineteenth-century novel *Frankenstein*—is the problem of understanding and dealing responsibly with these powerful technologies. To accept technology's power without also accepting the responsibility to use that power wisely is to invite death, devastation, and pollution—the monster's retaliation against its maker.

For any man to abdicate an interest in science is to walk with open eyes toward slavery.

JACOB BRONOWSKI,
PHILOSOPHER-SCIENTIST

Scientific activity is one of the main features of the contemporary world and, perhaps more than any other, distinguishes our times from earlier centuries.

SCIENCE FOR ALL AMERICANS,
A REPORT TO THE AMERICAN
ASSOCIATION FOR THE
ADVANCEMENT OF SCIENCE

I hesitate to use the word "God," but in my studies of the universe I have come to the conclusion that there is some purpose to it. The universe has organized itself in such a way as to become aware of itself. As conscious beings, we are part of that purpose.

PAUL DAVIES, PHYSICIST

Technology is the application of science to achieve useful human goals. This book often uses the single word *science* to refer to science and technology.

The dangers that face the world can, every one of them, be traced back to science. The salvations that may save the world will, every one of them, be traced back to science.

ISAAC ASIMOV,
SCIENTIST AND WRITER

We need people who can see straight ahead and deep into the problems. Those are the experts. But we also need peripheral vision and experts are generally not very good at providing peripheral vision.

ALVIN TOFFLER,
WRITER AND FUTURIST

Teach me your mood, O patient stars!
Who climb each night the ancient sky,
Leaving no space, no shade, no scars,
No trace of age, no fear to die.

RALPH WALDO EMERSON,
THE POET

This book focuses on that part of science called physics, and on its human connections. You have heard of most of the major sciences: biology, geology, chemistry, astronomy, physics, and others. When people ask me "What is physics?" I like to pick up something and drop it. Most things fall when you drop them. You can drop a rock, you can drop a frog, or you can drop other objects, and they all* fall. **Physics** is the study of these kinds of phenomena, phenomena that, like falling, are universal. Geologists study rocks and Earth's structure, biologists study frogs and other living organisms, while physicists study those general principles that rocks and frogs and everything else follow.

Please pick up this book and drop it. That's physics! It's about falling and about all the other general features of nature.

The remainder of this chapter is about what scientists do when they do science. How does science operate? We answer this with a significant case study of the actual operation of science: the early history of astronomy. We begin in Section 1.2 with some commonsense observations and conclusions about the night sky. Sections 1.3, 1.4, and 1.5 present three theories about the way the heavens are organized: the ancient Greek, Earth-centered view culminating in Ptolemy's theory; then Copernicus's postmedieval sun-centered theory; and Kepler's slightly later sun-focused theory. Section 1.6 discusses what we can learn about science's operations from this history, and Section 1.7 looks at the cultural implications of all of this.

1.2 OBSERVING THE NIGHT SKY

Science and technology are so pervasive that they are hard to fit into perspective. And yet precisely because science is pervasive, we need to understand it. This book's most important theme is the study of the nature of science itself.† How does science operate? What are its values? How valid are its conclusions? This topic is basic because science is more a method, a way of learning, than it is a body of knowledge.

The scientific method is often described as several activities that scientists sometimes practice: observing, hypothesizing, testing, and so forth. But such a list is superficial because it treats science as different from other human endeavors instead of capturing how science works in human culture. For a perspective on how science operates in real life, we will look at a historical example: the early history of astronomy. **Astronomy**, the scientific study of the stars and other objects in space, has usually been closely associated with physics.

Humans are fascinated by the heavens. The starry sky seems a more perfect place than our daily world. Life is full of clatter, the stars are serene; life is brief, the stars are forever. It is not surprising that ancient priests looked to the stars, for in addition to their beauty the stars also move in orderly, predictable patterns. Here was certain and timeless knowledge, a place, perhaps, of gods. And so for at least as long as there have been records to tell the story, we have looked

---

*Well, almost all fall. There are exceptions, such as helium balloons.
†Besides this theme, three others reappear throughout the book: the significance of post-Newtonian physics, the social impacts of physics, and energy.

to the sky for the time to plant and to reap, for omens of war and peace, for a sign regarding life's meaning, for our gods (Figure 1.1).

Although **astrology**, the belief that the positions of the stars significantly influence human affairs, has been an exploded superstition for two centuries, human scientific, religious, and aesthetic fascination with the stars may be greater today than ever (Figure 1.2).

In these electrically lit times, we sometimes fail to see the stars. On some clear night, try to follow the stars as they cross the sky. Find the moon, the seven stars of the Big Dipper, the North Star, any group of stars on the eastern horizon, and one or two stars on the western horizon (Figure 1.3, page 8). Observe all of these every 15 minutes for 1 or more hours. What happens? You should be able to see that the moon and stars move westward, that stars rise in the east and set in the west, that different stars maintain their positions relative to one another while moving as a group across the sky, that the North Star remains fixed, and that stars near the North Star move in circles around the North Star (Figure 1.4).

The Home Projects at the end of this chapter include many other astronomical observations that you can make without a telescope. You can observe that there are several small and unusually bright starlike objects that do not keep pace with the stars. When observed for a week or more, they can be seen to slowly shift their positions relative to the stars. These objects are called **planets**

FIGURE 1.1
Four thousand-year-old testimony to our reverence for and scientific awareness of the stars: the remains of Stonehenge, in England. Humans hauled the huge stones for more than 200 miles to make these monuments. These stones are the remains of a much larger structure used for religious purposes and to predict astronomical events, particularly solstices (longest and shortest days) and equinoxes (equal-length days and nights). Stonehenge perhaps also predicted eclipses of the moon, an impressive feat for people who did not use writing. Eclipses occur in an irregular and apparently random pattern that repeats itself only over a 56-year cycle. Even to have been aware that a repeated pattern exists required enormous dedication and attention to detail.

(a)

FIGURE 1.2
Our fascination with the stars seems greater than
ever. (a) *The Hubble Space Telescope.* (b) Model of
a four-telescope installation being built in Chile.
Each telescope contains a near-perfect mirror
8 meters in diameter.

(b)

(c)

FIGURE 1.2 (continued)
(c) Many telescopes receive nonoptical
signals from space. This large radio
telescope in Arecibo, Puerto Rico, receives
radio signals emitted by objects that pass
overhead as Earth rotates. (d) Drawing of the
neutrino telescope planned for installation in
Arkansas. The telescope "looks" downward,
through Earth, to receive subatomic particles
called neutrinos emitted by objects in space.
The neutrinos pass entirely through Earth at
the speed of light. A few neutrinos interact
with atoms near the telescope, causing these
atoms to emit particles called *muons* that in
turn emit light as they travel through the
telescope's purified water. This light is
detected underwater.

North Star

Little Dipper

Big Dipper

FIGURE 1.3
Look for these two constellations and the North Star in the night sky.

(*wanderers* in Greek). Five planets are visible without a telescope. The moon and the sun also move at a different pace from the stars.

From such observations, most people would conclude that the stars, sun, moon, and planets all are moving in circles around Earth, with their axis of rotation fixed in the direction of the North Star. Figure 1.4 is rather convincing evidence of this notion. This is the conclusion most observers drew centuries ago, and it is surely the conclusion that observers draw today unless they learn differently in school or elsewhere.

The observations and conclusions described here are typical of science's two main processes: observing, and conceptualizing or making theories. Science is not really different from a lot of other human thought. Whenever we observe our surroundings and develop ideas based on what we observe, we are acting somewhat like scientists.

FIGURE 1.4
Time-exposed photograph showing the "star trails" near the North Star. Observations such as this appear to provide convincing evidence that the stars move in circles around Earth.

1.3 ANCIENT GREEK THEORIES: AN EARTH-CENTERED UNIVERSE

At least as early as 3000 B.C., the Babylonians and Egyptians were aware of the differing motions of the stars, sun, moon, and the five planets then known. Beginning around 500 B.C., a few Greeks sought a new kind of understanding of these motions. They desired to go beyond the observed facts, to grasp the essentials of how the system worked.

All things are numbers.
PYTHAGORAS,
SIXTH CENTURY B.C.

Figure 1.5 indicates the early Greek concept of the heavenly architecture. In agreement with the observations described in the preceding section, this figure shows the heavenly objects circling a motionless Earth. Because the stars all keep pace with one another, the Greeks supposed that they all were attached to a single transparent invisible sphere centered at Earth's center and rotating around Earth once a day, carrying the stars with it. The sun, moon, and five visible planets, seven objects in all, each move along their own circle at rates a little different from that of the stars.

The Greeks imagined that each of these seven objects was attached to a large transparent sphere centered on Earth, one sphere for each object. Each of the seven spheres rotated at an unchanging (or **uniform**) rate around Earth roughly once each day. These spheres rotated at slightly different rates about the same axis through Earth's center.

A Greek philosophical–mathematical–religious group led by Pythagoras developed this picture of the universe. The Pythagoreans formed a secretive cult that believed passionately in the importance of abstract ideas. An idea is, in a sense, eternal. A real table, for example, eventually rots and turns to dust, but the idea of "table" or "tableness" seems eternal. Pythagoras believed that the

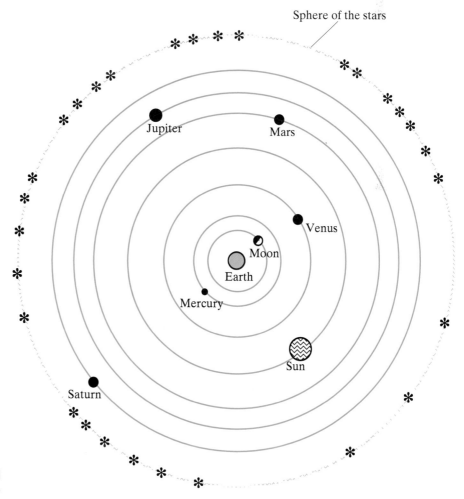

FIGURE 1.5
The earliest Greek conception, around 500 B.C., of the layout of the universe.

most perfect ideas were mathematical because they could be stated so precisely yet abstractly. The idea of a table is rather imprecise—a flat rock might be considered a table, or it might be just a rock. But mathematical ideas like a straight line, a circle, or the number five were precise, pure. For example, a circle is all of the points on a flat surface that are at the same distance from some fixed point on the surface.

Number is the ruler of forms and ideas, and the cause of gods and demons.

PYTHAGORAS

FIGURE 1.6
If you try to draw a circle, there will always be imperfections.

Let no one without geometry enter here.

INSCRIPTION OVER THE ENTRANCE TO PLATO'S ACADEMY, FOURTH CENTURY B.C.

■ **DIALOGUE 1**   To check your understanding of the preceding definition of a circle, try answering this question: What do we normally call the *fixed point* and the *distance* referred to in the definition? (Dialogue questions like this will appear throughout the book. Try to answer them as you come to them, before looking at the answer at the bottom of the page.)

Although this definition is precise, if you draw a circle that follows this definition, there will always be imperfections (Figure 1.6). Indeed, the Pythagoreans believed that it was the idea of a circle, rather than any particular representation of it, that was pure and eternal.

These mathematical mystics discovered that many features of the natural world could be described by mathematical ideas. The famous *Pythagorean theorem* is one example, and the simple numerical relationships between tones in the common musical intervals is another.* They believed that such ideas exist apart from human minds, that they are eternal and more real than any earthly object, and that the universe must be based on deep mathematical principles or "harmonies" analogous to the numerical relationships between the common musical intervals. They believed that nature was ultimately mathematical and that when one studies mathematics one studies the mind of God.

Other Greeks regarded all of this suspiciously. The Pythagoreans were persecuted and eventually banished. But their thinking had a deep influence on subsequent Greek philosophers such as Plato and Aristotle and on European and Western civilization.

Given their general beliefs, it's not surprising that the Pythagoreans sought a precise geometric scheme for the heavens. What geometrical forms could be more fitting for the stars than the sphere and the circle? After all, the sphere is the only perfectly symmetric (the same from all sides) shape in space, and the circle is the only perfectly symmetric shape on a flat surface. And as befits the timeless stars, circular paths have no beginning or end.

In line with their elegant and fitting picture of a universe made of eight Earth-centered spheres within spheres, these early Greeks had the startling notion that Earth itself was a sphere residing motionless at the center of the transparent spheres. The Pythagorean concept of a spherical Earth (although not their idea of a motionless Earth) survives to this day.

---

1. No looking down here until you've formed your own answer! The answers are: the center, the radius.

*The Pythagorean theorem states that in any triangle having a 90-degree or right angle, if you draw three squares, each one based on one of the triangle's three sides, the sum of the areas of the two smaller squares will equal the area of the larger square. As an example of the relationships between musical tones, if you create a musical tone by plucking a string and then precisely halve the string's length and pluck it again, the two notes you create will be exactly one octave apart. Other simple string ratios, such as 3 to 2 or 4 to 3, produce the other musical intervals that sound harmonious.

H O W   D O   W E   K N O W ?

All good science is based on observable evidence. Science is always skeptical, always asking, What is the evidence? It is a question that we, too, will ask frequently. So we will often interrupt the flow of ideas to ask, How do we know?

How do we know that Earth is a sphere? The evidence is fairly direct today (Figure 1.7), but what evidence did the ancient Greeks have? The Greek philosopher Aristotle, living two centuries after Pythagoras, stressed the importance of evidence. He gave many good observational reasons to believe that Earth is a sphere. For one thing, ships sink little by little below the horizon as they go out to sea, finally vanishing completely below the curve of Earth (Figure 1.8). For another, Greek travelers reported that in northern lands, the noontime sun is lower in the sky. For another, the shadow cast by Earth on the moon, as observed during an eclipse of the moon, is the shape that would be expected if both Earth and the moon were spherical.

FIGURE 1.7
"Spaceship Earth"; photograph of Earth from space. The general cultural impact of photos like this, showing Earth as a single freely moving ball in space, may be among the space program's most important benefits.

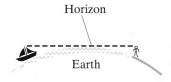

FIGURE 1.8
One piece of evidence that Earth's surface is curved like a sphere. As a ship sails out to sea, an observer on shore sees it sink little by little below the horizon.

But there was a problem with the Pythagorean transparent spheres theory, concerning the motion of the five observable planets. Because of the uniformly rotating transparent spheres, the theory implies, or "predicts," that each planet moves at a uniform rate around Earth. But careful observation shows that they do not. Instead, their rate of rotation, as seen from Earth, changes. Figure 1.9 diagrams this effect for a single planet such as Mars. The diagram is drawn relative to the background stars, and so it does not show the nightly rotation of Mars and the stars. Relative to the stars, Mars generally moves from west to east, but at a variable rate. Occasionally, Mars even changes directions and moves east to west relative to the stars, a phenomenon known as **retrograde motion**. Retro-

FIGURE 1.9
The motion of a planet such as Mars, relative to the background stars. Relative to the stars, Mars usually moves from west to east. In this illustration, Mars moves more slowly during July–August than it does during June–July. It slows to a stop by October and then reverses direction during October–December and regains its normal direction during December–February.

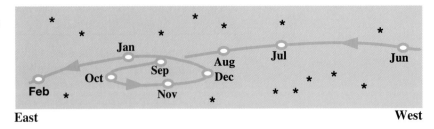

grade motion is the most obvious example of the fact that each planet moves nonuniformly as seen from Earth.

The Greek philosopher Plato, convinced of the Pythagorean belief in an elegant mathematical reality lying behind the motions of the heavens, challenged his students with the problem of finding a geometric scheme that would explain the observed motion of the planets. Plato and his students constructed a theory similar to the Pythagorean theory but more complex. They retained the uniformly rotating transparent spheres centered on Earth because they believed that the universe must be based on spheres and circles. To make their theory agree with such facts as retrograde motion, they imagined several unseen spheres for each planet, all centered on Earth, all "nested" within one another, with each sphere's axis of rotation fixed to the inside of the next outward sphere, and with all of these axes pointed in different directions and the single planet riding on the outermost sphere. This complex theory of **multiple transparent spheres** for each planet was favored by Plato's student Aristotle and by many others.

One Greek thinker, Aristarchus, proposed that it was the sun and not Earth that was at rest at the center of the universe, that Earth and the five planets moved in circles around the sun, and that Earth spun on its axis. It was a radical idea, and few astronomers took it seriously because it seemed absurd on two different grounds. First, our senses suggest to us that Earth is nothing like the heavens. So how could Earth be a planet like the heavenly planets? Second, it seems absurd to believe that Earth moves. It's too big! What immense force could be pushing it to keep it moving? If it does move, it seems that objects such as birds and clouds that are not attached to the ground should be left behind. If Earth spins on its axis, objects should be hurled off it in the way a stone is hurled from a rotating sling. These things were not observed, and so for reasons that made sense at the time, Aristarchus's theory was rejected. It would be nearly 2000 years before a sun-centered theory would again be seriously considered.

Although the theory of multiple transparent spheres correctly described the wandering of the planets, another problem arose. The Greeks noticed that during a planet's retrograde motion, it appeared brighter than at other times, as though it were closer to Earth during this time. Yet the theory of multiple transparent spheres had each planet on an Earth-centered sphere, so that each planet maintained a fixed distance from Earth. This contradicted the observations of varying brightness.

To explain the varying brightness of the planets, the Greeks proposed a theory that was rather different from the transparent spheres but that retained the uniform circular motion favored by all Greek thinkers. The basic idea was that each planet moved around Earth in a circle within a circle. As shown in Fig-

FIGURE 1.10
The orbit of Mars around Earth, according to the epicycle theory.

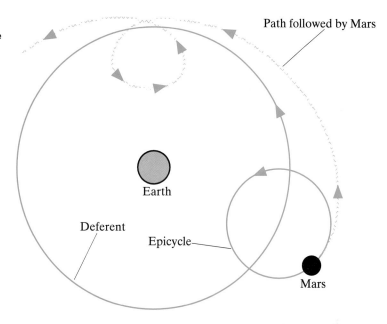

Path followed by Mars

Earth

Deferent

Epicycle

Mars

Having set ourselves the task to prove that the apparent irregularities of the five planets, the sun and moon can all be represented by means of uniform circular motions, because only such motions are appropriate to their divine nature. . . . We are entitled to regard the accomplishment of this task as the ultimate aim of mathematical science based on philosophy.

PTOLEMY

ure 1.10, a planet such as Mars moved uniformly around a circle whose center was on another circle that was centered on Earth. The small outer circle was called the planet's *epicycle*, and the inner circle centered on Earth was called the planet's *deferent*. The center of the epicycle moved uniformly along the deferent, so that Mars moved in two circles at the same time.

This produced a loop-the-loop orbit for each planet (Figure 1.10). In agreement with observation, the theory predicted that there would be occasional periods of retrograde motion (on the inside of the loops) and that the planet would be closest to Earth during retrograde motion and so should appear brightest. It was a satisfying picture, and it explained the observations. It was a good theory.

Figure 1.11 pictures the entire theory, although greatly simplified. In order to agree with the observations, each planet needed lots of epicycles—in fact, more than 80 "wheels within wheels" were needed. King Alfonso X commented that "If the Lord Almighty had consulted me before embarking upon the creation, I should have recommended something simpler." This theory was developed over a period of a few centuries and was finally refined and summarized in about A.D. 100 by Ptolemy, antiquity's greatest astronomer (Figure 1.12). In order to agree with the observations at that time, Ptolemy introduced two new ideas: the displacement or *eccentricity* of the centers and the *equant point* from which the motion appears uniform (Figure 1.13).* The details of these two ideas are not crucial here.

The important point is that Ptolemy devised a rather complicated, Earth-centered, mathematical theory that correctly predicted—to within the accuracy of the measurements then possible—the observed motions of the stars, sun, moon, and five known planets.

---

*None of these ideas were original with Ptolemy, but he was the first to put them all together in a consistent and quantitatively correct theory.

FIGURE 1.11
Ptolemy's Earth-centered epicycle theory (around A.D. 100) of the layout of the universe, according to which the five visible planets move on epicycles around Earth. The epicycles of the two innermost planets, Mercury and Venus, are centered on the line joining Earth to the sun.

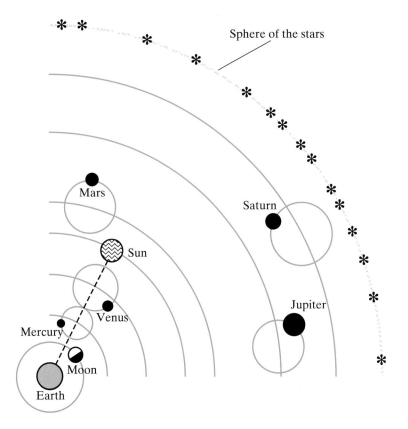

FIGURE 1.12
The ancient astronomer Ptolemy, A.D. 85–165. Using epicycles and many other theoretical devices, he prefected the Earth-centered theory of the layout of the universe.

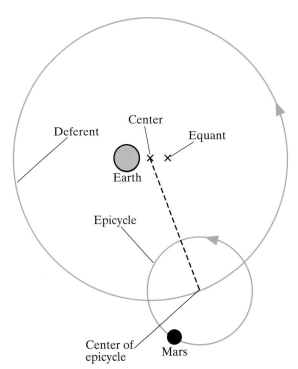

FIGURE 1.13
Explanation of the displaced center and the equant. Earth is displaced slightly from the center of the deferent. Furthermore, the center of the epicycle moves with unchanging speed, as seen from the equant and not from either the center or Earth.

Ptolemy checked his theory with many *quantitative* (numerical) measurements of the heavens. Because telescopes hadn't been invented yet, the measuring devices were long sighting rods with a scale to measure the angular position of a planet (or star). For example, one measured quantity was the number of angular degrees between the horizontal and the direction to the planet. The sighting devices were accurate to within about 0.2 degrees (recall that there are 360 angular degrees in a complete circle). Ptolemy's was the first theory to give such a complete, detailed, and quantitative account of all the known heavenly motions.

Ptolemy's theory predicted more details and was more accurate than any preceding theory. It survived, with modifications, for some fifteen centuries and was used by navigators, astronomers, and mystics such as astrologers. As the historian and philosopher of science Thomas Kuhn remarked, "For its subtlety, flexibility, complexity, and power the epicycle–deferent technique . . . has no parallel in the history of science until quite recent times. In its most developed form the system of compounded circles was an astounding achievement."

■ **DIALOGUE 2** The ancient Greeks believed that the stars and other astronomical objects shine by means of their own light. Can they have believed this of every astronomical object?

■ **DIALOGUE 3** Use Ptolemy's theory (Figure 1.11) to explain why the planet Venus often appears as the morning star (the last-seen "star" near the rising sun) or the evening star (the first-seen "star" near the setting sun).

1.4 COPERNICUS'S THEORY: A SUN-CENTERED UNIVERSE

The year A.D. 1543 is often taken as the birth year of modern science. In that year, an old man in Poland (Prussia, at that time), on his deathbed, signed the first printed copy of his life's work, *On the Revolutions of the Heavenly Spheres*. Nicolaus Copernicus (Figure 1.14)—astronomer, mathematician, linguist, physician, lawyer, politician, economist, and canon in a Catholic cathedral—had kept the manuscript locked up for thirty years, fearing the storm of criticism it would unleash.

With this book, the sun was setting on the medieval world. During the Middle Ages (about A.D. 500 to 1500 in Europe), philosophers such as St. Augustine

FIGURE 1.14
Renaissance astronomer Nicolaus Copernicus, 1474–1543. Finding Ptolemy's system to be "neither sufficiently absolute nor sufficiently pleasing to the mind," he devised a simpler theory. Copernicus's theory placed the sun at the center of the universe, with Earth moving around it. The odd idea that Earth moved and was a planet like the other planets met with much resistance because it conflicts with the intuitive notion that Earth is at rest at the center of things and because it conflicted with prevailing philosophies.

2. No. The moon's *phases* (new, crescent, quarter, and so on) show that it shines by means of reflected light from the sun.

3. The center of Venus's epicycle is fixed to the line joining Earth to the sun (Figure 1.11), which means that when viewed from Earth, Venus will never be seen far from the sun. Since Venus and the other planets are among the brightest astronomical objects and since Venus is never seen far from the sun, it is viewed as the morning star during those periods when it rises before the sun and as the evening star when it sets after the sun. This evening, or tomorrow morning, try to find Venus.

and Thomas Aquinas had linked Greek thought, including Ptolemy's astronomy, to Christian theology. But it was 1543 and the times were changing. During the past century the intellectual and artistic flowering known as the Renaissance had germinated in Italy and spread to all of Europe. Martin Luther had led a frontal assault on the Catholic church's authority. Christopher Columbus had made a memorable voyage.

These trends had a liberating effect on thoughtful minds. Copernicus and others were influenced by the new forms of art, new religious views, and exploration of faraway places. To begin with, Copernicus was uncomfortable with Ptolemy's theory. Not that Copernicus was a revolutionary. Quite the contrary: Copernicus's thinking resonated with the ancient views of Pythagoras. Copernicus objected to Ptolemy's theory on the grounds that with some eighty epicycles, plus all manner of eccentrics and equants, Ptolemy had strayed too far from the Pythagorean ideals. Ptolemy's system lacked the simplicity that scientists have always sought among nature's complex details. Copernicus put it this way:

> The planetary theories of Ptolemy and most other astronomers, although consistent with the data, seemed to present no small difficulty. For these theories were not adequate unless certain equants were also conceived; it then appeared that a planet moved with uniform motion neither on its deferent nor about the center of its epicycle. Hence a system of this sort seemed neither sufficiently absolute nor sufficiently pleasing to the mind.
>
> Having become aware of these defects, I often considered whether there could perhaps be found a more reasonable arrangement of circles, from which every apparent inequality would be derived and in which everything would move uniformly about its proper center, as the rule of absolute motion requires.

Note that like all thinkers since Pythagoras, Copernicus believed that the natural heavenly motions were both circular and uniform. He was adamant about this, stating that "the intellect recoils with horror" from any other suggestion and that "it would be unworthy to suppose such a thing [as noncircular motion] in a Creation constituted in the best possible way."

As an intellectual Renaissance man, Copernicus was more able than those before him had been to adopt a broad outlook and embrace a large part of nature. As Renaissance artists looked beyond Christian art and as Columbus looked beyond Europe, Copernicus looked beyond Earth itself to imagine it as an object in space, an object that he believed to be similar to other objects in space. To Copernicus, and to science since Copernicus, the ancient idea that the universe is centered on Earth seemed narrow-minded.

And so Copernicus was able to take a new point of view. He asked a broader question than had been asked before: What is the most elegant geometric scheme for the motion of the stars, sun, moon, five observed planets, *and Earth* that will fit the known measurements of the heavens? It was a crucial change of focus, a point of view that clashed with medieval intuitions and so had been inconceivable during the preceding 1000 years. Given this change of focus, Copernicus soon discovered "a more reasonable arrangement of circles," a theory in which the visible planets and Earth move in uniform circular motion around the sun and only the moon circles Earth (Figure 1.15). As in previous figures, Figure 1.15 pictures only the slow wandering of the planets and moon relative to the stars. Copernicus obtained the east-to-west daily motion of the stars, sun, moon, and planets (Figure 1.4) by allowing Earth to spin from west to east about

In the centre of everything rules the sun; for who in this most beautiful temple could place this luminary at another or better place whence it can light up the whole at once? . . . In this arrangement we thus find an admirable harmony of the world, and a constant harmonious connection between the motion and the size of the orbits as could not be found otherwise.
COPERNICUS

Whether a man is on the earth, or the sun, or some other star, it will always seem to him that the position that he occupies is the motionless center, and that all other things are in motion.
NICOLAS DE CUSA, BISHOP OF BRIXEN, 1450

We live on a great round wonder rolling through space.
WALT WHITMAN, AMERICAN POET

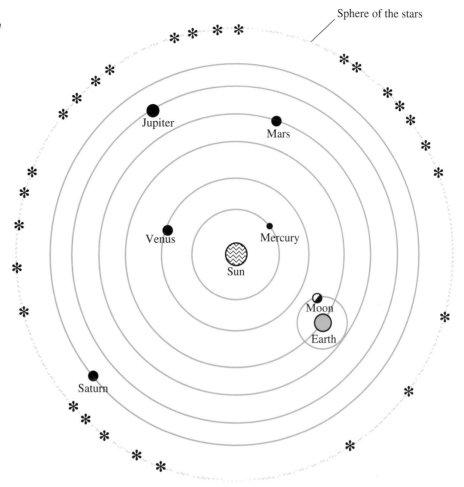

FIGURE 1.15
Copernicus's sun-centered theory of the layout of the universe, A.D. 1543. The diagram is simplified; the planets all move on epicycles, similar to those in Ptolemy's theory, but here there are far fewer epicycles. Furthermore, Copernicus used no equants.

a north–south line through Earth's center, rather than by allowing the sphere of the stars to rotate from east to west.

H O W   D O   W E   K N O W ?

Should we not attribute to God, the Creator of nature, that skill which we observe in common makers of clocks? For they carefully avoid inserting in the mechanism any superfluous wheel, or any whose function could be served better by another with a slight change of position.

GEORGE RHETICUS, AN EARLY SUPPORTER OF COPERNICUS'S IDEAS

There were still no telescopes in Copernicus's day, and data were still gathered with star-sighting devices. With properly chosen radii and rotation rates and eccentrics, Copernicus obtained quantitative agreement with the data. His theory explained many things, such as retrograde motion (see Figure 1.16 and Dialogue 5). But as Copernicus admitted, Ptolemy's theory was also "consistent with the data." Both theories agreed with the data.

Despite its agreement with the data, there were good observationally based objections to the new theory, like those that had earlier confronted Aristarchus's theory. Being so large, how can Earth move? What keeps it moving? Why aren't birds and clouds left behind? Why aren't objects hurled off Earth? Copernicus did not really have an answer. Instead, he pointed out that such problems loomed even larger for Ptolemy's great spinning sphere of stars

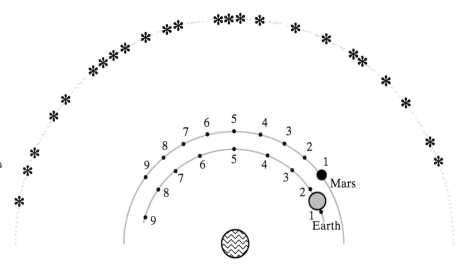

FIGURE 1.16
The Copernican theory's explanation of retrograde motion. As Earth passes another planet, such as Mars, the other planet appears to move backward as seen against the background stars, because of the rotation of the Earth-based observer's line of sight. You can demonstrate this by following the instructions in Dialogue 5. A similar effect occurs when you pass a car moving down a straight highway. Viewed against distant background trees and houses, the slower car appears for a few seconds to move backward, because of the rotation of your line of sight.

No one in his senses, or imbued with the slightest knowledge of physics, will ever think that the earth, heavy and unwieldy from its own weight and mass, staggers up and down around its own center and that of the sun; for at the slightest jar of the earth, we would see cities and fortresses, towns and mountains thrown down.

JEAN BODIN, SIXTEENTH-CENTURY POLITICAL PHILOSOPHER

There is perhaps no other example in the history of thought of such dogged, obsessional persistence in error, as the circular fallacy which bedeviled astronomy for two millennia.

ARTHUR KOESTLER, TWENTIETH-CENTURY WRITER AND HISTORIAN OF SCIENCE

than for Copernicus's much smaller spinning Earth. Why don't the stars fly off Ptolemy's spinning sphere? What keeps Ptolemy's sphere of stars moving? In making this argument, Copernicus was assuming that the stars were subject to natural laws like those operating on Earth. Nobody had looked at it in this way before. But the new view is a natural one if you believe, as Copernicus did, that Earth is not so different from other astronomical objects.

Such questions as What keeps Earth moving? were not answered for more than a century, not until Isaac Newton and others devised a radically new view of motion, an idea known as the law of inertia. Copernican astronomy started science moving toward the physics of Isaac Newton. Newton's physics arose partly because of such questions.

The first decisive blow against Ptolemy's theory and in support of a sun-centered theory did not come until Galileo introduced the telescope into astronomy, some 70 years after Copernicus's death.*Among other things, Galileo observed that Venus goes through phases similar to the moon's phases (new moon, crescent moon, quarter moon, and the like). This means that Venus shines not by its own light but by light reflected from the sun. In Ptolemy's theory, the center of Venus's epicycle must be fixed on the line joining Earth to the sun (see Figure 1.11), in order to explain the fact that Venus is never seen far from the sun. As shown in Figure 1.17, this means that we should never be able to see a "full-Venus" phase from Earth. On the other hand, the sun-centered theory predicts that we should be able to see a full-Venus phase whenever Earth and Venus are on opposite sides of the sun. Galileo observed that the phases of Venus included a full Venus.

Copernicus's theory, still brimming with epicycles and eccentrics, remained far from the Pythagorean simplicity of eight transparent spheres. Furthermore,

---

*Although Galileo did not invent the telescope, he was the first to make significant scientific use of it and the first to use it to study the heavens.

FIGURE 1.17
Ptolemy's theory predicted that an Earth-based observer would never see a "full" phase of Venus because Venus's epicycle lay between Earth and the sun. Copernicus's theory predicted that a nearly full Venus could be seen whenever Venus was on the far side of the sun in its orbit around the sun, as it is in Figure 1.15. But Galileo observed that the phases of Venus included a full Venus, thereby disproving Ptolemy's theory.

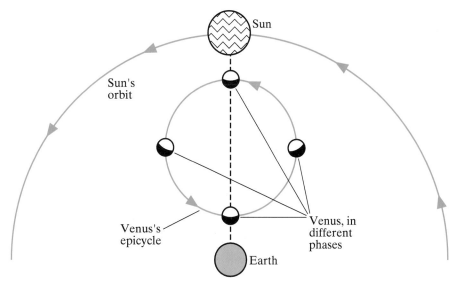

the observational data at that time gave no reason to prefer Copernicus's theory over Ptolemy's. Still, for Copernicus and for a few other intellectuals at the time, the new theory was an improvement because it dispensed with equants and with many epicycles. And the new planetary orbits were very close to being just simple circles around the sun, which means that for most purposes one could use a simple circular model like Figure 1.15 to explain and predict astronomical events (Dialogues 5 and 6). Ptolemy's theory, on the other hand, would be wildly inaccurate if the epicycles were left out. Copernicus's theory was "approximately simple," so it seemed closer to the truth.

■ **DIALOGUE 4** When we say that the sun "rises," what do we really mean, from the Copernican point of view?

■ **DIALOGUE 5** Use Copernicus's theory to explain retrograde motion: On Figure 1.16, draw lines of sight from Earth through Mars to the background stars during the period when Earth is catching up with and passing Mars. Number the nine points (1 through 9) at which the line of sight intersects the background stars. How would Mars appear to move, as viewed from Earth against the background stars? Still using Figure 1.16, what direction would Mars appear to move during this period if you viewed it from the sun?

■ **DIALOGUE 6** Use Copernicus's theory (Figure 1.15) to explain why Venus often appears as the morning star or the evening star (compare this with Dialogue 3).

---

4. We mean that Earth has rotated eastward around its axis, to bring the sun into view.

5. The line of sight rotates backward for a brief time during passing, so when viewed from Earth against the background stars, Mars appears to move "backward" (in other words, opposite to its usual direction of motion). Viewed from the sun, Mars would appear to continue moving "forward" during this period.

6. The orbit of Venus is inside Earth's orbit, closer to the sun than Earth's orbit. So when viewed from Earth, Venus can never appear far from the sun. As explained in Dialogue 3, this means that Venus will sometimes be seen as the morning star and sometimes as the evening star.

FIGURE 1.18
Tycho Brahe, 1546–1601. By making measurements of the planetary positions that were five times more accurate than were previous measurements, he overthrew two theories of the architecture of the heavens.

1.5 KEPLER'S THEORY: A SUN-FOCUSED UNIVERSE

Tycho Brahe (Figure 1.18), born three years after the death of Copernicus, loved the night sky. A skillful fund-raiser, he obtained financing from the king of Denmark to build a large astronomical observatory (Figure 1.19).

H O W   D O   W E   K N O W ?

Brahe's elegant sighting devices (Figure 1.20) were so accurate that his data are sometimes used today. Before Brahe, the best measurements had inaccuracies (possible errors) of at least 10 arc-minutes (an arc-minute is 1/60th of 1 degree). Brahe's measurements had inaccuracies of only 2 arc-minutes, a fivefold improvement.

When Brahe began his project, there were two competing theories of the universe: Ptolemy's Earth-centered theory and Copernicus's sun-centered theory. Despite their great dissimilarity, both theories agreed with the data known at that time. Would Brahe's measurements be able to distinguish between them and so determine which one was correct? For the next 20 years, Brahe cataloged accurate data on the positions of the sun, moon, and planets.

It soon became obvious that neither Ptolemy's nor Copernicus's theory agreed with Brahe's data. Each made predictions that differed from the observations by several arc-minutes—small differences but big enough for Brahe to detect. This meant that both theories were flawed.

FIGURE 1.19
Brahe's sextant for measuring the positions of the planets. Brahe's work was done without telescopes.

Just 18 months before Brahe died, the 29-year-old Johannes Kepler (Figure 1.21) managed to gain employment with the famous astronomer. Kepler was born to a ne'er-do-well father who abandoned his family when the boy was 17 and to a mother who was later tried for being a witch. As one historian describes it, "The boy was precocious above all in illness, being beset by small pox, headaches, boils, rashes, worms, piles, the mange, and worst of all for an aspiring astronomer, defective eyesight. His visual problems included double vision in one eye and myopia in both eyes."* The hardships of his youth seem to have toughened Kepler for the challenges to come. This philosopher-mathematician-astronomer-astrologer was deeply devoted to the Pythagorean notion of an elegant mathematical order in the cosmos, and he harbored another devotion that could only be described as sun worship. Regarding mathematical order, Kepler proclaimed:

> Why waste words? Geometry existed before the Creation, is coeternal with the mind of God, is God Himself . . . ; geometry provided God with a model for the Creation . . .

*Michael J. Crowe, *Theories of the World from Antiquity to the Copernican Revolution*, New York: Dover Publications, Inc., 1990.

In the first place, lest perchance a blind man might deny it to you, of all the bodies in the universe the most excellent is the sun, whose whole essence is nothing else than the purest light, than which there is no greater star; which singly and alone is the producer, conserver, and warmer of all things; it is a fountain of light, rich in fruitful heat, most fair, limpid, and pure to the sight, the source of vision, portrayer of all colours, though himself empty of colour, called king of the planets for his motion, heart of the world for his power, its eye for his beauty, and which alone we should judge worthy of the Most High God.

KEPLER, PROBABLY WRITTEN AT 22 YEARS OF AGE

FIGURE 1.20
An instrument that Brahe used for measuring the angular altitudes of the planets. The two wooden arms are joined together. The lower arm is placed horizontally, as determined by the string and weight hanging vertically from the upper end of the scale. The upper arm is raised by turning the screw until the two sights point toward the planet. The planet's angular position above the horizontal is then read from the graduated scale.

Regarding the sun, Kepler stated:

> The sun in the middle of the moving stars, himself at rest and yet the source of motion, carries the image of God the Father and Creator.... He distributes his motive force through a medium which contains the moving bodies even as the Father creates through the Holy Ghost.

Given his beliefs, it's not surprising that Kepler was the first professional astronomer to openly support the Copernican system, a theory whose beauty he contemplated with "incredible and ravishing delight." Kepler's words and thoughts convey the passion of the true scientist. This passion to know the universe is the driving force of science.

It is not surprising that such a man would seek out Brahe, who had made humankind's most intimate observations of the universe's architecture. Kepler deeply desired to read the mind of God, and Brahe's catalog might be the key. When he came to work with Brahe, Kepler had already spent many years pursuing the structure of the heavens, working on a beautiful scheme in which the six planets (including Earth) were attached to imaginary spheres, like the transparent spheres of the ancient Greeks but centered on the sun. He believed that the five intervals (gaps) between these six spheres were of such a size that these intervals could be exactly filled by using the surfaces of the five geometric shapes known as the *perfect solids* as "spacers" between the spheres.* He believed that the planets were arranged in such a way that this unseen collection of six spheres and five solid surfaces would precisely nest within one another, like boxes nested within other boxes, with the sun at the center (Figure 1.22).

H O W   D O   W E   K N O W ?

This idea first struck Kepler while he was teaching a class. He drew, for purposes unrelated to the layout of the planets, a blackboard diagram like that in Figure 1.23, in which an equilateral (equal-sided) triangle lies between two circles. He suddenly realized that the ratio of the sizes of the two circles was about the same as the ratio of the sizes of the orbits of Saturn and Jupiter. It was a real "eureka" moment for Kepler and no doubt an interesting moment for his students. The thought struck him that there were five intervals between the six planetary orbits and precisely five different perfect solids. Perhaps God had arranged things so that the perfect solids would just fit between the spheres of the planets! Kepler spent the next four years trying, unsuccessfully, to make this theory agree with the known data. Although he successfully pursued other theories after joining Brahe, Kepler always retained his enthusiasm for this idea, hoping to find further justification for it. But no further justification appeared. Kepler's alluring hypothesis was a scientific dead end.

One disproof of Kepler's theory today is the fact that there are nine planets, not six. But even in Kepler's day, the only actual data supporting the theory

---

*The perfect solids, or Pythagorean solids, are those flat-faced solid shapes whose faces are identical "regular polygons": two-dimensional (flat) figures bounded by straight lines of equal length. For example, a cube, bounded by six squares, is one of the perfect solids. A pyramid, bounded by four equilateral triangles, is another. The most complex of the five is the dodecahedron, bounded by twenty equilateral triangles. There are only five different perfect solids.

FIGURE 1.21
Johannes Kepler, 1571–1630. He contemplated the Copernican theory with "incredible and ravishing delight," although the scientific facts compiled by Brahe forced him to alter that theory in ways that would have displeased Copernicus. This scientist/mystic has been described by philosopher and novelist Arthur Koestler as the watershed between medieval science and modern science.

were that it approximately predicted a few of the planetary ratios and that there were five perfect solids, the same as the number of intervals between the six planets. This could easily be, and turned out to be, a meaningless coincidence. Kepler's pursuit of this scheme borders on **pseudoscience**—the dogmatic and irrational belief in an appealing idea that appears scientific but that is not supported by scientific methods.

Kepler was a convinced Copernican when he came to Brahe. But he found that Copernicus's theory did not fit Brahe's data, despite playing with many variations on Copernicus's theory. Brahe's data for Mars proved the most difficult to fit, even though Kepler tried reintroducing the equant device that Copernicus had so despised. The calculations were tedious, for Kepler needed to translate Brahe's Earth-based data into sun-based data: How would the planetary positions appear if seen from the sun? Kepler spent four years on this project, filling 900 notebook pages with finely handwritten calculations.

The project failed. The Copernican orbit coming closest to Brahe's data for Mars was still off by 8 arc-minutes. Before Brahe, an error of 8 arc-minutes could have been ascribed to observational error. But Kepler, toughened by the confrontation with his master's hard-won data, knew that neither observational error nor further tinkering would make uniform circular motion agree with the observed facts. Kepler rejected the Copernican theory.

A less passionate person would have given up. Worse yet, a less tough-minded person would have found a way to fudge the data to get them to agree

FIGURE 1.22
(a) Kepler's drawing of his model of the universe. The six planets move on six different spheres separated by the five perfect solids. The outermost sphere is Saturn's, then Jupiter's, and then Mars's. The remaining smaller spheres can be seen in (b), a detail of the central part of Kepler's illustration. The sphere of Mars is on the outside, then Earth, Venus, and Mercury, with the sun in the center.

(a)   **(b)**

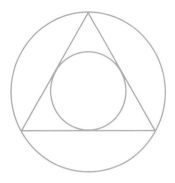

FIGURE 1.23
A blackboard diagram similar to this gave Kepler the original inspiration for his planetary theory based on the five perfect solids. In this diagram, two circles are separated by a triangle.

FIGURE 1.24
You can draw an ellipse with the help of a loop of string and two thumbtacks. The thumbtacks represent the two focuses.

I measured the skies, now the shadows I measure.
Skybound was the mind, earth-bound the body rests.

KEPLER'S EPITAPH,
COMPOSED BY KEPLER

with the Copernican preconceptions that Kepler had believed most of his life. But the ever-fervent Kepler, writing "on this 8-minute discrepancy, I will yet build a theory of the universe," began anew. He began studying planetary motions that, for the first time in history, were neither uniform nor circular. Copernicus, and all previous astronomers, would have been horrified.

Sixteen years later, Kepler found the precise and elegant scheme he sought. This theory is based on three geometrical principles, now called *Kepler's three laws*, describing the planetary orbits. The most significant law states that rather than moving in sun-centered circles, each planet moves in a sun-focused ellipse: an ellipse having the sun at one of its two "focuses." There is nothing at the other focus. Figure 1.24 shows how to draw an ellipse.* It is, roughly, a squashed circle. The planetary orbits are only slightly elliptical, which is why sun-centered circles come so close to fitting the observations.

The ellipse has just the kind of elegance Kepler had sought. He was elated:

> What sixteen years ago I urged as a thing to be sought, that for which I joined Tycho de Brahe . . . at last I have brought to light and recognize its truth beyond my fondest expectations. . . . The die is cast, the book is written, to be read either now or by posterity, I care not which. It may well wait a century for a reader, as God has waited six thousand years for an observer.

■ **DIALOGUE 7**   Who is the "observer" mentioned by Kepler in the preceding quotation? Kepler says that God has waited 6000 years. Why 6000?

■ **DIALOGUE 8**   Is a circle a particular example of an ellipse? Explain how to get a circle from the tack-and-string construction of Figure 1.24. Explain how to get a highly elliptical (very elongated) orbit from the tack-and-string construction.

1.6 THE SCIENTIFIC REVOLUTION: A DIALOGUE BETWEEN NATURE AND MIND

The dynamic interplay between observations and theories is the essence of science. Figure 1.25 illustrates this dialogue with nature.

Observation is the beginning of the scientific process. But a catalog of observed facts does not add up to an understanding of nature, any more than a telephone book adds up to an understanding of a city. To understand, literally to *stand beneath*, means to perceive a framework. A framework of scientific ideas is called a theory. In the development of astronomy, observations stimulated speculations that led to theories, and these theories in turn suggested new observations to check the theories and suggest new speculations. The key to science is the presence of both observation and theory.

---

*Here is the exact definition: An ellipse is all the points for which the sum of the distances of each point on the ellipse from two fixed points (the two "focuses") is constant. The construction shown in Figure 1.24 follows this definition.

7. The observer is Brahe. In Kepler's time, most people believed the universe to be just a few thousand years old: 6000 years is the figure that one gets from a literal reading of the Bible.

8. Yes, a circle is an ellipse for which the two focuses (the thumbtacks in Figure 1.24) are at the same point. For a highly elliptical orbit, move the thumbtacks far apart.

FIGURE 1.25
How we began to learn where we are in the universe.

A Summary of the Early History of Astronomy

Observations	Typical Dates	Theories
Stars, sun, moon, and planets are moving overhead.	3000 B.C.	
	500	Pythagorean theory: Earth-centered transparent spheres.
Each planet moves at a varying rate; retrograde motion.	400	Theory of multiple Earth-centered transparent spheres.
	300	Aristarchus's theory: sun-centered circles.
Heaven and Earth seem different; Earth seems motionless, apparently contradicting Aristarchus's theory.	200	
Planets are brighter during retrograde motion.	100	Theory of Earth-centered epicycles.
Detailed quantitative measurements show need for small corrections.	0	Ptolemy's theory: Earth-centered epicycles, equants.
	A.D. 100	
	1500	Copernicus's theory: sun-centered circles.
Brahe's accurate measurements disprove Ptolemy's and Copernicus's theories.		
	1600	Kepler's theory: sun-focused ellipses.
Galileo's telescopic observations disprove Earth-centered theories.		

The scientific method: doing one's damnedest with one's mind, no holds barred.

PERCY W. BRIDGMAN,
IN *REFLECTIONS OF A
PHYSICIST*

In science it often happens that scientists say, "You know that's a really good argument; my position was mistaken," and then they would actually change their minds and you never hear that old view from them again. They really do it. It doesn't happen as often as it should, because scientists are human and change is sometimes painful. But it happens every day. I cannot recall the last time something like that happened in politics or religion.

CARL SAGAN, ASTRONOMER
AND SCIENCE WRITER

Physical theory without experiment is empty. Experiment without theory is blind. It is the experimentalists who keep the theorists honest.

HEINZ PAGELS, PHYSICIST

Observation refers to the data-gathering process and has more or less the same meaning as measurement and experiment. A **measurement** is a quantitative observation, and an **experiment** is an observation that is designed and controlled by humans, perhaps in a laboratory.

The word *theory* comes from a Greek word meaning *to see*. A scientific **theory** is a framework of ideas that explains or unifies a group of observations. It is a way of understanding, a way of seeing, the observations. Theory has more or less the same meaning as principle, law, and model. A **model** is a theory that can be visualized, and a principle or law is one idea within a more general theory. The word *law* can be misleading because it sounds so absolute and certain. As we will see, no scientific idea is ever certain.

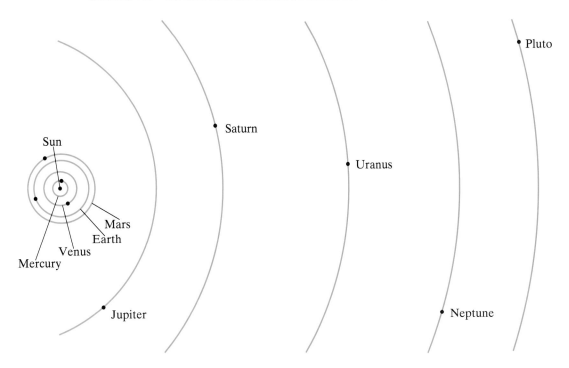

FIGURE 1.26
The arrangement of the solar system as it is now known. Uranus, Neptune, and Pluto are visible only with a telescope. The orbits are elliptical, although their ellipticity is too small to be visible in this diagram.

The whole of science is nothing more than a refinement of everyday thinking.
ALBERT EINSTEIN

Kepler's theory makes a good example. Figure 1.26 shows the general form of Kepler's model of the solar system (the sun and its planets) extended to include all nine known planets.

Kepler's theory explains all of Brahe's data and all preceding observations, in the sense that we can deduce (predict) all these observations from Kepler's theory. The theory also unifies all these data in a few principles such as the principle of elliptical orbits. Kepler's theory represents an enormous simplification or reduction of many observations into a few simple ideas.

But Kepler's theory does more than describe known data. It also predicts new observations that were unknown to Kepler. For example, when new planets (Uranus, Neptune, Pluto) were discovered, Kepler's theory predicted, correctly, that they too would move in elliptical orbits. A theory having no predictive value, that needs to be patched up to account for every new observation, is not worth much. For example, Ptolemy's theory could doubtless be amended with enough new epicycles to make it agree with all of Brahe's data, but the result would be a confusing patchwork with little ability to predict, say, the orbits of Uranus, Neptune, and Pluto.

Even more important, Kepler's theory is able to suggest further developments. Isaac Newton, born a few years after Kepler's death, made important use of Kepler's theory in his theories of force, motion, and gravity. Without Kepler, Newton's more sweeping theories would have been impossible.

Science is the belief in the ignorance of the experts.

RICHARD FEYNMAN, PHYSICIST

The great intellectual division of mankind is not along geographical or racial lines, but between those who understand and practice the experimental method and those who do not understand and do not practice it.

GEORGE SARTON, HISTORIAN OF SCIENCE

The aim of science is not to open the door to everlasting wisdom, but to set a limit on everlasting error.

BERTOLT BRECHT, PLAYWRIGHT, IN *THE LIFE OF GALILEO*

Science can purify religion from error and superstition, while religion can purify science from idolatry and false absolutes.

POPE JOHN PAUL II, 1988

I believe that ideas such as absolute certitude, absolute exactness, final truth, etc., are figments of the imagination which should not be admissible in any field of science. . . . This loosening of thinking seems to me to be the greatest blessing which modern science has given us. For the belief in a single truth and in being the possessor thereof is the root cause of all evil in the world.

MAX BORN, PHYSICIST

Physical concepts are free creations of the human mind, and are not, however it may seem, uniquely determined by the outside world.

ALBERT EINSTEIN

How valid, and how certain, is scientific knowledge? There are two common misconceptions about this. One is the belief that scientific knowledge is absolute, or certain. The other is the opposite belief, that scientific knowledge is just dubious guesswork.

Concerning the first misconception, that scientific knowledge is absolute, let's look at history. Ptolemy's theory correctly predicted the planetary observations during and after his time. So did Copernicus's theory. Both theories were useful reductions of the planetary motions to a few ideas, and both had predictive value. Both are correct today to within some 10 arc-minutes. They were, and are, good theories for many purposes.

But new observations led to improvements. Brahe's data contradicted both theories, opening the way for Kepler's theory. Kepler's theory had predictive value, predicting, for example, the phases of Venus and the shapes of the orbits of Uranus, Neptune, and Pluto. Did Kepler, then, discover the true motion of the planets?

Not necessarily. Tomorrow night, astronomers might discover that the planets have begun severely deviating from their elliptical paths! In fact, we will see below that the planets do deviate, although only slightly, from precise elliptical paths. It is always possible that new data will contradict any general theory. We cannot even be absolutely certain that the sun will rise tomorrow! Good science is always provisional, nondogmatic. Its ideas dangle by the slender thread of evidence, a thread that cannot prove a theory but that can disprove it with a single contradictory observation.

In fact, if a theory cannot be tested against observations, then it is not a scientific theory at all, because it tells us nothing about the observable universe. Scientific theories must be testable by observations that could conceivably contradict the theory. For example, a notion such as "undetectable alien creatures are living among us" is not a scientific statement, not because this notion seems odd (most scientific theories are odd), but because the creatures are said to be undetectable. Scientifically, this idea is not true and it's not even false. Being untestable, it is outside science.

Nonscientific ideas can, of course, have their own validity. "Beethoven's music is sublime," or "I love you," or "May God bless this home," can be meaningful statements, but they lie outside science.

Today, highly accurate planetary observations can test, and possibly disprove, Kepler's theory. Strictly speaking, Kepler's theory actually has been disproved by such observations! Observations made since Kepler's day show that the planets move along orbits that deviate slightly from precise ellipses. According to Isaac Newton's theories (see Chapters 3, 4, and 5), Kepler's elliptical orbits are caused by gravity acting between the sun and each planet. The main cause of the deviations from elliptical motion is gravity acting between the different planets and between each planet and its moons. These deviations cause the planets to wobble a little as they orbit the sun. Interplanetary dust and many other things also cause smaller deviations. Nevertheless, scientists have retained Kepler's theory because it is a good and useful approximation. Perhaps we should describe theories as *good* or *useful* rather than *true*.

Let's turn to the other misconception, namely the belief that theories border on dubious guesswork. As one example of this belief, some people who disliked the Copernican theory on religious or other grounds argued that because the

The gods did not reveal, from the beginning, all things to us; but in the course of time, through seeking, men find that which is better. But as for certain truth, no man has known it, nor will he know it; neither of the gods, nor yet of all things of which I speak. And even if by chance he were to utter the final Truth, he would himself not know it; for all is but a woven web of guesses.

XENOPHANES, GREEK
HISTORIAN,
SIXTH CENTURY B.C.

We have to live today by what truth we can get today, and be ready tomorrow to call it falsehood.

WILLIAM JAMES, PHILOSOPHER

We are trying to prove ourselves wrong as quickly as possible, because only in that way can we find progress.

RICHARD FEYNMAN

Science is built up with facts, as a house is with stones. But a collection of facts is no more a science than a heap of stones is a house.

JULES HENRI POINCARÉ,
SCIENTIST AND
MATHEMATICIAN, 1854–1912

When old age shall this generation
    waste,
Thou shalt remain, in midst of other
    woe
Than ours, a friend to man, to
    whom thou sayest,
Beauty is truth, truth beauty, —that
    is all
Ye know on earth, and all ye need
    to know.

JOHN KEATS, FROM *ODE ON A
GRECIAN URN*

idea that Earth circles the sun was a "mere theory," it need not be taken seriously. Today, people who dislike the theory of evolution attack it on similar grounds. Our culture often disparages the word *theory* as though a theory were just a wild guess. Perhaps this is a sign of anti-intellectualism ("against thought") in our culture. At any rate, theories—coherent mental frameworks for what we observe—are what science is all about.

The fact that theories are not absolutely certain is a strength, not a weakness, of science. Absolute certainty can foster dogmatism and a rigid inability to change what needs changing. As we noted, theories can be good, useful, fruitful, or compelling, but they are never certain.

A plausible suggestion that is not yet well confirmed is called a **hypothesis**. Kepler's hypothesis that the five perfect solids act as spacers between the six planetary orbits turned out to be false. But because Brahe's data confirmed his hypothesis of elliptical orbits, that hypothesis was elevated to the status of a theory.

The history of astronomy shows that science thrives on creativity. Creativity shows up in the elegant tools of Brahe and the inspired theories of Pythagoras and Kepler. It is one of nature's mysteries that these beautiful inventions actually turn out to produce a consistent picture of the universe. As Albert Einstein once put it, "The most incomprehensible thing about the world is that it is comprehensible."

Science's ideas are not so much discovered as created. Guided by Brahe's data, Kepler invented elliptical orbits. Ptolemy and Copernicus created two very different models, yet in their day, both were accurate. It is possible, even today, to conceive of the sun and planets as "really" going around a stationary Earth, although such a theory would need very many epicycles indeed and would require some unusual inventions to explain the phases of Venus.* Given a choice among theories that agree with the observations, scientists opt for the theory that is simplest, least arbitrary, most fitting.

Scientists generally believe in the Pythagorean ideal of a universe based on simple and elegant principles. Copernicus adopted a sun-centered theory over the hallowed Earth-centered theory because it was "pleasing to the mind." Scientists such as Kepler strove passionately to perceive such an elegant framework. When creating his theories, Einstein used to ask himself how he would have constructed the universe if he were God.

But science is more than elegant inventions. Theories must agree with the data. It is the synergistic interplay of theory and observation that leads us to a deeper understanding of nature. For example, Copernicus's theory helped inspire Brahe's observations, and both of these then led to Kepler's vastly improved theory, which in turn was the groundwork for Newton's grand theories.

All in all, the scientific method comes down to careful observation and the creation of ideas that correlate observations. The scientific way of observing and theorizing is not very different from our ways of coping with daily life. In science, as in life, we learn from experience and by thinking carefully about our experiences. It is a very human activity. To summarize:

---

*In order to maintain such a contorted theory, we would also need to change the law of inertia (Chapter 3) and thus nearly all of present-day science!

SCIENTIFIC METHOD
The dynamic interplay between experience (in the form of experiments and observations) and thought (in the form of creatively constructed theories and hypotheses that correlate experiences).

We will return frequently to the theme of how science operates.

■ **DIALOGUE 9**   Which theory, Ptolemy's or Copernicus's, proved fruitful for Kepler?

■ **DIALOGUE 10**   "Contrary to scientific opinion, I know that the universe was created just 200 years ago. Historical records, old-looking trees, fossils, and so forth that make Earth appear more than 200 years old are misleading. Actually, all these things were created just 200 years ago in order to fool the human race into believing the universe is more than 200 years old." Can you prove this statement wrong? Comment.

■ **DIALOGUE 11**   William claims to be absolutely certain of some idea. Can you then conclude that William's idea is right? Wrong? Crazy? Irrelevant? Is William being scientific? Unscientific?

■ **DIALOGUE 12**   "Because Darwinian evolution is only a theory, we need not take it seriously." Comment?

1.7 THE COPERNICAN REVOLUTION: DAWN OF THE MODERN AGE

The scientific age has its roots in two historical developments. One is the Pythagorean idea of *natural harmonies* or *natural law*, an idea that captivated influential Greek philosophers such as Plato and Aristotle and that then spread to Europe and the world. Its central premise is that the universe is organized in a framework of principles that can be uncovered by observation. The other historical development is the rejection of the "geocentric illusion" that humankind is central to the universe. Copernicus started this development, which is justly called the **Copernican revolution**. Its thrust is that we on Earth are not special; just as Earth is a planet similar to the other planets, nature is pretty much the same everywhere, differing perhaps in details at different places and times but always following the same general principles.

Others began thinking along these lines. It became apparent that the sun—considered by Copernicus and Kepler to be central to the universe—was a star like the other stars. We now know that the visible stars belong to a vast revolving aggregation of over 100 billion stars spread over a giant pancake-shaped re-

---

9. Kepler was led to his theory by starting from Copernicus's theory.
10. This is a nonscientific statement because it cannot be tested.
11. William is being unscientific.
12. Theories are precisely what we should take most seriously in science.

gion of space. Our sun is one of the stars in the outreaches of this aggregation, and it circles the center every 200 million years. But the center of this aggregation is not at the center of the universe, either. Instead, there are hundreds of billions of other similar aggregations of stars throughout the universe. And according to current theories, none of these aggregations is at the center, because the universe has no center (Chapter 11 describes this odd idea). This is the ultimate extension of Copernican astronomy.

Each of these giant aggregations of stars is called a **galaxy**. Ours is the **Milky Way galaxy**. Figure 1.27 shows a typical distant galaxy, one much like our own. If the galaxy in Figure 1.27 were our galaxy, our sun would be about two-thirds of the way out from the center toward the edge of the photograph. The cloudlike glow in the night sky that is called the Milky Way is our galaxy seen from our position within it. The glow comes from the stars in only a small, local portion of our entire galaxy. The center of our galaxy lies far beyond the visible Milky Way, in the direction of the constellation (group of stars) known as Sagittarius.

Telescopic surveys of galaxies give some indication of the size of our universe. Figure 1.28 is a photograph of a few of the galaxies in a distant region of the sky containing hundreds of galaxies. Figure 1.29 is a computerized plot of the locations of over a million known galaxies, still just a tiny fraction of the galaxies in the universe.

Copernicus sowed the seeds of many revolutions. Aristotle had developed an Earth-centered physics with two sets of natural laws, one for Earth and one for the heavens. Once Copernicus announced that Earth is a planet, Isaac Newton could unify the heavens and Earth in a new physics based on principles that acted uniformly throughout the universe. And just as Copernicus unified Earth with the other planets, Charles Darwin conceived an evolutionary biology that unified all life and included humankind as one species among many other

How vast those Orbs must be, and how inconsiderable this Earth, the Theatre upon which all our mighty Designs, all our Navigations, and all our Wars are transacted, is when compared to them. A very fit consideration, and matter of Reflection, for those Kings and Princes who sacrifice the Lives of so many People, only to flatter their Ambition in being Masters of some pitiful corner of this small Spot.

CHRISTIAN HUYGENS,
PHYSICIST,
AROUND 1690

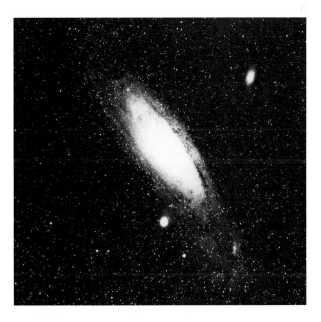

FIGURE 1.27
The Andromeda galaxy, photographed through a telescope. This is the nearest other large galaxy, outside our own Milky Way galaxy. Like our galaxy, Andromeda is made of billions of stars, each one somewhat similar to our sun, whirling around a bright star-filled center. Andromeda is nearly invisible to the unaided eye and lies far beyond the visible stars of our own galaxy. Since light takes more than 2 million years to reach here from there, you are looking at 2-million-year-old history.

FIGURE 1.28
This view of distant galaxies was taken with the Hubble Space Telescope [Figure 1.2(a)]. Nearly every object in the photograph is an entire galaxy, most of them so far away that the light takes many billions of years to reach here. The view covers only a narrow speck of sky 1/30th the diameter of the full moon, and reaches deeper into space than any previous visible image. Some of these galaxies formed less than 1 billion years after the *Big Bang* that created the universe some 12 billion years ago. Although this view shows a very small sample of sky area, it is considered representative of the typical distribution of galaxies in space because the universe, statistically, looks the same in all directions.

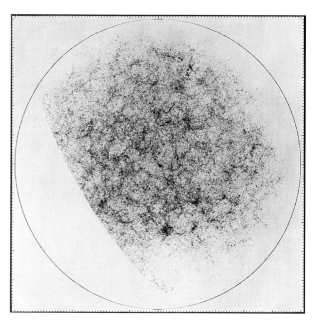

FIGURE 1.29
A computerized plot of the locations of more than a million known galaxies. Each tiny dot represents an entire galaxy! On such a large-scale plot, the pattern of galaxies looks like a thin film of cobwebs. The lower left-hand part of the plot is blank because this region of the distant universe is obscured by the gas and dust of our own Milky Way galaxy.

species. The Copernican/Newtonian conception of natural laws that apply democratically everywhere and to all people is an important component of the political transition from medieval authority to constitutional law and democracy.

Copernican astronomy was correctly perceived as revolutionary by religious and philosophical authorities. Ptolemy's system had been developed in parallel with Earth-centered Aristotelian physics, and Aristotle's thinking was a foundation of Catholic theology. The perfection of heaven, the imperfection of Earth, and humankind's centrality to God's plan for the universe were threatened by the loss of the Ptolemaic system. It is not surprising that Copernicus, prudent by nature, withheld publication of his ideas until his dying day.

Seventy years after Copernicus's death, the Catholic church pronounced his theory "false and erroneous," "altogether opposed to Holy Scripture," and "heretical." Science historians believe that during this period, science and religion fell out into two noncommunicating camps that still, today, feel they are at odds with each other. This situation is a far cry from the Pythagoreans who considered science and religion to be the same thing. Champions of the Copernican theory such as Kepler, Galileo, teacher and writer Giordano Bruno, and Coper-

Humanity has perhaps never faced a greater challenge; for by [Copernicus's] admission [that humanity is not the center of the universe], how much else did not collapse in dust and smoke: a second paradise, a world of innocence, poetry and piety, the witness of the senses, the conviction of a religious and poetic faith . . . ; no wonder that men had no stomach for all this, that they ranged themselves in every way against such a doctrine.

JOHANN WOLFGANG VON GOETHE, NINETEENTH-CENTURY GERMAN POET AND DRAMATIST

nicus's colleague and friend Rheticus were denounced and persecuted by religious authorities.*

The leaders of the Protestant Reformation were even more extreme in denouncing the new astronomy. The main Protestant objection was that the new theory ran counter to a "literal" reading of the Bible, in other words, a reading that interpreted the Bible as an accurate history and science text rather than, as many others argued, a spiritual work not meant as precise history or science. The Bible frequently mentions a moving sun and a fixed Earth, contrary to the Copernican theory. Even before publication of the new theory, the Protestant leader Martin Luther heard about Copernicus's ideas and violently condemned them for contradicting the Bible. In Luther's opinion, "The fool [Copernicus] will turn the whole science of astronomy upside down. But, as the Holy Writ declares, it was the sun and not the Earth which Joshua commanded to stand still."

SUMMARY OF IDEAS AND TERMS

Ideas and terms are listed here in the order of their appearance in the chapter, so that you can use this list as a brief summary of the chapter. At the end of the book, you will find a complete and alphabetized glossary that includes all of these ideas and terms.

Science The observation and theoretical understanding of the natural world.

Physics The branch of science that studies the most general principles underlying the natural world.

Astronomy The scientific study of the stars and other objects in space.

Astrology The belief, rejected by science for over two centuries, that the stars have a significant direct influence on human affairs.

---

* In 1984 the Vatican stated that church officials had erred in condemning Galileo and called for increased dialogue between science and religion. Then in 1992, the pope announced that the church had wrongly accused Galileo, laying the blame on seventeenth-century church authorities who interpreted the Bible too literally.

Planet To the Greeks, these were objects that looked like stars but that wandered, out of step with the stars. Today, we view planets as objects that orbit the sun in nearly circular orbits.

Solar system The sun and the objects that orbit the sun. It includes the nine planets and the moons that orbit many of the planets.

Uniform circular motion Motion in a circle at an unchanging or uniform speed.

Pythagoreans The earliest scientists. A society of ancient Greeks who believed that the fundamental principles of the universe were precise mathematical ideas.

Pythagorean theory of the universe The earliest Greek theory. The stars, sun, moon, and each of five planets circle Earth on uniformly spinning, transparent, Earth-centered spheres to which each is attached.

Retrograde motion A temporary change in the direction that a planet moves relative to the stars, as seen from Earth.

Multiple transparent spheres theory An extension of the Pythagorean theory to account for the planets' "wandering." Each planet has several transparent, Earth-centered spheres, with the planet riding on the outermost sphere.

Aristarchus's theory A sun-centered theory that was rejected because it seemed to conflict with everyday observations.

Ptolemy's theory An Earth-centered theory in which the planets move in circles within circles, or loop-the-loops. It was a good theory that survived for 1500 years.

Copernicus's theory A sun-centered theory, similar to Aristarchus's. The planets, including Earth, circle the sun, and Earth spins on its axis.

Tycho Brahe's observations Highly accurate data on planetary positions that disproved both Ptolemy's and Copernicus's theories.

Kepler's theory The planets orbit the sun in ellipses with the sun at one focus. This theory agrees with Brahe's observations.

Pseudoscience The dogmatic and irrational belief in an appealing idea that appears scientific but that is not supported by scientific methods.

Theory An idea, or several related ideas, that explains or unifies a group of observations. A **model** is a theory that can be visualized. A **principle** or **law** is a single idea, often within a larger theory.

Observation The fact-gathering process. A **measurement** is a quantitative observation, and an **experiment** is a controlled observation.

Hypothesis An educated guess, a tentative theory.

Scientific method The dynamic interplay between experience (in the form of experiments and observations) and thought (in the form of creatively constructed theories and hypotheses that correlate experiences).

Copernican revolution The rejection of the idea that humankind is central to the universe.

Galaxy A large aggregation of stars. Most galaxies, such as our own **Milky Way**, have a disklike, pancake shape and revolve about their centers.

REVIEW QUESTIONS

Each chapter has review questions, exercises, and problems at the end. Review questions go over many of the essentials; you can find the answers by looking in the chapter. Exercises ask you to draw on your understanding of the material to answer new conceptual (nonquantitative) questions not directly answered in the chapter. The answers to the odd-numbered exercises are in the back of the book. The many exercises allow your instructor flexibility; you will probably want to tackle only some of them. Problems are similar to exercises but they are quantitative, and occasionally algebraic, rather than conceptual. Answers to the odd-numbered problems are given at the back of the book.

SCIENCE AND THE NIGHT SKY

1. What two reasons does this chapter give for studying science?
2. What is physics?
3. Distinguish astronomy from astrology.
4. What astronomical objects can you normally see in the night sky? Describe their motion as seen from Earth.

ANCIENT GREEK ASTRONOMY

5. What did the Pythagoreans believe, and how did these beliefs influence the development of science?

6. According to the earliest Greek theory, the planets orbit Earth in uniform circular motion. In what way does this theory disagree with simple observations made without telescopes?
7. Give two observational reasons for believing that Earth is a sphere.
8. How does Ptolemy's theory explain the retrograde motion of the planets and the fact that planets are brighter during retrograde motion?
9. Did Ptolemy's theory agree with the quantitative observations known in Ptolemy's time? How were these observations made?

COPERNICUS'S IDEA

10. "Copernicus rejected Ptolemy's theory because it disagreed with the data, and he proposed a new sun-centered theory that did agree with the data." True or false? Explain.
11. Use Copernicus's theory to explain the retrograde motion of the planets and the fact that they are brighter during retrograde motion.
12. Why did Copernicus propose his theory?
13. State at least one plausible argument *against* the notion that Earth moves around the sun.
14. How did new telescopic evidence decisively disprove Ptolemy's theory?

KEPLER'S IDEA

15. What is pseudoscience, and in what way does Kepler's earliest theory of the planetary orbits border on pseudo-science?
16. "Kepler was attracted to Copernicus's theory because the known data supported that theory." True or false? Explain.
17. Describe Brahe's work and its effect on the theories of Copernicus and Ptolemy.
18. What aspect of Kepler's theory would have horrified all previous astronomers?
19. According to Kepler's theory, what geometric shape fits the planetary orbits?

THE SCIENTIFIC AND COPERNICAN REVOLUTIONS

20. What is the most characteristic and significant feature of science?
21. Describe several characteristics of a good scientific theory.
22. Can a scientific theory be proved (can we show that the theory is certainly true)? Can it be disproved? Explain.
23. Strictly speaking, Kepler's theory has been disproved. What has been found wrong with it? Why, then, do we still use it?
24. How does a hypothesis differ from a theory?
25. Distinguish between the Copernican theory and the Copernican revolution.
26. In what sense can evolutionary biology be said to be "Copernican"?

HOME PROJECTS

There is nothing like hands-on experience in science. Try some of these at home.

1. On a clear night, follow the stars across the sky. Look at the moon, the seven stars in the big dipper (Figure 1.3), the North Star (Figure 1.3), any group of stars that starts on the eastern horizon, and one or two stars that start on the western horizon. Try to observe them every 15 minutes for least 1 full hour. Describe your observations.
2. Observe the moon each night for a few weeks. Begin with a new crescent moon in the west, near the setting sun. Look at the moon at the same time each evening after sunset. On paper, draw its appearance (crescent, quarter, or whatever) each night and its position relative to the setting sun. After about two weeks, you will no longer see the moon just after sunset. Where is it? When does it rise and where? Continue recording your nightly observations until the moon has returned to its initial position near the setting sun. How many days did the complete process take? Explain why the moon had the appearance it did on different nights. Draw a diagram showing the relative positions of the sun, moon, and Earth when the moon is new; when it is quarter full; and when it is full.
3. Find one of the visible "outer planets" (preferably Mars, but Jupiter or Saturn will do) that is high in the night sky at some convenient time such as after sunset. A member of your local astronomy or physics department might help you locate one. On one night, observe the planet every 15 minutes for an hour or more. During that time interval, does the planet appear to stay in step with the east-to-west motion of the stars? Does the planet move along the zodiac—the path followed by the sun and the moon? Follow the planet during a longer time by observing it once each night for a few weeks. Does the planet remain in step with the stars, or do your observations show that it moves relative to the stars? About how many days does it take for the planet to noticeably shift its position relative to the stars? Does it move east to west faster than the stars do or slower? Is it following its normal motion or its retrograde motion?
4. Find Venus. It is often either the morning or evening "star." A member of your local astronomy or physics department might help locate it. Verify that Venus is a planet and not a star, by noting its changing position relative to the stars over a few weeks. Your observations of Venus should verify that it stays close to the sun in the sky. What is the explanation of this observation, according to Ptolemy's theory (Figure 1.11) and Copernicus's theory (Figure 1.15)?
5. Together with members of your class, lay out a scale-model solar system. Begin with a ball, about 23 centimeters in diameter (the size of a bowling ball), to represent the sun. Then lay out nine objects to represent the planets, as shown in the accompanying table. The model stretches over 983 meters, nearly a kilometer. We could choose a smaller scale, but then planets such as Earth and Mercury would be even smaller than a pinhead. The solar system is essentially all empty space.

To represent	—use an object of about this diameter		—placed this many paces (meters) from the preceding object
Sun	23 cm	(bowling ball)	
Mercury	1 mm	(pinhead)	10
Venus	2 mm	(peppercorn)	8
Earth	2 mm	(peppercorn)	7
Mars	1 mm	(pinhead)	13
Jupiter	2.4 cm	(chestnut)	92
Saturn	2.0 cm	(acorn)	108
Uranus	9 mm	(peanut)	240
Neptune	8 mm	(peanut)	271
Pluto	0.5 mm	(pinhead)	234

FOR DISCUSSION

The discussion questions are meant to stimulate thinking about questions that have no specific correct answer. They can be used for class discussion, individual essays, or individual thought.

1. Discuss the similarities and differences between a scientific theory such as Kepler's and a work of art such as a symphony.
2. In light of the history in this chapter, speculate on what might have happened if Copernicus had never lived. Would we still believe today that Earth is at rest at the center of the universe? Would somebody else have proposed a similar theory? If so, then when might this have happened—just a few years after 1543 (when Copernicus published his theory), a century after 1543, or several centuries after 1543? When considering this, keep these historical details in mind: Aristarchus, Brahe, the Renaissance.
3. List 10 ways in which science and technology have directly or indirectly improved your life. Now list 10 ways in which they have worsened your life.

EXERCISES

Odd-numbered answers are in the back of the book.

SCIENCE AND THE NIGHT SKY
1. How can you tell, from direct observation alone, whether or not a particular object in the sky is a planet?
2. Draw a diagram showing the positions of Earth, the moon, and the sun at new moon, crescent moon, nearly full moon, and full moon.
3. Are the stars in Figure 1.4 circling clockwise or counterclockwise? A time-lapse photograph made in the Southern Hemisphere, looking toward the South Pole, would also show the stars moving in a circle around a fixed point in the southern sky. Would the stars in the southern view be circling clockwise or counterclockwise?

ANCIENT GREEK ASTRONOMY
4. Describe an unaided-eye observation you could make to disprove the theory that the planets orbit Earth in a simple uniform circular motion.
5. Describe an unaided-eye observation you could make to disprove the theory that the planets orbit Earth attached to transparent spheres that rotate in a complicated fashion but that are always centered on Earth.
6. In seeking an explanation of retrograde motion, why didn't the Greeks just allow the planets to change their speed and direction of motion as the planets moved along circular paths around Earth, instead of resorting to circles within circles?

COPERNICUS'S IDEA
7. Is it possible that on some evenings, the planet Mars is the evening star? Is this very likely? (See Figure 1.15.)
8. Use Copernicus's theory to predict whether Mars goes through moonlike phases. Do we ever see a "full Mars"? A "new Mars"?
9. It is possible, but difficult, to see the planet Mercury with the unaided eye. How, then, would you go about finding it?

KEPLER'S IDEA
10. Which aspects of Kepler's theory would Copernicus have liked? Disliked?
11. Would Kepler's theory have agreed with the data available in Ptolemy's time? In Copernicus's time?
12. Did Brahe's data *prove* that planets move in ellipses? Explain.
13. Is there anything in Kepler's theory that resembles the displaced centers of Ptolemy and Copernicus?

THE SCIENTIFIC AND COPERNICAN REVOLUTIONS
14. What is the most important and characteristic feature of science?
15. Can two different theories both be true in the sense that at some particular time in history, they correctly predicted the known data? Defend your answer with a historical example.
16. "If Earth is curved, it must have a spherical shape, because a sphere is the most perfect curved solid form." Does an aesthetic argument like this have any place in science?
17. A sensationalist tabloid "news"paper carries this headline: "SCIENTISTS PREDICT THAT THE UNIVERSE AND EVERYTHING IN IT WILL DOUBLE IN SIZE AT THE BEGINNING OF THE NEXT NEW YEAR!" Is this a testable hypothesis? If so, how could you test it, and if not, why not? Is this good science, bad science, or neither?
18. What should be your attitude toward claims by proponents of astrology, dianetics, extrasensory perception (ESP), a 6000-year-old universe, creationism (belief in the literal truth of Genesis), visitations by extraterrestrials, the Bermuda Triangle, and pyramid power?
19. Aristotle, a careful observer of living organisms, wondered where the material that contributes to the growth of a plant comes from. He hypothesized that all of it comes from the soil. Based on your knowledge of biology, do you consider this hypothesis to be correct? Propose an experiment to test this hypothesis.

20. Some people believe that plants will grow better if they talk to them. Is this a testable hypothesis? If so, propose an experiment to test it.

21. "Certain people are gifted with extrasensory perception (ESP), such as the ability to move material objects with their own minds. However, ESP is so delicate that every attempt to verify it always destroys it." Is this a scientific hypothesis?

22. Isaac Newton predicted that because of its spinning motion, Earth would bulge out near the equator and be flattened near the poles. In 1735 the French Academy of Sciences sent an expedition to the Arctic to measure the exact shape of Earth. When they returned, reporting the predicted results, the philosopher Voltaire mocked them with the following couplet:

    To distant and dangerous places you roam
    To discover what Newton knew staying at home.

    Was Voltaire's sarcasm justified? Why or why not?

23. Consider the *flat Earth hypothesis*. Give evidence for this hypothesis. Give evidence against it.

PROBLEMS

Odd-numbered answers are in the back of the book.

1. Since there are some 100 billion stars in a typical galaxy, and since there are at least 100 billion galaxies in the known parts of the universe, how many stars are there in all? Write this number out.

2. Figure 1.29, representing 1 million galaxies, is only a tiny fraction of the 100 billion galaxies in the known parts of the universe. How tiny? Write the answer out as a decimal number.

3. An *astronomical unit* (AU) is the distance from Earth to the sun. The radius of the orbit of Mars is about 1.5 AU. As Earth and Mars orbit the sun, what is their greatest and least distances apart, measured in AU?

4. A *light year* (LY) is the distance light travels in one year. Our nearest neighboring star is 4 LY away. Using the fact that light gets here from the sun in 8 minutes, how many AU (preceding problem) is it to our nearest neighboring star?

5. The astronomical object known as the Crab Nebula is the remnant of an exploded star. The explosion was seen, by the Chinese, in A.D. 1054. However, the Crab Nebula is about 3500 LY (preceding problem) distant from Earth. In what Earth year did the star actually explode?

CHAPTER

Atoms
the nature of things

We turn now from stars to atoms. One of science's key principles is that everything is made of imperceptibly small particles. This explains an extraordinary range of observations, and it deeply influences our culture's perception of reality. As we will learn in later chapters, science profoundly changed its view of these *atoms* during the twentieth century, a development crucial to one of this book's four themes:* the significance of post-Newtonian physics.

Section 2.1 presents the 2500-year-old idea that everything is made of small particles. Section 2.2 discusses the atom as chemists have understood it for the past 200 years and distinguishes atoms from molecules. Following an excursion in Section 2.3 into metric units and large and small numbers, Section 2.4 explores a tiny portion of the tremendous explanatory power of the atomic idea. Section 2.5 describes the amazingly small size of atoms, and Section 2.6 considers the philosophical implications of the atomic idea. Section 2.7 looks ahead to compare this chapter's model of the atom with two more modern models presented in later chapters. Finally, Section 2.8 explores several examples of chemical reactions, or recombinations of atoms.

2.1 THE GREEK ATOM: THE SMALLEST PIECES

The ancient Greeks produced an astonishing number of profound and original thinkers. These philosophically minded people wanted to think their way to the bottom of everything. Because of their prointellectual attitude, this small group of people, during just two centuries centering on the fifth century B.C., laid the foundations for most of the Western world's great ideas. Among the things they thought about was the nature of matter: material substances such as wood, cotton, sausage, ice, water, soil, and gold. They speculated about the underlying unity that they believed lay behind the different substances. What do sausage and gold have in common? What is matter?

Leucippus and his student Democritus sharpened their focus on this question with a "thought experiment," an imagined experiment that seemed possible in

*The three other themes are the processes of science, the social context of physics, and energy.

principle but difficult to carry out in practice.* Suppose, they argued, you cut a piece of gold in half, and then cut one of the halves in half, and so forth. How far could you continue making such divisions? Either the divisions could go on forever, or there would be a limit at which no further divisions would be possible. That is, matter is either *continuous*, divisible without limit, or else it is made of particles that cannot be divided. The first alternative seemed absurd to them. Matter, they concluded, is made of imperceptibly small, "a-tomic" (Greek for *not divisible*), particles. They called these smallest particles **atoms**. †

■ **DIALOGUE 1** Would you classify this idea as scientific fact, experimental observation, hypothesis, theory, principle, or law? [Dialogue questions like this appear throughout the book. Try to answer them as you come to them, before looking at the answer at the bottom of the page.]

This turned out to be an influential and useful idea. We will call it:

> **THE ATOMIC THEORY OF MATTER**
> All matter is made of tiny particles, too small to be seen.

This idea is a good example of a scientific principle or law or theory: an idea or group of ideas that explains a broad range of observations. In Leucippus's and Democritus's time, the atomic idea had not yet been confirmed by observations; rather, it was an educated guess or hypothesis. When observations confirmed this idea during the nineteenth and twentieth centuries, it then became an established theory (or principle or law). A general idea like this cannot, however, be called a fact, no matter how often observations may have confirmed it, because we cannot observe every possible material object to prove for certain that everything really is made of atoms. Theories are never certain.

What do your thumbnail, beer, the Brooklyn Bridge, an acorn, Saturn, the Amazon River, and the period at the end of this paragraph have in common? Each is made of atoms, the unity beneath the diversity of materials and the kind of underlying unity that the Greeks loved and that scientists seek.

> H O W   D O   W E   K N O W ?

Science's power comes from its insistence on evidence. So we frequently ask, How do we know? This "show me" skepticism characterizes much of science.

How do we know everything is made of small unseen particles?

If, in some cataclysm, all of scientific knowledge were to be destroyed, and only one sentence passed on to the next generations of creatures, what statement would contain the most information in the fewest words? I believe it is the atomic hypothesis (or the atomic fact, or whatever you wish to call it) that all things are made of atoms—little particles that move around in perpetual motion, attracting each other when they are a little distance apart, but repelling each other upon being squeezed into one another. In that one sentence . . . there is an enormous amount of information about the world.

RICHARD FEYNMAN, PHYSICIST

---

*The Greek philosopher Epicurus later incorporated and developed the ideas of Leucippus and Democritus. During the first century A.D., the Roman poet Lucretius elaborated the ideas of the Greek atomists in his long poem *On the Nature of Things*.

†*Atom* is used in a slightly different sense today. Today, *atom*, or *chemical atom*, refers to the smallest particle of a chemical element. This chemical atom is actually divisible, being made of smaller, subatomic parts: electrons, protons, and neutrons. In fact, the protons and neutrons are themselves made of quarks. As far as we can tell today, electrons and quarks are the ultimate, smallest particles, of which the others are made. These smallest possible parts are what the Greeks meant by an atom.

1. In Leucippus's and Democritus's time, this was a hypothesis, but today it is an established theory. It is incorrect to call a general idea, such as this one, a fact or observation.

Although the ancient Greeks had no direct microscopic evidence for atoms, Democritus did have some ingenious indirect evidence. He argued, for instance, that since we can smell a loaf of bread from a distance, small bread particles must break off and drift into our noses; these bread particles were, he proposed, related to the bread's atoms. This is still an acceptable explanation of odors today (see Section 2.4). Medieval Christians rejected the atomic theory because they associated it with atheism. But Islamic scholars preserved Greek culture, including the atomic theory, during the Middle Ages, and they then reintroduced these ideas to the Christian West. In this way, the atomic principle became part of the background for the physics of Galileo, Newton, and others. But because the idea had no direct observational support, many scientists did not believe in atoms.

The first specific evidence for atoms was discovered around 1800 by the chemist John Dalton. Dalton discovered that when certain substances combine chemically to form other substances, they always combine in simple ratios by weight. For example, when hydrogen and oxygen combine to form water, the ratio of the weights of the two substances is always 1 to 8.

The occurrence of such simple ratios is not easy to understand based on the hypothesis that matter is infinitely divisible. But if matter is made of atoms, then there is a simple explanation: If, for example, 1 atom of hydrogen and 1 atom of oxygen have a simple weight ratio, and if these atoms always combine in simple ratios to create water, the weight ratios of the hydrogen and oxygen in water will be simple numbers also. Today we know that individual atoms of hydrogen and oxygen have a weight ratio of 1 to 16 and that it always takes 2 hydrogen atoms for every 1 oxygen atom to form water. So we can see, today, why the weight ratio should be 1 to 8.

So the atomic theory explains the simple ratios that Dalton observed. But does this prove the theory? The answer is no! It is possible that atoms do not exist and that there is some other explanation, or even no explanation, for the simple ratios. A general idea cannot be proved by observations, but it can be made more plausible.

A few decades later, a botanist, Robert Brown, using a microscope, observed that tiny pollen grains suspended in liquid move around erratically (Figure 2.1), even though the liquid itself had no observable motion. His first hypothesis was that the grains were alive. But lifeless dust grains suspended in liquid executed the same erratic dance, disproving the hypothesis. Hypotheses and theories can be disproved, although they cannot be proved.

What unseen cause lay behind this *Brownian motion*? Toward the end of the nineteenth century, it was suggested that submicroscopic (smaller than could be seen with a microscope) motions of atoms caused Brownian motion. The idea was that atoms were in constant motion and that the dust grains were jostled by atoms (Figure 2.1).

This hypothesis got strong support in 1905 from an unknown young physicist named Albert Einstein. Einstein's contribution was on the theoretical, not the observational, side. He used an already established theory to calculate the details of the jostling of particles such as dust grains—particles that are much larger than atoms—when bombarded randomly by moving atoms.* He made

FIGURE 2.1
Brownian motion. This erratic path is typical of a small particle such as a single dust grain, suspended in water, observed under a microscope. The atomic theory explains this behavior by the ceaseless, rapid, random motion of water molecules. Although the molecules are far too small to be seen even under a microscope, the effect of their impact on a dust grain can be seen in the erratic motion of the dust grain.

---

*That theory was the kinetic theory of heat, the theory that atoms move because of their thermal energy. Some of these topics will be discussed in Chapters 6 and 7.

several quantitative (numerical) predictions, such as the rate at which a collection of dust grains should spread out, or **diffuse**, because of this random jostling in a liquid. Predictions like this could be, and soon were, checked by measurements. The measurements agreed with Einstein's predictions. It was difficult to dispute this quantitative evidence. Either unseen atoms really were the cause of Brownian motion, or else Einstein's calculations were extremely lucky in just happening to give all the right numbers. Since Einstein's work, scientists have not questioned the atomic theory.

■ **DIALOGUE 2** An individual sulfur atom has twice the weight of an individual oxygen atom. When sulfur and oxygen combine to form sulfur dioxide, one sulfur atom is required for every two oxygen atoms. What is the weight ratio of sulfur and oxygen in the formation of sulfur dioxide? What is the weight ratio in the formation of sulfur *tri*oxide, having one sulfur atom for every three oxygen atoms?

▌ 2.2 ATOMS AND MOLECULES

Think of all the different material substances around you: this paper, your shirt, your hair, and so forth. You could list hundreds. Nineteenth-century chemists found that most of them could be *decomposed* into a much smaller number of simpler substances but that this small number could not be further decomposed by chemical means. Any process that changes a single substance into two or more other substances is called a **chemical decomposition** of the original substance.

For example, water can be decomposed* into two distinct substances, called hydrogen and oxygen, neither of which are anything like water. But hydrogen and oxygen turn out to be among that small (fewer than 100) group of substances that nineteenth-century chemists could not decompose. No matter how they tried to decompose hydrogen, it remained hydrogen, and the same was true for oxygen.

Apparently, these roughly 100 substances that cannot be chemically decomposed are particularly fundamental. They are called **chemical elements**, or simply elements.

By studying the weight ratios just discussed, Dalton and others soon recognized that the different elements corresponded to different kinds of atoms. The idea was that each element was made of only one kind of atom and that elements differed because their atoms differed. But water is *compounded* of two kinds of atoms, hydrogen and oxygen, which is why you can decompose water but not oxygen or hydrogen.

Today, 112 elements are known. About 90 occur naturally on Earth and about 20 others are made in laboratories. The table on the inside back cover of

2. [Are you reading this before forming your own answer? If so, do you exercise by watching somebody else jog? Exercise your mind by providing your own input to the dialogues!] 1 to 1. 2 to 3.
*By passing an electric current (Chapter 8) through water.

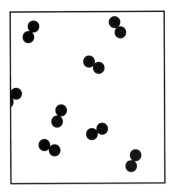

FIGURE 2.2
A simplified drawing of hydrogen gas, magnified 50 million times. Each molecule of hydrogen is made of two hydrogen atoms. Each molecule moves rapidly in a nearly straight line, changing direction only when it collides with another molecule or with the container wall. This and other microscopic drawings view only a tiny region within a much larger container.

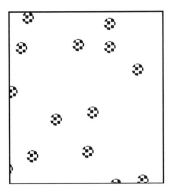

FIGURE 2.3
A simplified drawing of helium gas, magnified 50 million times. Each molecule of helium is simply an unattached atom of helium.

this book gives the name and standard abbreviation for each element and numbers them. Each number, known as the element's **atomic number**, represents a particular kind of atom. Higher atomic numbers correspond (with some exceptions) to heavier atoms. We will discuss the physical meaning of an atom's atomic number in Section 8.7.

Scientists found that certain groups of elements have similar chemical properties. If we list these groups having similar properties vertically and also list the elements in order of increasing atomic number, the result is the *periodic table* shown on the inside of the back cover. This table is a nice example of the regularities that scientists find in natural phenomena. When many elements had still not been discovered, scientists used the periodic table as a predictive device, by noting the table's unfilled gaps and searching for elements with properties that just fit those gaps.

What about all the other substances, those made of more than one element? A pure substance, such as pure water (with no impurities like salt or dirt mixed in), that is made of more than one element is called a **chemical compound**.

Imagine dividing a cup of water into smaller and smaller amounts. If the water is pure, you will always get just water—not something else such as salt or dirt. Working downward in size, you will eventually arrive at the smallest particle of water. In pure water, every one of these smallest particles must be a particle of water, and so they should be identical. And every particle must contain atoms of both hydrogen and oxygen, because we know that water can be chemically decomposed into these elements. This smallest particle that still has the characteristics of water is called a *molecule* of water.

This reasoning shows that every pure chemical compound must be made of tiny particles that are identical and that are themselves made of two or more kinds of atoms attached together into a single identifiable unit. Such a particle, the smallest particle of a compound that still has the characteristics of that compound, is called a **molecule** of that compound.

Some elements are made of two-atom molecules. For example, a molecule of hydrogen gas is made of two hydrogen atoms (Figure 2.2).* Helium gas, on the other hand, is made of individual helium atoms (Figure 2.3). The common forms of oxygen gas and nitrogen gas are made of two-atom molecules, and air is made primarily of these two kinds of molecules (Figure 2.4).

Chemists can deduce a compound's molecular structure by means of chemical decomposition and experiments like Dalton's. Such experiments show, for example, that the water molecule is made of two hydrogen atoms and one oxygen atom (Figure 2.5).

We represent compounds and elements by abbreviated formulas. For example, water is represented by H_2O, where the subscript 2 belongs to the symbol preceding it and indicates the number of atoms of that type in each molecule. Hydrogen gas is represented by H_2, oxygen by O_2, and helium by He.

Molecules can get pretty complicated. You might expect the molecules of life, such as your protein and DNA molecules, to be highly complex because they must carry the coded information needed to organize your body. Biological molecules are among the most varied and complicated known. Hemoglobin, the

---

*However, outside Earth's atmosphere nearly all the universe's hydrogen is in the "atomic" (single-atom) form rather than the "molecular" (two-atom) form, because the universe began with atomic hydrogen and remains mostly in that form.

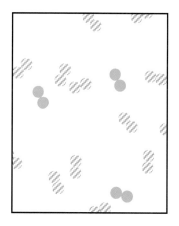

FIGURE 2.4
A simplified drawing of air, magnified 50 million times. Air is a mixture mostly of nitrogen (striped), and oxygen (blue) molecules, both two-atom molecules. About 80% of all air molecules are nitrogen and about 20% are oxygen.

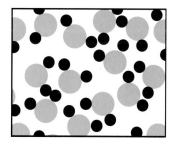

FIGURE 2.5
A simplified drawing of liquid water. As in the preceding figures, all the molecules are in rapid motion. In a liquid the molecules are in close contact and slide past one another. Each water molecule is made of one oxygen atom (blue) and two hydrogen atoms (black).

protein responsible for the red color of blood, has the formula $C_{3023}H_{4816}O_{872}N_{780}S_8Fe_4$. DNA molecules contain millions of atoms and vary from one individual to the next. DNA carries the instructions that make you you.

■ **DIALOGUE 3**   Name four other elements that have chemical properties similar to those of chlorine. Of these five similar elements, which is made of the lightest-weight atoms?

■ **DIALOGUE 4**   The simple sugar known as glucose has the chemical formula $C_6H_{12}O_6$. What elements, and how many atoms of each, does a glucose molecule contain?

■ **DIALOGUE 5**   What is the chemical formula for carbon dioxide (one carbon and two oxygen atoms)? Carbon monoxide (*mono*-oxide)? Methane (carbon and four hydrogens)? Sulfur dioxide? Carbon tetrachloride ("tetra" means "four")?

▌ 2.3 METRIC DISTANCES AND POWERS OF 10

Unfortunately, U.S. students must learn two systems of measurement: the nationally used "English" system based on feet and pounds and so forth, and the metric system that everybody else uses. Because not even England uses it any longer, we will refer to the feet-and-pounds system as the "English" (in quotes) system.

The basic metric distance unit is the **meter** (abbreviated m). It's about 39 inches, a little over 3 feet. Table 2.1 lists other metric distances and relates them to the meter. The most important are the *kilometer* (km), the *centimeter* (cm), and the *millimeter* (mm). Table 2.2 lists four common prefixes that can be attached to any metric unit: **mega-** (M), **kilo-** (k), **milli-** (m), and **micro-** (μ). For example, the kilowatt is 1000 watts, and the megawatt is 1 million watts (we'll later see what a watt is).

For handling large and small numbers, a technique known as **powers of 10** is useful. A power of 10 means 10 raised to some power. For instance, 10^2 means

TABLE 2.1

Metric distances

Name of unit	Distance	Conversion to "English" units
Kilometer (km)	$1000 \text{ m} = 10^3 \text{ m}$	1 km = 0.62 mi, 1 mi = 1.6 km
Meter (m)		1 m = 3.3 ft, 39 in, 1 ft = 0.30 m
Centimeter (cm)	$0.01 \text{ m} = 10^{-2} \text{ m}$	1 cm = 0.39 in, 1 in = 2.5 cm
Millimeter (mm)	$0.001 \text{ m} = 10^{-3} \text{ m}$	
Micrometer (μm)	$0.000001 \text{ m} = 10^{-6} \text{ m}$	

---

3. Fluorine, bromine, iodine, astatine (in the same column in the periodic table with chlorine). Fluorine is the lightest.

4. 6 carbons, 12 hydrogens, 6 oxygens.

5. CO_2, CO, CH_4, SO_2, CCl_4.

TABLE 2.2

Metric prefixes

Mega (M)	one million, 10^6
Kilo (k)	one thousand, 10^3
Milli (m)	one-thousandth, 10^{-3}
Micro (μ)	one-millionth, 10^{-6}

10×10, which equals 100, and 10^5 means $10 \times 10 \times 10 \times 10 \times 10 = 100,000$. The superscript is the power to which 10 is raised, the number of 10s that are multiplied together.

For example, the solar system's diameter (distance across) is 12,000,000,000,000 m. You can write this as $1.2 \times 10,000,000,000,000$ m or as 1.2×10^{13} m. Each multiplication by 10 moves the decimal point one place to the right, so to write out 1.2×10^{13}, you begin with 1.2 and move the decimal point 13 places to the right. The number in front (the 1.2) is usually written as a number between 1 and 10. If there is no number in front, you can think of a 1 in front; for instance, 10^5 is the same as 1×10^5.

Negative powers are used for small numbers. For instance, 10^{-2} means $1/10^2$, which equals 1/100, or 0.01, and 10^{-5} means $1/10^5 = 0.00001$. The minus sign indicates that the power of 10 is to be divided into 1. For example, the diameter of an atom is about 0.000 000 000 11 m, which can be written as 1.1×10^{-10} m. Since each division by 10 moves the decimal point one place to the left, to write 1.1×10^{-10}, you begin with 1.1 and move the decimal point 10 places to the left.

Thousand (10^3), *million* (10^6), *billion* (10^9), and *trillion* (10^{12}) all represent various powers of 10. Similarly, thousandth (10^{-3}), millionth (10^{-6}), and so forth represent negative powers of 10.

To multiply two powers of 10, just add their powers. For instance, $10^2 \times 10^5 = 10^{2+5} = 10^7$, and $10^2 \times 10^{-5} = 10^{2+(-5)} = 10^{-3}$. The numbers in front of the power of 10 must be grouped together first, before multiplying. For example,

$$(1.5 \times 10^2) \times (3 \times 10^5) = (1.5 \times 3) \times (10^2 \times 10^5) = 4.5 \times 10^7$$
$$(1.5 \times 10^2) \times (3 \times 10^{-5}) = (1.5 \times 3) \times (10^2 \times 10^{-5}) = 4.5 \times 10^{-3}$$

To divide two powers of 10, subtract the denominator's power from the numerator's power. For instance, $10^2 / 10^5 = 10^{2-5} = 10^{-3}$, and $10^2 / 10^{-5} = 10^{2-(-5)} = 10^7$. Numbers in front of the powers of 10 are grouped together first, before dividing them.

You can calculate all sorts of amazing things with powers of 10. For example, the solar system's diameter divided by an atom's diameter (the ratio of the two diameters) is

$$1.2 \times 10^{13} \text{ m}/1.1 \times 10^{-10} \text{ m} = (1.2/1.1) \times (10^{13}/10^{-10}) = 1.1 \times 10^{13-(-10)}$$
$$= 1.1 \times 10^{23}$$

This number, 110,000,000,000,000,000,000,000, is the number of atoms you would have to line up side by side in a row in order for them to stretch across the solar system.

Powers of 10 are something that all of us need to be able to handle. You need them just to begin to understand the U.S. national debt!

■ **DIALOGUE 6** (a) Convert 6 miles to yards (there are 1760 yards per mile). To inches. (b) Now convert 6 kilometers to meters. To centimeters. Would you rather do this sort of thing in "English" or metric units?

■ **DIALOGUE 7**   The universe is only seconds old, a million trillion seconds old, in fact. Write this number in ordinary notation and in powers of 10. The diameter of an atomic nucleus is about a hundredth of a trillionth of a meter. Write this in powers of 10.

2.4 THE ATOM'S EXPLANATORY POWER: THE ODOR OF VIOLETS

What are smells? This is a good question to introduce us to the power of the atomic theory and of careful observation.

Observation shows that we are surrounded by an invisible substance, air. You know it's there because you can feel the wind. The atomic theory tells us every material substance, anything that you can pick up or touch, is made of atoms. It's reasonable to suppose that air is a material substance, too, because you can feel it blow on you. A careful measurement would show that air has weight, further confirming our hypothesis. We conclude, from the atomic theory, that air is made of atoms. As we know from the Brownian motion experiment, the atoms in a liquid move all the time, even when the liquid appears to be motionless. So it seems reasonable to suppose that air molecules are in constant motion, too, even in still air.

Consider the odor of violets. Among a violet's various molecules, there must be some that make it smell the way it does. Chemists have learned how the violet's odor molecule is strung together (Figure 2.6). Scientifically, at least, that's what the smell of violets is—those molecules. In order for a violet's smell to spread out, odor molecules must break loose from the violet. Once in the air, moving air molecules knock odor molecules around; it's just like the Brownian motion of dust particles in water. This random jostling causes the odor molecules to spread out, to *diffuse*, in all directions in the air. Eventually, they reach your nose. Think about it the next time you smell something.

The atomic theory links human-scale or **macroscopic** phenomena that we can see around us to phenomena at the unseen **microscopic** level. The microscopic perspective is especially helpful in understanding the states of matter. Water, for example, comes in three states: ice, liquid, and steam. We call these the **solid state**, **liquid state**, and **gas state** of water. Nearly every substance can exist in any one of these three states. At the macroscopic level, the three states of matter are distinguished by their shapes. In a closed container, a solid maintains its shape; a liquid spreads out over the bottom; and a gas fills the volume. How do they differ at the microscopic level? A little thought, guided by the atomic theory and by some simple macroscopic observations, can answer this. As I hope you will discover throughout this book, it's amazing what careful thought guided by simple observations can accomplish!

Because solids maintain a fixed shape, their molecules must be locked into a fixed arrangement. Because solids are difficult to compress into a smaller volume, their molecules must be crowded against one another. The precise arrangement is determined by the ways in which the substance's molecules push and pull on one another when they get close together. If you have ever seen a large number of balls tossed one by one into a big box (Figure 2.7) or gunshot (BBs)

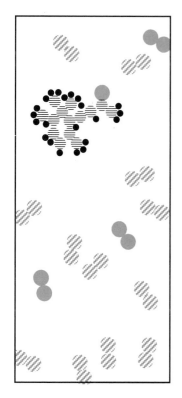

FIGURE 2.6
The odor of violets, in air. The odor-of-violets molecules are made of carbon (horizontal stripes), hydrogen (black), and a single oxygen atom (blue).

6. (a) 6 miles = 10,560 yards = 380,160 inches. (b) 6 km = 6000 m = 600,000 cm.

7. 1,000,000,000,000,000,000 seconds, or 10^{18} seconds; 10^{-14} m.

FIGURE 2.7
If you toss a large number of balls one by one into a box, the balls often lock into an orderly pattern. For similar reasons, the atoms in many solid materials lock into an orderly *crystal pattern* at the microscopic level.

FIGURE 2.8
An unseen microscopic order lies behind the beautiful symmetries seen in macroscopic crystals such as these hexagonal quartz crystals.

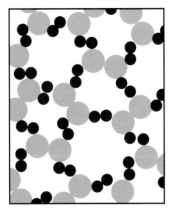

FIGURE 2.9
Ice. Compare this diagram of solid water with that of liquid water in Figure 2.5. In the solid state, atoms vibrate around their average position in the crystal pattern, but they do not migrate throughout the material, as water molecules do in the liquid state.

filling a small container, you can guess that molecules tend to lock into an orderly pattern that repeats itself throughout the solid. Figure 2.7 is, in fact, a reasonable representation of the arrangement of molecules in many solid substances. Orderly molecular patterns are responsible for the regular surfaces and beautiful symmetries seen in macroscopic crystals (Figure 2.8). Figure 2.9 shows the microscopic six-sided crystal pattern of ice, and Figure 2.10 shows the macroscopic snowflake crystals that are formed from it.

Because liquids have no fixed shape, their molecules must not be rigidly attached to one another. But liquids are about as difficult to compress into a smaller volume as are solids, so we expect that a liquid's molecules are crowded together about as closely as are a solid's molecules. This reasoning leads us to a microscopic picture of a liquid that is similar to a jumbled bowlful of marbles that can be molded into different shapes depending on the shape of the bowl. The molecules in a liquid can slide past one another and are free to migrate throughout the liquid. Most liquids take up slightly more volume than do the solids of the same substance, because their molecules need room to move around. But water is an exception to this rule; because of its open crystal structure, ice takes up slightly more volume than does liquid water (Figure 2.9).

Because gases can be compressed into a much smaller volume, their molecules must be widely separated. Gas molecules dart back and forth, bouncing off the container's walls or colliding and rebounding from one another. From this microscopic picture, we would expect that a gas should press outward against any surface it contacts. This outward press, or **gas pressure**, is caused by a continual torrent of gas molecules hitting the surrounding surfaces. It's as though hundreds of baseballs were thrown at a wall, pressing the wall backward. You can see the effect of gas pressure when you fill a balloon or tire with air. A balloon's elastic material is pressed outward by trillions of gas molecules hitting the walls every second.

Figure 2.11 pictures the differences between solids, liquids, and gases.

A complete absence of air and all other forms of matter is called a **vacuum**. A perfect vacuum is impossible to achieve in any ordinary macroscopic volume on Earth, but it is not difficult to achieve a partial vacuum, in which the container holds far less air than it would if filled with air at its normal density. A good vacuum in a laboratory still contains a trillion molecules in every cubic centimeter! But in space the large regions between galaxies are nearly perfect vacuums—the molecules are some 2 meters apart!

FIGURE 2.10
The hexagonal symmetry we see in snowflakes mirrors the hexagonal symmetry of the underlying crystal structure (compare this with Figure 2.9).

When you heat a substance, what happens to its molecules? With our understanding of solids, liquids, and gases, we are in a position to answer this important question. Consider an air-filled balloon with its top tied so that no air enters or leaves. What happens if you heat or cool it? Try it! Put the balloon in the freezer. Now hold it over boiling water. What happens? The balloon expands as the air warms inside it. Returning to our microscopic picture of a gas, we see that as the air in the balloon warms, the molecules inside must bounce harder off the inner walls in order to cause the balloon to expand. This means that, as the air warms, the molecules must move faster.

Experiments now amply confirm this connection between warmth or temperature on the one hand and molecular motion on the other. This is not only true in gases such as air, but also in liquids and in solids. That is, molecules are always in random (or disorganized) motion, whether in a gas, liquid, or solid, and those random motions get faster as the temperature increases. Gas molecules dart around faster and liquid molecules slide more rapidly past each other as the temperature increases. In solids, each molecule vibrates randomly around a fixed position, and these vibrations increase as the temperature increases.

The connection between warmth and molecular motion is so close that scientists consider warmth (or temperature) and molecular motion to be actually the same thing—two aspects of the same phenomenon. That is, the phenomenon known macroscopically as *warmth* is, at the microscopic level, molecular motion. Let's summarize this idea in the following way:

> **THE MICROSCOPIC INTERPRETATION OF WARMTH**
> At the microscopic level, temperature (warmth) is the random, or disorganized, motion of a substance's molecules. This motion cannot be directly observed macroscopically but is observed instead as temperature or warmth. When temperature increases, microscopic motion also increases.

The atomic theory unifies the odor of violets, the three states of matter, chemical compounds, gas pressure, temperature, and much more. It is a good theory.

■ **DIALOGUE 8**   Give several observations that confirm that we are surrounded by air.

■ **DIALOGUE 9**   Can you suggest an experiment that would show that air has weight?

■ **DIALOGUE 10**   What if, in addition to its random molecular motion, all the air molecules in some volume of air had an overall *collective* motion, all of them moving, say, eastward. Could this collective motion be observed macroscopically? What would we call it?

---

8. Trees bending in the wind, leaves blowing in the wind, air entering and leaving your nose, wind blowing a candle flame, an air-filled balloon.

9. Weigh two identical rigid containers, one containing air and one that has had some of its air removed. If air has weight, the air-filled container should weigh a little more.

10. It can be observed as wind.

(a)

(b)

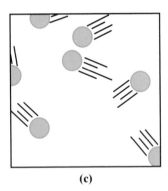

(c)

FIGURE 2.11
The (a) solid, (b) liquid, and (c) gas states of matter, viewed microscopically.

FIGURE 2.12
Chains of thorium atoms, as detected by a scanning electron microscope.

2.5 THE SMALLNESS OF ATOMS: WE ALL ARE BREATHING ONE ANOTHER

The most convincing evidence for atoms is to see one. But it is very difficult to see atoms with ordinary light, even with the best optical microscopes. The reason lies in the nature of light. Light is a wave, similar in some ways to water waves on the surface of a pond. The wavelength of light, the distance from one crest to the next, is very small—10 to 100 times smaller than the smallest visible dust particle. But a single atom is much smaller still—5000 times smaller than the wavelength of light. To visualize this, imagine that light has a wavelength of 5 meters. On this scale, an atom would be a speck just 1 millimeter across!

H O W   D O   W E   K N O W ?

Before 1970, Brownian motion was probably the closest we had come to seeing atoms. The object actually seen, through a microscope, is a tiny dust particle darting this way and that, but we can infer from its sudden changes in direction that atoms are bumping it.

In 1970, scientists developed a more direct way to detect individual atoms: the scanning electron microscope. It shoots a steady stream, or beam, of tiny material particles called electrons at the object to be detected. Your television set sprays an electron beam across the inside of the screen's face. An electron beam is fundamentally different from a light beam because electrons are particles of matter—material substance having weight—whereas light is not made of matter. During the 1920s, physicists discovered that every particle of matter, such as an electron, has a certain kind of wave associated with it. These waves are called *psi waves.**

Scanning electron microscopes use electron psi waves, having wavelengths thousands of times smaller than the wavelength of light, to detect individual atoms. As a powerful yet extremely thin electron beam scans, or sweeps by, an individual atom, the beam's psi wave is disturbed. The disturbed psi wave is then monitored by devices that collect and record the patterns made by the electrons. The result is an image such as Figure 2.12, a photograph of a scanning electron microscope image of a string of thorium atoms. It is the first clear image of individual atoms.

The scanning tunneling microscope, developed in 1983 and also based on electron psi waves, operates on an even smaller scale. A tiny probe, shaped like a sharp pencil tip a few atoms wide, scans the surface of a specimen. Electrons sense the shape of the surface by moving between the probe and the surface in a process known as *quantum tunneling* (Figure 2.13). The results are processed by a computer and displayed as a three-dimensional image of the atomic details of the surface (Figure 2.14).

Atoms were detected using light waves in a 1994 experiment in which a scanning tunneling microscope was used in a manner that caused the atoms in a solid surface to emit light in a very precise pattern. This light was then detected by a

---
*There is more about electrons in Chapter 8, more about light in Chapters 8 and 9, and more about psi waves in Chapters 13 and 14.

FIGURE 2.13
The tip of a scanning tunneling microscope and a small portion of the specimen being scanned.

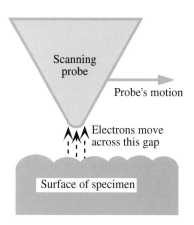

Scanning probe

Probe's motion

Electrons move across this gap

Surface of specimen

FIGURE 2.14
A scanning tunneling microscope image of a tiny part of the surface of a silicon crystal, showing the location of individual atoms. The surface shows a few areas containing impurities.

photomultiplier—a device in which light causes an electric current because of a phenomenon known as the *photoelectric effect* (see Chapter 13). In another experiment, an isolated atom was suspended in vacuum by electric and magnetic fields, and then placed in the path of a laser beam. The laser beam caused the single atom to emit light that could be seen with the naked eye.

We can even use the scanning tunneling microscope to perform the thought experiment that Leucippus and Democritus could only imagine 2500 years ago. In addition to detecting individual atoms, the probe tip can pick up individual atoms and drag them from place to place. In 1990, scientists picked up single atoms of xenon gas and rearranged them.* In effect, they divided xenon gas into the smallest particles of xenon, into atoms, just as Leucippus and Democritus had supposed was possible. These scientists positioned 35 xenon atoms to spell out the name of their laboratory (Figure 2.15). Democritus would have been delighted!

FIGURE 2.15
Thirty-five individual xenon atoms have been manipulated into position by the tip of a scanning tunneling microscope. The distance between atoms in the pattern is about 10^{-9} m, or 10 times the width of a single atom.

Atoms are among the smallest of all objects (Figure 2.16). The nucleus at the atom's center (see Section 2.7) is about 100,000 times smaller. The electron, also found within the atom, is known to be smaller than the smallest distance yet measured, which makes it at least 100,000 times smaller than the nucleus. Since it might in fact have zero size, the electron does not appear in Figure 2.16.

At the other end of the scale, the galaxies are among the largest objects known. Larger still are clusters of galaxies, forming relatively thin "sheets" of galaxies that can individually stretch across as much as 1% of the entire known universe. The largest structures ever detected are the ripples in the so-called cosmic background radiation, the faint afterglow of the great explosion that created the universe. These ripples stretch across as much as two-thirds of the known universe.

Standing somewhere in the middle, somewhere between the atoms and the stars, is humankind (Figure 2.16).

*Xenon atoms do not combine easily with other atoms, making them easy to manipulate.

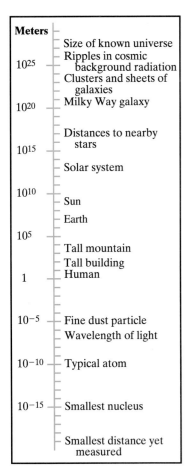

FIGURE 2.16
The range of sizes in the universe.

Atoms are pretty small. If you put a million of them side by side, the lineup would be no longer than the period at the end of this sentence. The head of a pin contains more than 10^{18} atoms. One breath of air, about 1 liter (1000 cm^3, about a quart), contains more than 10^{22} atoms.

Now, 10^{22} happens to be about the number of liters of air in the Earth's atmosphere, too, which leads to an interesting conclusion. Any particular parcel of air, such as the liter of air you will exhale in your next breath, mixes throughout Earth's atmosphere during the next few years. This means that of the air you exhaled a few years ago in any particular breath, about one or two atoms are now in every liter of air on Earth and inside the lungs of every person on Earth! And one or two atoms breathed out by every person on Earth, in any particular breath, are in your lungs now. One or two atoms from George Washington's first breath, from his dying breath, and from every other breath he ever took, are in your lungs right now. So are atoms from each of the breaths of all the other people who have ever lived. We all are breathing one another.

Atoms are ageless. Earth's atoms have been here since Earth formed 5 billion years ago, and very few have changed during that time.* It is only the connections between atoms that have changed. A particular oxygen atom might be part of a nerve cell in your brain today, part of an atmospheric water molecule a century from now, and part of a tree a century after that. "Your" atoms, the ones in your body, have just been borrowed from the air, from the breaths of every person who ever lived, from Earth, borrowed by the acts of breathing and eating, to be given back perhaps soon, perhaps later, to be given back entirely when you die.

MAKING ESTIMATES

About how many times have you exhaled during your life? Of all the atoms you have exhaled during your life, about how many will your class instructor inhale during his or her next breath?

Such questions look difficult at first glance. Although it would be impossible to answer them precisely, it is surprisingly easy to make a rough estimate. In estimating very large numbers like the number of times you have exhaled, an estimate to the nearest power of 10 is usually good enough. In other words, is the answer closer to 10, or 10^2, or 10^3, and so on? Such estimates are called *order-of-magnitude estimates*.

We will make such estimates of all sorts of things throughout this book, but do not expect to get a single correct answer. Although different people make different estimates, all of them should be in the same ballpark.

How many exhales in a lifetime? Here is my estimate (yours may be different): By timing my breathing, I find I take about 12 breaths per minute (note how **per** is used—it always means "in each"). To get the number of breaths per year, multiply 12 breaths per minute by 60 minutes per hour, then by 24 hours per day, then by 365 days per year: $12 \times 60 \times 24 \times 365$. Since we are making only

Every time you drink a glass of water, you are probably imbibing at least one atom that passed through the bladder of Aristotle. A tantalizingly surprising result, but it follows [from the simple] observation that there are many more molecules in a glass of water than there are glasses of water in the sea.

RICHARD DAWKINS,
ZOOLOGIST, OXFORD
UNIVERSITY

*Only those relatively few atoms that have been involved in radioactive decay, fission, or fusion have changed into different types of atoms.

a rough estimate, we might as well round off these numbers so that they will be easy to multiply: Make it $10 \times 60 \times 25 \times 400$.

$$10 \times 60 \times 25 \times 400 = (1 \times 6 \times 25 \times 4) \times 10^4$$
$$= 600 \times 10^4 = 6 \times 10^6 \text{ (6 million) breaths per year}$$

I am 64 years old, or "about 60" (I like this better than "about 70"). So the number of times I have exhaled is about

$$60 \times 6 \times 10^6 = 360 \times 10^6 = 3.6 \times 10^8$$

or about 400 million exhales. If you are younger, your answer will be smaller.

Now, how many of my exhaled atoms will some other person—you, for example—inhale in your next breath? Recall that every liter of air contains about one atom from every one of my exhaled breaths (except for recent breaths that have not yet had time to mix throughout Earth's atmosphere). So you will inhale about one atom from every one of my exhaled breaths in your next breath. Since I have exhaled some 400 million times, you will inhale about 400 million of my exhaled atoms in your next breath! And in your next breath, you also will inhale some 400 million atoms from the exhaled breaths of each person living on Earth and from each person who has ever lived on Earth. And all of this will only be a small fraction of the total number of atoms you will inhale on that breath. It's something to consider when you take a breath.

■ **DIALOGUE 11 MAKING ESTIMATES**   About how many millimeters thick is one sheet of paper? *Hint*: Roughly how thick is a 500-sheet package of typing paper?

■ **DIALOGUE 12 MAKING ESTIMATES: THE SIZE OF $1 TRILLION**   If you stacked up a trillion dollars in new $100 bills, about how many kilometers high would the stack be? *Hint*: Assume that they stack like typing paper, and see the previous dialogue.

2.6 ATOMIC MATERIALISM: ATOMS AND EMPTY SPACE

From 1550 to 1700, the revolutionary ideas of Copernicus, Kepler, and others became widespread, and consequently, to educated people Earth was no longer the motionless central focus of the universe. Humans became passengers on one planet among many, inhabitants of an impersonal mechanical universe (Figure 2.17). The new point of view, coupled with the new freedom of ideas in Renaissance Europe, stimulated an advancing scientific tide. The high point was Newtonian physics, the ideas about motion, force, and gravity developed by Isaac Newton (1642–1727) and others. Newtonian physics dominated science during

---

11. A 500-sheet package of typing paper is about 4 to 6 cm (a few inches) thick, say 5 cm. So the thickness of one sheet is about $5/500 = 0.01$ cm = 0.1 mm (1/10 of 1 mm).

12. The number of $100 bills needed is $10^{12}/100 = 10^{10}$. The thickness of one bill (previous dialogue) is 0.01 cm = 10^{-2} cm = 10^{-4} m. The height of the stack is $10^{10} \times 10^{-4}$ m = 10^6 m = 10^3 km = 1000 km (about 600 miles)!

the eighteenth and nineteenth centuries and is still influential. Its influence extended far beyond science to the general culture of the times, and it deeply affected the way that people thought about themselves, their society, and their role in the universe. Although the foundations of Newtonian physics have been superseded today by new theories, Newtonian ideas continue to dominate our general culture.

In summing up his scientific career, Isaac Newton once stated, "If I have seen farther than others, it is by standing on the shoulders of giants." Two such giants, René Descartes (1596–1650) and Galileo Galilei (1564–1642), strongly influenced the philosophical underpinnings of Newton's physics. Together, Descartes, Galileo, and Newton were the leading founders of science as we know it today.

Although there was very little evidence at that time for the atomic theory, the idea of atoms underlies much of the work of Descartes, Galileo, and Newton. It was an idea that went pretty deep, a philosophical idea. Democritus stated it in a far-reaching form:

> By convention sweet is sweet, by convention bitter is bitter, by convention hot is hot, by convention cold is cold, by convention color is color. But in reality there are atoms and empty space. That is, the objects of sense are supposed to be real, and it is customary to regard them as such, but in truth they are not. Only the atoms and empty space are real.

This goes considerably further than the atomic theory. Democritus is saying that not only matter but *everything* is made of atoms. And he goes beyond this to say that atoms are all there is—only atoms are real. Sweetness, bitterness, hot,

FIGURE 2.17
One way of picturing humankind's shift in psychological perspective, when medieval science gave way to the new science of Copernicus and Newton. The cozy pre-Newtonian universe was replaced by a vast impersonal mechanical universe. This woodcut was made during the nineteenth century, long after the transition had taken place.

cold, color, and all other sense impressions are only ways of speaking. They are not real. Only atoms moving in empty space are real.

According to this view, when you say "the water is hot" or "the shirt is red," you really mean that the atoms in the water and shirt and in your body are moving in a certain way. There really is no such thing as hot or red—there are only atoms. And when you say that "the painting is beautiful," you really mean only that the painting's atoms cause your brain's atoms to move in a certain way, a way that ultimately causes your vocal chords to form the words "the painting is beautiful." It is all purely mechanical. The beauty that you think you perceive in the painting is not real, and your thinking is not real either; only the atoms are real. And so it is with all sense impressions, all feelings: Beauty, ugliness, love, hate, and so forth are really only atoms moving mechanically in your brain or in the external world.

According to Descartes, sense impressions are merely *secondary qualities*, qualities that exist not in the real universe of atoms but rather only in our minds. The real universe contains only the atoms and their physical properties, such as weight and size. These are the *primary qualities*. According to Descartes, science is the study of the primary realm, and so its chief task is to explain natural phenomena by means of atoms. Although Newtonian physics does not require this point of view, it is quite congenial to it.

But such a view—concerning what is real and what is less real—goes beyond what can be observed and verified, and so it lies outside science. It is a philosophical view. The view of the Greek atomists and the associated views of the founders of modern science are one version of **materialism**: the philosophical view that matter is the only reality and that everything is determined by its impersonal, machinelike workings. Not that Descartes, Galileo, or Newton were materialists themselves—it was psychologically impossible for an educated person of that time to maintain a purely materialistic position. They all subscribed to nonmaterialistic religious ideas, in particular a belief in God. But in the new scientific philosophy there was little room for the God of the Middle Ages, a God who is continually and intimately involved in the world. Instead, these thinkers believed in a creator-God, a God who created the universe and set it into motion, who established the physical laws of the universe once and for all, and who then made his presence felt only by maintaining those laws.

Newtonian physics is quite compatible with a purely materialist philosophical position. As we will see (Chapters 3, 4, and 5), Newtonian physics explains an astonishing range of phenomena, in ways that are compatible with the idea that the universe is an impersonal clocklike machine whose parts are atoms. Since Newton, many later ideas have pointed in the same direction. We have seen in this chapter that heat is the motion of atoms. We will see in Chapter 9 that light and color are caused by vibrations of atoms. According to modern biochemistry, human sense impressions are connected with chemical changes in certain molecules in the human brain. It certainly seems that every observable phenomenon is connected with the arrangement or motion of atoms. Is it true, then, that the atoms and their motions is all there is?

For every good philosophical idea, there are always good arguments on the other side. One of the twentieth century's greatest scientists, the physicist Niels Bohr, put it this way: "The hallmark of a profound idea is that its converse is also profound." And so it is with atomic materialism. Respectable arguments on the other side include the following:

All nature then, as it exists by itself, is founded on two things: There are bodies and there is void in which these bodies are placed and through which they move about.
ROMAN POET LUCRETIUS, ABOUT 50 B.C.

All these things being considered, it seems probable to me that God in the Beginning formed Matter in solid, massy, hard, impenetrable, movable Particles, of such Sizes and figures, and with such other Properties, and in such Proportion to space, as most conduced to the end for which he formed them.
ISAAC NEWTON, 1704

1. Science always starts from evidence. All evidence comes ultimately from sense impressions, whether it is the smell of bread or the visual appearance of an electron microscope image. Materialism, by arguing that atoms are the primary reality and that sense impressions are only secondary manifestations of atoms, has the cart before the horse. The direct evidence is that "this smell exists," and "this visual image exists." It is the theoretical idea of atoms that is secondary to this direct evidence, rather than the other way around.

2. Considered as a philosophical worldview, materialism is a view that is rooted in science. But science is only one way of viewing reality. Other views—religious, aesthetic, intuitive, and psychological—have equal claim to being real.

3. Considered as a scientific principle, atomic materialism is only tentative, just as all scientific ideas are tentative. For example, we cannot know for certain that everything is really made of atoms, because we have not yet observed everything.

4. During the past century, scientists have found that most of the fundamentals of Newtonian physics are incorrect.*

As we will see throughout the last half of this book, post-Newtonian physics can be interpreted to have philosophical implications quite different from those of Newtonian physics. Whereas Newtonian physics is congenial to materialism, recent theories seem neutral or perhaps even uncongenial to materialism.

Lucretius is the truly Roman heroic poet; his heroes are the atoms, indestructible, impenetrable, well-armed, lacking all qualities but these; a war of all against all, the stubborn form of eternal substance. Nature without gods, gods without a world.

KARL MARX

2.7 THREE ATOMIC MODELS: GREEK, PLANETARY, AND QUANTUM

Science's way of working back and forth between observations and theories comes with a kind of guarantee of success. If an observation agrees with a theory, that's fine for that theory. And if an observation disagrees with a theory, that's too bad for that theory, but it's fine for science because science makes its greatest progress by repairing, or replacing, disproved theories.

The historic evolution of the atomic theory of matter is a good example. Science's notion of the nature of atoms has changed several times. In the original Greek model of the atom, an atom was an unchangeable, single object like a small and rigid pea. The Greek atom was also Descartes's and Galileo's and Newton's way of looking at the atom, and it got considerable experimental support from the nineteenth-century discoveries of elements, compounds, and Brownian motion.

Around 1900, experiments with electricity (Chapter 8) contradicted the Greek model of the atom. Electricity had been studied throughout the nineteenth century. Nobody suspected that an entirely new model of the atom would be needed to explain these experiments until, in 1897, scientists discovered a new, very lightweight, "electrified" particle. It was the first discovery of a

---

*Newtonian physics is only approximately correct over only a limited range of phenomena. Outside that limited range it is not even approximately correct.

FIGURE 2.18
An atom of helium and an atom of
carbon, according to the
planetary model of the atom. The
small black dots are electrons in
orbit around the nucleus, and the
blue circles and white circles
represent protons and neutrons in
the atom's nucleus. This is not
drawn to scale! The nucleus
should be 100,000 times smaller
than the electron orbits, and most
physicists believe that the electron
has no size at all.

particle that weighed less than an atom. It was, apparently, one part of the so-called a-tom (remember, the word means *indivisible*). This particle was the **electron**.

A few years later, in 1911, physicists discovered that an atom is itself nearly entirely empty space and that nearly all of an atom's material substance resides in a tiny central core, or **nucleus**. Experiments indicated that each atom also contained electrons moving through the large empty regions outside the nucleus. Scientists developed the theory that electrons orbit the nucleus much as planets orbit the sun. Later, scientists learned that the nucleus itself is made of two kinds of subnuclear particles, **protons** and **neutrons**. Figure 2.18 pictures an atom of the element helium and an atom of the element carbon, according to this **planetary model of the atom**. The planetary model got support from experiments in nuclear physics, chemistry, and electricity. Niels Bohr developed a version of the planetary model that explained certain previously inexplicable observations of the light that atoms can emit.

During the 1920s, new experiments involving electrons contradicted the planetary model and, in fact, contradicted Newtonian physics itself. An entirely new theory of matter was needed to explain the new results. The planetary model was inconsistent with this new *quantum theory* (Chapters 13 and 14), and so scientists developed a new model, the **quantum theory of the atom**. To date, no disagreements have been found between experiments and the quantum theory of the atom.

There is a funny thing about the quantum theory of the atom: It cannot be visualized or drawn. In fact, it is a fundamental principle of quantum theory, known as the *uncertainty principle* (Chapter 14), that any picture of the microscopic world, such as Figure 2.18, must contain features that are experimentally wrong. Briefly, the microworld is not visualizable! This is why we speak of the quantum theory rather than the quantum model of the atom: A model means a visualizable theory.

There is another funny thing about the atom: Not only is it mostly empty space, it is in a sense all empty space! Today's theories represent the electron as truly a point particle—simply a geometric point, occupying no volume, residing at the center of the electron's *force field* (force fields are discussed in Chapter 9). Furthermore, it is now fairly firmly established that neutrons and protons are themselves made of still-smaller particles called *quarks* (Chapter 18), with each neutron being made of a particular combination of three quarks and each proton being made of another particular combination of three quarks. Like the electron, a quark is thought to be a point particle—a geometric point in space surrounded by its force field. Since quarks and electrons occupy no volume at all, 100% of an atom's volume is occupied only by force fields. So it seems that an atom is all empty space, filled only with force fields and geometric points at the centers of those fields.

So our understanding of the atom has evolved through at least three different theories. At each stage, new experiments disproved the old theory, and scientists invented a broader theory that explained both the old and the new observations. But scientists did not discard the Greek and planetary models, despite their shortcomings, because these models are useful within their proper range. For instance, we can use the Greek atom to explain all sorts of things like air pressure and the odor of violets. We need not resort to the planetary atom or the quantum atom to explain these things, because the atom's internal structure and its quantum nature are irrelevant to these phenomena. Restricted to its proper range, the Greek atom is a fine theory. Once again, we see that theories are best described as useful rather than true.

■ **DIALOGUE 13** Since the quantum theory of the atom agrees with every experiment to date, can we now say that this theory is certainly correct?

2.8 CHEMISTRY AND LIFE: WHAT DID ATOMS EVER DO FOR YOU?

You can get atoms to do fantastic things by connecting them in sufficiently subtle ways. It is possible, for example, for a pile of atoms to acquire additional atoms from its environment, to move itself from one place to another, to respond to external events such as the presence of particular molecules or light or warmth, and to create copies of itself. We would call such a pile of atoms *alive*. Indeed, you are just such a pile of atoms. The pile of atoms that is you has an especially surprising property: It is aware of itself, and in this scientific age it is even aware that it is a pile of atoms.

The element that gives life its powerful abilities is carbon, which is plentiful on Earth and connects readily to a variety of other plentiful elements. Other elements that are abundant in biological molecules are oxygen, hydrogen, and nitrogen (remember "COHN").

Any rearrangement of molecules into new molecular forms is called a **chemical reaction**. In the remainder of this section, we will study three chemical reactions that are important in everyone's life: burning, respiration, and photosynthesis.

It is instructive to present **burning*** within its historical context. It was once believed that fire was one of the substances of which things are made and that a burning object released the fire that it already contained. The prevailing theory was that this intangible and nonmaterial substance—fire—had weight and carried its weight away when any object burned. Around 1780, Antoine Lavoisier studied burning more closely. He accurately weighed all of the materials involved when an object burned, including the gases used and given off. The prevailing theory predicted that the weight should decrease, since fire supposedly carried away some of the weight. But Lavoisier showed that there was no net change in weight. This was one of the earliest experiments supporting the idea that everything is made of indestructible atoms, and it initiated the modern science of **chemistry**—the study of the properties and transformations of substances (chemical compounds).

The key to the new science of chemistry was the concept that chemical reactions are rearrangements of atoms that are themselves changeless and indestructible. It followed that the total amount of matter involved in any chemical reac-

---

13. No, general scientific theories always are tentative.
*Burning is one form of *combustion*, which means any chemical reaction that generates heat.

tion is the same before and after the reaction. This idea, known as the **conservation** (preservation) **of matter**, or the conservation of mass, was assumed to extend to every physical process. It was and is a useful theory, especially in chemistry, but experiments have proved it wrong. Although it is a good approximation in chemical reactions, in other situations it is entirely wrong.

Most burnable substances are derived from biological materials that contain carbon and perhaps hydrogen, along with other elements. As simple experiments demonstrate (see the Home Projects), burning requires air, and something is removed from the air during burning.

Air is not a pure substance but is a mixture of molecules of many different substances. Nitrogen and oxygen dominate: Nearly 80% of air's molecules are nitrogen (N_2), about 20% are oxygen (O_2), and 1% are single argon atoms. All the other gases added together total less than 1%. These "trace gases" include all sorts of compounds. Some, such as water vapor (H_2O) and helium (He), are natural. Others, such as carbon dioxide (CO_2) and ozone (O_3), come from both natural and industrial sources. And still others, such as carbon monoxide (CO) and complex chlorofluorocarbon and hydrocarbon molecules, come almost entirely from industry.

For burning, the crucial component needed from air is oxygen. Carbon from the burning substance combines with oxygen from the air to form carbon dioxide. We abbreviate the preceding sentence symbolically as

$$C + O_2 \rightarrow CO_2$$

The plus sign means *combines with*, and the arrow means *changes into*. If the fuel contains hydrogen, it too combines with oxygen to form water vapor. For example, methane gas, CH_4, is the simplest of the hydrocarbon (hydrogen and carbon) fuels. It burns in air to form carbon dioxide and water vapor*

$$CH_4 + O_2 \rightarrow CO_2 + H_2O$$

An important feature of any chemical reaction is its energy balance. For now, we use the important word *energy* to mean either of two things: heat or the ability to move things around. As you know, heat—perhaps from friction or a burning match—is needed to start a substance burning. Once it starts, the burning reaction itself creates more than enough heat to maintain itself, so excess heat is given off. Including heat, the reaction formula for burning a typical fuel such as methane is

$$CH_4 + O_2 \rightarrow CO_2 + H_2O + \text{excess heat}$$

Let us now turn to biology. Animals get their bodily material and their energy from the food they eat and the air they breathe. Your blood absorbs carbon-

---

*This formula is not quantitatively "balanced." For example, there are four Hs on the left, but only two on the right. The balanced formula is $CH_4 + 2O_2 \rightarrow CO_2 + 2H_2O$. Since we are not interested here in how much of each compound enters into a reaction, we omit the numerical coefficients.

based molecules from food, and oxygen from air, and ferries them all over your body. When they arrive at, say, your thumb, they enter a biological cell there. In a process known as **respiration**, certain parts of the cell use these substances to create biologically useful energy. In a typical case, a simple sugar called glucose reacts with oxygen to create carbon dioxide and water:

$$C_6H_{12}O_6 + O_2 \rightarrow CO_2 + H_2O + \text{useful energy}$$

As you can see, this is similar to burning. Animal life is a slow burn. So it is not surprising that it generates biologically useful energy, just as burning generates heat. In respiration, part of this energy goes into making a high-energy molecule known as ATP, and the rest appears as heat. ATP is the energy carrier in animals. It can remain in storage, or it can move from place to place within the cell. It can be used for all sorts of things, such as bending your thumb. As you can see from the reaction formula, respiration generates water and carbon dioxide as waste products that are excreted in your sweat, urine, and exhaled breath.

Plants and animals have different strategies for generating useful energy. Plants use energy directly from the sun, whereas animals gather energy by eating plants and other animals that have stored it. Plants gather carbon dioxide and water from their surroundings and put them together to form high-energy carbohydrates (carbon compounded with water) such as glucose. From the formula for respiration, you can see that this process in plants is exactly the opposite of respiration in animals! Since respiration generates useful energy, we expect the opposite reaction in plants to require an input of energy. Plants have worked out a complicated process that gets this needed energy from the sun. Since animals consume oxygen, we also expect that plants must generate oxygen. The overall reaction is

$$CO_2 + H_2O + \text{solar energy} \rightarrow C_6H_{12}O_6 + O_2$$

It is called **photosynthesis** (putting together by light).

Both the food and the energy that animals use can be traced to plants. We depend on plants not only for food and fuel but also for the crucial component of the very air we breathe. Earth's oxygen comes from photosynthesis. Without plants, animals would soon be out of food, fuel, and breath.

Plants and animals complement each other quite nicely in the balance that evolution has worked out on Earth. The carbon dioxide exhaled by animals is taken in by plants for photosynthesis, and the oxygen released by plants is inhaled by animals for respiration. The two kingdoms of life recycle each other's primary waste products.

■ **DIALOGUE 14**   Is there another element, other than carbon, that could be the key element in forming living organisms elsewhere in the universe? *Hint*: In what table should you look to find an element whose chemical properties are similar to carbon's?

---

14. In the periodic table, beneath carbon, is silicon. It is quite abundant on Earth, in sand.

"OF COURSE THE ELEMENTS ARE EARTH, WATER, FIRE AND AIR. BUT WHAT ABOUT CHROMIUM? SURELY YOU CAN'T IGNORE CHROMIUM."

SUMMARY OF IDEAS AND TERMS

Ideas and terms are listed here in the order of their appearance in the chapter, so that you can use this list as a brief summary of the chapter. At the end of the book, you will find a complete and alphabetized glossary that includes all of these ideas and terms.

Atomic theory of matter   All matter is made of tiny particles, too small to be seen.

Diffusion   The spreading out of a collection of particles due to their random jostling by one another or by other neighboring particles.

Chemical decomposition   Any process that changes a single substance into two or more other substances.

Chemical element   A substance that cannot be chemically decomposed, of which 112 are known. They are listed in the **periodic table**, along with their **atomic number**. An **atom** is the smallest particle of an element.

Chemical compound   A pure substance that can be chemically decomposed. Its smallest particle is a **molecule**, two or more atoms connected into a single unit. All of a compound's molecules are identical.

Meter (m)   The basic metric unit of distance. About 39 inches. Other distances: kilometer (km), millimeter (mm). The most common metric prefixes are **mega-** (M), **kilo-** (k), **milli-** (m), and **micro-** (μ).

Powers of 10   10 raised to some positive or negative power. Powers of ten are used to express very large or very small numbers.

Microscopic   Too small to be seen with the unaided eye, as opposed to **macroscopic**, or visible to the unaided eye.

Solid, liquid, and gas   The three states of matter. Nearly every substance can exist in any of the three states. A solid's molecules are locked closely together in a regular pattern, a liquid's molecules are close together but not fixed in position, and a gas's molecules are far apart and move around rapidly.

Gas pressure   The outward push caused by gas molecules hitting the walls of the gas's container.

Vacuum   An absence of atoms and other material particles.

The microscopic interpretation of warmth   Temperature is associated with disorganized, or random, microscopic motions that are not visible macroscopically.

Per   In each.

Newtonian physics   The ideas about motion, force, and gravity developed by Isaac Newton and others around 1700.

Materialism   The philosophy that only matter is real and that everything is determined by its impersonal workings. The early Greek atomists were atomic materialists. Newtonian physics is compatible with this philosophy.

Models of the atom   The **Greek atom** is a tiny indestructible object, like a small and rigid pea. The **planetary atom** is made of parts, including a tiny central **nucleus** containing **protons** and **neutrons**, and one or more **electrons** orbiting far outside the nucleus. The **quantum theory of the atom** is based on the post-Newtonian quantum theory.

Chemistry   The study of the properties and transformations of substances (chemical compounds).

Conservation of matter   The idea that the total amount of matter involved in any chemical reaction, or in any other physical process, is the same after the reaction as it was before. This is also called conservation of mass. Although it is a useful theory, especially in chemistry, experiments have proved that it is only a useful approximation in chemical reactions, and entirely wrong in many other situations.

Air   A mixture of several chemical compounds: N_2 (about 80%), O_2 (about 20%), argon (about 1%), and a smattering of trace gases.

Chemical reaction   A rearrangement of atoms into new molecular forms. **Burning** creates heat by combining oxygen from air with a fuel such as carbon or hydrogen. **Respiration** in animals combines oxygen and carbon-based molecules such as glucose to generate useful energy, along with carbon dioxide and water. **Photosynthesis** in plants combines atmospheric carbon dioxide and water to form carbohydrate molecules such as glucose, along with oxygen.

REVIEW QUESTIONS

Each chapter has review questions, exercises, and problems at the end. Review questions go over many of the essentials. You can find the answers by looking in the chapter. Exercises ask you to draw on your understanding of the material to answer new, conceptual (nonquantitative) questions not directly answered in the chapter. The many exercises allow your instructor flexibility; you will probably want to tackle only some of them. Problems are similar to exercises but they are quantitative, and occasionally algebraic, rather than conceptual. Answers to the odd-numbered exercises and problems are given at the back of the book.

THE ATOMIC IDEA
1. What macroscopic evidence is there for atoms?
2. What light-based microscope evidence is there for atoms?
3. What experiment did the ancient Greek atomists imagine doing, and what did they believe the result would be?
4. An experiment such as the Greeks (previous question) imagined was actually carried out recently. Describe it.
5. Which is bigger, an atom or the wavelength of light? A little bigger or a lot?

ATOMS AND MOLECULES
6. From the microscopic point of view, what is the difference between an element and a compound?
7. From a macroscopic point of view, what is the difference between an element and a compound?
8. What is the difference between an atom and a molecule?

9. Why is the periodic table arranged in the way that it is?
10. If you chemically decompose water, will you get anything like water? What will you get?

THE ATOM'S EXPLANATORY POWER
11. Describe the microscopic process by which perfume gives off an odor that you can smell some distance away.
12. How do solids, liquids, and gases differ macroscopically? Microscopically?
13. Which is easiest to compress: solids, liquids, or gases? Why?
14. Is a perfect vacuum ever attained on Earth, over a volume as large as 1 cubic centimeter? Elsewhere?
15. What is the microscopic difference between hot water and cold water?

ATOMIC MATERIALISM AND ATOMIC MODELS
16. Name several things that people ordinarily regard as real but that, according to atomic materialism, are not real.
17. Describe the philosophy of materialism.
18. Give arguments for the materialist philosophy.
19. Give arguments against the materialist philosophy.
20. Name three different models of the atom. Describe two of them.

CHEMISTRY AND LIFE
21. What is meant by a *chemical reaction*?
22. Name three different chemical reactions.

23. Is air a single substance (a single compound)? Describe its chemical composition.
24. Describe an experiment involving burning that supports the notion of conservation of matter.
25. In what types of experiments is the conservation of matter correct to a very good approximation?
26. Describe the following reactions: burning, respiration, and photosynthesis.

HOME PROJECTS

There is, literally, nothing like "hands on" experience in science. Try some of these at home!

1. The diffusion game. Draw a row of 11 squares, making each square large enough for a penny. Number and mark them as in Figure 2.19. Stack 4 pennies on square 0. Now flip another coin, and move the top penny one place to the right (to +1) or to the left (to −1), depending on whether the flip was heads or tails. Do the same with the other three pennies, flipping each time. You have now completed one *diffusion step*. Take another diffusion step, moving each penny one place rightward or leftward, depending on the coin flip. Take several diffusion steps. Do the pennies diffuse outward from the center? How many diffusion steps does it take before a penny "diffuses" to your nose or to Joe's nose? Who smelled the pennies first? What are the similarities and differences between this process and the physical diffusion of an odor discussed in this chapter?

FIGURE 2.19
The diffusion game.

2. Evidence that *hotter* means *faster molecules*. Blow up a balloon and tie off its top. Measure its diameter by placing it between two bricks or other movable vertical surfaces. Now put it in your freezer for a minute or so, and quickly remove and measure again (you could perhaps measure it inside the freezer, but this wastes electricity). Hold it over boiling water for a minute or so, and quickly measure again.

3. You can use your results from the preceding project, along with a thermometer, to check one of the gas laws. This law, also called Charles's law, states that the fractional change in volume is proportional to the fractional change in temperature. Since the volume of a sphere is proportional to the cube of its diameter, the fractional change in volume is just $(\Delta D/D)^3$, where D represents the balloon's diameter (any units) and ΔD represents the change in its diameter (same units). The fractional change in temperature (the change in temperature divided by the temperature) must be expressed in *absolute* temperature units. The absolute temperature is the ordinary Celsius temperature plus 273°. If you happen to live in a country that is still using degrees Fahrenheit, then the absolute temperature is the Fahrenheit temperature plus 460°. For safety, be sure you are using a thermometer that goes to the boiling temperature.

4. You may have heard that warm water freezes faster than cold water does. Try it and see, starting with identical ice trays of cold and warm water.

5. Light two identical candles, and invert a large jar over one. What happens? Explain. Now get two identical large jars, take a deep breath of the air from one, and exhale back into it. Quickly put the jars over the candles. Does one candle go out before the other? Explain what you observe. Simultaneously place a liter jar over one candle and a half-liter jar over the other. Does the candle in the liter jar burn twice as long?

FOR DISCUSSION

The discussion questions are meant to stimulate thinking about questions that have no specific "correct" answer. They can be used for class discussion, individual essays, or individual thought.

1. What problems would be created if the United States switched to the metric system? What would be the advantages?

2. Would you know that there is such a thing as "air" if nobody had ever told you about it?

3. In what respects do you agree or disagree with the atomic materialist philosophy described in this chapter? Defend each of your opinions. Points to consider: What happens at the microscopic level when you have a sense impression, for instance, when you smell perfume or touch a warm object? Is that all that happens? What if,

instead of a smell or touch, your sense impression had been of love, anger, beauty, or religious feelings?

4. Social commentators often note that modern society is materialistic. What does this use of the word *materialism* mean? Is this *social* materialism the same thing as the *philosophical* materialism described in this chapter? Are social and philosophical materialism at all related?

EXERCISES

THE ATOMIC IDEA

1. Is the atomic theory known, for certain, to be true?
2. Carbon atoms are about 25% lighter than oxygen atoms (the ratio of their weights is 3 to 4). What is the weight ratio of the carbon and oxygen that go into the formation of carbon monoxide? Answer the same question for carbon dioxide.
3. A carbon atom is 12 times heavier than a hydrogen atom. If methane (CH_4) is chemically decomposed, what will be the ratio of the weights of the resulting carbon and hydrogen?
4. A carbon atom is 12 times heavier than a hydrogen atom, and an oxygen atom is 16 times heavier than a hydrogen atom. If glucose ($C_6H_{12}O_6$) is chemically decomposed, what will be the ratio of the weights of the resulting elements?

ATOMS AND MOLECULES

5. How many atoms are in a molecule of H_2SO_4 (sulfuric acid)?
6. How many atoms are in the alcohol molecule C_2H_5OH?
7. Which of these is a pure compound, which is an element, and which is neither: helium gas, carbon dioxide, polluted water, $C_6H_{12}O_6$, gold, steam?
8. Which of these is a pure compound, which is an element, and which is neither: pure water, oxygen gas, liquid mercury, H_2SO_4, U, air?
9. Suppose you obtained the smallest single particle of each of the following substances. In which cases would this particle be a molecule made of more than one atom, and in which cases would it be a single unattached atom: pure water, oxygen gas in the form found in Earth's atmosphere, H_2SO_4, U, He, carbon dioxide, H_2, H?
10. Helium is an inert gas, meaning that it does not readily enter into chemical reactions with other substances. List five other substances that you would expect to also be inert gases.
11. Chlorine has a strong tendency to combine with a single hydrogen atom to form HCl. Look in the periodic table and list at least three other elements that you would expect to combine with hydrogen the way that chlorine does.
12. Consider a pure chemical substance A. Suppose that it can be chemically decomposed into two other pure substances B and C. Can we then conclude that B and C *must* be elements? That B and C *must* be chemical compounds? That A *must* be a chemical compound?

THE ATOM'S EXPLANATORY POWER

13. Why can't you observe Brownian motion in easily visible objects such as floating bits of paper?
14. What is the chemical formula for the odor of violets (see Figure 2.6)?
15. A dog follows an escaped convict's trail by putting its nose to the ground. Explain this from a microscopic point of view.
16. If air is put into a sealed container that is then compressed (reduced in volume), what do you predict will happen to the air pressure on the container walls? Explain this from a microscopic point of view.
17. If air is put into a sealed container and heated, what do you predict will happen to the air pressure on the container walls? Explain this from a microscopic point of view.
18. If a balloon is partially filled with air (so that it isn't fully expanded), sealed, and then heated, what do you predict will happen to the balloon? Explain this from a microscopic point of view. What if the balloon is cooled instead?
19. Why is it so difficult to remove the lid from a vacuum-sealed jar?
20. Suppose that you observe tiny pollen grains floating in still air enclosed in a glass bottle, in Brownian motion. What would happen if you increased the amount of air? What would happen if you heated the air?
21. Why does the air pressure in a tire increase as you add air? Why does the air pressure in a tire increase as you heat the tire?

THE SMALLNESS OF ATOMS

22. Put these in order from lightest to heaviest: water molecule, oxygen atom, raindrop, hydrogen atom, glucose molecule, electron, DNA molecule.
23. Put these in order from lightest to heaviest: H_2 molecule, methane molecule, fine dust particle, hemoglobin molecule, proton, glucose molecule.
24. How old are a baby's atoms? Are they older than an old person's atoms? What about a baby's DNA molecules?

CHEMISTRY AND LIFE

25. What is the chemical reaction formula for burning hydrogen gas in air? What substance is created by this reaction?
26. For safety, gas-filled balloons are filled with helium instead of hydrogen. What does this tell you about the behavior of helium in the atmosphere?

27. Gasoline is a hydrocarbon fuel. What are the main compounds created when gasoline burns in a car engine?
28. NO_x (nitrogen oxide and nitrogen dioxide) is one pollutant from automobiles. What elements must combine to form NO_x?
29. Referring to the preceding exercise: Gasoline contains neither oxygen nor nitrogen. So where must these elements come from when NO_x is formed in car engines?
30. Are there any molecules in your body that you could claim are "your" molecules, unique to your body and unlike any other molecules in the universe?

PROBLEMS

ATOMS AND MOLECULES

1. On the simplifying assumption that oxygen and carbon atoms have the same weight, how many tons of carbon dioxide gas are formed when one ton of coal burns (coal is nearly pure carbon)?
2. Referring to the preceding problem: In a typical large coal-fed electrical generating plant, a ton of coal is burned every 10 seconds. About how many tons of carbon dioxide enter the atmosphere every hour from such a plant?

METRIC DISTANCES AND POWERS OF 10

3. Write as ordinary numbers: 10^9; 10^{-6}; 3.6×10^{13}; 5.9×10^{-8}.
4. Write in powers of 10 notation: 3 trillion; five-thousandths; 730,000,000,000,000; 0.000 000 000 082.
5. The distance to the moon is 384,000 km. Express this using powers of 10. How far is this in meters? In millimeters?
6. Our Milky Way galaxy contains perhaps 400 billion stars. Suppose that 0.05% (i.e., 0.0005) of these stars have planetary systems containing at least one Earthlike planet (one that could conceivably support Earthlike life). Express these two numbers using powers of 10, and then multiply them together to find how many Earthlike planets there are in our galaxy. Express this number in words (thousands or millions, etc.).

THE SMALLNESS OF ATOMS

7. Making estimates. According to Dialogue 11, one sheet of paper is about 0.1 mm thick. An atom is about 10^{-10} m across. About how many atoms thick is one sheet of paper?
8. Making estimates. The average weight, per atom, of the atoms in your body is about 10^{-26} kg (2×10^{-26} pounds). About how many atoms are there in your body?
9. Making estimates. The smallest dust particle visible to the unaided eye measures about 0.05 mm across. About how many *atoms* across is this? In other words, if we line up atoms side by side, about how many would it take to make a line of atoms 0.05 mm long?
10. Making estimates. Referring to the preceding exercise: Assume the small dust particle is shaped like a cube. About how many atoms does it contain?

PART TWO

THE
NEWTONIAN
UNIVERSE
▼
a clockwork
kingdom

How Things Move

CHAPTER 3

Galileo asks the right questions

When you look around, your eye falls on this book, a flower, your foot. What are these things made of? It's a natural question if you want to study your physical environment. We saw in Chapter 2 that the ancient Greeks, the earliest people to seriously ask such questions about their physical surroundings, answered that things are made of atoms.

You notice that the book falls, the flower sways, your foot taps. Why, and how, do things move? Again, the ancient Greeks asked first. And this is what we ask, in this and the following two chapters.

The Greek philosopher and scientist Aristotle (Figure 3.1) developed the earliest theory of motion. His theory was intuitively plausible, and had some observational support, but post-Renaissance (1600–1700) European scientists such as Galileo and Newton found Aristotle's theories inadequate. They discarded Aristotelian physics in favor of powerful new ideas, proposed by Isaac Newton and others, that then dominated science for three centuries. Beginning in 1900, science again changed its fundamental view of motion, when the relativity and quantum theories altered most of Newtonian physics.

Chapters 3, 4, and 5 discuss Newtonian physics. Although these ideas have been superseded today, they continue to be useful for understanding the way the macroscopic world around us works. Most important, Newtonian views have retained their powerful cultural influence.

After a look at Aristotelian physics (Section 3.1), this chapter discusses Galileo's ideas about motion itself as distinct from the causes of motion. Section 3.2 presents Galileo's objections to Aristotelian physics and the experimental background for the law of inertia, the foundation of Newtonian physics, and Section 3.3 examines this law. Sections 3.4 and 3.5 explore the two ideas needed to describe motion, namely velocity and acceleration. Section 3.6 applies all of this to a familiar and fundamental phenomenon, falling.

Looking ahead, Chapter 4 discusses the causes of motion, more accurately stated as the causes of the changes in motion. Chapter 5 discusses one such cause: gravity. Throughout, we focus on those aspects of Newtonian physics that are importantly related to our lives and our times.

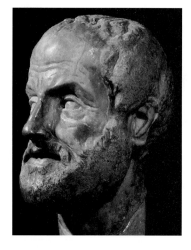

FIGURE 3.1
The Greek philosopher and scientist Aristotle, 384–322 B.C. He developed the earliest theory of motion. His theory, which agreed with our common intuitive notions, was replaced by the far less intuitive theory of Newton and, during the twentieth century, by the even less intuitive quantum theory.

3.1 ARISTOTELIAN PHYSICS: A COMMONSENSE VIEW

Aristotle's physics agrees with most people's commonsense feelings about motion. But these plausible notions are precisely the ones that Newtonian physics discarded. Because Aristotelian physics is so ingrained in our intuitions, we must look at it closely to see why, and where, it is wrong.

In his science, Aristotle was no armchair philosopher. Keenly observant of his surroundings, he is regarded as the father of biology because of his observations of plants and animals. In observing motion, Aristotle noticed that some motions maintain themselves without assistance but that others could be maintained only by an outside agent. One example of motion that Aristotle considered to be unassisted: When you push a rock off a ledge, it falls toward the ground with no obvious assistance. Aristotle regarded motion that could maintain itself in this manner to be **natural motion**.

In addition to solid objects falling, Aristotle perceived three other sorts of natural motion: liquids falling or running downhill, air rising, and flames leaping upward. These four represented to Aristotle the four natural motions that could occur on Earth. He believed that natural motion occurred because everything on Earth was made of four different substances or elements—earth, water, air, and fire—and that each of these was striving to reach a different natural place. Although **Aristotle's elements** are different from the chemical elements (Chapter 2), the word *element* is used in a similar way today to mean the substances of which everything is made.

Each Aristotelian element has its own type of natural motion and its own natural place toward which it strives: The element earth moves downward because Earth's center is its natural resting place (educated Greeks knew that Earth was a sphere). Water flows downward because its natural place is just above earth. Air rises because the atmosphere is its natural place, and fire leaps upward because its natural place is above the air.

Aristotle believed that all objects were composed of these four elements and that an object's natural motion was determined by how much of each element the object contained. So a rock sank in water because rocks were mostly made of earth; wood floated because it contained air in addition to earth; and warm air rose because it contained fire.

All these natural motions are directly downward or upward. Horizontal motions seem different. When you pull a cart along a road, throw a stone horizontally, or push a box along the floor, your activity maintains the motion: You keep the cart moving; the stone moves only because you threw it; you must push the box. Pushes and pulls are needed to keep the cart, the ball, and the box moving. Aristotle believed these pushes and pulls were needed because objects must be forced to behave contrary to their own natures, in other words, contrary to their natural motion. Aristotle classified all such forced motion as *violent motion*, meaning that an external push or pull was needed to maintain it.

He believed that all motion on Earth was either natural or violent, but he perceived in the heavens an entirely different kind of motion. He believed that the moon, sun, planets, and stars were made of a fifth element, *ether* (from the Greek word for *to kindle or blaze*), not found on Earth. Unlike the four earthly elements, ether had no weight and was incorruptible (unchangeable, eternal). Perfect in every way, ether's natural place was in the heavens, and it naturally moved in perfect circles around Earth. This third kind of motion was called *celestial motion*.

This theory explained a wide range of observations with a few general principles. It was a fitting and useful theory and gained wide acceptance, partly because of its plausibility. It does seem to us that a rock falls all by itself, that a push is needed to maintain horizontal motion, and that motion in the heavens really is different from motion on Earth. But we will see that Newtonian physics contradicts all of these ideas.

Aristotle's theory supported other Greek ideas about Earth's place in the universe. Earth was believed to be at rest, with the stars circling it. Aristotle's physics explained why: Earth could not move because it was made mostly of the element earth, whose natural resting place was at Earth's center. The stars moved in circles because that was ether's nature. Objects on Earth could not move in the way that the stars did, because there was no ether on Earth. Aristotle's physics explained why Earth was so different from the heavens.

Medieval Christian philosophy found Greek physics and astronomy quite congenial. An Earth-centered universe, an Earth made of corruptible elements, humans made of corruptible mortal elements, a natural resting place for everything, a perfect heaven made of incorruptible ether in eternal celestial motion—all of these agreed nicely with church theology. Until about 1600, an Earth-centered astronomy and Aristotle's physics were central to the philosophy and culture of European civilization.

Beginning in about 1550, Copernicus and others introduced new astronomical theories. New theories of physics were not far behind. Descartes, Galileo, and others began rethinking Aristotelian physics around 1600.*

■ **DIALOGUE 1**   How would Aristotle's theory explain these facts: People float better in water when their lungs are full of air; an air-filled balloon rises when heated; liquid mercury sinks in water?

Odd as it may seem, most people's views about motion are part of a system of physics [Aristotle's] that was proposed more than 2000 years ago and was experimentally shown to be inadequate at least 1400 years ago.

I. BERNARD COHEN,
SCIENCE HISTORIAN

3.2 HOW DO WE KNOW? DIFFICULTIES WITH ARISTOTELIAN PHYSICS

Aristotelian physics explained a lot, but it had its weaknesses. You can demonstrate some of these for yourself. Drop a piece of notebook paper to the ground. Now crumple it into a tight ball and drop it again. Does it fall faster?[†] It does fall noticeably faster, and this is not easy to explain on the basis of Aristotelian physics. The crumpled paper is still the same paper, with the same amount of Aristotelian earth in it. It should fall no faster than the flat sheet does.

Aristotle would argue here that the flat sheet falls slower because in plowing through so much of the element air, it carries along enough air that it now seeks the

---

*Post-Aristotelian ideas had actually been brewing for many centuries, although most educated people continued to believe Aristotle's physics until Newton's time.

1. Full lungs contain more of the element air, so they stay better above the element water; heated air contains the element fire, making it rise; liquid mercury contains a lot of the element earth.

†Please do these simple experiments when they are suggested, to reinforce your learning. There's nothing like observing the real thing!

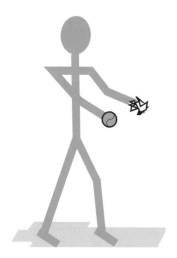

FIGURE 3.2
Hold a baseball and a wadded-up piece of paper above the ground, and drop them simultaneously. What does Aristotelian physics predict? What do you observe?

I do not feel obliged to believe that the same God who has endowed us with sense, reason, and intellect has intended us to forgo their use.

GALILEO

FIGURE 3.3
A feather dropped in vacuum falls as fast as a stone.

ground less strongly. To get around this argument, try dropping two objects that have the same shape—so that they plow through equal amounts of air—but that have very different weights: a block of wood and a block of iron of the same shape, or a baseball and a tightly crumpled piece of newspaper of about the same size (Figure 3.2). What does Aristotelian physics predict? What do you find?

According to Aristotelian physics, the heavier object, which contains much more earth, should fall noticeably faster. But it does not. If you do this experiment carefully and from a high place such as a second-story window, you might detect that the heavier object actually does fall a little faster, but not a lot faster as is predicted by Aristotle's physics. Today's theories predict that for two objects of the same shape, the lighter one will fall slower because of **air resistance**—the resistance to the motion of an object through the air due to the object's collisions with numerous air molecules. According to today's theories, air resistance also explains why a flat sheet of paper falls slowly.

One can test the hypothesis that these small differences in falling are due to air resistance by letting two objects fall in a vacuum, perhaps in a container from which the air has been removed. One then finds that the light and heavy objects fall at precisely the same speed. A feather dropped in a vacuum falls as fast as a stone, in clear contradiction of Aristotelian predictions (Figure 3.3)!

Aristotle's concept of violent motion also has problems. If you shoot an arrow, it can travel a great distance horizontally while hardly slowing down. A brief strong push from the bowstring starts it, but what external assistance keeps it moving once it is released from the bow? Aristotle himself had difficulty reconciling this sort of example with his own theory, and later scientists had similar difficulties.

Galileo Galilei (Figure 3.4) was a brilliant, cocky Italian who supported Copernican astronomy, issued sarcastic opinions about Aristotelian physics, and generally annoyed the authorities. His writings eventually earned him a visit by the Catholic inquisition, which "persuaded" the now elderly man to "confess" and then confined him to house arrest for his remaining 10 years of life. Even under house arrest, the irrepressible Galileo pursued his experiments and wrote a large physics book.

To focus his thinking, Galileo imagined the following experiment: Let a ball roll down an incline. Its speed will increase. Now give the ball a starting push and let it roll up an incline. It slows down (then stops and rolls back down). Suppose we make the inclines nearly horizontal (Figure 3.5). If you have ever let a ball roll down a very slight incline, you know that it is likely to slow down and come to rest, even though it is going downhill. Galileo understood that this slowing is due to the roughness of the incline and ball. Today we call it **friction**. Galileo's crucial step was to idealize the experiment by neglecting, at least in his mind, the effect of friction.

Galileo saw that neglecting friction, the ball would speed up on any downward incline, no matter how slight, and would slow down on any upward incline. Then he took another characteristically brilliant step: He considered the "limiting case" of slight inclines, namely, a perfectly horizontal surface. On a frictionless horizontal surface, the ball could neither speed up nor slow down because the horizontal surface was intermediate between downhill and uphill. Galileo concluded that in absence of friction, a ball that once started rolling on a horizontal surface would roll forever. This radically contradicted Aristotle's theory, which stated that continued pushing or pulling was needed to maintain violent (that is, horizontal) motion.

FIGURE 3.5
A smooth ball on a smooth incline always (a) speeds up going down and (b) slows down going up, even for a very slight incline. In the limiting case (c) of a perfectly smooth and level surface, the ball should keep going forever once it has started rolling.

FIGURE 3.4
Galileo Galilei, 1564–1642. He helped overthrow Aristotelian physics, helped formulate the law of inertia that formed the foundation for Newtonian physics, made many important astronomical discoveries that provided support for the new Copernican view of the universe, and did the first detailed experimental and theoretical work in the study of motion. But his most important contribution might have been his development and support of scientific methodology, *the way of science*: the notion that we learn not from authority but rather from experience and rational thought.

The Holy Spirit intended to teach us in the Bible how to go to heaven, not how the heavens go.
GALILEO

From the Divine Word, the Sacred Scripture and Nature did both alike proceed.... Nor does God less admirably discover himself to us in Nature's actions than in the Scripture's sacred dictions.
GALILEO

Galileo's methods have been crucial to science ever since. They included

- *Experiments*, designed to test specific hypotheses. Although careful observation dates back at least to Aristotle, Galileo was the first to refine this process with controlled experiments to test specific hypotheses.
- *Idealizations* of real-world conditions, to eliminate (at least in one's mind) any side effects that might obscure the main effects.
- *Limiting the scope of the inquiry* by considering only one question at a time. For example, Galileo did not make Aristotle's mistake of trying to understand both why horizontal motion occurs and how it occurs. Aristotle believed that horizontal motion occurred because of external assistance (which is why it occurs), so that an unassisted horizontally moving object should slow down and stop (that is how it occurs, says Aristotle). But Galileo looked more carefully at only the question of how horizontal motion occurred and found that in the absence of friction an object would not slow down.
- *Quantitative methods*. Galileo went to great lengths to measure, for example, the motion of bodies. He understood that a theory capable of making quantitative predictions was more powerful than one that could make only qualitative, descriptive predictions, because quantitative predictions were more specific and could be experimentally tested in greater detail.

Galileo was the first person to practice what we recognize today as the **scientific method**: the dynamic interplay between experience (in the form of experiments and observations) and thought (in the form of creatively constructed theories and hypotheses). This notion that we learn not from authority but rather from experience and rational thought is what make Galileo's work, and science itself, so powerful and enduring.

3.3 THE LAW OF INERTIA: THE FOUNDATION OF NEWTONIAN PHYSICS

With experiments and arguments such as Galileo's, scientists eventually arrived at a profound non-Aristotelian insight. It involved an extreme idealization: Suppose you could get away from the effects of friction and air resistance *and also gravity*. It isn't easy to imagine such a thing, because gravity is such an unchang-

Oh, my dear Kepler, how I wish that we could have one hearty laugh together! Here at Padua is the principal professor of philosophy, whom I have repeatedly and urgently requested to look at the moon and planets through my glass, which he perniciously refuses to do. Why are you not here? What shouts of laughter we should have at this glorious folly! And to hear the professor of philosophy at Pisa labouring before the Grand Duke with logical arguments, as if with magical incantations, to charm the new planets out of the sky.

GALILEO, IN A LETTER TO KEPLER, COMMENTING ON LIBRI, TEACHER OF PHILOSOPHY AT PADUA, WHO REFUSED EVEN TO LOOK INTO GALILEO'S TELESCOPE IN ORDER TO VIEW THE NEWLY DISCOVERED MOONS OF JUPITER

Libri did not choose to see my celestial trifles while he was on Earth; perhaps he will do so now he has gone to Heaven.

GALILEO'S FURTHER COMMENT ON LIBRI, WHO DIED SOON AFTER THE TELESCOPE INCIDENT

ing part of every person's daily existence that we don't even notice its presence. It's hard to imagine the absence of something that we aren't even aware of in the first place! French philosopher and scientist René Descartes was the first to arrive at the new insight, with contributions from Galileo.

What if you could turn off gravity? This question would have been meaningless to Aristotle. To Aristotle, there was no such thing as gravity. Things just fell, by themselves, because that was the nature of things. But the question had meaning from the enlarged post-Copernican perspective that saw Earth as one among many planets. One could imagine that on other planets, gravity might be very weak, that in outer space—far from all planets and stars—gravity might have practically no effect. One could extrapolate to a situation of zero gravity, just as Galileo had extrapolated to an absence of friction.

If you released a stone in midair and there were no gravity, the stone would not fall. It would hang in midair. And if you flicked the stone with your finger to start it moving and there were no gravity or friction or air resistance, *it would coast in a straight line with no change in speed, forever*! This was Descartes's new insight.

This idea was an invention of the human mind, like all scientific theories. But it was not a free invention of the mind; like all scientific theories, it was conditioned on observation of the natural world. Partly, it was a redefinition of *natural motion* and *externally influenced motion*. Not only friction but also gravity were now to be regarded as external influences rather than something that bodies provide for themselves. It was a new way of looking at nature, much as Copernican astronomy was a new way of looking at the heavens.*

Descartes is saying that if it were moving to begin with, an unassisted object will keep moving without external assistance. And an unassisted object that was at rest to begin with will stay at rest; it will hang in midair, for instance. It is a strange idea. At first, our intuition rejects it. Because gravity, friction, and air resistance are all around us, our intuition tells us that objects can keep moving only if they are pushed or pulled.

As a way of hanging on to our intuitive notion that something must assist an object if it is to keep moving, we give a word to an object's ability to keep moving: *inertia*. We also apply this word to an object's ability to stay at rest. In other words, an object's **inertia** is its ability to keep moving and to remain at rest. So all objects have inertia. All objects have the ability to keep moving and to remain at rest. This use of the word "inertia" doesn't really explain anything; inertia is simply a word that stands for the unexplainable fact that bodies do keep moving.

We summarize all of this as:

> **THE LAW OF INERTIA**[†]
> A body that is subject to no external influences (also called external forces) will stay at rest if it was at rest to begin with and will keep moving if it was moving to begin with; in the latter case, its motion will be in a straight line at an unchanging speed. In other words, all bodies have inertia.

---

*Historians refer to such a shift of fundamental viewpoint as a *paradigm shift*.

[†]This is often called Newton's first law, even though Descartes invented it, because Newton listed it first among his three basic principles of motion. We refer to these three principles as the law of inertia, Newton's law of motion, and the law of force pairs, rather than by their more common but less accurate, less descriptive, and more boring titles: Newton's first law, Newton's second law, and Newton's third law.

Of all the intellectual hurdles which the human mind has confronted and has overcome in the last fifteen hundred years the one which seems to me to have been the most amazing in character and the most stupendous in the scope of its consequences is the one relating to the problem of motion.

HERBERT BUTTERFIELD, HISTORIAN, REFERRING TO THE LAW OF INERTIA, IN *THE ORIGINS OF MODERN SCIENCE*

It is possible to see the law of inertia in action by observing motion that is horizontal (to eliminate the effect of gravity) and nearly frictionless. For example, an object that coasts on a cushion of air or other gas moves with practically no friction. Figure 3.6a is a multiple-flash photograph of such a moving coaster, made in a completely dark room with a camera whose shutter stayed open, using a rapidly flashing light to illuminate the coaster only briefly at several equally spaced times.

As you can see by checking the meter stick next to the coaster, the coaster is moving in a straight line at an unchanging speed: It moves the same distance (as far as it is possible to measure this in the photo) in each time interval. Although careful measurement would show that air resistance slows the coaster slightly, this example is close enough to Galileo's ideal case that if you view this demonstration in a lab, it will give you an intuitive feel for the law of inertia.

Space furnishes lots of nice examples. **Space** refers to those regions of the universe outside Earth and outside other astronomical objects, where "Earth" means the solid and liquid and gaseous Earth, including the atmosphere. The atmosphere is so thin at 10 kilometers (6 miles) high that it can no longer sustain life, although air resistance still affects spacecraft at this altitude. The atmosphere is so thin above 100 kilometers that the drag (air resistance) on satellites is nearly negligible. Figure 3.7 puts these altitudes into perspective. As you can see, even 100 kilometers high is still relatively near Earth. Beyond this lies space.

When astronauts traveled to the moon, their spacecraft's rocket engines first boosted (pushed) them up and into orbit around Earth. Then they fired their rocket engines for a few minutes to leave Earth's orbit and start toward the moon. Then they shut their engines off and coasted for three days, covering about 400,000 kilometers (more than a quarter of a million miles) to the moon. The spacecraft became a long-distance coaster and a great example of the law of inertia.

(a)

FIGURE 3.6
(a) Multiple-flash photo of the motion of a glider on a smooth surface, viewed from above. (b) The glider at rest on the smooth surface.

(b)

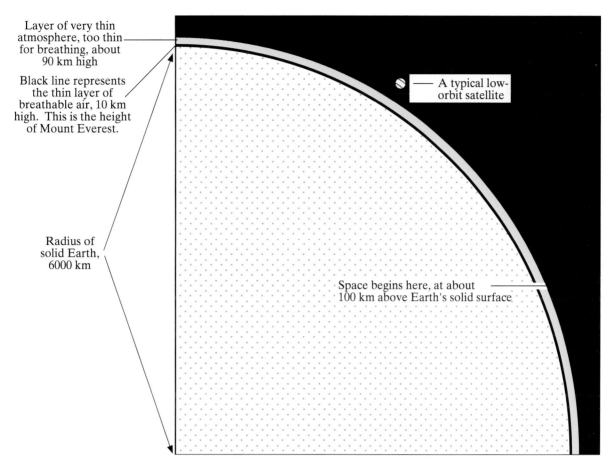

Layer of very thin atmosphere, too thin for breathing, about 90 km high

Black line represents the thin layer of breathable air, 10 km high. This is the height of Mount Everest.

Radius of solid Earth, 6000 km

A typical low-orbit satellite

Space begins here, at about 100 km above Earth's solid surface

FIGURE 3.7
The solid Earth and Earth's atmosphere and space, in perspective.

However, this coasting was not entirely free of external influences. Although there is no significant air resistance in space, there is still significant gravity unless the spacecraft is extremely far from all large bodies such as Earth and its sun and moon. Gravity has a very long-range effect. For example, at one-sixth of the distance to the moon, gravity is still 1% as strong as it is on Earth's surface—a much-reduced effect but still not negligible. Figure 3.8 demonstrates this. The spacecraft to the moon actually slowed during the first part of its journey because of the pull of Earth's gravity and then sped up during the last part because of the pull of the moon's gravity.

A big difficulty for a sun-centered astronomy, such as Copernicus proposed around 1550, was the problem of how Earth could keep moving with nothing to push it. This problem perplexed Copernicus, and it gave his opponents powerful ammunition.

The law of inertia has the answer, but it came a century too late to help Copernicus. Earth is a coaster in space! Like the astronauts coasting to the moon, Earth coasts around the sun. It keeps going because there is nothing in space to stop it. There are other questions that one can still ask about Earth's motion: Why does it move in a circle rather than in a straight line at unchanging speed? What started it moving in the first place? These questions cannot be an-

FIGURE 3.8
At one-sixth of the distance to the moon, gravity still has 1% of the effect it has on Earth's surface: much reduced but still not negligible.

Space begins here

60,000 km, one-sixth of the distance to the moon. At this distance, the effect of Earth's gravity is reduced to 1% of its normal value.

Radius of Earth, 6000 km

swered by the law of inertia, which tells us only why Earth keeps moving, not why it moves in a circle or how it got started. But here is some indication of the answers: The sun's gravity bends Earth's path into a circle, and Earth started moving because the gas and dust from which it was formed were already moving (see Chapter 5).

Another difficulty for a moving-Earth theory of astronomy is that it seems as though birds, clouds, and baseballs in flight should be left behind the moving Earth. Since they are not attached to the ground, how can they keep up with Earth as Earth moves through space? The answer is that birds, clouds, and baseballs have inertia, too. A bird standing on the ground is participating in Earth's 30-kilometer-per-second (70,000 miles per hour) coaster ride around the sun, so when the bird takes to the air, it is already moving at this speed around the sun, and the law of inertia says that there is no reason for it to stop. It coasts right along with Earth.

■ **DIALOGUE 2**   According to Aristotelian physics, what outside influences act on a stone while it falls? According to Newtonian physics?

■ **DIALOGUE 3**   Suppose that because of unknown causes, the sun suddenly "stood still," in other words, suddenly stopped moving across the sky as seen by an Earth-based observer. From the viewpoint of Copernican astronomy, what would this mean? Because of Earth's daily spin, a typical point on Earth's surface moves around Earth's center at about 1600 kilometers per hour (1000 miles per hour). So what would happen to people and houses at the moment the sun "stood still"?

3.4 MEASURING MOTION: SPEED AND VELOCITY

Sometimes **quantitative**, or numerical, methods are needed to get at nature's deeper secrets. At other times, however, quantitative details are superfluous, and **qualitative**, or nonnumerical, descriptions are preferable. For a clear under-

---

2. None. Gravity and air resistance.

3. It would mean that Earth stopped spinning. Because of the law of inertia, people and houses would then immediately start sliding across the Earth's surface, at 1600 km/hr. Eventually, friction would bring them to rest. There would also be 1600 km/hr winds across Earth's surface (10 times faster than strong hurricane winds); clouds would move at 1600 km/hr across the sky; and so forth. Most objects would slide for about 2 minutes, across a distance of about 25 kilometers (15 miles) and would be heated by friction to nearly the boiling point of water.

standing of motion and related ideas like force and energy, we need to think both qualitatively and quantitatively. And quantitative methods are needed to understand important practical matters such as the world's energy problems.

Scientists like to specify their measurements in terms of just a few basic quantities. To describe motion, only two are needed: distance and time. For example, suppose we want to describe quantitatively the motion photographed in Figure 3.6 using only the meter stick shown and a clock. Suppose the flasher flashes steadily at 0.40 second intervals and that we start the clock at the first flash, when the coaster is at the 10-centimeter (cm) mark. Then successive flashes occur at 0.40 seconds (s), 0.80 s, 1.20 s, 1.60 s, 2.00 s, and 2.40 s. Table 3.1 tabulates these data. The distances are estimated to the nearest millimeter (0.1 cm) from Figure 3.6.

How can we use these data to describe how fast the coaster is moving? In other words, how can these data tell us what a speedometer attached to the coaster would read? Inspection of Table 3.1 shows that during each 0.40 s time interval the coaster travels 14.1 cm. A speedometer tells you the distance traveled per (in each) unit of time, such as the number of kilometers per hour or the number of centimeters per second. To choose easier numbers, suppose you walked 6 kilometers in 2 hours. The number of kilometers in each hour would be 6 divided by 2, or 3 kilometers per hour. This is the distance divided by the time. So a speedometer attached to the coaster should register

$$\frac{14.1 \text{ cm}}{0.40 \text{ s}}$$

or 35.2 centimeters per second. We call this the coaster's **speed**. We abbreviate it as 35.2 cm/s, where the divide sign (/) is to be read "per" or "in each."

But there is a problem with defining speed in this way. Suppose you ride your bicycle 24 kilometers in 3 hours. According to this definition, your speed would be 24 km / 3 hr = 8 km/hr. But surely you did not maintain exactly this speed during every minute of the 3-hour trip. Rather, the 8 km/hr is an overall speed, actually the speed you would have had to maintain in order to make the entire trip at an unchanging speed. We call this the **average speed** for the trip. It is only in the special case of an unchanging speed that the average speed is really the speed during the entire time interval.

But a car's speedometer gives a value at every instant of time. What does a speedometer read? A speedometer is based on wheel rotation rates. If you look closely at the way it operates, you will find that it actually reads only an average speed during some time interval, but that this time interval is very short—so short that the speed is practically unchanging during this time interval.* This average speed during a time interval so short that the speed hardly changes is called the **instantaneous speed**. It is defined in the same way that average speed is, but with the understanding that the time interval is short. It is what

TABLE 3.1

Positions and times for coaster of Figure 3.6

Clock time (s)	Position (cm)
0.00	10.0
0.40	24.1
0.80	38.2
1.20	52.3
1.60	66.4
2.00	80.5
2.40	94.6

---

*A car speedometer is based on the effects created by a magnet that rotates when the car's wheels rotate. It usually gives a good approximation of the instantaneous speed. But if the wheel rotation rate changes very quickly, for instance, during sudden braking, the speedometer cannot respond fast enough to measure the instantaneous speed and so gives only an average. Every measured speed is really an average speed during some time interval.

a speedometer reads. We use the unmodified word *speed* to mean "instanta-neous speed," the kind of speed you read on a speedometer, and we will use "av-erage speed" when that is what we mean.

Quantitative statements, like our definition of average speed, are often easier to grasp if written as an abbreviated formula:

$$\text{average speed} = \frac{\text{distance traveled}}{\text{duration of time interval}}$$

This can be further abbreviated using symbols. We can choose the symbols to suit ourselves. We will use s, d, and t to stand for the average speed during some time interval, the distance traveled during that time interval, and the duration of that time interval. Then the formula is

$$s = \frac{d}{t}$$

Formulas like this are not essential to physics! The idea, not the formula, is essential. The formula can be stated in words, without symbols. And many of the most important principles, such as the law of inertia, are best stated without for-mulas. In fact, if you do use a formula, be sure that you can first state it com-pletely in words, because otherwise you could easily use the formula incorrectly. For example, the t in the speed formula's denominator is not just any arbitrary time—it means something very specific: the duration of the time interval during which the object traveled the distance indicated in the numerator. It's easy to get carried away by formulas in physics and to think that you understand when all you've done is remember a formula.

When you travel, it makes a difference which direction you move. Jogging at 10 km/hr northward will get you to a different place than will jogging at 10 km/hr westward. Speed and direction of motion occur together so frequently in physics that it is useful to have a separate word for the combination. We use the word **velocity** to mean *speed and direction*. The words *speed* and *velocity* are in-terchangeable in everyday language, but in physics they are not. The distinction might seem small, but small distinctions can be crucial.

Test your understanding of speed and velocity by trying these dialogue questions.

■ **DIALOGUE 4**   A car driving through a city travels 12 km in half an hour. A bicycler sprints for 1 minute at a steady 30 km/hr. Which has the higher average speed, the car or the bicycler?

■ **DIALOGUE 5**   Are the following speeds increasing, decreasing, or unchang-ing? (a) A car covers longer and longer distances in equal times. (b) A car takes longer and longer times to cover equal distances. (c) A car covers equal dis-tances in equal times. (d) A car covers equal distances in shorter and shorter times. (e) A car takes equal times to cover equal distances. (f) In equal times, a car covers shorter and shorter distances.

---

4. The average speed of the car = 12 km / 0.5 hr = 24 km/hr, which is slower than the bicycler.
5. (a) Increasing. (b) Decreasing. (c) Unchanging. (d) Increasing. (e) Unchanging. (f) Decreasing.

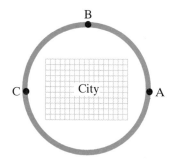

FIGURE 3.9
Illustration for Dialogue 7.

■ **DIALOGUE 6** Two bicyclers pass each other on a straight road, one moving north and the other south. Could they possibly have the same velocities? The same speeds?

■ **DIALOGUE 7** The automobile beltway around many cities is approximately circular. Suppose that you start driving at point A on the east side of the beltway and drive counterclockwise at an unchanging 90 km/hr (Figure 3.9). (a) As you pass point B on the north side, what is your speed? Your velocity? (b) As you pass point C on the west side, what is your speed? Your velocity? (c) Did your speed change? (d) Did your velocity change?

3.5 MEASURING MOTION: ACCELERATION

The law of inertia says that a body that feels no external forces must either move in a straight line at an unchanging speed, or remain at rest if it were at rest to begin with. But motion in a straight line at an unchanging speed is motion at an unchanging velocity, and the condition of remaining at rest is also a condition of unchanging velocity (the velocity remains zero). So we can state the law of inertia more concisely:

> **THE LAW OF INERTIA** (more concise form)
> A body that is subject to no external forces must maintain an unchanging velocity.

Now suppose that there are external forces. How will they affect an object's motion? It's not too difficult to guess the answer once you understand the law of inertia: External forces must cause changes in velocity. As we will see in the next chapter, this idea is central to Newtonian physics.

So changes in velocity are crucial to understanding Newtonian physics. Any object whose velocity is changing is said to be **accelerated**. The next dialogue will exercise your thinking about this idea. Remember: An object will be accelerated only if its velocity is changing, and velocity refers to the combined instantaneous speed and direction.

FIGURE 3.10
Illustration for Dialogue 8(e).

FIGURE 3.11
Illustration for Dialogue 8(f).

■ **DIALOGUE 8** During a trip, a car executes several different kinds of motion. For each case, state whether the car is accelerated: (a) moving along a straight level road at a steady 70 km/hr, (b) moving along a straight level road while speeding up from 70 km/hr to 80 km/hr, (c) moving along a straight level road while slowing down from 80 km/hr to 50 km/hr, (d) rounding a curve at a steady 50 km/hr, (e) moving uphill along a straight incline at a steady 50 km/hr (Figure 3.10), (f) rounding the top of a hill at a steady 50 km/hr (Figure 3.11), (g) stop-

---

6. No. Yes.

7. (a) 90 km/hr, 90 km/hr toward the west. (b) 90 km/hr, 90 km/hr toward the south. (c) No. (d) Yes.

8. (a) No. (b) Yes. (c) Yes. (d) Yes, because the direction of motion is changing, so the velocity is changing. (e) No, because neither the instantaneous speed nor the instantaneous direction of motion is changing. (f) Yes, because the direction of motion is changing. (g) Yes. (h) Yes.

ping (coming to rest) along a straight level road, and (h) starting up from rest along a straight level road.

We have seen how to describe velocity quantitatively in terms of two measured quantities: distance and time. What about acceleration? To answer this, imagine a car moving north along a straight level highway. Suppose it speeds up, say from 70 km/hr to 82 km/hr. Its change in speed is then 12 km/hr. Imagine how this would feel to you if you were in the car. It would make a difference to you how fast this change took place. For instance, if it took place over an entire hour, you would hardly notice it but if it occurred during one-tenth of a second, you could wind up with a whip-lashed neck! So it is the rate at which the speed changes—the amount of speed change per second—that we are interested in.

So we need to know the time interval during which the 12 km/hr speed change occurred. Suppose that this time interval is 8 s. Then the amount of speed change per second is

$$\frac{12 \text{ km/hr}}{8 \text{ s}}$$

or 1.5 *kilometers per hour per second*. These units tell us that in every second the speed changes by 1.5 km/hr. We will write this as 1.5 (km/hr)/s.

This useful quantity, which measures the rate of speeding up, is called the *acceleration* of the car. But remember that an object is said to be accelerated whenever its velocity changes and that the velocity changes not only when the object speeds up but also when it slows down or changes direction. This means that an object's **acceleration** is its change in velocity (and not simply the change in speed) divided by the time to make the change. Written as a formula,

$$\text{acceleration} = \frac{\text{change in velocity}}{\text{time to make the change}}$$

Although this definition is useful in physics, it has the drawback of not entirely agreeing with the popular meaning of acceleration. The scientific meaning includes not just speeding up (the common meaning of acceleration) but also slowing down (commonly called deceleration) and changing direction. Definitions of words like *acceleration* in physics are arbitrary, in the sense that nature does not tell us that we must define these words in any particular way. Physicists define their words for maximum convenience.

The distinction between velocity and acceleration is important and often misunderstood. *Velocity* refers to motion itself—an object has a velocity whenever it is moving. But *acceleration* refers only to changes in velocity.

The law of inertia says that in the absence of external forces, objects move with unchanging velocity. But *unchanging velocity* is the same thing as *unaccelerated*, which gives us yet another, very concise, way to state the law of inertia:

THE LAW OF INERTIA (most concise form)
A body that is subject to no external forces must be unaccelerated.

You might wonder exactly how you would measure the change in velocity in those cases in which the direction changes. We will take this up in Chapter 5. Here, our quantitative examples will stick to objects speeding up along a straight line. Then the change in velocity is numerically the same as the change in speed, and the acceleration is just the change in speed divided by the time to make that change.

 DIALOGUE 9   Give examples of an object that has (a) a high velocity and low acceleration, (b) a low velocity and high acceleration, and (c) an unchanging speed but high acceleration.

■ **DIALOGUE 10**   Let us return to the car speeding up along a straight road. Find its acceleration when (a) it speeds up from 70 to 82 km/hr in 4 s instead of in 8 s, (b) it speeds up from 70 to 82 in 16 s, (c) it speeds up from 70 to 94 km/hr in 8 s, and (d) it speeds up from 70 to 76 km/hr in 8 s.

3.6 GALILEO'S LAW OF FALLING

We have seen that simple experiments demonstrate that if air resistance is negligible, any two objects dropped together will fall together. We summarize this as:

> **GALILEO'S LAW OF FALLING**
> If air resistance is negligible, then any two objects that are dropped together will fall together, regardless of their weights and regardless of the substances of which they are made.

This is a pretty general statement. It applies to any two objects. Each object could be anything: a stone, shoe, frog, person, feather, balloon, even a helium-filled balloon, just as long as air resistance is negligible. For a feather, balloon, or helium-filled balloon, you need a nearly complete vacuum before air resistance becomes truly negligible, but once you do have a vacuum, you will find that even a helium-filled balloon that floats upward in air falls just like a stone. As Galileo put it, "a grain of sand falls as rapidly as a grindstone." Scientists get excited about ideas like this that apply to such a wide variety of situations.

Galileo's law of falling tells us that we can learn about the fall of any object by studying the fall of just one particular object, because all objects fall in the same way. How does your book fall?

---

9. (a) Any fast-moving object whose speed or direction of motion is changing slowly, such as a high-velocity bullet that is gradually slowing down; (b) any slow-moving object that is quickly speeding up, such as a car starting up rapidly; (c) any object turning a sharp corner at a high but unchanging speed.

10. (a) 3.0 (km/hr)/s, (b) 0.75 (km/hr)/s, (c) 3.0 (km/hr)/s, and (d) 0.75 (km/hr)/s.

You can learn a lot just by looking

Hold this book, flat, above the floor, and let it go.

—This is a pause, for dropping your book.

What is the numerical value of the book's speed at the instant you release it? While you are holding it, the book's speed is zero, so its speed at the instant of release must also be zero. But does its speed remain zero? Obviously not. At the beginning of the motion, the speed increases from zero to something bigger than zero. So the book must accelerate, at least at the beginning.

Now drop your book from about half a meter above the floor. Pick it up and drop it again from about 2 meters above the floor. (Maybe you should do this with some other book if you want to preserve this one!) On a hard floor, you should hear the book hit noticeably harder the second time. So the book must have been moving faster at the end of the second drop. This tells us that books get going faster during 2 meters of fall than during half a meter. You could repeat this experiment for other heights, with similar results demonstrating that objects that fall farther attain a higher speed at the end of the fall. Applied to the fall of a single object, this tells us that the farther an object falls, the faster it moves. In other words, neglecting air resistance, falling objects accelerate all the way down.

The preceding two paragraphs demonstrate the power of careful observation, especially the observation of everyday occurrences like a falling object. As the Yankee catcher and famous sage Yogi Berra once said, "You can learn a lot just by looking."

Figure 3.12 shows how falling can be measured quantitatively. It is a multiple-flash photo (described in Section 3.3) of a falling object. A billiard ball falls past a 2-meter stick while being photographed at several equally spaced times. The time interval between photos is 1/30 s. You can see the ball's acceleration: The images get farther and farther apart.

Figure 3.13 is an idealized diagram of a ball falling through a larger distance, neglecting air resistance. A real object falling farther than about 20 m is strongly affected by air resistance because air resistance becomes stronger as an object moves faster. We will, however, neglect the effect of air resistance for now, in order to focus on the effects due to gravity alone. So Figure 3.13 shows how an object would fall if dropped from a high building onto an airless Earth. An object like this whose falling is influenced by gravity alone, and not by air resistance or other influences, is said to be in **free fall**. The ball in the figure is in free fall for more than 5 s.

Suppose you measure distances downward from the release point and that a speedometer attached to the ball measures the ball's speed as it falls. At the instant of release, you start a clock that tells you the time elapsed since the ball was dropped. Figure 3.13 gives some of the data you would get in this idealized experiment. In order to emphasize the pattern, the actual experimental data are rounded.*

FIGURE 3.12
A multiple-flash photo of a falling billiard ball. The position scale is in centimeters, and the bulb flashed every 1/30 s.

---

*The rounding-off process introduces a 2% error. More precise distances are 4.9 m, 19.6 m, 44.1 m, 78.4 m, and 122.5 m. More precise speeds are 9.8 m/s, 19.6 m/s, 29.4 m/s, 39.2 m/s, and 49 m/s. Using these more accurate numbers, the calculated acceleration is $(9.8 \text{ m/s}) / (1 \text{ s}) = 9.8 \text{ m/s}^2$.

Time	Approximate distance	Approximate speed
0 s	0 m	0 m/s
1 s	5 m	10 m/s
2 s	20 m	20 m/s
3 s	45 m	30 m/s
4 s	80 m	40 m/s
5 s	125 m	50 m/s

FIGURE 3.13
A freely falling ball dropped from the top of a tall building. The effect of air resistance is neglected in this illustration.

You can see, in three different ways, that the ball accelerates: First, the drawing shows the stretching out noted in Figure 3.12. Second, the speed is greater at the end of each successive second, so the ball is moving faster and faster. Third, if you look closely at the distance data, you will see that the distances covered during each successive time interval grow larger and larger. For instance, the distance covered during 0 to 1 s is 5 m; the distance covered during 1 to 2 s is 15 m; the distance covered during 2 to 3 s is 25 m; and so forth. These distances (5 m, 15 m, 25 m) stretch out, so the ball must be accelerating.

How big is this acceleration? Recall that the acceleration during any time interval is the change in velocity during that time interval divided by the duration of that time interval. Since this is just straight-line motion at increasing speed, we can replace *velocity* in this prescription with *speed*, as discussed in the preceding section. Now look at each successive 1-second time interval. All the changes in speed are precisely the same! During 0 to 1 s, the speed change is 10 m/s. During 1 to 2 s, it is also 10 m/s, and so forth. So calculating the acceleration for any one of these time intervals,

$$\text{acceleration} = \frac{\text{change in speed}}{\text{time to change}} = \frac{10 \text{ m/s}}{1 \text{ s}}$$

which is 10 meters per second per second, or 10 (m/s)/s, usually abbreviated as 10 m/s^2. This means that the speed changes by 10 m/s in every second. The units can be confusing because both times are measured in seconds. Time must appear twice because acceleration is a rate of change of speed and speed is a rate of change of position. Two rates of change are involved.

Since we get 10 m/s^2 for every time interval, the acceleration is the same all the way down. And Galileo's principle of falling tells us that it must be the same for every freely falling object. The numerical value of this **acceleration due to gravity** is, more accurately, 9.8 m/s^2. When accuracy is important, you should use 9.8 rather than 10 m/s^2. The precise measured value varies by as much as 0.5% at different places in the world because of variations in Earth's shape and in the material of which Earth's crust is made. For example, it is 9.832 m/s^2 at the North Pole, 9.803 in New York City, and 9.782 in Panama, near the equator.

There is a pattern in the speed data. At times of 0 s, 1 s, 2 s, 3 s, 4 s, and 5 s, the speed (in m/s) is 0, 10, 20, 30, 40, 50, and so on. Just by looking at this string of numbers, it is easy to guess what should come next: 60. The reason this can be guessed is that there is a recognizable pattern here. Patterns are what scientists look for in nature. Kepler (see Chapter 1) searched for years through Brahe's data before finding the elliptical pattern in the data. Your expectation that the sun will rise tomorrow is also based on a pattern in nature.

One way to describe this pattern is with proportionalities. One quantity is **proportional to** a second quantity if doubling the first means you must double the second, tripling the first means you must triple the second, and so forth. In general, whatever you multiply the first quantity by, you must also multiply the second quantity by. Looking at the speed data, doubling the time (from 1 to 2 seconds, say) results in doubling the speed (from 10 to 20 m/s), and similar proportionalities hold throughout the data. So, for free fall:

speed is proportional to the time; $s \propto t$

The symbol $\propto$ means *is proportional to*.

Similarly, speed is proportional to the time for any object that starts from rest and moves in a straight line with unchanging acceleration. For example, an object dropped onto the surface of the moon falls freely (there is no air resistance on the moon, because there is no atmosphere) with an unchanging acceleration of 1.7 m/s^2. So a falling object reaches a speed of 1.7 m/s at the end of 1 s of falling, 3.4 m/s at the end of 2 s, 5.1 m/s at the end of 3 s, and so forth. As on Earth, the speed is proportional to the time; for instance, the speed at the end of 3 s is three times greater than the speed at the end of 1 s.

The pattern in the position data is not as easy to recognize. To make it easier to find, let's express the position data in multiples of the position at 1 s, in other words, in 5-meter units. The data tell us that at 2 s, the position (the total distance fallen) is 4 of these units, at 3 s it is 9 units, at 4 s it is 16 units, and at 5 s it is 25 units. So the positions, measured in 5-meter units, are 0, 1, 4, 9, 16, and 25. What is the next number in this sequence? In other words, what is the pattern?

To recognize the pattern, note that each of the numbers is a perfect square: 0^2, 1^2, 2^2, 3^2, 4^2, 5^2. What comes next? 6^2, or 36 of our 5-meter units! To get the distance in meters, multiply this by 5: $5 \times 36 = 180$ m.

This pattern can also be described quantitatively with proportionalities. The distances, in meters, are **proportional to the square of** the time. This means that doubling the time multiplies the distance by four, tripling the time multiplies the distance by nine, and so forth. In general, whatever number you multiply the time by, you must square this number and then multiply it by the distance. For instance, if you multiply the time by 3 (from 1 to 3 seconds, say), then you must multiply the distance by 3^2, or 9 ($9 \times 5 = 45$ m). So for free fall:

distance fallen is proportional to the square of the time; $d \propto t^2$

Again, this is a proportionality that is valid for any case of unchanging acceleration. For instance, distance is proportional to the square of the time for an object falling freely onto the moon.*

■ **DIALOGUE 11**   How much farther does a freely falling object fall in three times as much time? In four times as much time? Would your answers be the same if the object were falling freely onto the surface of Mars?
■ **DIALOGUE 12**   Answer "true" or "false" to each of the following statements regarding free fall: (a) The total distance covered keeps increasing during the fall. (b) The distance covered during each second keeps increasing. (c) The speed keeps increasing. (d) The change in speed during each second keeps increasing. (e) The acceleration keeps increasing.

---

*With further analysis, we could arrive at two general formulas for the speed and position of any object that starts from rest and moves in a straight line with unchanging acceleration:
$$s = at, \qquad d = (\tfrac{1}{2}) at^2$$
where a represents the acceleration. For freely falling objects on Earth, $a = 9.8$ m/s^2.
11. Nine times as far. Sixteen times as far. Yes, both answers would be the same on Mars.
12. (a) True. (b) True. (c) True. (d) False. (e) False.

SUMMARY OF IDEAS AND TERMS

Natural motion Unassisted motion. According to Aristotelian physics, falling is one form of natural motion, and horizontal motion is always unnatural, or violent. According to Newtonian (and current) physics, any motion at a constant velocity is natural motion, and falling is not natural motion.

Aristotelian elements Earth, water, air, fire, and ether.

Difficulties with Aristotelian physics Contrary to Aristotelian predictions, heavy and light objects often fall at the same speed, and horizontally moving objects would move forever if there were no external forces.

Air resistance The resistance to the motion of an object through the air due to the object's collisions with numerous air molecules.

Friction The backward pull that slows an object that is sliding on a surface, caused by the roughness of the surface or of the sliding object.

Scientific method The dynamic interplay between experience (in the form of experiments and observations) and thought (in the form of creatively constructed theories and hypotheses that correlate experiences).

Inertia An object's ability to stay at rest or to maintain an unchanging velocity when acted on by no external forces.

Law of inertia A body that is subject to no external forces maintains an unchanging velocity. More briefly: All bodies have inertia.

Space The universe outside Earth and its atmosphere and outside other astronomical bodies.

Average speed An object's average speed during a time interval is the distance traveled during that time interval divided by the duration of that time interval, $s = d/t$.

Instantaneous speed, or **speed** The average speed during a time interval that is so short that the speed hardly changes. Speedometers measure this. Measured in meters per second (m/s).

Velocity The combined instantaneous speed and direction of motion.

Acceleration An accelerated object is one whose velocity is changing. Quantitively, the acceleration is the change in velocity during a time interval divided by the duration of that time interval. It can be measured in (km/hr)/s, or in $(m/s)/s = m/s^2$.

Galileo's law of falling Neglecting air resistance, any two objects dropped at the same speed fall at the same speed, regardless of their weights.

Free fall Falling that is influenced only by gravity and not by air resistance or other influences.

Acceleration due to gravity The acceleration of any freely falling object. On Earth this is about 10 m/s² or, more precisely, 9.8 m/s².

Proportionality One quantity is proportional to a second quantity if the first is always multiplied by whatever the second one is multiplied by. One quantity is **proportional to the square of** a second quantity if the first is always multiplied by the square of whatever the second is multiplied by. The symbol ∝ means *is proportional to*.

Quantitive description of free fall For an object that starts from rest and then falls freely: $s \propto t$, $d \propto t^2$. These proportionalities are correct for any motion that starts from rest and maintains an unchanging acceleration in a straight line.

REVIEW QUESTIONS

ARISTOTELIAN PHYSICS

1. Describe the four kinds of motion that Aristotle considered to be natural on Earth.
2. Give two examples of violent motion.
3. According to ancient Greek thought, in what fundamental way does Earth differ from the heavens?
4. Give an example of a motion that contradicts Aristotelian physics.
5. According to Aristotelian physics, why does a stone fall when it is released above the ground? According to Newtonian physics?
6. Describe at least one principle of Aristotelian physics that seems intuitively plausible but that Newtonian physics rejects.

THE LAW OF INERTIA

7. What does it mean to say that an object has inertia?
8. What does the law of inertia say about the velocity of a body that is subject to no external influences? What does this law say about such a body's acceleration?
9. Which of the following describes how high you must go before you will first reach "space": Anywhere above the ground, about 100 meters, about 1 kilometer, about 100 kilometers, beyond the moon, beyond the solar system?
10. Is there air in space?
11. Give at least one example that is very nearly unassisted motion at a constant velocity.

VELOCITY

12. Describe how you could use a clock and a meter stick to measure a moving object's speed.

13. When we say "5 centimeters per second," what does the "per" mean?
14. What is the difference between speed and average speed? In what circumstances are they the same?
15. Can you give an example in which the speed is unchanging but the velocity changes? If so, give one.
16. Can you give an example in which the velocity is unchanging but the speed changes? If so, give one.

ACCELERATION

17. A car speeds up along a straight line. Describe how you could use clocks and meter sticks to measure its acceleration.
18. How is acceleration related to velocity?
19. If an object's position is changing, can we be certain that it has a nonzero velocity? Can we be certain that it has a nonzero acceleration?
20. If an object is moving in a circle at an unchanging speed, is it accelerated?
21. If an object is slowing down, is it accelerated?

FALLING

22. An object is released above the ground and falls freely. At which of these positions during the fall is its velocity greatest? The top, the midpoint, or a point near the bottom? At which position is its acceleration greatest?
23. What is the meaning of the phrase "acceleration due to gravity"? What is its approximate value on Earth?
24. What does "speed is proportional to time" mean?
25. In twice the time, does a freely falling object fall (from rest) twice as far, at twice the speed, or both?

HOME PROJECTS

1. Drop two sheets of notebook paper to the ground at the same time, one of them open and the other crumpled up tightly. Drop the crumpled sheet and a stone at the same time. Drop a pebble and a large stone at the same time. Explain the results.
2. Get a small piece of notebook paper, lay it on top of a large book, and drop the book. Now hold the book and the paper in different hands and drop the two at the same time. Try to explain the results, and then compare your explanation with the answer in the footnote.*

---

*When the sheet of paper is placed on top of the book, it keeps up with the book because the book sets the surrounding air into motion at the book's velocity. Since the sheet of paper is falling in *moving* air, the sheet no longer needs to push through the air as it falls. This effect is called *drafting*. Bicycle racers know about this.

3. Measure the speed of an ant. Use a ruler (metric, preferably) and a watch with a second hand. Try to find an ant that moves in a straight line.
4. Check the accuracy of your car's speedometer, using the kilometer or mile markers posted along many highways. Have a friend measure the time (it isn't safe to do this alone), using a watch with a second hand. Calculate the speed.
5. Measure various speeds, such as your top bicycling speed, running speed, and hopping-on-one-foot speed.

6. Measure the time for your car to accelerate from 0 to 50 km/hr (30 mi/hr). For safety, have a friend do the timing. Find the car's acceleration in (km/hr)/s or (mi/hr)/s. Convert your result to m/s² by using the facts that 1 km = 1000 m, 1 mi = 1610 m, and 1 hr = 3600 s. An acceleration of 1 g means 9.8 (about 10) m/s². What is your car's acceleration in g's?

FOR DISCUSSION

1. "Meteoroids" are pebble-sized to boulder-sized rocks, similar in many ways to Earth rocks, moving through space. A typical meteoroid moves at a nearly unchanging speed in a nearly straight line. How might Aristotelian physics have been affected if Aristotle had known about this? How might Newtonian physics have been affected if Descartes, Galileo, and Newton had known about this? (Meteoroids were not known to science until about 1800.)

2. Many intuitively plausible beliefs conflict with science. For example, the belief that the sun moves around a stationary Earth conflicts with current astronomy. List several other intuitively plausible beliefs that conflict with either Copernican astronomy or Newtonian physics or both.

EXERCISES

ARISTOTELIAN PHYSICS
1. You roll a ball. It soon rolls to a stop. How would Aristotle interpret this? How would Galileo interpret it?
2. What explanation would Aristotle's physics give for these facts? Dry wood floats and water-logged wood sinks; an air-filled balloon falls more slowly than does a partly deflated balloon; and smoke rises.

THE LAW OF INERTIA
3. Most meteoroids—pebble-sized to boulder-sized rocks in space—have been moving for billions of years. What, if anything, keeps them moving?
4. If you ride on a smooth fast train at an unchanging speed and throw a baseball upward inside the train, will the baseball then get left behind, and will it come down toward the rear of the car? Explain.
5. If a ball is moving at 20 m/s and no forces ever act on it, what will its speed be after 5 seconds? After 5 years?
6. Do you suppose that the photo of Earth shown in Figure 1.7 was taken from a low-orbit satellite (see Figure 3.7).
7. In order to experimentally verify the law of inertia, would you need to be able to measure time? Weight? Distance?
8. When a moving bus comes rapidly to a stop, why do the riders who are standing suddenly lurch toward the front of the bus?

SPEED AND VELOCITY
9. Can you drive your car around the block at a constant velocity?

10. Mary passes Mike from behind while bicycling. As she passes him, do the two have the same velocity? Do the two have the same speed?
11. Mary is bicycling straight north at 15 km/hr, and Mike is bicycling straight south at 15 km/hr. As they pass each other, do they have the same velocity? Do they have the same speed?
12. Figure 3.14 represents a multiple-flash photo of two balls moving to the right. The figure shows both balls at several numbered times. The flash times are equally spaced. Which ball has the greater acceleration? The greater speed? The greater velocity? Does either ball pass the other, and if so, when?

FIGURE 3.14

ACCELERATION
13. A French TGV train cruises on straight tracks at a steady 290 km/hr (180 mi/hr). What is its acceleration?
14. Is the "motion sickness" that some people get in a car actually due to motion per se or to something else? Describe one form of motion that would not make people sick.

15. When you drive a car, might you depress the accelerator pedal without actually accelerating? Might you be accelerating without having your foot on the accelerator? Explain.

16. One car goes from 0 to 30 km/hr. Later another car goes from 0 to 60 km/hr. Can you say which car had the greater acceleration? Explain.

17. Figure 3.15 represents a multiple-flash photo of two balls. Describe each ball's motion. Does either ball pass the other? When? Do they ever have the same speed? When?

FIGURE 3.15

18. In each of the following cases, is the motion accelerated or not accelerated? (a) A rock falling freely for 2 m. (b) A meteoroid (a rock in space) that is so far from all planets and stars that gravity is negligible. (c) An artificial satellite orbiting Earth at a steady 30,000 km/hr. (d) The moon. (e) An ice-skater coasting on smooth ice, neglecting friction and air resistance.

19. Can a slow-moving object have a large acceleration? Can a fast-moving object have a small acceleration?

20. How many devices in a car are designed to cause acceleration?

21. For an unassisted (unforced, or isolated) moving object, which of the following quantities change, and how do they change: distance (from the starting point), speed, velocity, acceleration?

22. An automobile moves along a straight highway at an unchanging 80 km/hr. During the motion, which of the following quantities change, and how do they change: distance (from the starting point), speed, velocity, acceleration?

23. A ball rolls down a straight ramp. During the motion, which of the following quantities change, and how do they change: distance (from the starting point), speed, velocity?

FALLING

24. Multiple choice: Two metal balls are dropped from a cliff at the same time. They are the same size, but one weighs twice as much as the other. The time to reach the ground will be (a) about twice as long for the heavy ball; (b) about twice as long for the light ball; (c) about the same for both; (d) considerably longer for the heavy ball, but not necessarily twice as long; (e) considerably longer for the light ball, but not necessarily twice as long.

25. Figure 3.13 shows that a falling object has a speed of 10 m/s at $t = 1$ s. So it seems that it should move 10 m in 1 s. Yet the data say that the object moves only 5 m in 1 s. What is wrong here?

26. Figure 3.16 represents a multiple-flash photo of a falling ball. Neglect air resistance. At which point, A or B, is the ball's acceleration larger? At which point is its velocity larger?

FIGURE 3.16

27. By how much does a freely falling object's speed increase during its third second of fall (from $t = 2$ s to $t = 3$ s after release)? During its fourth second of fall?

28. For an object that is freely falling to Earth, which of the following quantities increase during the fall: distance (from the starting point), speed, velocity, acceleration?

29. As an object falls freely, what will be its speed at the end of the first second? second second? third second?

30. As an object falls freely, what will be its acceleration at the end of the first second? second second? third second?

PROBLEMS

VELOCITY

1. In Figure 3.14, suppose the large divisions on the measuring rod are centimeters and that the time-intervals each have a duration of 0.20 s. Find the speed of each ball.

2. Find the average speed of a jogger who jogs 3 km in 15 min. Give your answer in km/hr.

3. Worldwide sea levels are currently rising at about 2 mm per year, probably because of global warming. At this rate, how long will it be before sea levels have risen by 0.5 m?

4. It takes light about 8 minutes to travel here from the sun. Given that the speed of light is 300,000 km/s, how far is it to the sun?

5. It is 3.8×10^8 m to the moon. How long does it take a radar beam, traveling at the speed of light (300,000 km/s), to get from Earth to the moon and back?

6. You wish to travel from downtown New York City (NYC) to downtown Washington DC (DC), a distance of 330 km. You consider two options: train and plane. The high-speed train takes 1.5 hr, plus 30 min in stations. The airplane flies the 330 km (airport to airport) in just 30 min, but the drive to the NYC airport takes 30 min, you must arrive 30 min before departure time, the plane waits 15 min for takeoff, and it takes 45 min to get your luggage and drive into DC. Find the train's track speed, the plane's flying speed, the total travel time for each option, and the overall average speed for each option.

7. You drive from New York City to Washington DC, by car. You drive in traffic for the first hour at an average 50 km/hr. You cover the next 250 km in 3.0 hr, and then drive the remaining 30 km into Washington in 30 min. Find your total time and average speed.

ACCELERATION

8. A bicycler increases her speed along a straight road from 3 m/s to 4.5 m/s, in 5 s. Find her acceleration.

9. A car accelerates from 0 to 100 km/hr in 10 s. Find its acceleration. Drag-racers can get to 400 km/hr from rest in 5 s. How big is this acceleration?

10. A car starts from rest and maintains an acceleration of 4.5 (km/hr)/s for 5 s. How fast is it going at the end of the 5 s?

11. A car is moving at 30 km/hr. The driver then presses harder on the accelerator, causing an acceleration of 2.25 (km/hr)/s, which she maintains for 4 s. How fast is the car going at the end of the 4 s?

FALLING

12. An astronaut on another planet, one that has no atmosphere, drops a rock off a cliff. How much faster is the rock moving at the end of 3 s, as compared with 1 s? How much farther (measured from the release point) does the rock fall in 3 s, as compared with 1 s?

13. Neglecting air resistance, would the answers to the preceding exercise be different if the rock is dropped on Earth? Neglecting air resistance, would the *distance fallen* in 3 s on Earth be likely to be the same as the distance fallen in 3 s on the other planet?

14. You drop a rock off a cliff and note that it hits the ground below in 6 s. How high is the cliff, assuming that the air is so thin that air resistance can be neglected during the entire fall? How fast is the rock moving when it hits the ground, still assuming negligible air resistance? Unless the air is extremely thin, air resistance will not be negligible. What does this tell you about your calculation of the cliff's height: Is the cliff actually higher, or is it lower, than your calculated answer?

15. You drop a rock down a well and hear a splash 3 s later. As Charlie Brown would say, the well is "three seconds deep." But how many meters deep is it, assuming that air resistance is negligible and that the time for sound to travel back up the well is also negligible? How fast is the rock moving when it hits the water, under the same assumptions?

16. A car speeds up from rest, along a straight highway. Its acceleration is unchanging. How much farther (measured from the starting point) does it get in 10 s, as compared with 1 s? How much faster is it moving in 10 s, as compared with 1 s?

17. On the planet Mars, a free-falling object released from rest falls 4 m in 1 s and is moving at 8 m/s at that time. How fast would such an object be moving after 2 s? 3 s? How far would such an object fall in 2 s? 3 s?

CHAPTER 4

Why Things Move As They Do

> *Nature and nature's laws lay hid in night;*
> *God said, "Let Newton be," and all was light.*
> ALEXANDER POPE, EIGHTEENTH-CENTURY BRITISH POET

The world changed in 1687. In that year, Isaac Newton (Figure 4.1) published his *Mathematical Principles of Natural Philosophy*. To take just one example, by making Descartes's and Galileo's law of inertia the foundation of his work, Newton undermined our intuitive view of how things move, a view made explicit by Aristotle and accepted by all educated people for 2000 years.

The Newtonian world is surprisingly simple. Using only a few key concepts and principles, Newton was able to give clear, quantitative explanations for motion on Earth and in the heavens. Newton's theory explains the orbits of planets and moons, Kepler's laws of planetary motion, the motion of comets, falling objects, weight, ocean tides, Earth's equatorial bulge, the motion of an automobile, and the stresses on a bridge. It was an unparalleled expansion and unification of our understanding of nature.

Newtonian habits of thought influenced far more than physics and astronomy. Not only scientific fields such as chemistry and biology but also history, the arts, economics, government, theology, and philosophy were shaped by the general way in which the new physics portrayed the universe. For example, the ideals of inalienable human rights that inspired the American and French revolutions stemmed largely from a populace steeped in a Newtonian culture of universal natural law that applied equally to all people, to kings as well as commoners.

Unchallenged for more than two centuries, Newtonian physics worked so well for so long that it was eventually regarded as absolute truth. The very word *understand* came to mean *to explain in terms of Newtonian physics*. Most importantly, people eventually took for granted many subtle Newtonian habits of mind that had profound cultural and philosophical implications but that were unstated, unexamined, and unconsciously assumed by scientists, philosophers, and others.* Eventually, everyone from the person in the street to the profes-

FIGURE 4.1
Isaac Newton, 1642–1727. His *Mathematical Principles of Natural Philosophy*, summarizing his, Descartes's, and Galileo's studies on the motion of material objects on Earth and in the heavens, may well be the single most important book in the history of science.

---

*Two classic historical studies of physical science from the early Greeks through Newton examine the transition to the new worldview. The very title of Arthur Koestler's *The Sleepwalkers* (Universal Library and The Macmillan Company, New York, 1963) refers to the philosophically naive manner in which the new view developed. E. A. Burtt's *The Metaphysical Foundations of Modern Science* (originally published in 1932; re-issued by Humanities Press, Atlantic Highlands, NJ, 1980) is a close examination of the history and implications of these unstated philosophical assumptions.

I do not know what I may appear to the world; but to myself I seem to have been only like a boy playing on the seashore, and diverting myself in now and then finding a smoother pebble or a prettier shell than ordinary, whilst the great ocean of truth lay all undiscovered before me.

NEWTON

If I have seen further it is by standing upon the shoulders of Giants.

NEWTON

sional scholar assumed that Newtonian physics represented the single true framework for all human knowledge. Although Newtonian physics has been superseded by relativity theory and quantum theory, there appears to be no widely accepted philosophical framework for the new physics, and old Newtonian cultural habits remain. In order to help provide you with the tools to examine these cultural habits, one of the themes of this book is the significance of post-Newtonian physics.

Newton's physics starts from only a few concepts and principles. We have already discussed two Newtonian concepts, velocity and acceleration, and two principles, the law of inertia and the law of falling. Newton's other key concepts are force and mass, examined in Sections 4.1 and 4.2. His other key principles are the law of motion (Sections 4.3 and 4.4), the law of force pairs (Section 4.5), and the law of gravity (Chapter 5). In Section 4.6, we use these ideas to look at the motion of a device that has had a remarkable social impact: the automobile.

4.1 FORCE: WHY THINGS ACCELERATE

The preceding chapter used the word *force* synonymously with *external influence*. Now we need to be more specific. Like velocity and acceleration, there is no predetermined way that we must necessarily define *force*. We want to define this word in the most useful way.

One example of an object that experiences no external influences is a body that is isolated in distant space, far from all planets and stars. The law of inertia predicts that such a body will be unaccelerated. But what about a body that is not isolated, for example, a stone in space with a comet nearby? We often find it convenient to use examples of objects in space, because earthly effects such as friction and gravity tend to obscure the basic principles of motion. It was precisely such confusion that kept the scientific world thinking in Aristotelian terms for 2000 years. An air coaster riding on a cushion of air on a horizontal surface (Figure 3.6) is another convenient example, because friction is negligible and gravity does not affect the horizontal motion. A smoothly rolling ball is a pretty good example, for the same reasons.

How can we tell whether or not a nearby comet "influences" a stone in space? Physicists answer this question by using the law of inertia as a definition of what we mean by an *influence*: We say that the comet influences, or *exerts a force on*, the stone if the comet causes the stone to accelerate.* This is the way we use the word *force*. That is, we say that a body exerts a **force** on another body whenever the first body causes the second body to accelerate.

There are many misconceptions about the word *force*. The operational term here is *exert a force on*. Force is an action, rather than a thing: A body cannot be a force. And force is not a property of a body: A body cannot possess force. Instead, a body can *exert a force on* another body. It would be best to use *force*

---

*If you are saying to yourself that this reasoning sounds circular, you are right, for the law of inertia is partly a way of defining natural motion and force. But experiments, like the one shown in Figure 3.6, are needed to demonstrate the consistency of the definitions. The law of inertia is part definition, part natural law.

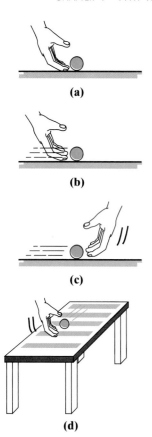

FIGURE 4.2
A push (a) starts a ball moving and (b) speeds a ball up. (c) A brief pat from in front slows a ball down. (d) A brief push from the side changes the direction of motion of a moving ball.

FIGURE 4.3
Both you and Ebenezer are exerting a force on the ball, but the ball is not accelerating. You exert a force on the ball whenever you would cause it to accelerate if no other forces were acting. In the case pictured, Ebenezer's force on the ball prevents your force from causing it to accelerate.

only in a pure-verb sense, as in "the comet forces the stone," but unfortunately this is not done.*

Some examples (Figure 4.2): If a ball is at rest on a table and you push it with your hand, the ball will start moving. If a ball is already moving across a table and you push it briefly from behind, the ball will speed up. If a ball is moving across a table and you "pat" it briefly and lightly from in front, toward the rear, it will slow down. If a ball is moving across a table and you push it briefly and lightly sideways, perpendicular to the direction of motion, it will change directions. In all four cases, your hand push accelerated the ball. In fact, a little experimentation with a smooth hard ball rolling on a smooth hard surface (better yet, with an air glider) shows that you cannot push the ball without accelerating it.† Since we defined a force as any influence that causes an acceleration, we conclude that every hand push is a force.

The hand push is probably the most familiar example of a force. When you use the word *force*, it is useful to keep in mind that a force is similar to a push.

Pulling is another example. Starting with the ball at rest on the table, you can clasp it and pull it toward you, accelerating it into motion. So pulls, as well as pushes, are examples of forces. Instead of pulling the ball with your hand, you could attach a string and pull the string, which pulls on the ball to accelerate the ball. So strings, when they pull, exert forces.

Only a single force is exerted on the ball in these examples. Now suppose that you and your friend Ebenezer both push on a ball in opposite directions (Figure 4.3). If you and Ebenezer adjust your pushes, you can get them to balance so that the ball remains at rest. Yet you both are pushing on the ball. Even though the ball is unaccelerated, we will still say that you exert a force on the ball and that Ebenezer does, too. In cases like this, involving more than one force, we say that a body exerts a force on another body if, in the absence of the other forces, the first body would cause the second to accelerate.

You could tap a ball with a hammer instead of pushing or pulling it with your hand. Since the hammer tap accelerates the ball, it exerts a force on the ball. Try tapping a motionless or moving ball from various directions yourself, and observe it carefully. Exactly when is it accelerating?

—This is a pause for finding a ball and a hammer and trying this.

The ball accelerates only during the brief fraction of a second that the hammer is actually touching it. So the hammer exerts a force on the ball only during this fraction of a second. After the tap, the ball moves at a constant velocity, so there is no force exerted on it. So the moving ball does not *have force* or *carry force along with it*. To exert a force is to perform a certain action.

Friction and air resistance also are forces. If you briefly shove a book and let go so that it slides across a table, the book will slow down as it slides. We know that some force must cause this slowing, because the law of inertia says that the speed would be unchanging if there were no force on the book. If we glide an air coaster across a table, the air coaster will not slow down. So the force that slows the book must result from the contact between the book and the tabletop. It

---

*Scientists like to turn their verbs into nouns, so rather than the verb *to force*, physicists persist in using *to exert a force on*. This bad habit has caused centuries of confusion for students, but it seems impossible to break.

†We are assuming that only a single push is exerted at a time. Two simultaneous pushes could cancel each other, as we will discuss later.

must be exerted by the tabletop on the bottom of the book. This force, by one surface on another surface, is called **friction**.

A fast bullet moving horizontally through air slows down a little, so there must be a force on the bullet. Since there is no slowing when a bullet moves through a vacuum, we conclude that this force is due to the air around the bullet. It is caused by air molecules that the bullet hits as it travels and is called **air resistance**. The effect is similar to friction: The bullet slides through the air in somewhat the same way that a book slides across a table.

As we know from the preceding chapter, an apple falling freely to the ground accelerates all the way down. Since the apple accelerates, there must be a force on it. It is commonly called the **force of gravity**. But remember that forces are always actions by one object on another object. Gravity is a force on the apple, but what is this force exerted by? The answer must be that this force is exerted by Earth. The experimental evidence for this is that no matter where you go on Earth, a falling apple always accelerates downward, toward Earth's center. The direction of its fall is determined by the position of Earth's center.

You can think of a gravitational force as a pull, although not a human, muscular pull. Forces do not require living creatures to exert those forces. The gravitational force is one example: It is a pull by Earth on nearby bodies. There is an interesting difference between the gravitational force and the other forces we've looked at. Hand pushes, hand pulls, hammer taps, string pulls, friction, and air resistance all are contact forces: forces exerted by an object that is touching another object. The gravitational force by Earth on a falling apple is different, because Earth is not actually touching the apple while it falls. Air is touching the apple, but we could imagine removing the air, or we could take the apple above Earth's atmosphere and then drop it, and the apple would still fall. The gravitational force acts at a distance.

As a final example, if a magnet is brought near another magnet, the second magnet will accelerate toward or away from the first, depending on the orientation of the two magnets. So the first magnet exerts a force on the second, a force that can apparently be either attractive (a pull toward the first magnet) or repulsive (a push away from the first magnet). This *magnetic force* is another force that acts at a distance.

4.2 Connecting Force and Acceleration

The thrust of Newton's theory of motion is that *forces cause accelerations*. It is a surprising, nonintuitive idea, as we have seen. Our Aristotelian intuitions tell us that outside influences are needed to keep a thing moving, that is, that forces cause *velocities*.

Newton formulated the specific relation between force and acceleration. To follow his reasoning, suppose that you put a smooth ball on a smooth table and tap it once with a hammer. As we know, it accelerates during the tap. Experience shows that the more strongly you tap, the faster it will be moving after the tap. We conclude that stronger forces cause larger accelerations.

To make this relation quantitative, we need to do the experiment using an easily measurable type of force. We could do this, for example, by using stretchable springs rather than hammer taps to exert the forces. We then find that if we attach a spring to an object and pull it with an unchanging force (unchanging

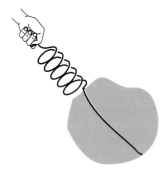

FIGURE 4.4
Imagine pulling on a stone in distant space by means of a spring attached to the stone.

FIGURE 4.5
What happens to the stone of Figure 4.4 when you double the force acting on it?

amount of stretch in the spring), and if there are no other forces except the spring force, then the object will maintain an unchanging acceleration. This shows that an unchanging force causes an unchanging acceleration. It isn't easy to do this experiment, because gravity and friction are usually present in addition to the spring's force. One way to get away from gravity and friction would be to pull a stone through outer space by means of an attached spring (Figure 4.4). Another technique that you might use in a physics laboratory is to attach a spring to a coaster gliding horizontally on a cushion of air (Figure 3.6).

If we now double the force by attaching two springs instead of one, the acceleration will be doubled, too (Figure 4.5). And a tripled force causes a tripled acceleration, and so forth. We conclude that an object's acceleration is proportional to the total force exerted on it. In symbols,

$$a \propto F$$

Now imagine exerting forces on two different objects, one of them heavier than the other. You will find that they respond differently to similar forces. For instance, if you give equal taps to a light ball and a heavy ball, you will find that the light ball accelerates into faster motion than does the heavy ball. So the light ball gets a larger acceleration during the tap.

It is useful to extend the concept of inertia to this situation. Recall that an object's inertia is its ability to maintain its velocity. In the preceding experiment, the heavier object changes its velocity the least. We will say that the heavy ball has more inertia than the light one, using the word **inertia** to mean a body's degree of resistance to acceleration.

To examine this notion of inertia more carefully, imagine that all experiments occur in distant space, far from gravity and other outside influences. Objects that are beyond the influence of gravity are weightless; they would not weigh down on your hand if you held them.

Experiments show that objects that are tapped by a hammer in space accelerate by the same amount that they accelerate when given the same tap on Earth (assuming that friction is negligible in the experiment on Earth). This is an important and surprising point. Even though an object in distant space has no weight, just as much force is needed to accelerate it. Although objects in space have no weight, they retain all of their inertia.

So if you tap a heavy stone and a light stone in space, the heavy one accelerates a little and the light one accelerates a lot. But now we cannot say that this is due to their different weights, because neither stone weighs anything in space. How can we describe the difference between the two stones in space, without resorting to the concept of weight? One answer is that one stone—the heavier one on Earth—has more matter in it, more atoms, for instance. But it is difficult to see how to make this notion of *quantity of matter* quantitative without resorting to the notion of weight—which has no meaning in space.

So we come back to the concept of inertia. An object's inertia can be measured even when the object is in distant space and has no weight, for instance by tapping it and seeing how much it accelerates. When the concept of inertia is made quantitative in this way, it is called **mass**. That is, the mass of an object is its amount of inertia. In the next section we will find that this definition of mass agrees with our intuitive notion that two objects have the same mass if they balance on a balance beam.

To establish a measurement scale for mass, we choose one particular object, called the **standard kilogram**, and define its mass to be one **kilogram**, abbrevi-

FIGURE 4.6
The U.S. National Standard Kilogram no. 20, an accurate copy of the International Standard Kilogram kept at Sèvres, France. It is stored inside two bell jars.

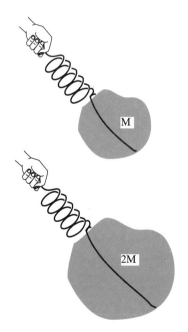

FIGURE 4.7
What happens when a stone having twice the mass is pulled by the same force?

ated kg (Figure 4.6). There are good duplicates of it in most physics laboratories. Any object having the same inertia as the standard kilogram has a mass of one kilogram. And any object having the same inertia as 2 kilograms bundled together is said to have a mass of 2 kilograms. We can define the mass of any object in this way, by comparing its inertia with that of 1 or more kilograms (or with a half of a kilogram or some other fraction).

Now, imagine using identical forces to pull two stones through space, one stone having twice the mass of the other (Figure 4.7). How do their accelerations compare? Upon doing this experiment, we find that the stone containing twice as much mass has *half* as big an acceleration. And three times as much mass yields one-third as much acceleration, and so forth. The acceleration is divided by whatever you multiplied the mass by. We express this by saying that an object's acceleration is **proportional to the inverse** of its mass. Since the inverse of a number is 1 divided by that number, we can write this proportionality as a $a \propto 1/M$.

We can put our two proportionalities, $a \propto F$ and $a \propto 1/M$, together to read

$$\text{acceleration} \propto \text{force/mass}; \qquad a \propto F/M$$

This says that an object's acceleration is proportional to the force on it divided by its mass.

We have not yet mentioned any measurement scale for force. The unit of force, called the **newton** (abbreviated N), is defined as the amount of force that can give a 1 kg mass a 1 m/s^2 acceleration. This means we can write the proportionality as an equality:

$$\text{acceleration} = \text{force/mass}; \qquad a = F/M$$

This formula gives the acceleration in m/s^2, provided we express the force in N and the mass in kg.

In the "English" system, force is measured in the familiar (in the United States) pounds. One newton is a little less than a quarter of a pound. Think of a quarter-pound of margarine. The "English" unit of mass happens to be called the *slug*. Sounds terrible. You've probably never heard of it, and you can immediately forget it. We can only hope that the United States will someday be able to forget the entire "English" system!

4.3 NEWTON'S LAW OF MOTION: CENTERPIECE OF NEWTONIAN PHYSICS

We need to discuss two other facets of Newton's law of motion. First, what happens when more than one force acts on an object, perhaps pushing or pulling it in different directions? All the forces will affect the object's motion, so we need to somehow include all of them. When the individual forces act in the same direction—as was the case for the two springs attached to a stone (Figure 4.5)—their effects must add up, and the overall effect is the sum of the individual forces. But what if the individual forces are in different directions?

For example, what if two forces act in opposite directions on the same object? If the two forces are of equal strength (Figure 4.3), you know from experience that the object will not accelerate. So the overall effect must be zero. This suggests that we should subtract the two forces when they are in opposite directions. This suggestion is correct. If two forces act in opposite directions, they will have the same overall effect as a single force that is in the direction of the larger force and is equal to the difference between the larger and the smaller force.

To summarize: If two or more forces act in the same direction, their overall effect, which we will call the **net force**, is the same as that of a single force equal to their sum. If two forces act in opposite directions, the net force is in the direction of the larger force and is equal to the difference between the larger and the smaller force.

We will not be concerned with finding the net force when the individual forces are neither in the same nor in opposite directions.

For instance, suppose that you push your book along a tabletop with a force of 10 N and that the tabletop exerts a 3-N frictional force on the book (Figure 4.8). Then the net force on the book is 7 N. These 7 newtons represent the net, overall effect of the external environment pushing and pulling on the book. It is this 7-N net force, and not just your 10-N pushing force, that accelerates the book.

The second point we need to consider is the direction of the acceleration. Recall that an acceleration is a change in velocity. Since velocity has a direction, it is not surprising that acceleration should have one, too. So far, the only accelerations we have discussed quantitatively were ones in which an object was speeding up along a straight line. In this case, the velocity has an unchanging direction, and the velocity changes by becoming larger, so that the change in velocity is forward (Figure 4.9). That is, the direction of the acceleration is forward.

What about an object moving along a straight line and slowing down? In this case, the velocity gets smaller, so the change in velocity is backward (Figure 4.10). This means that the acceleration is backward, opposite to the velocity. In the next chapter, we will discuss the direction of the acceleration in cases in which the object does not move along a straight line.

Since an object's acceleration is determined by the net force on it, it seems plausible that the direction of this acceleration should be the same as the direction of the net force. Simple experiments verify this: If you give a motionless ball a brief hammer tap, it will accelerate into motion along the direction in which you tapped, so the acceleration is along the direction of the force. If the ball is already moving and you tap it from behind, it will speed up; the acceleration is forward, again along the direction of the force. And if you tap a moving ball lightly

FIGURE 4.8
How strong and in what direction is the net force on the book?

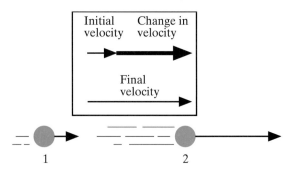

FIGURE 4.9
When an object speeds up along a straight line, its change in velocity is along the direction of the motion.

FIGURE 4.10
When an object slows down along a straight line, its change in velocity is opposite to the direction of the motion.

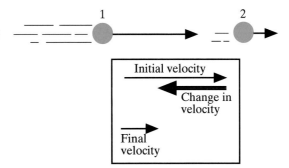

from in front, the ball will slow down; this is a backward acceleration, which again is along the direction of the force. We save the case of a sideways force, which is neither forward nor backward, for the next chapter.

In summary:

NEWTON'S LAW OF MOTION*
An object's acceleration is determined by the net force exerted on it by its environment, and by the object's mass. Quantitatively, the acceleration is proportional to the net force divided by the mass. Furthermore, the direction of the acceleration is the same as the direction of the net force. In symbols,

$$\text{acceleration} \propto \frac{\text{net force}}{\text{mass}}; \qquad a \propto \frac{F}{M}$$

If force is measured in newtons, mass in kilograms, and acceleration in m/s², the proportionality becomes an equality:

$$\text{acceleration} = \frac{\text{net force}}{\text{mass}}; \qquad a = \frac{F}{M}$$

■ **DIALOGUE 1**   An old slow car moves at a steady 10 km/hr down a straight highway, while another car speeds by at a steady 120 km/hr. Which car has the greatest net force on it?

■ **DIALOGUE 2**   You push your 2-kg book along a tabletop, pushing it with a 10-N force. If the book is greased so that friction is negligible, what is the book's acceleration? Another, ungreased, book also has a mass of 2 kg, and you also push it with a 10-N force, but now there is a 4-N frictional force. Find the net force on this book, and the book's acceleration. What object exerts the frictional force?

---

*Often called Newton's second law.

1. Neither. Each car has zero net force on it.

2. First book: $a = F/M = 10/2 = 5$ m/s². Second book: $F = 10 - 4 = 6$ N; $a = F/M = 6/2 = 3$ m/s². The table (or tabletop) exerts the frictional force.

■ **DIALOGUE 3** You are in distant space. A giant boulder, many times larger than you, is at rest in front of you. You tap the boulder lightly with a small hammer. Describe the motion of the boulder during and after the tap. If the boulder were on Earth, at rest on level ground, would the experimental result be the same? If not, why not?

■ **DIALOGUE 4** You are in distant space. You hold two blocks of metal. They look and feel identical when touched, but you have been informed that one is made of lightweight aluminum and the other of heavy lead. How can you determine which one is which? Can you determine how much more massive the lead block is than the aluminum block? Explain.

4.4 WEIGHT: GRAVITY'S FORCE ON A BODY

We omitted gravity from the preceding section. Now we put it back into the picture.

As we know (Section 4.1), Earth exerts a gravitational force on objects that are falling to the ground. We use the word *weight* for this force. An object still has weight even when it is not falling, for instance, when it is at rest on the ground.

As we will discover in the next chapter, the sun, moon, planets, stars, and all other astronomical bodies exert gravitational forces, too. It is useful to extend the meaning of the word *weight* to include all such possibilities. In other words, the **weight** of an object refers to the net gravitational force exerted on it by all other objects.

Weight and mass are related concepts, but they certainly are not the same thing. An object's weight is the force on it due to gravity, while its mass is its quantity of inertia. Like all forces, weight is measured in newtons, while mass is measured in kilograms. An object's weight depends on its environment; for instance, an object's weight is different when it is near the sun than when it is near Earth. But an object's mass is a property of the object alone and not of its environment.

As one example, the acceleration of a stone falling onto the moon's surface is only 1.6 m/s², one-sixth of the acceleration due to gravity on Earth. Since Newton's law of motion says that acceleration is proportional to net force and since gravity is the only force exerted on the falling stone, the force exerted by gravity on this stone must be only one-sixth of the force exerted by gravity on the same stone on Earth. In other words, the stone's weight on the moon is one-sixth of its

---

3. It accelerates during the tap. After the tap, it moves away at a slow, unchanging velocity. On Earth, the boulder would not accelerate because when you tap the rock, the ground exerts a frictional force in the other direction, making the net force (caused by the hammer tap and ground) zero.

4. Exert a force on each block, for instance by giving each a small tap with your finger. The one that accelerates more is aluminum. To find the ratio of their masses, exert the same force on each block, and measure their accelerations. The ratio of the masses is the inverse of the ratio of the accelerations; for instance, if the aluminum's acceleration is three times larger, then its mass must be three times smaller.

weight on Earth. But the stone's mass is the same in both places, since its inertia doesn't change. The stone's mass would be the same even in distant space, where its weight would be essentially zero.

If a stone and a baseball are held above Earth, dropped, and allowed to fall freely (no air resistance) to the ground, Galileo's law of falling tells us that their accelerations will be the same (9.8 m/s²). Suppose that the stone and the baseball happen to have the same mass. Then Newton's law of motion tells us that the forces on the two objects are the same because their accelerations are the same and their masses are also the same. But this force is just the force of gravity; in other words, it is the weight of each object. So their weights are equal.

This is a fairly plausible conclusion and also a very important one. It says that *two objects of equal mass also have equal weight.* So you can compare masses by comparing weights, for instance on a balance beam (Figure 4.11). Any object that balances a kilogram has a mass of 1 kilogram, for example.

Since weight is a force, it can be measured in newtons. In "English" units, weight is measured in pounds.

The **metric ton**, or **tonne** (it's always spelled this way), is 1000 kilograms. It is useful for larger masses. The similar "English" unit—the "English ton" or simply *ton*—is 2000 pounds. It is not a unit of mass but, instead, a unit of force. On Earth, the mass of a ton is about 900 kilograms, so a ton is a little less massive than a tonne. As you can see, things can get confusing if we try to keep up with both metric and "English" units, so we will confine ourselves to metric.

As a common example, consider this textbook resting on a table. Suppose the book weighs 12 N; in other words, suppose that the gravitational force by Earth on the book is 12 N. This force has a downward direction. But the book is obviously not accelerating downward through the table. Since the book's acceleration is zero, Newton's law of motion tells us that the net force on it must also be zero. So there must be an upward force of 12 N acting on the book, to balance the downward weight. The table must exert this force, because the book will fall if we suddenly remove the table.

It may seem strange that an inanimate object could exert a force. Why should a table push on a book? The tabletop doesn't seem to be doing anything. A microscopic view is enlightening. The upward force is exerted by the atoms in the table on the atoms in the book.* When the book presses against the tabletop, the tabletop is squeezed down and slightly deformed. That is, the book squeezes the atoms in the tabletop. And like a squeezed spring, the atoms then push upward against the book (Figure 4.12). The direction of this force by the tabletop is directly away from the surface, perpendicular to it. A force similar to the upward force by the table on the book is exerted when any object touches a solid surface. We will call any such force a **perpendicular force**,† because it is always perpendicular to the surface.

Figure 4.13 shows the forces exerted on the book. Each force is represented by an arrow. A **force diagram** like this can help in analyzing forces and motion. When you draw a force diagram, show every one of the individual forces acting on whatever object is of interest. Show each force as an arrow pointing in

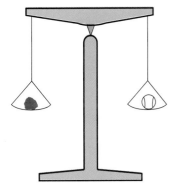

FIGURE 4.11
You can compare masses by comparing weights, for instance on a balance beam.

FIGURE 4.12
An explanation of the perpendicular force by a table on a book.

---

*More precisely, this force is an electric force by the electrons in the table's atoms on the electrons in the book's atoms. Electrons repel one another strongly when they get very close together.

†Many textbooks call it a *normal force*, because *normal* sometimes means *perpendicular*.

FIGURE 4.13
The forces exerted on a book at rest on a table.

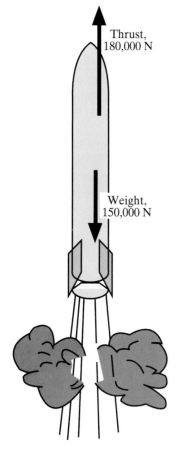

FIGURE 4.14
The forces exerted on a rocket during liftoff. Note that the diagram shows only the individual forces and not the 30,000-N net force. Adding another arrow, labeled "net force," would only confuse the analysis.

the direction in which that force pushes or pulls on the object, and name each force.

As another example, suppose that a rocket at liftoff weighs 150,000 N and has a mass of 15,000 kilograms and that the rocket engines exert a 180,000-N "thrust" force on the rocket (Figure 4.14). (We'll learn more about the thrust force in the next section.) What net force accelerates the rocket, and how big is the rocket's acceleration?

The net force is 180,000 N – 150,000 N = 30,000 N upward. It is only this 30,000 N that actually accelerates the rocket. To find the acceleration, divide the net force by the mass (Newton's law of motion): 30,000 N/15,000 kg = 2 m/s^2.

■ **DIALOGUE 5**   (a) Suppose an engine fails and the rocket develops a thrust of only 165,000 N. Find the net force and the acceleration. (b) Suppose that, at the end of the countdown on the launch pad, the rockets turn on but with a thrust of only 150,000 N. What happens? (c) What if the thrust were only 120,000 N? (d) What if the rockets didn't exert any thrust at all?

■ **DIALOGUE 6**   (a) Would it be easier to lift this book on Earth or on the moon? (b) Neglecting friction, would it be easier to set this book into horizontal motion at 5 m/s on the Earth or on the moon? (c) Neglecting friction, would it be easier to set this book into horizontal motion at 5 m/s on Earth or in distant space?

■ **DIALOGUE 7**   A standard kilogram in your physics lab weighs about 10 newtons (2.2 pounds). Find its mass and weight on the moon, where the acceleration due to gravity is one-sixth of its value on Earth. Find its mass and weight in distant space.

4.5 THE LAW OF FORCE PAIRS: YOU CAN'T DO JUST ONE THING

H O W   D O   W E   K N O W ?

Try these: Slap a tabletop with your hand. Grasp the edge of a table and pull hard on it. Now push hard against the table. Find two marbles, place one at rest, and "shoot" the other marble at it.

When you slap a table, it slaps back, as you can feel when it stings your hand. This slap by the table is a force, because it accelerates your hand (by stopping your hand).

5. (a) 15,000 N, 1 m/s^2. (b) The net force on the rocket is zero, so $a = 0$. The rocket will remain poised on the launch pad, not pressing down on the launch pad but also not taking off. (c) The rocket remains at rest, $a = 0$. Now the launch pad must exert a 30,000-N upward force in order to make the net force zero. (d) The rocket remains at rest, $a = 0$, the launch pad exerts 150,000 N (upward), net force = 0.

6. (a) On the moon. (b) The same in both places. (c) The same in both places.

7. On the moon: mass = 1 kg, weight = 1/6 × 10 N = 1.7 N. In space: mass = 1 kg, weight = 0.

When you pull on a table, the table pulls you toward it. And when you push on the table, the table pushes you away. These are forces exerted by the table on you.

When the marbles collide, the shooter marble exerts a force on the second marble, as you can see from the fact that the second marble accelerates into motion. But the second marble exerts a force on the first marble, too, as you can see from the fact that the first marble's velocity changes.

These experiments indicate that whenever one object exerts a force on a second object, the second exerts a force on the first: Forces always come in pairs, called **force pairs**.

Do things still work out this way even if the two objects are not touching? Find a pair of magnets and investigate this. Place the magnets on a smooth surface and hold them at rest with their poles near each other but not touching. When you release them, they both accelerate. So each exerts a force on the other. Even for two separated objects, forces come in pairs.

Touch your friend's face. Your hand, too, is touched, by your friend's face. Again, forces always come in pairs. You cannot touch without being touched.*

Slap lightly on a tabletop. Now slap hard. The table slapped back harder the second time, right? This observation gives us quantitative information about force pairs: When one member of a force pair grows bigger, so does the other. In fact, quantitative experiments show that the two members of any force pair have the same strength. If one of them is, say, 3 N, the other one will be 3 N too.

The directions of the two forces in a force pair are not the same, however. In fact, our examples show that they are opposite in direction. For instance, when you pull a table toward you, the table pulls you toward it (Figure 4.15).

Newton recognized this idea about force pairs as a key physical principle. We summarize it as follows:

> **THE LAW OF FORCE PAIRS†**
> Forces always come in pairs: Whenever one body exerts a force on a second body, the second exerts a force on the first. Furthermore, the two forces are equal in strength but opposite in direction.

FIGURE 4.15
When you pull or push on a table, it pulls or pushes on you in the opposite direction.

(a) Pulling on the edge of the table.

(b) Pushing on the table.

---

*Thanks to Paul Hewitt, author of *Conceptual Physics*, 8th ed. (Boston: Scott, Foresman, 1998), for this nice way of putting it.

†Often called Newton's third law.

FIGURE 4.16
Force pairs.

Force by table
on your hand

Force by your
hand on table

(a) Slapping a table.

Force by S
pole of light
blue magnet
on N pole of
dark blue
magnet

Force by N pole of
dark blue magnet
on S pole of light
blue magnet

(b) Opposite poles of two magnets
attracting each other.

Force by N
pole of light
blue magnet
on N pole of
dark blue magnet

Force by N pole
of dark blue
magnet on N
pole of light
blue magnet

(c) Like poles of two magnets
repelling each other.

Force by table
on book

Force by
book on
table

FIGURE 4.17
Another force pair.

Figures 4.16 and 4.17 show several examples of force pairs. These diagrams are not complete force diagrams. Figure 4.14, for example, is a complete force diagram for a rocket—it shows *all* the forces acting on a *single* object, namely, on the rocket. On the other hand, each illustration in Figures 4.16 and 4.17 shows forces on two objects without showing all of the forces on either of the two objects.

Figure 4.17 is the example of a book on a table (Figure 4.13). Although Figures 4.13 and 4.17 look similar, the difference between them is important: In Figure 4.13, the downward force is the gravitational force by Earth on the book; all the forces in this figure are exerted on the book. In Figure 4.17, the down-

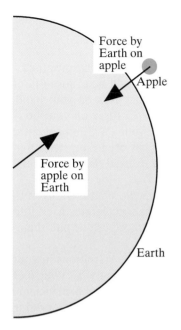

FIGURE 4.18
Earth and a falling apple: Which one exerts the larger force? (Answer: They are the same.)

ward force is by the book on the table. Microscopically, this force is exerted by the atoms in the book's lower surface on the atoms in the tabletop. This force is repulsive, away from the book. If you lay a book on the palm of your hand, you can feel this downward force by the book's pressing on the atoms of your hand.

Figure 4.18 illustrates an interesting point. Since Earth exerts a gravitational force on an apple, the law of force pairs says that the apple must exert a gravitational force on Earth! Furthermore, the strengths of these two forces are equal: If Earth exerts a 2-N force on the apple (in other words, if the apple weighs 2 N), then the apple also will exert a 2-N force on Earth. This might seem surprising. Why haven't we noticed this force, by objects such as apples, on Earth? Why doesn't Earth accelerate toward the apple?

The answer is that an apple causes only a slight acceleration of Earth because Earth's mass is so large. You can't use an apple to noticeably accelerate a planet. Large astronomical objects, however, can noticeably accelerate a planet. For example, Earth accelerates noticeably in response to the motions of the moon.

■ **DIALOGUE 8**   For each of the following forces, describe the other member of the force pair. (a) The force a pitcher exerts on a baseball while throwing it. (b) A bat hitting a baseball. (c) A baseball hitting a catcher's mitt. (d) An apple's weight. (e) Your weight. (f) Your arms pulling on a box as you slide it across the floor. (g) A rope pulling forward on a water skier.

■ **DIALOGUE 9**   A 12-N book rests on the palm of your hand. Describe the two forces that act on the book. How strong is each force, and in what direction does each act? Are these two forces equal but opposite to each other? Are these two forces part of one force pair?

■ **DIALOGUE 10**   Referring to the preceding dialogue, suppose you accelerate the book into upward motion. Describe the individual forces that act on the book during the acceleration. What can you say about the strengths of each of these two forces? Are they equal but opposite to each other?

■ **DIALOGUE 11**   (a) A large truck collides head-on with a small car. Which force is bigger: the force by the truck on the car or the force by the car on the truck? (b) A moving car hits a stationary car from behind. Which is bigger, the force by the moving car on the stationary car or the force by the stationary car on the moving car?

8. (a) The baseball pushing backward on the pitcher's hand. (b) The ball hitting the bat. (c) The mitt stopping (exerting a backward force on) the baseball. (d) The apple exerting an upward gravitational force on Earth. (e) Your body exerting an upward gravitational force on Earth. (f) The box pulling backward on your arms. (g) The skier pulling backward on the rope.

9. The two forces acting on the book are the weight of the book and the force by your hand. The weight is 12 N downward, and the hand force is 12 N upward. They are equal but opposite. They are not members of a single force pair.

10. As in Dialogue 9, the two forces are the weight of the book and the force by your hand. The weight is 12 N downward, and the hand force is larger than 12 N upward. These two forces are not equal, and they are opposite in direction. As in Dialogue 9, they are not members of the same force pair.

11. (a) The law of force pairs says that the two forces are the same size (same strength). (b) Ditto.

4.6 NEWTON MEETS THE AUTOMOBILE

Simple experiments demonstrate that you cannot get anywhere by pushing or pulling on yourself. Try pulling on your nose. You might pull your nose out of joint, but you won't go anywhere because your nose pulls back on your hand, and both pulls are on your body, so they result in zero net force on your body (Figure 4.19). If you want to accelerate, something in your environment must exert a force on you. It is for this reason that Newton's law of motion states that an object's acceleration is determined by the net force exerted on it *by its environment*.

This presents us with an interesting dilemma when we consider any self-propelled object such as an automobile or an animal that accelerates itself into motion. How can they get themselves going if things cannot push or pull themselves into motion?

Try this: Stand up and take one step, noting carefully the sensations in your legs, especially along the bottom of the foot that is accelerating you. Your foot pushes backward against the floor.* You can demonstrate more convincingly that your foot actually pushes backward when you walk or run by accelerating rapidly from rest on a dusty dirt road. Your feet push dust backward, showing that they exert a backward force on the road.

The law of force pairs tells us that if you push backward on the ground, the ground must push forward on you. This push by the ground on you is the force that propels you forward! It is one example of a **drive force**: the forward force exerted by the environment on a self-propelled object .[†]

FIGURE 4.19
You can't get anywhere by pulling on your nose.

---

*If you are barefoot. Otherwise, your foot pushes backward against your shoe, which in turn pushes backward against the floor.

[†]*Self-propelled* means that the energy (Chapter 6) to propel the object comes from within the object itself. But the force to propel it comes from the outside.

Most drive forces are actually particular cases of the force of friction. The backward push by your foot against the floor is a frictional force and so is the forward push—the drive force—by the floor against your foot. If you were standing on a smoothly frozen pond and you tried to quickly accelerate from rest to a fast run, you would soon find that friction (some people call it *traction*) is required to provide the forward drive force! There is, however, one important difference between sliding friction and frictional drive forces: while sliding friction acts backward, frictional drive forces act forward, along the direction of motion.

The automobile is a fine example of Newtonian principles. More important, motor vehicles are one of the technologies that have drastically reshaped the social fabric of the modern world. Like all powerful technologies, the car has important social pros and cons. It gives us unparalleled freedom of movement; it has revamped our cities; it uses most of our petroleum, creates much of our pollution, and is the leading cause of death of Americans under 35. It will come in for a lot of discussion in this book.

Before reading further, try listing, or drawing, the forces exerted on a car while it is driving along a straight level road.

As shown in Figure 4.20, one force acting on a car driving along a straight level road is the gravitational force, or weight of the car, exerted downward. A second force is the perpendicular force, exerted upward by the road on the car. These two forces act vertically. Since a car on a level road has no acceleration in the vertical direction, the net vertical force must be zero. So these two vertical forces must be of equal strength.

The three horizontal forces relate more directly to the car's motion. Two backward **resistive forces** act on the car: The air through which the car moves exerts the force of air resistance, and the contact between tires and road creates another backward force known as **rolling resistance**. Rolling resistance is caused by the squeezing or compression of the tire where the rubber meets the road. The force that the road must exert to cause the compression turns out to be partly backward, retarding the car's motion. To look at it another way, the repeated squeezing of the tire represents a loss of the car's energy (Chapter 6). This is most pronounced in flexible, air-filled tires. Hard tires rolling on a hard

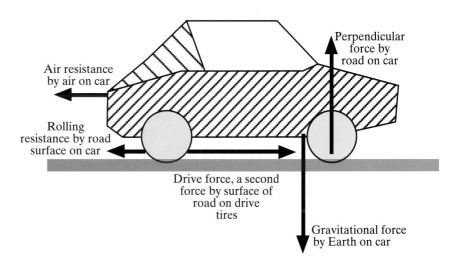

FIGURE 4.20
The five forces on an automobile.

smooth surface, such as steel wheels on steel tracks, reduce rolling resistance to a minimum—one reason trains are more energy efficient than cars.

The four forces discussed so far act even on a car that is coasting with its engine shut off. If these are the only forces on the car, then the net force must be backward, so the acceleration is backward, thus the car slows down.

When a car is driving instead of coasting, there is an additional force on it: a drive force. This force is not exerted by the engine on the car, because things cannot push themselves. Instead, the engine causes the drive wheels to turn; the drive wheels push backward against the road; and the *road* pushes forward against the drive wheels. This push by the road is the drive force.

If the car moves at an unchanging velocity, there is no acceleration. Newton's law of motion tells us that the net force must then be zero, which means that the five forces shown in Figure 4.20 must balance. In this case, the forward drive force must equal the sum of the strengths of the two resistive forces. Note that "no net force" does not mean "no drive force." Even with no acceleration, a drive force is needed. The car speeds up whenever the drive force is larger than the sum of the two resistive forces and slows down whenever the drive force is smaller than the sum of the two resistive forces.

Most other drive forces are similar to this: A bicycle's drive tire pushes backward on the road, and the road pushes forward on the tire. A swimmer pushes backward on the surrounding water, and the water pushes forward on the swimmer. A motorboat's propeller pushes backward on the water, and the water pushes forward on the propeller. A prop-driven airplane's propeller pushes backward on the surrounding air, and the surrounding air pushes forward on the propeller, just like water pushing forward on a motorboat's propeller. A jet airplane pushes air backward, too. As a jet engine moves through air, the air flowing into its front end heats by combustion with jet fuel, and the heated gas expands and rushes out of the back end at high velocity. The jet engine pushes air backward, but by a different mechanism than in a prop-driven airplane.

One nice thing about space travel is that there are no resistive forces in space, so you don't need a drive force to keep going. Your spaceship keeps going because of the law of inertia. But if you want to give your spaceship an acceleration—a change in direction for instance—then you have a problem. It is difficult to get the surroundings to exert a drive force on your spaceship because there is nothing around to push against!

You would have a similar problem if you were stranded in the middle of a smoothly frozen pond. If the ice were absolutely smooth, you could not walk off it because, with no friction, you could not push on the ice. How could you get off? You could fan the air, pushing air backward in the way that a swimmer pushes water backward. That would work. But suppose that, as in space, there were no air.

Suppose that you had something along with you that you could throw away. Your physics book, for example. Or a shoe. You could take off a shoe and throw it away! While throwing, you would push on the shoe, so it would push in the other direction on you, so your body would accelerate away from the shoe. If you now let go of the shoe (if you hung onto it, the shoe would slow your body to a stop), you would have acquired a velocity. So you would slide along the pond. If you wanted to slide faster, you could either throw your shoe away faster or throw away your other shoe, too.

This is the principle of **rocket drive**. Rockets take along their own material just to have something to push against. Shoes would work (Figure 4.21), but they aren't very practical. The rocket fuel for the U.S. space shuttle's main rocket engines is hydrogen and oxygen, stored as low-temperature liquids. When combined, their combustion produces steam, which streams rapidly out of the back end of the engine, pushing the shuttle forward.

■ **DIALOGUE 12**   A car weighing 10,000 N moves along a straight level road at an unchanging 80 km/hr. Air resistance is 300 N, and rolling resistance is 400 N. Find the net force on the car and the strength and direction of the perpendicular force and the drive force.

■ **DIALOGUE 13**   Magnetic forces can levitate railroad trains a short distance above the tracks, making friction practically negligible. Suppose such a

FIGURE 4.21
Shoe power.

---

12. Zero, 10,000 N upward, 700 N forward.

13. There would be a forward drive force during a relatively brief acceleration period while leaving New York; no drive force would be needed during most of the trip; and a backward force would slow the train to a stop in Chicago. *Note*: Although there are plans for maglev trains, there are no plans for evacuated tunnels.

"maglev" train runs inside an evacuated (emptied of most air) tunnel, from New York City to Chicago. If friction and air resistance are negligible, during what parts of the trip would a drive force or other external horizontal force act on the train? Discuss the direction of the external horizontal force.

SUMMARY OF IDEAS AND TERMS

Force A body exerts a force on another body whenever the first body causes the second body to accelerate. A force is an action by one body on another; it is not a thing or a property of a body. Every force is similar to a push or a pull.

Force of gravity, or **gravitational force** The downward pull by Earth on objects in Earth's vicinity. In Chapter 5, we will learn that not only Earth, but also every other material object exerts a gravitational force on every material object.

Newton (N) A unit of force. The amount of force that can accelerate a 1-kg mass by 1 m/s^2.

Net force The total, overall force on an object. The net force due to two forces acting in the same direction is the sum of the two. The net force due to two forces acting in opposite directions is the difference between the two and acts in the direction of the stronger force.

Inertia A body's ability to stay at rest or to maintain an unchanging velocity. Quantitatively, a body's inertia is its degree of resistance to acceleration, or its **mass**.

Mass A body's mass is its amount of inertia and also its quantity of matter. We find a body's mass by comparing its inertia with the inertia of a standard kilogram.

Kilogram (kg)   A unit of mass. The mass (or inertia) of the object known as a **standard kilogram**. Any object that has the same inertia as the standard kilogram has a mass of 1 kilogram.

Metric ton, or **tonne**   1000 kilograms or about 2200 pounds.

Proportional to the inverse   A quantity is proportional to the inverse of another quantity if the first is proportional to (equal to some number times) 1 divided by the second quantity.

Newton's law of motion   An object's acceleration is proportional to the net force exerted on it by its surroundings and is proportional to the inverse of its mass. The direction of the acceleration is the same as the direction of the net force, $a \propto F/M$. In the appropriate units (m/s^2, newtons, kilograms), $a = F/M$.

Weight   The weight of an object is the net gravitational force exerted on it by all other objects.

Resistive force   Any force that acts on a moving body in a direction opposite to the body's motion. **Friction** is the resistive force by a surface on an object sliding across that surface. **Air resistance** is the resistive force that air molecules exert on an object moving through the air. **Rolling resistance** is the resistive force by a surface on a rolling object.

Perpendicular force   The force, perpendicular to a solid surface, that is exerted by any solid surface on any object touching it.

Force diagram   A diagram that shows all of the individual forces acting on an object. Each force is shown as an arrow pointing in the direction of that force.

Force pair   The two forces that two bodies exert on each other.

Law of force pairs   Forces always come in pairs: Whenever one body exerts a force on a second body, the second exerts a force on the first. The two forces are equal in strength but opposite in direction.

Drive force   The force, exerted on a *self-propelled* (self-energized) object, that accelerates it or keeps it moving against resistive forces. Most drive forces are due to friction, but unlike sliding friction they act in the forward direction.

Rocket drive   A particular type of drive force. When material is thrown away from, or ejected from, any vehicle, the material exerts a reaction force back on the vehicle because of the law of force pairs. When this reaction force is used as a drive force, it is called rocket drive.

REVIEW QUESTIONS

FORCE
1. How can we tell whether or not a body is exerting a force on another body?
2. Can an object have force? Can an object exert a force? Can an object be a force? Can an object feel a force?
3. List at least six specific examples of forces. Try to list examples that are significantly different.
4. What is a resistive force? Give two examples.
5. Give two examples of forces that act at a distance.

NEWTON'S LAW OF MOTION
6. What does Newton say that forces cause? What does Aristotle say?
7. What do we mean when we say that one object has "more inertia" than another object does?
8. When you move an object from Earth to the moon, does its inertia change? Does its weight change? Does its mass change? Does its amount of matter change? Does its acceleration differ while falling freely? Does it respond differently to a force of 1 N?
9. Forces of 8 N and 3 N act on an object. How strong is the net force if the two forces have opposite directions? The same directions?
10. Is an object's acceleration always in the same direction as its velocity? If not, give an example in which it is not. Is an object's acceleration always in the same direction as the net force on the object? If not, give an example in which it is not.
11. As you increase the net force on an object, what happens to its acceleration? What if you double the net force? As you increase the mass of an object (for example, by gluing additional matter to it), what happens to its acceleration? What if you double the mass?

WEIGHT
12. What is weight? Is it the same as mass? If not, what is the difference?
13. Describe a simple way to determine, in a lab, whether two objects have equal masses. Would this method work in distant space? What would work in distant space?
14. Find the gravitational force on a 1-N apple. Would it still be a 1-N apple if we took it to the moon?
15. Draw a force diagram showing the forces on an apple at rest on a table. Find the net force on the apple.
16. Draw a force diagram showing the forces on a rocket during liftoff. Which force is largest? What is the direction of the net force?
17. Where is it easiest to lift your automobile: on Earth or on the moon? Where is the automobile's mass larger?

LAW OF FORCE PAIRS
18. Describe several experiments demonstrating that forces come in pairs.
19. Do you exert a gravitational force on Earth? How do you know? What direction is this force?
20. Describe the other member of the force pair for each of the following forces: the perpendicular force on a book, the weight of an apple, the force by a bat against a baseball, the force by a baseball hitting a catcher's mitt.

DRIVE FORCES

21. Describe four examples of drive forces.
22. Draw a force diagram showing the forces on a car driving along a straight level road. How would this force diagram be altered if the car were coasting? What if the car were braking?
23. How does propeller drive differ from jet drive?
24. How does a car's drive force compare with the resistive forces when the car maintains a constant speed? When the car is speeding up? Slowing down?

25. When a car moves at constant speed along a straight road, is the drive force zero? Is the net force zero? Is the acceleration zero? Is the velocity zero?
26. What is the main difference between rocket drive and other drive forces such as propeller drive and automobile drive?

HOME PROJECTS

1. Experience the meaning of an unchanging force and an unchanging acceleration. Pull a smoothly rolling wagon along a smooth level sidewalk, with a spring or a strong rubber band stretched by an unchanging amount. You will have to move faster and faster—you must accelerate! Try increasing the mass by putting an object, or a friend, in the wagon. For the same amount of stretch, is the acceleration larger or smaller with the larger mass?
2. Place a spool of thread on a table, as shown in Figure 4.22a. According to Newton's law of motion, what should the direction of the spool's acceleration be when you pull on the thread? Try it. Now turn the spool over, as shown in Figure 4.22b. Intuitively, what do you believe will happen when you pull on the thread? What does Newton's law of motion predict? (*Hint*: What is the direction of the net force?) Try it. You might not have expected this result. The experiment does agree, however, with Newton's law of motion. Nature, and Newton's theory of motion, can be surprising!

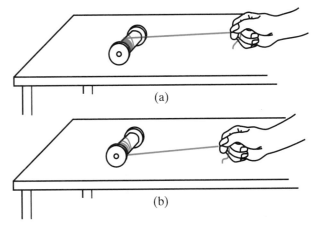

FIGURE 4.22
Try this.

EXERCISES

FORCE

1. Is any force exerted on you when you speed up along a straight line? When you slow down along a straight line? How do you know?
2. Is any force exerted on you while you move in a circle at unchanging speed? How do you know?
3. A smooth ball rolls on a smooth table. Initially, no horizontal forces are exerted on the ball. Then you bring a magnet near the rolling ball, but you are not sure whether the magnet actually exerts a magnetic force on the ball. How can you tell whether or not the magnet is exerting a horizontal force on the ball?
4. Does a high-velocity bullet contain force? Does a stick of dynamite contain force?

NEWTON'S LAW OF MOTION

5. You place your book on a table and hit it horizontally with a hammer, strongly but briefly. Do not neglect friction. Describe the motion of the book, beginning from just before you hit it with the hammer. Describe the direction and strength of the net force on the book during the entire motion.
6. If you exert a force on an object and then exert three times as strong a force on the same object, what (if anything) can you say about the object's acceleration during the exertion of each force?
7. A ball weighing 8 N is thrown straight upward. Disregarding air resistance, find the direction and strength of the net force on the ball as it moves upward. What is the direction of the ball's acceleration? Are the net force and the acceleration in the same direction in this case? Can they ever be in different directions?
8. An object moves with unchanging velocity. Does it then have no forces acting on it? Explain. Does it have no net force acting on it?
9. An object is at rest. Does it then have no forces acting on it? Explain. Does it have no net force acting on it?
10. When you stand on the floor, does the floor exert a force on your feet? In which direction? Why, then, don't you accelerate in that direction?

11. You push on a solid concrete wall. Is your push the only horizontal force on the wall? How do you know? What can you say about the net force on the wall?

12. A car starts up from rest, moving along a straight highway with an acceleration of 1 m/s^2. A second car comes racing past at a steady 120 km/hr. Which car has the larger net force acting on it?

13. An 80-tonne truck and a 1-tonne car maintain a steady speed of 80 km/hr on a straight highway. Which vehicle has a larger net force exerted on it? Which vehicle has the larger perpendicular force exerted on it?

14. An 80-tonne truck and a 1-tonne car maintain a steady speed of 80 km/hr on a straight highway. Which vehicle has the larger drive force exerted on it? The larger air resistance force? The larger net force?

WEIGHT

15. Roughly, what is your weight in newtons?

16. Which has the greater mass, a tonne of feathers or a tonne of iron? Which has the greater weight? Which has the larger volume?

17. Would you rather have a hunk of gold whose weight is 1 N on the moon or one whose weight is 1 N on Earth—or wouldn't it make any difference?

18. Would you rather have a hunk of gold whose mass is 1 kg on the moon or one whose mass is 1 kg on the Earth—or wouldn't it make any difference?

19. Find the strength and direction of the net force on an apple weighing 2 N, neglecting air resistance, in each of the following cases: The apple is held at rest in your hand. The apple is falling to the ground. The apple is moving upward, just after you threw it upward.

20. An apple is accelerated upward by your hand. Which is larger, the apple's weight or the upward force by your hand? What if you accelerate the apple downward while it is in the palm of your hand? What if you lift the apple at an unchanging velocity? What if you lower the apple at an unchanging velocity?

21. Would it be easier (in other words, would it require less thrust and less rocket fuel) to lift a rocket off the moon's surface than off Earth's surface? Why?

22. An astronaut on the moon picks up a large rock. Would it be easier, or harder, or neither, for him to pick up the same rock on Earth?

23. An astronaut on the moon kicks (horizontally) a large rock. What if she kicked the same rock on Earth: Neglecting frictional effects, would it hurt her foot more, or less, or just as much?

LAW OF FORCE PAIRS

24. "Planet Earth is pulled upward toward a falling boulder with just as much force as the boulder is pulled downward toward Earth." True or false? Why?

25. "Planet Earth is pulled toward a falling boulder with just as much acceleration as the boulder has as it moves toward Earth." True or false? Why?

26. A large truck breaks down on the highway and receives a push back into town by a small car. The car speeds up while pushing the truck. While speeding up, the car exerts a forward force on the truck. But does the truck exert any force on the car? If so, is this force weaker or stronger than the force that the car exerts on the truck?

27. A car collides head-on with a large truck. Which vehicle exerts the stronger force? Which has the larger force exerted on it? Which experiences the larger acceleration?

28. When a rifle fires, it accelerates a bullet along the barrel. Explain why the rifle must recoil.

29. A 2-N apple hangs by a string from the ceiling. Describe the two forces on the apple. How strong is each of these forces? Do these forces form a single force pair? If not, then for each force, describe the other member of that force's force pair.

30. A horizontally moving bullet slows down. Is anything exerting a force on it? How do you know? Is it exerting a force on anything? How do you know?

31. I push you away from me. Do you also push (exert a force on) me? Which force is stronger—the force by me on you or the force by you on me?

DRIVE FORCES

32. Since the law of inertia states that no force is needed to keep an object moving at an unchanging velocity, why does a car need a drive force to keep moving?

33. While driving your car on a straight level road, you slam on your brakes. Draw a force diagram of the car during braking. What is the direction of the net force? Draw a force diagram for a car that is coasting without braking. In which of the two cases is the net force stronger?

34. Why is it easier to pedal a bicycle with hard high-pressure tires, as compared with soft balloon tires?

35. When you hold your foot on a car's accelerator pedal, is the car necessarily accelerating? Could it be accelerating? Could it have a forward acceleration? Could it have a backward acceleration?

36. There are three acceleration devices on any car. What are they, and what kinds of accelerations does each one give to the car?

37. If a jet airplane were above Earth's atmosphere, could it then accelerate? What about a rocket-driven plane?

PROBLEMS

NEWTON'S LAW OF MOTION

1. A 3-kg rock rests on the ice. You kick it, exerting a 60-N force. Find the rock's acceleration, assuming that there is no friction. Still assuming no friction, what will be the rock's acceleration after your foot is no longer in contact with the rock? Will the rock have a (nonzero) *velocity* at this time?

2. In the preceding problem, assume now that a frictional force of 6 N acts on the rock whenever it is moving across the ice. Find the net force on the rock and the rock's acceleration. What can you say about the net force on the rock after your foot is no longer in contact with it?

3. Ned the skydiver weighs 600 N. How large must be the force of air resistance acting on Ned in order that Ned will maintain an unchanging speed while falling through the air? What would be Ned's acceleration if there were no air?

4. What would a skydiver's acceleration be if air resistance was half as large as the skydiver's weight? What if air resistance were as large as the skydiver's weight?

5. How much force must a pitcher exert on a 0.5-kg baseball in order to accelerate it at 50 m/s^2?

6. Find the force acting on a 0.01-kg bullet as it is accelerated at 1 million m/s^2 (100,000 g's!) down a rifle barrel.

7. A car weighing 8000 N moves along a straight level road at a steady 80 km/hr. The total resistive force on the car is 500 N. Find the net force on the car, the acceleration of the car, and the drive force.

8. The car of the preceding problem has a mass of 800 kg. The driver pushes down on the accelerator, increasing the drive force to 2100 N. All other forces on the car remain unchanged. Find the net force on the car, the acceleration of the car, and the directions of both.

9. The driver of the car of the preceding two problems lets up on the accelerator, reducing the *drive* force to only 100 N. All other forces are unchanged. Find the *net* force on the car, the acceleration of the car, and the directions of both.

10. A freely falling apple has a weight of 1 N. Earth's mass is 6×10^{24} kg. How strong is the force by Earth on the apple? How strong a force does the apple exert on Earth? How big is the apple's acceleration? Find the acceleration that the apple would cause Earth to have if the apple was the only object exerting a force on Earth.

THE LAW OF FORCE PAIRS

11. Wearing frictionless roller skates, you push horizontally on a wall with a force of 50 N. How hard does the wall push on you? If your mass is 40 kg, what is your acceleration?

12. Your friend (mass 80 kg) and you (mass 40 kg) are both wearing frictionless roller skates. You are at rest, behind your friend. You push on your friend's back, with a force of 60 N. How hard does your friend's back push on you? What is your acceleration? What is your friend's acceleration?

13. A small car having a mass of 1000 kg runs into an initially stationary 60,000-kg 18-wheeled truck from behind, exerting a force of 30,000 N on the truck. How big, and in what direction, is the force that the truck exerts on the car? Find the car's acceleration. Is this a "speeding up" or a "slowing down" type of acceleration? Find the truck's acceleration. Is this of the "speeding up" or "slowing down" variety?

14. You press downward with a 100-N force on a brick weighing 40 N that rests on a table. With what force, and in what direction, does the brick press against your hand? Draw a force diagram similar to Figure 4.20, with an arrow representing each force acting *on the brick*. How big is the net force on the brick? Now find the force (how big and in what direction) that the table exerts on the brick. How hard is the brick pressing down against the table?

CHAPTER 5

The Universe According to Newton

And from my pillow, looking forth by light
Of moon or favouring stars, I could behold
The antechapel where the statue stood
Of Newton with his prism and silent face,
The marble index of a mind for ever
Voyaging through strange seas of Thought, alone.
WILLIAM WORDSWORTH, PRELUDE (BOOK III), 1850

The law of inertia might be history's most fruitful scientific idea. Besides unifying natural motion on Earth and in the heavens, undermining Aristotelian views, promoting the idea of universal natural law, and leading to Newton's law of motion, it also led Newton to look in a new way at one specific kind of force: gravity. This is the subject of most of this chapter.

Gravity is perhaps the most universal force, because it is all around us on Earth and also accounts for most motions observed in the heavens. Gravity is so all-pervading that we are hardly aware of it, and this historically made it difficult to conceptualize. Prior to the inertial view of motion, it was believed that every body containing the element "earth" had a natural tendency to seek out Earth's center, in the way that a thirsty person seeks out water. External influences—forces—were not needed to explain why objects fall: They fell because that was what they "wanted" to do.

The inertial view is that bodies "want" to maintain their velocity. Descartes, more than anyone else, first conceived of this new view of motion. It was a conceptual shift comparable to Copernicus's shift to a sun-centered view. Newton then built on Descartes's idea.

If you believe that bodies have inertia, then you must ask why an apple, held at rest in midair and then released, falls. The inertial view made this question possible. Newton's answer applied to more than apples. The answer reaches beyond the moon and demonstrates convincingly that in the heavens and on Earth, the same forces are at work. We live in one natural universe, not two.

Section 5.1 presents the general idea of Newton's law of gravity, and Section 5.2 gives the specifics along with several examples. Section 5.3 applies these ideas to understanding the origin of the sun and Earth and the future final gravitational collapse of the sun. Section 5.4 tells the tale of the final violent collapse of stars that are more massive than the sun and the exotic objects that result from the collapse. Sections 5.5 and 5.6 return to our theme of comparing Newtonian and post-Newtonian physics: Section 5.5 looks at some broad philosophical implications of Newtonian physics, particularly the general worldview known as the mechanical universe. Section 5.6 places Newtonian physics within a broader scientific perspective, noting the limitations of Newton's theories in light of the more recent relativity and quantum theories.

5.1 THE IDEA OF GRAVITY: THE APPLE AND THE MOON

At age 22, Isaac Newton had just completed his bachelor of arts degree at Cambridge University in England. He had been invited to remain there, but the school then closed for 18 months because of an epidemic of the plague, and so the graduate returned to his family's farm. During those 18 months Newton laid the foundations for a theory of gravity, a theory of light, and calculus.

Some say that greatness is partly a matter of timing. Newton lived at a time that was scientifically and culturally ripe for a new view of the universe. The scientific foundations had been laid by Copernicus, Brahe, Kepler, Descartes, and Galileo. We have seen that the inertial view of Descartes and Galileo leads naturally to Newton's law of motion. The ideas surrounding the law of motion, combined with the astronomical ideas of Copernicus and Kepler, then led Newton to the law of gravity. Newton truly stood, as he himself said, "on the shoulders of giants." But by the same token, it took the supreme inventive genius of a Newton to put all this together.

As Newton recounted it late in life, he invented the central idea of his law of gravity during that stay on his family's farm. The idea came to him when an apple fell from a tree while he could see the moon in the sky. Beyond the fact that both are more or less round, it is difficult to think of two more dissimilar objects than an apple and the moon. One is on Earth, the other is in the heavens; one quickly rots, the other seems eternal; one falls to the ground, the other stays aloft. Yet where others saw difference, Newton saw similarity. Great science often comes from just such unifying insights.

Let us trace Newton's thinking. Figure 5.1 shows an apple falling toward the ground, accelerated by Earth's gravitational pull. The directions of the apple's velocity and acceleration and of the gravitational force all are downward toward Earth's center, as shown. The moon's motion is quite different. The direction of its velocity is around Earth rather than toward its center. But we are interested in the forces on each, and according to the inertial view, forces cause accelerations, not velocities. So the forces on the two could be similar, despite the dissimilarity of the velocities. How do the forces compare?

Aristotle would say that no force is needed to make the moon move in a circle because that is its natural motion, but the new inertial view is that in order for the moon to deviate from straight-line motion, a force must act on it. What is the direction of this force? If the moon were at point A in Figure 5.2 and if no force acted on it, it would keep moving in a straight line from A to B. But instead it moves around to point C. The moon does two things simultaneously: Its inertia keeps it moving forward and would carry it from A to B, but it is simultaneously pulled inward so that it also "falls" from B to C. The force required to

One had to be a Newton to see that the moon is falling, when everyone sees that it doesn't fall.

PAUL VALÉRY, FRENCH POET AND PHILOSOPHER, 1871–1945

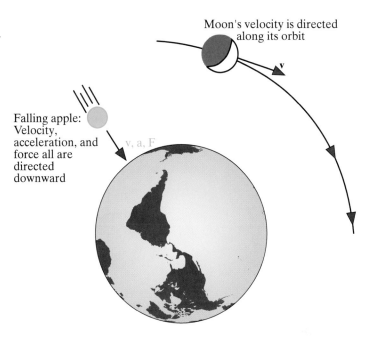

Moon's velocity is directed
along its orbit

v

Falling apple:
Velocity,
acceleration, and
force all are
directed
downward

v, a, F

pull the moon inward—so that it arrives at C rather than B—is directed toward Earth's center, just like the force on an apple. Newton hypothesized that this force had the same source as the force that pulled an apple downward: Earth's gravitational attraction.

Since the force on the moon is inward, Newton's law of motion tells us that the moon's acceleration must be inward, too. Figure 5.3 shows in another way that the moon's acceleration is inward: In order for the moon to maintain a circular motion, its velocity must keep changing toward the inside. So the change in velocity—in other words the acceleration—is toward the center. Although the velocity of the apple and the velocity of the moon are quite different, the accelerations of both are directed toward the center.

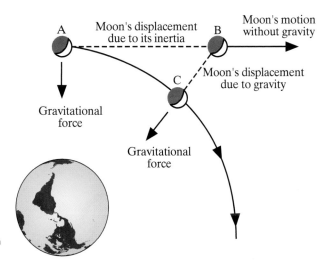

A

Moon's displacement
due to its inertia

B

Moon's motion
without gravity

C

Moon's displacement
due to gravity

Gravitational
force

Gravitational
force

FIGURE 5.2
The moon is held into its orbit by an
inward-directed force.

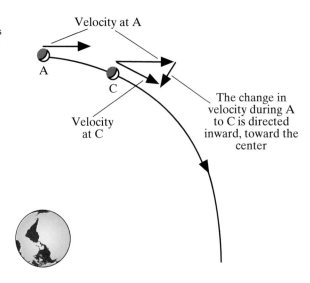

Velocity at A

Velocity
at C

The change in
velocity during A
to C is directed
inward, toward the
center

Newton offered another argument, one that helps us understand why the moon and other satellites stay up. If you throw an apple horizontally, it will follow a curved path as it falls to the ground (Figure 5.4). If you throw the apple faster, it will go farther before hitting the ground. And if you throw it fast enough, it might "fall" around a large part of Earth's surface before striking the ground (Figure 5.5).

If the apple is launched at such a high speed that the curvature of its path just matches Earth's curvature, it will fall all the way around. In other words, it goes into orbit. The required speed is about 8 km/s, or 29,000 km/hr. This is what any orbiting satellite, including the moon, does.

The force that shapes the moon's path is gravity—the same gravity that pulled the apple to the ground that day on Isaac Newton's family's farm.

It was an imaginative leap, in more ways than one. It was difficult to believe that anything at all was pulling on the moon, much less that it could be the same force that pulled on an apple. Most difficult of all was the notion that the gravitational force could reach out across nearly half a million kilometers of empty space (the distance was known in Newton's time) to pull on the moon. It is easy to see that things exert forces on one another when they are in direct contact, but a force that acts across so much empty space seems like magic.

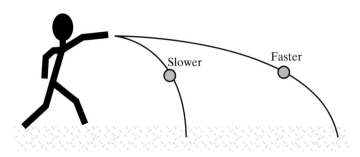

Slower

Faster

FIGURE 5.5

Falling around Earth. If you throw an apple fast enough, it will fall around a large part of Earth's surface or even go into orbit. A diagram like this appears in Newton's notebook.

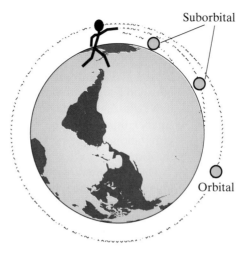

Suborbital

Orbital

■ **DIALOGUE 1**   Figures 5.4 and 5.5 show five possible paths for an apple that has been thrown horizontally. Assume that air resistance is negligible. For each path, draw three arrows—labeled *f, a, v*—attached to the apple that show the directions of the gravitational force on the apple, the acceleration of the apple, and the velocity of the apple.

■ **DIALOGUE 2**   A 2-N apple falls from a tree. While it is freely falling, what are the strength and direction of the force on it? What are the numerical value and the direction of the apple's acceleration?

■ **DIALOGUE 3**   Continuing Dialogue 2, suppose we throw the apple horizontally, as shown in Figures 5.4 and 5.5. While the apple is in each of the five positions shown in the figures, what is the strength and the direction of the force on it (neglect air resistance)? Would the apple's mass still be the same as it was in Dialogue 2? Since the net force and the mass are the same as they were in Dialogue 2, what must the numerical value and the direction of the apple's acceleration be?

5.2 THE LAW OF GRAVITY: MOVING THE FARTHEST STAR

Since Earth's gravitational pull holds the moon into its orbit, it is reasonable to suppose that all satellites—bodies in orbit around larger astronomical bodies—are held in their orbits by the gravitational pull exerted by the larger body. Since the planets are satellites of the sun, Newton's insight regarding the moon also resolves the old question of why the solar system moves as it does! The planets keep moving forward because of the law of inertia, and the sun's gravitational pull bends their orbits into ellipses (elongated circles). Similarly, the moons of

---

1. *f* and *a* point directly toward Earth's center, and *v* points along the path of motion.

2. 2 N downward. 10 m/s^2 downward.

3. 2 N downward. The mass would still be the same. Since the mass and force are the same as in Dialogue 2, Newton's law of motion tells us the acceleration is the same: 10 m/s^2 downward.

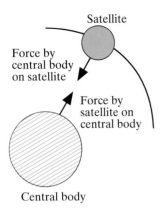

Satellite

Force by
central body
on satellite

Force by
satellite on
central body

Central body

FIGURE 5.6
An astronomical object with a
satellite. The two bodies could
be Earth and the moon, the sun
and Earth, the sun and Mars,
Earth and an artificial satellite,
and so forth. Each of the two
bodies pulls on the other.

What makes planets go around the
sun? At the time of Kepler some
people answered this problem by
saying that there were angels
behind them beating their wings
and pushing the planets around an
orbit. . . . The answer is not far
from the truth. The only difference is
that the angels sit in a different
direction and their wings push
inwards.

RICHARD FEYNMAN, PHYSICIST

Force by
apple on
book

Force by book
on apple

FIGURE 5.7
Even ordinary-sized objects
exert gravitational forces on
one another. Your physics
book exerts a force on an
apple, and vice versa. It's a
small force, but forces like this
have been measured.

the planet Jupiter are held in their orbits by Jupiter's gravitational pull. And because of the law of force pairs, every satellite must also pull back on the larger body. The gravitational force pulls on both bodies (Figure 5.6).

But why would gravity act only between astronomical bodies and their satellites? For instance, it seems plausible that there should be a gravitational force between Earth and Mars. Such a force between planets had not been noticed yet in Newton's day, but Newton supposed that this was only because it was so much smaller than the force by the sun on the planets. Likewise, there should be a gravitational force between any two astronomical bodies, even between the farthest stars.

But why should gravity be restricted to astronomical bodies? Why shouldn't a gravitational force be exerted between smaller objects on Earth—oranges, rocks, and so forth? Your physics book, for instance, should exert a gravitational pull on an apple, and vice versa (Figure 5.7). You won't notice this force, but that is only because the force between such objects is very small.

So Newton reasoned that the gravitational force is universal; it is exerted between every pair of objects throughout the universe. This is the central idea of Newton's theory of gravity.

Newton understood the importance of quantitative methods. Although his basic insight was qualitative, its expression in a quantitative form led to powerful explanations and predictions. Quantitatively, we expect that the gravitational attraction between two objects must be stronger when the objects' masses are larger, because an apple's weight is larger when its mass is larger (double the mass, for example, by replacing the one apple by two apples joined together, and you double the weight). And since widely separated objects do not attract each other very strongly, the gravitational force should get smaller when the distance between the objects gets larger.

Newton put all this together (see How do we know? at the end of this section for how he did this) and came to the following conclusions:

NEWTON'S LAW OF GRAVITY
Between any two objects there is an attractive force that is proportional to the product of the two objects' masses and proportional to the inverse of the square of the distance between them:

$$\text{gravitational force} \propto \frac{(\text{mass of 1st object}) \times (\text{mass of 2nd object})}{\text{square of distance between them}}$$

$$F \propto \frac{m_1 \times m_2}{d^2}$$

If mass is expressed in kilograms, distance in meters, and force in newtons, then this proportionality becomes

$$F = 6.7 \times 10^{-11} \frac{m_1 \times m_2}{d^2}$$

For our first example, consider your own weight, in other words, the gravitational force exerted by Earth on you. The law of gravity tells us that this force is

Pick a flower on Earth and you move the farthest star!

PAUL DIRAC, PHYSICIST

proportional to your mass times Earth's mass, which means that the force is proportional to each of the two masses separately. So doubling your mass would double your weight, tripling your mass would triple your weight, and so forth—which seems quite plausible. But the law of gravity also tells us that if you imagined that somehow Earth's mass were doubled (without, however, changing its size), this also would double your weight; halving Earth's mass would halve your weight; and so forth. You can reduce your weight without dieting or exercising: Simply reduce Earth's mass!

What if you altered both masses? For instance, suppose you tripled your mass while simultaneously doubling Earth's mass. Since the force is proportional to the product of the two masses, this would multiply your weight by 6. As another way of looking at this, take it in steps: If you first tripled your mass, this would triple your weight; if you then doubled Earth's mass, this threefold weight would be doubled, giving an overall multiplication by 6.

What happens when the distance between Earth and you is changed? In fact, exactly what do we mean by the *distance between the objects* in a case like this? Does the distance from Earth to your body mean the distance from the near side of Earth (the ground beneath your feet), from the far side, from the center, or from some other point? And to what point in your body should we measure the distance?

Newton worked through a lot of mathematics to answer this—in fact he needed to invent *integral calculus* to answer it. Newton's answer was that the distance between the "centers" of the two bodies is the correct distance to use when applying the gravitational force law to two extended bodies. In the case of a body such as Earth that has an obvious center, distance is measured from that center. For other bodies, such as your own, the distance should be measured from the body's *balance point*—the point at which the body would be balanced under the force of gravity. But because your body is so small compared with the distance from Earth's center to your body, it matters little which point you choose within your body.

Suppose that you travel away from Earth. Since the gravitational force is proportional to the inverse of the square of this distance, the increased distance makes the force decrease—another way to reduce your weight! For instance, your weight at the top of Mount Everest, nearly 10 km above sea level, is 0.3% less than at sea level. If your weight is normally 600 N (135 lbs), it will be 598 N (134.5 lbs) at the top of Mount Everest. Your weight reduction is greater at an altitude of a few hundred kilometers, where low-orbit artificial satellites travel. For example, at a 200-km altitude, your weight would be reduced by 6%, so a person normally weighing 600 N would weigh only 560 N. Now you're really losing weight (but not mass!).

Moving to still higher altitudes, suppose you are 6400 km—1 Earth radius—above the ground. What is your weight? The proportionalities in the law of gravity make this an easy question. If you rise 1 Earth radius above the ground, the distance from Earth's center to your body will have doubled from its normal value, so the square of the distance is multiplied by 4. Since the force is proportional to the inverse of the square of the distance, the force is divided by 4. Your weight is now one-fourth of normal.

Figure 5.8 is a graph of your weight at various distances, in Earth radii, measured from Earth's center. No matter how far you are from Earth, the gravitational force by Earth on your body will never reach precisely zero. But very far

It is only by bringing into the open the inherent contradictions, and the metaphysical implications of Newtonian gravity, that one is able to realize the enormous courage—or sleepwalker's assurance—that was needed to use it as the basic concept of cosmology. In one of the most reckless and sweeping generalizations in the history of thought, Newton filled the entire space of the universe with interlocking forces of attraction, issuing from all particles of matter and acting on all particles of matter, across the boundless abysses of darkness.

ARTHUR KOESTLER, AUTHOR AND PHILOSOPHER, IN *THE SLEEPWALKERS*

FIGURE 5.8
A graph of your weight at various
distances from Earth's center. The same
graph applies to the weight of any object.

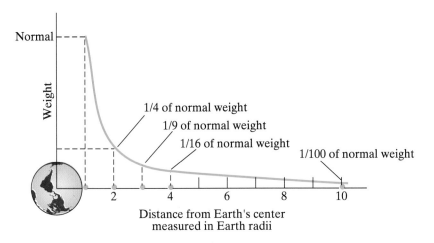

Normal

Weight

1/4 of normal weight

1/9 of normal weight

1/16 of normal weight

1/100 of normal weight

2   4   6   8   10

Distance from Earth's center
measured in Earth radii

away, the force becomes very small. For example, at 10 Earth radii, your weight is 1% of your normal weight.

It's a good thing for us that the force of gravity declines at larger distances in just the way it does. If the gravitational force declined a little faster, the planets would not move in ellipses but would instead spiral into the sun, and we would not be here to ask about things like gravity. If, on the other hand, the gravitational force declined a little more slowly, the gravity from distant stars would dominate the gravity from Earth, and again we would not be here.

The law of gravity can be used to calculate the gravitational attraction between any pair of objects, from apples and books to stars and moons. For example, the force between 1 kilogram and another kilogram 1 meter away is found by putting 1 kg, 1 kg, and 1 m into the gravitational force formula. The answer is 6.7×10^{-11} newtons, or 0.000,000,000,067 newtons! It is no wonder that the gravitational force between ordinary objects is difficult to detect. The delicate experiments needed to measure such tiny forces could not be performed until about a century after Newton's work. When they were performed, they verified Newton's predictions.

The situation inside an orbiting satellite seems paradoxical. Judging from Figure 5.9, you would feel weightless in an orbiting satellite, at any altitude. But we have seen that if the satellite is in low orbit, your weight is actually only a little less than normal. Why, then, would you feel weightless? To answer this, let's imagine a somewhat similar situation (Figure 5.10): Suppose you are in an elevator and the elevator cable breaks. The elevator is then in free fall, and so are you. After the cable breaks, your feet no longer press down against the floor. If you try to press your feet against the floor, you will simply push yourself away from the floor. You are apparently weightless, but, because we have defined weight as the gravitational force on an object, you are not really weightless.

You would feel weightless in an orbiting satellite for the same reason that you would feel weightless in a freely falling elevator. As we saw in the preceding section, the satellite falls freely around Earth. You are falling freely around Earth too, regardless of whether you are inside the satellite or outside in space. Since both you and the satellite are just falling around Earth, you have the sensation of weightlessness. Your body behaves as though it were removed from the effects of gravity, but you are not really weightless.

(a)

FIGURE 5.9
Space travelers feel weightless when they are in orbit and at any other time that they are "falling" freely through space. (a) Balancing. (b) Floating. (c) Spacewalking.

(b)   (c)

I'm apparently weightless.

You're apparently in trouble.

FIGURE 5.10
Falling freely in a freely falling elevator.

H O W   D O   W E   K N O W ?

Since the forces between ordinary objects on Earth were too small for Newton to measure, how did he verify his hypothesized law of gravity? The dependence on mass was not hard to deduce. Because an object's weight is proportional to its mass (for instance, two identical apples tied together surely have twice the weight of one, so doubling the mass doubles the weight), Newton reasoned that the force of gravity must be proportional to each of the two masses.

But what about the dependence on distance? Newton knew that the distance to the moon is about 60 times larger than Earth's radius (Figure 5.11). Newton's hypothesized law of gravity then implies that an object at the moon's distance should experience a force that is 3600 (the square of 60) times smaller than the force on the same object on Earth. So (since acceleration is proportional to force) the acceleration of an object at this distance should be 3600 times smaller than the acceleration of an object falling to Earth. In other words, Newton's hy-

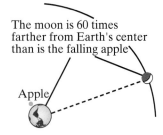

The moon is 60 times farther from Earth's center than is the falling apple

Apple

FIGURE 5.11
The moon is 60 Earth radii away from the center of Earth.

pothesis implies that the moon's acceleration should be $(1/3600) \times 9.8$ m/s², or 0.0027 m/s². But from the observed facts that the moon is 60 Earth radii from Earth and that the moon takes 27 days to complete a circle around Earth, Newton could calculate directly that the moon's acceleration (due to its circular motion) actually is 0.0027 m/s². Newton's hypothesis agreed with observation.

Gravity is one of nature's most fundamental forces. Physicists have found that all of nature's forces can be explained in terms of only four fundamental forces acting at the microscopic level. Gravity is one of these four. The others are the electromagnetic force (Chapter 8), and two forces whose main effects show up only within the atomic nucleus (Chapter 15).

■ **DIALOGUE 4**   Suppose that you were in distant space, far from all planets and stars and that you placed an apple and a book at rest in front of you, separated by about 1 m. What effect, if any, would gravity have on the apple or the book? What effect would your own presence have on this experiment? How could you reduce your own influence on the experiment?

■ **DIALOGUE 5**   Do you weigh less in a high-flying jet plane? Can you decrease your mass in this way? How can you decrease your mass? When people say that they are "overweight" and want to reduce, what do they really want to lose—weight or mass?

■ **DIALOGUE 6**   Calculate your weight at an altitude of 2 Earth radii. At an altitude of 3 Earth radii. Calculate your weight at a distance of 10 Earth radii from Earth's center. At a distance of 100 Earth radii.

5.3 GRAVITATIONAL COLLAPSE: THE BIRTH AND DEATH OF THE SOLAR SYSTEM

Gravity builds stars, and it builds planets. Stars are made mostly from an extremely diffuse gas, mostly hydrogen atoms, that is spread throughout the universe. In some regions, this material happens to be gathered slightly more densely into great gas clouds. These clouds are the spawning grounds for stars (Figure 5.12). Because of the gravitational pull between all bits of matter, all gas and dust in space tends to aggregate. If by chance there is a place within a gas cloud where matter is gathered more thickly than it is elsewhere, this place can become a center of gravitational attraction for other matter. If enough material

4. The apple and the book move very slowly toward each other, speeding up as they get closer. Your own body causes the apple and book to move toward you. You can reduce this effect by viewing the apple and book from a larger distance. (If undisturbed, a 1-kg apple and a 1-kg book started 1 m apart would come together in about 26 hours.)

5. Yes. No. You can reduce your mass by eating less, exercising, or cutting off one of your ears (not recommended!). Mass.

6. 1/9 of your normal weight; 1/16 of normal; 1/100, or 0.01 (1%) of normal; 1/10000, or 10^{-4}, of normal.

FIGURE 5.12
Star birth. These eerie, dark, pillarlike structures are columns of cool interstellar hydrogen gas and dust that are also incubators for new stars. They are part of the Eagle Nebula, a nearby star-forming region in our own galaxy. This region is "only" 7000 light-years away (i.e., it takes light 7000 years to get here from there). The tallest pillar (left) is about 1 light-year long from base to tip. This distance is about 800 times larger than the distance across our solar system.

falls into this center, it can become a new star. In this manner, stars create themselves. This falling together of matter due to gravitational attraction is called **gravitational collapse**.

This is how the sun and Earth were born. The solar system, including the atoms in your body, began as a diffuse, cold cloud of gas and dust in space, a cloud thousands of times larger than our present solar system. About 5 billion years ago, partly by chance, matter gathered more thickly in one part of the cloud than elsewhere. This clump of matter drew surrounding material toward it, making the clump more massive, which increased its pull so that it gathered even more material, which made the clump even more massive, and so forth. This self-reinforcing process continued until a large part of the original cloud had collapsed, forming a giant ball of gas larger than our solar system. Our ball of gas continued pulling inward on itself. As gas collected, new gas was pulled inward at higher and higher speeds. Atoms at the center collided at greater and greater speeds. In other words, the gas at the center heated up.

Every gas cloud has a slight overall spinning motion, simply because of the net effect of its chaotic flowing and swirling. As our gas cloud contracted, this spinning increased, just as a figure skater spins faster and faster as she brings her outstretched arms into her sides. Because of continued contraction, this spinning

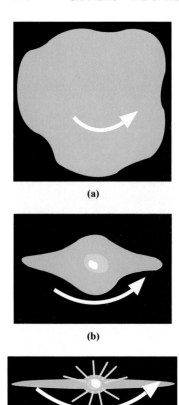

FIGURE 5.13
In (a), a slowly rotating gas cloud begins to contract because of its own gravity. In (b), the cloud rotates faster and flattens, and a hot central region begins forming. In (c), the sun forms in the cloud's center, surrounded by a rotating disk of gas. This disk will evolve into the sun's planets.

became rapid enough to flatten the outer regions of the gas ball into a thin disk, much as a wad of dough can be flattened into a pizza shape by spinning it (Figure 5.13). Some of the gas in the outlying flattened region eventually rotated fast enough to go into orbit around the larger central ball of gas. Because it was orbiting, this material was left behind as the center collapsed. The flattened outer region continued orbiting while cooling, condensing, and aggregating into clumps that became Earth and the other planets.

As the warming sun got hot enough to glow, light streaming outward from the warming sun swept away the dust and gas that had filled the solar system. And then there was light on Earth (Figure 5.14).

The large central ball of gas continued collapsing and heating until the center reached million-degree temperatures. New things happen at such high temperatures: Atoms collide so violently that their electrons are stripped off, leaving a gas made of bare nuclei and electrons. The violently colliding nuclei occasionally stick together, a process known as **nuclear fusion** (Chapter 16). Nuclear fusion creates lots of heat, and the pressure from this heat then prevents the ball of gas from collapsing further. A balance is created between the inward pull of gravity and the outward pressure of the hot gas, much like the balance in a balloon between the tension of the inward-pulling rubber and the pressure of the outward-pushing air in the balloon.

When it initiated nuclear fusion about 5 billion years ago, the sun turned itself on and became a normal, self-sustaining star.

All stars turn themselves on in roughly the same manner that the sun did. It's a process that is going on all the time, all over our galaxy and all over the universe. The starry sky is not static. Stars are born, have a lifetime, and die, like everything else.

Once the sun stopped collapsing, it settled into a middle age that has been going on now for nearly 5 billion years. The long-term stability of this period made it possible for atoms on one of the sun's planets to gather into highly complex and structured molecular forms that eventually evolved into life-forms and then, much later, into ourselves (Chapter 12). Humanlike creatures have existed on Earth for only about the past 4 million years, a thousandth of the sun's long history. Like the rest of the solar system, we came from the universe.

The sun's fuel for the fusion process is the hydrogen gas from which the sun is mostly made. Over billions of years, the supply of hydrogen must gradually deplete until, about 5 billion years in the future, it will no longer support nuclear fusion. Then the sun will enter old age. With no heat source, the sun will begin to collapse gravitationally, and this collapse will again heat the sun until its core gets hot enough to ignite new, hotter nuclear reactions involving elements other than hydrogen. The renewed heating will cause the outer portion of the sun to expand enormously, beyond the orbits of the two innermost planets, Mercury and Venus, and perhaps as far as Earth. Its heat will vaporize (turn into gas) any life remaining on Earth. Enveloped in the sun's gases, Mercury, Venus, and perhaps Earth will slow down and spiral into the sun's center, where they will be vaporized.

These nuclear reactions will eventually run their course. Then gravity will reassert itself for the final time. Without a nuclear heating source, there will be little to stop the sun from falling inward on itself. Certainly the interatomic forces that

FIGURE 5.14
An imaginary view of the newborn sun during formation of small rocky bodies that will soon coalesce to form planets. Dust clouds partially obscure the sun. (Painting by William K. Hartmann, author of *Astronomy: The Cosmic Journey* [Belmont, CA: Wadsworth, 1982])

hold up solid matter against outside pressures on Earth are far too puny to stand up against the enormous inward pull of gravity in the final collapse of a star.

The sun will squeeze itself far inside its present boundaries and far inside the volume it would have if it were made of ordinary solid material, squashing its atoms out of recognizable existence until all that remains at the microscopic level will be a solid, densely packed expanse of bare nuclei and unattached electrons. At this point, the collapse will be permanently stopped by an effect known as *quantum exchange forces* between the electrons.*

The sun's burnt-out "corpse" will be hot, solid, and small—about Earth's size. The sun's entire mass will be packed into one-millionth of its present volume! The sun's material will be extraordinarily compact, with many tonnes packed into each cubic centimeter. A thimbleful of this material would collapse a strong steel table on Earth! Gravitational heating will heat the sun enormously during its final collapse, but once the collapse ends there will be no further source of heating. This remnant of the sun will glow brightly for a while and then slowly dim like a dying ember.

A star the size of Earth? When the first astronomical body of this type was discovered, in 1862, astronomers found these figures too staggering to be believed and thought there must be an error in their observations. But two other small stars were soon discovered, and now more than 200 are known. Because of their white-hot glow and their size, these compact stars are called **white dwarfs**. They are believed to be remnants of collapsed stars.

---

*Quantum exchange forces have no explanation within Newtonian physics. Quantum exchange forces between electrons are far stronger than the ordinary repulsive electrical forces between electrons. These ordinary electrical forces are what maintains the solidity of normal solid matter.

Detailed quantitative theories of the evolution of stars predict the scenario just sketched. Observations of stars in the various evolutionary stages described and observations of Earth's oldest rocks, the moon, moon rocks, meteorites, other planets, other moons, and the sun itself all support these theories.

The natural place to look for star births is among thick gas clouds in space. When the *Hubble Space Telescope* searched the dense gas cloud known as the Eagle Nebula, they found thousands of newly minted stars (Figure 5.12). Just as the theory predicts, nearly all of these new stars were wrapped in disks of dust and gas, disks that are expected eventually to coalesce into planets.

■ **DIALOGUE 7**   If the sun were to collapse tomorrow to become a white dwarf, but without any explosions or expansion that would directly affect Earth, would the gravitational force by the sun on Earth be any different from what it is now? Would Earth's orbit be any different? Would you expect life on Earth to be affected?

5.4 GRAVITATIONAL COLLAPSE: THE DEATHS OF MORE MASSIVE STARS

A star's life cycle is determined primarily by how much mass it has. A star needs at least 10% of the sun's mass in order to get hot enough to initiate nuclear fusion and become a star in the first place. All stars massive enough to initiate nuclear fusion go through a middle age that is similar to the sun's present state. Then when its hydrogen fuel has been used up, it enters its final phase. Stars having masses up to three times the sun's mass go through a final phase similar to the sun's, ending their lives as white dwarfs.

But a quite different fate awaits more massive stars—those with more than about three times the sun's mass. Like the sun, these more massive stars use up their hydrogen fuel and then contract at the center. But because of the star's larger mass, the contraction is stronger, and its center gets hotter. The high temperatures initiate a wide range of nuclear reactions that eventually turn the star's small central core into solid iron. Iron continues forming until the inner core becomes so massive that it cannot hold itself up. The solid iron core then suddenly collapses during a time interval of just 1 second. This unimaginably violent process jolts the entire star, ripping it apart. Most of the star explodes into space, an event called a **supernova explosion**.

Only 10 to 20% of the original star remains after the explosion. No further nuclear reactions can occur in this remnant, so there is little to oppose the inpull of gravity. The remainder of the star enters its final collapse.

My suspicion is that the universe is not only queerer than we suppose, but queerer than we can suppose.
JOHN B. S. HALDANE, BRITISH GENETICIST, 1856–1928

---

7. No, because the sun's mass would be unchanged. Earth's orbit would not be affected, because the gravitational force by the sun on Earth would not change. Life on Earth would cease, however, because the sun's radiation would be radically altered.

If the original star is very massive, perhaps 10 or 20 solar (sun) masses, the final collapse is so strong that electron exchange forces are too weak to stop it. There is only one other known force that can stop the collapse. It is called the *neutron exchange force*. It is similar to the electron exchange force, except that it occurs between neutrons. Viewed at the microscopic level, the collapse not only squashes atoms out of existence, it also squashes electrons out of existence by forcing them to merge with nuclei. This turns each nucleus into a collection of neutrons, and it turns the entire star into an object that resembles a giant nucleus made of neutrons. This object is called a **neutron star**.

Nuclear physicist J. Robert Oppenheimer, who later became the leader of the team that developed the atomic bomb (Chapter 16), predicted neutron stars in 1938. None was discovered until 1967, when Jocelyn Bell, a sharp-eyed astronomy graduate student in England, discovered a source of radio waves in space that sent out "beeps" or "pulses" every 1.3 seconds. Project scientists thought at first that she might have discovered a radio beacon from an extraterrestrial civilization. But another was soon discovered, and by now more than 500 such sources are known, with a wide range of pulsation rates. There is little question about their identity: They are neutron stars.

A neutron star is an exotic object. Although more massive than the sun, it is only a few kilometers across with a billion tonnes of material packed into each cubic centimeter. On Earth, a tiny speck of this material would weigh as much as a large, fully loaded highway truck! Like a spinning ice-skater bringing in her arms, a neutron star spins faster and faster during its final collapse, until the entire star spins at speeds that are unimaginable for such a massive object. The fastest stars spin hundreds of times every second. This spinning combines with magnetic effects to create the rapid pulses of visible light and radio signals observed from Earth, the signals that Bell discovered in 1967. As seen from Earth, the entire star appears to flash on and off many times every second. Figure 5.15 is a sequence of photographs of a neutron star, showing two visible flashes. The supernova explosion that created this neutron star was seen and recorded on Earth in 1054. For a few days the light from the explosion was brighter than the planet Venus. Today, this star is called the Crab nebula because of the crablike shape of its dispersed, nebulous halo of gases blown into space by the explosion.

Gravitational forces are enormous on a neutron star's surface. The star's radius is only about 10 kilometers, which is 10^5 (100,000) times smaller than the sun's radius. Yet the star is more massive than the sun. Newton's law of gravity tells us that the weight of an object on the surface of a star is proportional to the inverse of the square of the star's radius. If a collapsing star's radius becomes 10^5 times smaller, the weight of an object on its surface will become $10^5 \times 10^5 = 10^{10}$ (10 billion) times larger.

What about stars that are even more massive than those that collapse to form neutron stars? When a star of more than about 30 solar masses runs out of fuel, the ensuing collapse is so strong that no known force can stop it. According to current theories, it collapses into a single point! Its matter—its atoms and subatomic particles—is squeezed out of existence. The star retains its mass, however, and so it retains its gravitational influence on the space around it.

Gravitational forces become unimaginably large near such a star. If you had the misfortune to get too close, you could not escape, because gravity won't let anything escape. If you were within a few hundred meters of the central point and you tried to throw an object away from the center, you would need to throw

Horatio: O day and night, but this is wondrous strange!
Hamlet: And therefore as a stranger give it welcome. There are more things in heaven and Earth, Horatio, Than are dreamt of in your philosophy.
SHAKESPEARE, HAMLET, *ACT I SCENE V*

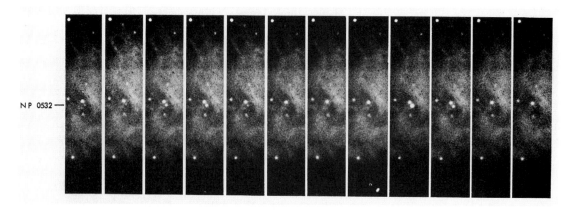

NP 0532 —

FIGURE 5.15
A sequence of photographs of the neutron star at the center of the Crab nebula. Portions of a surrounding gas cloud, the remnant of the supernova explosion that created the neutron star, can be seen. This sequence, which lasted a total of about 1/20 second, includes two flashes. The first flash occurred during frames 3 and 4, and the second occurred during frames 9 and 10.

it faster than the speed of light in order for the object to escape the star. But as Chapter 10 explains, objects cannot be thrown faster than the speed of light. Nothing can get out of the immediate vicinity of the star's central point, not even light itself.* Such an object is called a **black hole**.

HOW DO WE KNOW?

Observations of black holes

We can detect black holes by means of their gravitational influence on things around them. The first probable black hole, Cygnus X-1, was discovered in 1972. This object is thought to be a double star, two stars orbiting around each other, held together by their gravitational attraction. One star is a visible giant star, and the other is an unseen compact (in other words, far smaller than a normal star) object. By observing its gravitational effect on the visible star, the compact object's mass can be deduced to be some 15 solar masses. Since theories indicate that a compact object of more than 3 solar masses can only be a black hole, astronomers believe that Cygnus X-1 is a black hole. Satellites in orbit around Earth detect X rays from Cygnus X-1 that further confirm that it is a black hole. Apparently the invisible object's gravitational pull is drawing gases from the visible star and accelerating them down into and around the black hole, a process that tears apart the gas atoms and causes them to emit X rays that we can observe (Figure 5.16).

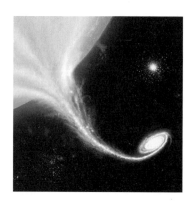

FIGURE 5.16
In this artist's conception, a black hole pulls matter from a companion star and accelerates it into a hot, X ray–emitting disk before slowly swallowing it.

*More precisely, quantum theory allows black holes to emit subatomic particles, but this effect is negligible for collapsed stars. This effect is expected to be important for low-mass black holes, although such small black holes have never been observed and may not exist.

It is difficult to believe that objects as exotic as black holes actually exist. Scientists don't go out of their way to invent strange ideas like this. To the contrary, they always look for the most natural, least strange, explanation of the observed data. For example, people once found it strange that Earth could orbit the sun, but astronomers such as Copernicus found that this was the most natural idea that could account for the data. In the same way, astronomers today find that a black hole is the most natural explanation of Cygnus X-1. If it is not a black hole, then this object is not compact, or it does not have a mass larger than 3 suns, or compact objects of greater than 3 solar masses do not always collapse to become black holes. Astronomers find it easier to believe that Cygnus X-1 is a black hole than to believe any one of these three options. Science is conservative, always preferring the least-odd conclusion possible.

Using X rays to identify possible black holes and then studying these candidates by more direct means, several objects have now been discovered that are believed to be black holes resulting from the gravitational collapse of a star.

Much more massive black holes are thought to exist at the centers of many galaxies. In some galaxies, confirmation of a central black hole comes from telescopic observations of the speeds of individual stars near the center. These stars are orbiting the center, indicating that something at the center is holding them into orbits. Measurements of the stars' orbital speeds enables the mass of the central object to be deduced from the law of gravity; faster speeds indicate a larger mass at the center. For example, observed orbital speeds as high as 1000 km/s near the center of our own Milky Way galaxy imply a black hole with a mass of about 2.6 million suns lurks at the center.

Such observations lead astronomers to conclude that many galaxies have black holes at their centers and that the distant and powerful galaxylike objects called *quasars* are powered by massive black holes.

■ **DIALOGUE 8**   Suppose that Earth collapsed to half of its present radius. How much would you weigh? What if Earth collapsed from its present 6000 km radius to only 60 km?

5.5 THE NEWTONIAN WORLDVIEW: A DEMOCRATIC, MECHANICAL UNIVERSE

During the sixteenth and seventeenth centuries, the new sun-centered astronomy and inertial physics ushered in not only new science but also a new philosophical and religious view that we will call the **Newtonian worldview.*** It is one of the most significant consequences of Newtonian physics. Even though Newtonian scientific ideas have been superseded today, the Newtonian worldview

The nature of the mind and soul is bodily; for when it is seen to push the limbs, rouse the body from sleep, and alter the countenance and guide and turn about the whole man, and when we see that none of these effects can take place without touch nor touch without body, must we not admit that the mind and the soul are of a bodily nature?

LUCRETIUS, ROMAN POET AND FOLLOWER OF DEMOCRITUS, FIRST CENTURY B.C.

---

8. Four times your present weight. The 60-km radius is 0.01 (1/100) of the original radius, so your new weight is $100 \times 100 = 10,000$ times your present weight.

*Newton was only one of many scientists and philosophers who contributed to these beliefs. Whereas Newton was the main scientific contributor to the new worldview, Descartes and Galileo contributed more to its philosophical formulation.

Just as the eye was made to see colors, and the ear to hear sounds, so the human mind was made to understand, not whatever you please, but quantity. Therefore, quantity is the primary feature of things, prior to the other categories [such as colors, sounds, and other sense impressions].

KEPLER

I am much occupied with the investigation of the physical causes [of the motions of the solar system]. My aim is to show that the heavenly machine is not a kind of divine, live being, but a kind of clockwork . . . insofar as nearly all the manifold motions are caused by a most simple, magnetic, and material force, just as all motions of a clock are caused by simple weight.

KEPLER

When anyone tells us that he sees color in a body or feels pain in one of his limbs, this is exactly the same as if he said that he there saw or felt something of which he was entirely ignorant, or that he did not know what he saw or felt. . . . When they [color or pain or other sense impressions] are judged to be certain things subsisting beyond our minds, we are wholly unable to form any conception of them.

DESCARTES

. . . which will seem in no way strange to those who, knowing all the parts which are in the body of each animal, consider this body as a machine.

DESCARTES

retains its influence. To properly appreciate this worldview, we will look at it in its historical context.

The pre-Newtonian or traditional worldview combined medieval Christianity, the Earth-centered astronomy of the ancient Greeks, and Aristotle's physics. Central to the traditional view is the idea of purpose, or future goals. Purpose is important to all of us in our individual lives, as many of our actions are directed toward some future goal. It is natural for people to extend this notion of purpose to human existence in general and to the physical universe. During most of human history, and certainly during the Middle Ages in Europe, popular culture and religion and science have been united in the belief that there is a purpose not only for individual human actions but also for humankind as a whole, for Earth, for the planets, for falling objects, and for everything else that happens. Furthermore, it was believed that the universe's larger purposes were tied to human purposes, that humankind represented the purpose of all creation.

Ancient Greek astronomy supported this traditional view. Earth was thought to be motionless at the center of things, with the sun, planets, and stars circling it for our benefit. Aristotle's physics also supported the traditional view. The five Aristotelian elements all had goals that explained their natural motions. For instance, heavy objects moved downward because they wanted to get to their natural place—their goal—at Earth's center. Everything had its natural place in a hierarchy of places and purposes. It was a cosmology—an understanding of the universe—that was well attuned to the era's hierarchical social structure, consisting of a God-ordained king surrounded by a few land-holding nobles surrounded in turn by many land-working serfs and peasants.

Astronomy and physics since the Middle Ages have fundamentally contradicted Earth-centered astronomy and Aristotelian physics. In the sixteenth century, Copernicus placed the sun, not Earth, at the center. Kepler replaced the planets' "natural" circles with ellipses. Descartes declared that there was only one kind of natural motion, namely, motion at an unchanging velocity, the same in the heavens as on Earth. According to the law of inertia, bodies move not because they have any goal but simply because there is nothing to stop them. The hierarchy of natural places, the notion that Earth is special, the centrality of humankind, and the scientific basis for universal purposes—all these were swept away.

It is not by chance that many early stirrings for religious and political freedom were occurring at about this time. Once the hierarchical cosmology began to crumble, it was no longer obvious to people that they should follow the old hierarchical cultural habits. The new science established universal natural laws, rather than particular human beings or religious beliefs, as the ultimate framework for human behavior. Thus sixteenth-century religious reformers such as Martin Luther (who, however, never accepted Copernican cosmology) were freed, culturally, to challenge medieval Christian traditions. Seventeenth-century social reformers such as philosopher John Locke could develop a theory of human behavior based on the notions of natural law and experimental evidence that had been in the air since Copernicus. And eighteenth-century political reformer Thomas Jefferson could draw up a Declaration of Independence that threw off the divine rights of the King of England and that is shot through and through with the concept of "unalienable rights" flowing directly from "the Laws of Nature and of Nature's God" to all people as the basis for human equality. Thus does our understanding of the universe determine, on quite a deep level, our religion, our social order, and our politics.

Without the living animal, I do not believe that odours or tastes or sounds are anything else than names—they have truly no other existence than in use; I say that I am inclined sufficiently to believe that heat is of this kind, and that . . . if the animate and sensitive body were removed, heat would remain nothing more than a simple word.

GALILEO

I am induced by many reasons to suspect that they [the phenomena of nature] all may depend upon certain forces by which the little parts of bodies . . . are either mutually impelled to one another and crowd together according to regular figures, or are repelled and recede from one another.

NEWTON

I shall express what I call general nature by "cosmical mechanism," i.e., [composed] of all the mechanical affections (figure, size, motion, etc.) that belong to the . . . great system of the universe.

ROBERT BOYLE, PHYSICIST AND CHEMIST, 1627–1691

Mathematical and mechanical principles are the alphabet, in which God wrote the world.

ROBERT BOYLE

The universe is like a rare clock, such . . . that the engine being once set moving, all things proceed according to the artificer's first design, and the motions . . . do not require the peculiar interposing of the artificer, or any intelligent agent employed by him, but perform their functions . . . by virtue of the general and primitive contrivance of the whole engine.

ROBERT BOYLE

[Men are] engines endowed with wills.

ROBERT BOYLE

Galileo emphasized that he sought only to describe how things behave, not why they behave as they do. He was not concerned with a physical phenomenon's purpose. Analysis—the new technique of separating phenomena into their simplest components and studying those components—was one of Galileo's tools in this endeavor. This led to a focus on the simplest and smallest components of matter: atoms.

And so **atomism**—the idea that nature can be reduced to the motions of tiny material particles—underlay the thinking of Galileo and others who developed the new physics. For example, in a view remarkably similar to Democritus's view (Chapter 2), Galileo stated:

I find myself impelled . . . to conceive that in its own nature [any piece of matter] is shaped in such and such a shape, that in relation to others it is large or small, that it is in this or that place, in this or that time, that it is in motion or remains at rest. . . . But that it must be white or red, bitter or sweet, sounding or mute, of a pleasant or unpleasant odor, I do not perceive my mind forced to acknowledge. . . . Hence I think that these tastes, odors, colors, etc., on the side of the object in which they seem to exist, are nothing else but mere names, and hold their residence solely in the sensitive body [that is, in the human or animal observer]; so that if the animal were removed, every such quality would be abolished and annihilated.

Newton noted:

It seems probable to me that God in the beginning formed matter in solid, massy, hard, impenetrable, movable particles, of such sizes and figures, and with such other properties, and in such proportion to space, as most conduced to the end for which he formed them; and that these primitive particles being solids are incomparably harder than any porous bodies compounded of them, even so hard as never to wear or break in pieces. . . . If at any time I speak of light and rays as coloured, I would be understood to speak not philosophically and properly, but grossly, and according to such conceptions as vulgar people in seeing all these experiments would be apt to frame. For the rays to speak properly are not coloured. . . . [C]olours in the object are nothing but a disposition to reflect this or that sort of rays more copiously than the rest; in the rays they are nothing but their dispositions to propagate this or that motion into the sensorium [the brain], and in the sensorium they are sensations of those motions under the forms of colours.

In the Newtonian worldview, the fundamental physical reality is atoms. For example, a napkin's redness is due to the motions of atoms in the napkin and in the observer's eye and brain. A napkin is not really red in itself. In the traditional view, human experience—feelings, colors, and so forth—had primary significance. A red napkin did not merely appear to be red, it actually was red, and precisely because it was perceived by humans as red.

Newton, Galileo, and Descartes were participants in the religious culture of their time and firm believers in God. What place could be found for God within the new science? Descartes reconciled the new science with traditional religion by assuming that there was another reality in addition to the physical world. This other reality was a spiritual world parallel to, but independent of, the physical world. The material world was made of matter and operated according to the inflexible laws of nature, in just the same way that an impersonal machine operated. In this material universe, there was no color or human feeling, although God was needed to maintain the physical laws and to start things moving. The

An image or colour is but an apparition unto us of the motion, agitation, or alteration, which the object worketh in the brain, or spirits, or some internal substance of the head.

THOMAS HOBBES,
PHILOSOPHER, 1588–1679

For light and colour, and heat and sound, and other qualities which are commonly called sensible, are not objects, but phantasms in the sentients. It is no more true that fire heateth, therefore it is itself hot; than that fire causeth pain, therefore it is itself in pain.

THOMAS HOBBES

Thus mind will be nothing but the motions in certain parts of an organic body.

THOMAS HOBBES

Look around the world, contemplate the whole and every part of it: You will find it to be nothing but one great machine, subdivided into an infinite number of lesser machines, which again admit of subdivisions to a degree beyond what human senses and faculties can trace and explain.

DAVID HUME, PHILOSOPHER
AND HISTORIAN, 1711–1776

[It is unbelievable] that all nature, all the planets, should obey eternal laws, and that there should be a little animal, five feet high, who in contempt of these laws, could act as he pleased, solely according to his caprice.

VOLTAIRE, FRENCH
PHILOSOPHER AND WRITER,
1694–1778

They may say what they like; everything is organized matter.

NAPOLEON BONAPARTE,
1769–1821

other world, the spiritual one, was the realm of human thoughts and feelings and communication with God. This notion of two independent realities is called **dualism**.

By putting all human qualities into a nonphysical spiritual realm, dualism lent support to the notion that physical reality consisted of atoms following the impersonal laws of nature. In the physical universe, the true realities, or **primary qualities**, were assumed to be impersonal characteristics such as position, time, shape, volume, mass, and the motions of atoms. Human sense impressions such as color and beauty were assumed to be **secondary qualities** that were not part of the physical world but were merely reflections of the primary qualities. Thus did science and philosophy relegate human concerns to a shadowy secondary role in a physical universe.

Galileo's introduction of scientific experimentation fit this trend nicely. Traditionally, scientists had observed the universe as it existed around them. Because Galileo was the first to conduct prearranged experiments, he was also the first to isolate natural phenomena from influences that tended to cloud the main point of his experiments. For example, Galileo sometimes removed—insofar as possible—friction and air resistance. Naturally, Galileo also wanted to remove human influence from his experiments in order to study nature in itself. This idea—that experimental results should not be influenced by humans, that they should be **objective**—has proved crucial to scientific progress.

It was assumed that objectivity actually was possible, at least in principle. If everything happened because of the motions of atoms and if these motions were determined by physical laws, then what was real did not depend on humans. The physical universe would do whatever it would do, regardless of human thoughts or feelings. It should be possible to observe the workings of the universe, uninfluenced by humans. In other words, it was assumed that humans did not interact with the physical universe in any essential way.

According to this view, there is little room for God in the day-to-day workings of the material universe. In the traditional view, God is continually and actively present throughout the universe, continually endowing all things with purpose. Descartes and Galileo believed that God was needed to establish the laws of nature and to start the universe moving but that once started, the whole thing would run itself.*

In the new view, God is, at most, an uninvolved observer.

A machine, especially a delicate and finely tuned machine such as a clock, is an excellent analogy for the Newtonian worldview. Once the owner has started them, clocks run themselves according to their own operating principles. The founding fathers of physics envisioned the universe as a clockwork mechanism whose operating principles were the laws of nature and whose parts were atoms. Once started, the universe ran itself. No further intervention was needed. Because of the machinelike quality of Newtonian physics, physicists refer to Newton's theories of motion and gravity as Newtonian *mechanics*.

---

*With a few exceptions, Newton also believed that God did not intervene in the universe. On certain occasions, namely, in situations for which Newton himself could not find a scientific explanation, he believed that God momentarily intervened. This view—that every phenomenon that cannot be explained by science at any particular time requires an intervention by God—has been referred to as a *God of the gaps* philosophy. But as science closes the gaps, such a philosophy becomes less and less tenable.

I never satisfy myself until I make a mechanical model of a thing. If I can make a mechanical model I can understand it. As long as I cannot make a mechanical model all the way through I cannot understand.

LORD KELVIN, NINETEENTH-CENTURY BRITISH MATHEMATICIAN AND PHYSICIST

That Man is the product of causes which had no prevision of the end they were achieving; that his origin, his growth, his hopes and fears, his loves and his beliefs, are but the outcome of accidental collocations of atoms; . . . all these things, if not quite beyond dispute, are yet so nearly certain, that no philosophy which rejects them can hope to stand.

BERTRAND RUSSELL, PHILOSOPHER AND MATHEMATICIAN, 1872–1970

Brief and powerless is Man's life; on him and all his race the slow, sure doom falls pitiless and dark. Blind to good and evil, reckless of destruction, omnipotent matter rolls on its relentless way.

BERTRAND RUSSELL

It's a material world.

MADONNA

Now I a fourfold vision see,
And a fourfold vision is given to me;
'Tis fourfold in my supreme delight
And threefold in soft Beulah's night
And twofold Always. May God us keep
From Single vision & Newton's sleep!

WILLIAM BLAKE, 1757–1827, POET, REBEL AGAINST THE MECHANICAL SINGLE VISION, OR LINEAR THINKING, OF NEWTON

According to Newtonian physics, every physical system is entirely predictable, like a perfectly operating clock. For example, Newtonian physics can predict precisely how far a freely falling object will fall in any specified amount of time.

This clocklike predictability of nature has surprising implications. To understand them, imagine a simple, self-contained collection of atoms that move and interact in accordance with Newtonian physics. Suppose we specify the precise positions and velocities of every atom at one particular time, such as noon today. Then, according to Newton's theory of motion, the entire future behavior of this collection of particles can be precisely predicted, for all time.

But the Newtonian view is that the entire universe is just such a self-contained collection of atoms. Thus, the future is entirely determined by what all the atoms of the universe are doing right now or at any other time. Once started, the clockwork universe is required to do precisely what it has done and what it will do. Since, according to the Newtonian worldview, every thought or feeling that enters your head is reducible to the motion of atoms within your brain and elsewhere, all of your thoughts, feelings, and actions are entirely predetermined and predictable. You never choose to scratch your nose, for example—the laws of nature choose for you. You might believe that you choose, but this, too, this believing that you choose, was chosen for you by the laws of nature.

This loss of free will was expressed most directly by the French scientist Pierre-Simon Laplace (1749–1827):

All events, even those which by their insignificance do not seem to follow the great natural laws, are indeed consequences of these laws as surely as is the rising and setting of the sun. We must therefore consider the present state of the universe as the result of its earlier state and as the cause of that which is to come. An intelligence which at a given instant knew all the forces acting in nature and the positions of every object in the universe—if endowed with a brain sufficiently vast to make all the necessary calculations—could describe with a single formula the motions of the largest astronomical bodies and those of the lightest atoms. To such an intelligence, nothing would be uncertain; the future, like the past, would be an open book.

Beginning in the seventeenth century and continuing into the twentieth century, these ideas influenced all educated people in Europe and elsewhere. Newtonian physics was so successful that the associated philosophy, the Newtonian worldview, was accepted with little question. These ideas were absorbed with little realization that they even were being absorbed.

There are reasons today to question both the Newtonian worldview and Newtonian physics (see Chapter 2 and the next section). Nevertheless, it would be surprising if these views were not still influential today. A person's worldview tends to be absorbed automatically, thoughtlessly, as part of the cultural air of the times. It seems likely that the Newtonian worldview remains active, especially among people who have never heard of Isaac Newton, let alone Descartes or Galileo.

It is for you, valued reader, to determine to what extent the Newtonian worldview is valid, whether it retains a significant influence, and what difference it might make.

5.6 BEYOND NEWTON: LIMITATIONS OF NEWTONIAN PHYSICS

Tested repeatedly during the eighteenth and nineteenth centuries, Newtonian physics stood up in quantitative detail to every challenge. In fact, it was so powerful and accurate that scientists began to accept it as true in an ultimate, absolute sense. To some, the word *understand* came to mean to explain in terms of Newtonian physics. But science is never absolute. Even though a scientific principle has been confirmed repeatedly, it hangs always by the slender thread of new experiments.

During the last two decades of the nineteenth century, experimental results began appearing that proved impossible to reconcile with Newtonian physics. Newtonian physics was found to be incorrect in certain situations.

During the first few decades of the twentieth century, physicists invented three new theories to account for nature's behavior in those situations in which Newtonian physics was incorrect. These three theories are called **special relativity**, **general relativity**, and **quantum theory**. To date, at least, scientists have found no exceptions to any of these three theories.

Newtonian physics has been found to give incorrect predictions in four types of situations: when speeds are very high, when gravitational forces are very strong, when distances are very large, and when distances are very small. Briefly, special relativity deals with high speeds; general relativity deals with strong gravitational forces and with large distances; and quantum theory deals with small distances.

Experiments show that Newton's law of motion and deeper Newtonian views concerning time and space break down at high speeds. The disagreement between Newtonian theory and experiment is not noticeable at slow speeds, but the errors become progressively worse as the speeds increase. The non-Newtonian effects cannot be observed for automobiles, bullets, jet planes, or even orbiting satellites moving at some 10 km/s. But at 30,000 km/s (roughly around the world in 1 second!), Newtonian predictions are off by 1%. At 290,000 km/s, nearly the speed of light, typical Newtonian predictions are incorrect by 1500%—a factor of 15! Scientists didn't notice these non-Newtonian effects for 200 years because they had never studied such fast-moving objects, at least not in enough detail needed to observe the effects. The theory of special relativity gives correct predictions at all speeds, both low and high. The predictions of special relativity become indistinguishable from those of Newtonian physics whenever the speeds are considerably less than the speed of light. Today, physicists think of special relativity as the "correct" (but never absolute!) theory, while Newtonian physics is regarded as a good approximation to reality.

Similarly, experiments show that Newton's law of gravity, Newton's law of motion, and Newtonian views about time and space, are incorrect for objects subjected to very strong gravitational forces and also over very large distances. For example, non-Newtonian effects are measurable, but small, for the orbit of the innermost planet, Mercury, which feels strong gravitational forces by the sun. Non-Newtonian gravitational effects are quite significant near neutron stars and black holes and for physical systems that range over large portions of the observable universe. The theory of general relativity gives correct predictions for all these situations and is regarded as the correct theory of gravity. The predictions of general relativity become indistinguishable from those of Newtonian physics whenever gravitational forces are not too strong and distances are not too large.

Finally, experiments show that Newton's law of motion and subtle but pervasive Newtonian views concerning predictability, cause and effect, and distant effects are incorrect for very small objects, objects whose sizes are comparable to that of one molecule or atom. Newtonian predictions are significantly in error when applied to individual atoms or molecules. However, when large numbers of atoms come together to form a macroscopic object such as a baseball, Newton's theory does give correct predictions for that object's overall motion. Quantum theory gives correct predictions for objects of all sizes, from microscopic to macroscopic. For ordinary macroscopic objects, quantum theory's predictions become indistinguishable from those of Newtonian physics.

Figure 5.17 is one way of indicating, graphically, these limits of validity of Newtonian physics.* The vertical axis represents speed. Special relativity predicts that objects cannot move faster than the speed of light, 300,000 km/s, so these speeds are forbidden. The horizontal axis shows distance, or the size of individual objects. Quantum theory predicts that objects cannot be both small and slow moving, in other words, that small objects must move rapidly. So there is another forbidden region in the small-size, low-speed corner of the diagram.

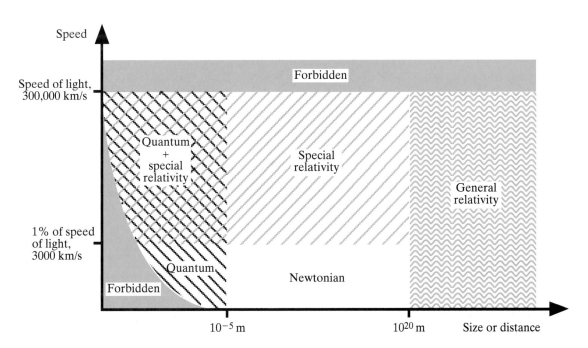

FIGURE 5.17
Newtonian physics breaks down for very small objects, very large objects, and very fast objects. Newtonian physics also breaks down for strong gravitational forces, such as those near a neutron star or a black hole. The diagram is only schematic and approximate.

---

*Thanks to Douglas Giancoli, the author of several physics textbooks, including *Physics: Principles with Applications* (Englewood Cliffs, NJ: Prentice Hall, 1991), for suggesting diagrams of this type.

In summary: Newtonian physics is correct for common phenomena on Earth, but it breaks down for phenomena that are far from this "normal" range. The quantum and relativity theories apply throughout the entire range of the phenomena observed to date, and these theories reduce to Newtonian physics when applied to common phenomena on Earth.

SUMMARY OF IDEAS AND TERMS

Newton's law of gravity   Between any two objects is an attractive force proportional to the product of the two objects' masses and proportional to the inverse of the square of the distance between them: $F \propto m_1 \times m_2/d^2$.

Satellite   A body in orbit around a larger astronomical body. Inertia keeps satellites moving, and the gravitational force exerted by the central body holds satellites in their orbits.

Weightlessness   Bodies are weightless only when they are far from all other bodies. Bodies in orbit around Earth are not weightless, but they seem weightless because they are falling freely around Earth.

Gravitational collapse   Matter falling together because of gravity.

Star birth   Stars form from collapsing gas clouds. Gravitational collapse heats the gas, which initiates nuclear fusion in the center, which stops the collapse.

Nuclear fusion   The process in which two nuclei stick together to form a single nucleus. This fuels the stars.

Supernova explosion   An explosion of an entire star.

White dwarf   A planet-sized, compact star. When stars of about the sun's mass run out of fuel, they flare up and then collapse to become white dwarfs.

Neutron star   A compact star a few kilometers in diameter, resembling a giant nucleus made of neutrons. It spins rapidly, sending out radio beeps and light flashes. When massive stars (10 to 20 solar masses before collapse) run out of fusion fuel, they explode, and the remaining remnant collapses to become a neutron star.

Black hole Any object that has gravitationally collapsed into a single point. Nothing can escape from its vicinity. When very massive stars (greater than 30 solar masses before collapse) run out of fuel, they explode, and the remaining remnant collapses to become a black hole. Giant black holes having millions of solar masses are thought to exist at the centers of many galaxies.

Newtonian worldview The philosophical notions associated with Newtonian physics, especially the mechanical universe, and the democratic ideals implied by universal natural laws.

Newtonian physics and democracy All humans are fundamentally equal, because all are ultimately governed by the same universal natural laws. Newtonian attitudes toward natural law pervade the strivings toward democracy of the past several centuries, and are particularly evident in the documents establishing the United States.

Atomism The notion that nature can be reduced to the motion of tiny material particles.

The mechanical universe The philosophical view, accompanying Newtonian physics, that the impersonal, clocklike workings of material atoms, as prescribed by natural laws, determine everything else. Thus the universe's future behavior is entirely determined by its state at any single time, and there is no free will.

Dualism Descartes's idea that there are two realities, physical and spiritual. In the physical realm, the real or **primary qualities** are objective, impersonal phenomena such as the motion of atoms. Human sense impressions are considered to be **secondary qualities**, caused by the primary qualities.

Objectivity An experiment is objective if its outcome is not influenced by humans. The Newtonian worldview assumes that perfect objectivity is possible, at least in principle.

Limitations of Newtonian physics Newtonian physics gives incorrect predictions for (1) fast-moving objects, (2) strong gravitational forces or large distances, and (3) small objects. **Special relativity, general relativity,** and **quantum theory** do give correct predictions for each of these three classes of phenomena, respectively.

REVIEW QUESTIONS

THE IDEA OF GRAVITY
1. What is the direction of a falling apple's velocity? Of its acceleration?
2. What is the direction of the moon's velocity? Of its acceleration?
3. Does Earth exert a force on the moon? What is its direction? How would the moon move if this force did not act?
4. In what ways are the moon and a falling apple similar? In what ways do they differ?

THE LAW OF GRAVITY
5. Does this book exert a gravitational force on your body?
6. What would happen to this book's weight if you managed to double Earth's mass? What if, instead, you doubled the book's mass? What if you doubled both?
7. In order to use Newton's law of gravity to calculate your weight, what data would you need?
8. If you were orbiting Earth in a satellite 200 km above the ground, would you be weightless? Would your weight be as large as it is when you are on the ground? Would you feel weightless? Explain.

GRAVITATIONAL COLLAPSE
9. What caused the sun to get hot? What keeps it hot today?
10. Describe the process that formed the planets.
11. Since gravity pulls inward on the material in the sun and since the sun is made only of gas, why doesn't the sun collapse?
12. Are there places in our galaxy where stars are being born?

13. Name the process and also the substance that fuels the sun.
14. What will happen to the sun after it runs out of fuel?
15. Name and describe the object into which the sun will evolve after it runs out of fuel.
16. What causes different stars to evolve differently?
17. All stars eventually evolve into one of three types of objects. Name them. What kinds of stars evolve into each of the three types of objects?
18. Describe a neutron star.
19. Describe a black hole. Since nothing can come out of a black hole, how can we detect it?

THE NEWTONIAN WORLDVIEW
20. List some ways in which ancient Greek astronomy and Aristotelian physics support the traditional philosophical and religious worldview.
21. List some of the ways in which Copernican and Newtonian science are less supportive of the traditional worldview.
22. How is Newtonian physics related to democracy?
23. According to the Newtonian worldview, is a red napkin really "red"? Explain.
24. List several ways in which, according to the Newtonian worldview, the universe is similar to a clock.

BEYOND NEWTON
25. For what kinds of phenomena is Newtonian physics incorrect? Why did it take so long to discover such exceptions?
26. List the three theories that give correct predictions for the situations in which Newtonian physics is incorrect.

HOME PROJECT

In an elevator, discover the experience of reduced apparent weight. As the elevator comes to rest while going upward or starts into motion while going downward, you should feel light. Is your weight really less when you experience this? This sensation of decreased weight is due to the elevator's downward acceleration, just as an orbiting astronaut's sensation of weightlessness is due to the satellite's downward acceleration while the satellite orbits Earth. What, if anything, can be done to give you the sensation of complete weightlessness inside the elevator? What motions of the elevator make you feel heavier than normal? What is the direction of the acceleration in these cases? Do you feel lighter, or heavier, or neither, when the elevator moves at an unchanging speed?

FOR DISCUSSION

1. Give at least one argument for the proposition that the Newtonian worldview is a valid view of reality. Now give at least one argument on the other side. What is your own view and why? List some ways in which the Newtonian worldview has influenced our culture, either in the past or today.
2. In view of the discussion of determinism and free will in Section 5.5, do you have free will? Support your view with at least two specific reasons (also see Section 2.6).
3. Discuss the meaning of ethics in a supposedly deterministic universe. If all actions really were predetermined, could we then talk meaningfully of what a person "should" do or denounce people for immoral actions? Could there be any right or wrong?

EXERCISES

THE IDEA OF GRAVITY

1. Does Earth's gravity pull more strongly on a block of wood or on a block of iron having the same size? Which one falls faster when dropped (neglect air resistance)?
2. Does Earth's gravity pull more strongly on a block of wood or on a block of iron having the same mass?
3. What is the magnitude (strength) and direction of the gravitational force on you right now?
4. When you crumple a sheet of paper into a tight ball, does its mass change? Does its weight change?
5. Are you in orbit around (falling around) Earth's center? Is there anything around which you are in orbit?
6. The moon is in orbit around two objects simultaneously. Which two? (Actually, there is a third—our galaxy's center.)
7. Do you exert a gravitational pull on people around you? Do they exert a gravitational pull on you?
8. Do you exert a gravitational force on Earth? If so, how large is it, and in what direction is it?
9. How far does Earth's gravitational influence extend?
10. As a spacecraft travels from Earth to the moon, does it ever entirely leave Earth's gravitational influence?
11. An orbiting satellite weighs 8000 N. How big and in what direction is the gravitational force on it? How big and in what direction is the gravitational force by the satellite on Earth?
12. What would be the approximate orbital period (time for one complete orbit) for an apple placed in orbit around Earth at the moon's distance from Earth?
13. If gravity suddenly shut off right now, what would be the shape of Earth's orbit? What about the moon's orbit?
14. What force (if any) keeps the planets moving along their orbits?

THE LAW OF GRAVITY

15. Which is larger, the gravitational force by Earth on the moon or the gravitational force by the moon on Earth?
16. How strongly and in what direction does Earth pull on a 1-N apple? How strongly and in what direction does the apple pull on Earth?
17. If you somehow decreased Earth's mass, with all other factors unchanged, would this affect your weight? How?
18. If you somehow increased Earth's radius, without changing its mass, would this affect your weight? How?
19. List at least three bodies that have a detectable (measurable) gravitational effect on Earth's motion.
20. The giant planet Jupiter is about 300 times more massive than Earth. It seems, then, that an object on Jupiter's surface should weigh 300 times more than it weighs on Earth. But it actually weighs only about 3 times as much. Explain.
21. If you were in a freely falling elevator and you dropped your keys, they would hover in front of you. Are the keys falling? Are the keys weightless?
22. If gold were always sold by weight, could you make money buying gold at one altitude above the ground and

selling it at a different altitude? Where would you want to buy—at a high altitude or a low altitude?

23. Would you weigh more in Denver or in Los Angeles? Why?
24. Is there any net force (Chapter 4) acting on the moon?
25. Is the moon accelerated? If so, in what direction is the acceleration? In what direction is the moon's velocity?
26. Suppose that a heavy- and a light-weight satellite are put into low orbits around Earth. Could you tell, by observing the shape or speed of the two orbits, which satellite was the heavy one?
27. Suppose that two satellites are put into orbit, one around Earth and one around the moon, and suppose that the radii of the two orbits (the distance from the center of Earth and the moon) are the same. From the knowledge that Earth's mass is larger than the moon's mass, can you make any predictions about the speeds of the two orbits?
28. Communications satellites must be in *geosynchronous orbits*. That is, they must remain above a fixed point on Earth's surface, enabling sending and receiving antennas to be aimed at a fixed point overhead. What, then, must be a communication satellite's orbital period (the time for one complete orbit around Earth)?

GRAVITATIONAL COLLAPSE
29. Will Earth ever collapse to become a black hole? Why? Will the sun?
30. The orbits of all nine planets lie approximately in the same flat plane. Why?
31. If Earth collapsed to a smaller radius, how (if at all) would this affect your weight?
32. If Earth expanded to a larger radius, how (if at all) would this affect your weight?

BEYOND NEWTON
33. What theory or theories would be needed to predict the behavior of an atom moving at half the speed of light?
34. According to the most widely accepted scientific theory of the creation of the universe, the universe during the first few moments (much less than 1 second) of its existence was extremely hot, was full of densely packed matter, and was very tiny—smaller than an atom. What theory or theories would be needed to explain what was happening during these first few moments?

PROBLEMS

THE LAW OF GRAVITY
1. Suppose that the gravitational force between an apple and an orange placed a short distance apart is one trillionth (10^{-12}) N. What would the force be if the distance was doubled? Halved? Tripled? Quartered?
2. Referring to the previous exercise, what would the force be if the mass of the apple was doubled? Tripled? What if the mass of the apple was tripled *and* the mass of the orange was quadrupled?
3. Referring to the previous exercise, what would the force be if the mass of the apple was doubled, the mass of the orange was doubled, and the distance between them was doubled?
4. If Earth collapsed to one-tenth of its present radius, how much would you then weigh?
5. If Earth expanded to 10 times its present radius, how much would you then weigh?
6. Earth's mass is 6.0×10^{24} kg, and its radius is 6.4×10^6 m. Use Newton's law of gravity to calculate the weight of (i.e., the force of gravity on) a 1-kg object lying on Earth's surface.
7. The moon's mass is 7.4×10^{22} kg, and its radius is 1.7×10^6 m. Use Newton's law of gravity to calculate the weight of a 1-kg object lying on the moon's surface. If you did the preceding problem, then compare the two answers.
8. A certain neutron star has a mass of 4.0×10^{30} kg (twice the sun's mass) compressed into a sphere of radius only 10,000 m (10 km). Calculate the force of gravity on a 1-kg object lying on the surface of this neutron star. If you did Problem 6, compare answers. Find the force on a cubic centimeter of water, whose mass is 1 gram, lying on the surface.
9. By how much would your weight change if you were in orbiting space lab territory? To answer this, find the weight of (i.e., the force of gravity on) 1 kg at Earth's surface and also in low Earth orbit, about 200 km above Earth's surface (see Figure 3.7). The data you will need are given in Problem 6. Find the percentage difference between the two answers.
10. Find the force by the moon on Earth. Their masses are 7.4×10^{22} kg and 6.0×10^{24} kg, and it is 3.8×10^8 m between their centers. How large is the force by Earth on the moon? In what direction is each of these two forces?
11. Find the force by the sun on Earth. Their masses are 2.0×10^{30} kg and 6.0×10^{24} kg, and it is 150 million kilometers between their centers. How large is the force by Earth on the sun? In what direction is each of these two forces? If you did the preceding problem, compare answers.
12. Making estimates. Earth's mass is about 100 times the moon's mass, and Earth's radius is about 4 times larger than the moon's radius. From this information, use the law of gravity to estimate how much more an object weighs on Earth, as compared with its weight on the moon.
13. Making estimates. Estimate the gravitational force, in newtons, that you exert on a person standing near you.

6

Conservation of Energy

you can't get ahead—

Human cultures are defined largely by their uses of nature's energy resources. Civilization itself is nearly synonymous with the organized use of solar energy. Although humanlike creatures evolved a few million years ago, it was not until the organized use of solar energy to grow food, just a few thousand years ago, that humans settled into villages and developed a sense of history, progress, and civilization.* The earliest village culture was an *agri*culture.

Despite the impact of energy on our lives, it is not easy to say what *energy* means. This is not surprising, for it took science two centuries following Newton's work to understand energy. Energy enters physics as an abstract idea, something apparently quite different from concrete reality, such as the chair you are sitting on. Nevertheless, we will see in later chapters that to modern physicists, nothing is more real than energy. In fact, the chair you are sitting on is made of energy.

The power of scientific ideas lies in their ability to explain and unify a variety of natural phenomena. At this, energy excels. And unlike many scientific principles, the principles of energy appear to be correct in all situations. Whether we observe huge cosmological structures, subnuclear particles, or an automobile on Main Street, the laws of energy apply correctly in every detail. In view of the limitations of even such a grand theory as Newtonian physics (see Section 5.6), the sweep of the energy principles is remarkable.

Maybe it is more than coincidence, then, that energy has a vital social context. As noted above, the organized use of solar energy formed the physical basis for the earliest human village culture. Currently, the chemically altered remains of ancient life known as **fossil fuels**—coal, oil, and natural gas—energize our industrial culture.

The coal-fueled steam engine stimulated the **industrial revolution** that began about 1750. Steam engines first pumped floodwater out of mines and later drove industrial machines, transported coal, and powered railroad locomotives, ships,

---

*Specifically human forebears, or *hominids*, evolved in Africa and are dated by radioactive and other methods at 3 million to 5 million years old. Anatomically modern humans, *Homo sapiens*, evolved from these forebears and are dated at 100,000 years ago in Africa. The agricultural revolution—the transition from nomadic hunter-gatherers to farmers—occurred first in the Middle East around 9000 B.C.

and the first automobiles. The economic and social consequences of these steam engines were profound. Because the new machines of production were expensive, they were accessible mainly to the rich. Workshops away from home grew into factories. Even unskilled workers could tend the new machines, unlike traditional craftspeople whose skills required a long apprenticeship. Thus surplus farm workers and children could be employed. Consequently, in Europe and North America, the nineteenth century was marked by increased productivity, the capitalistic organization of industry, and a shift of population from farms to cities. The political ideologies of the twentieth century grew out of the economics of the industrial revolution. The industrial revolution continues. It is spreading all over the world and to new industries such as computers.

As the industrial revolution transformed the planet, it stimulated scientists during the nineteenth century to understand the principles of energy. This chapter develops one of these principles, conservation of energy. This principle appears in Section 6.5, following the development in Sections 6.1 through 6.4 of the concepts of work and energy needed to understand this principle. Everything that happens in the universe involves an energy transformation of one sort or another. Section 6.6 studies several examples of energy transformations. Section 6.7 looks at a related idea of great practical importance: power, or the rate (per unit of time) of transforming energy. The next chapter explores the second grand principle of energy, the second law of thermodynamics.

Energy is one of the four recurring themes of this book. It will be used throughout to analyze all sorts of physical systems.

■ **DIALOGUE 1** Since there has never been an observed violation of them, are the principles of energy then true for certain?

6.1 WORK: USING A FORCE TO MOVE SOMETHING

Energy is based on the notion of work. In common language, you do work whenever you exert yourself to perform a task. The physicist's definition is a refinement of this common usage. In physics, work is done whenever an object is pushed or pulled through a distance. For instance, you do work on this book when you push it across the table. The agent doing the work need not be a person. For instance, a magnet does work on a paper clip when the magnet pulls the clip toward the magnet. More precisely, object A (a person or any other thing) does work on object B if A exerts a force on B while B moves. You do work on this book when you lift it. Earth does work on the book when the book falls. Notice that work is always done by a specific object on a specific object. And notice that both force and motion are needed in order for work to be done.

■ **DIALOGUE 2** Is either person in Figure 6.1 doing any work on the brick wall? Can you "work up" a sweat while pushing on a wall, without doing any physics-type work?

FIGURE 6.1
Is either person doing any work on the brick wall?

---

1. No! All scientific ideas are tentative.

2. No, because neither person is moving the wall. Yes, you can work up a sweat pushing on a wall without doing any actual work on the wall. (At a microscopic level, your muscles are doing work on one another, although this microscopic work is done within your body, not on the wall.)

FIGURE 6.2
It takes twice as much work to lift
two books.

FIGURE 6.3
It takes twice as much work to push
one book twice as far.

FIGURE 6.4
Scottish physicist James Prescott Joule.
His experiments in the 1840s helped
unravel the confusion surrounding
heat and so led to the first clear
understanding of energy.

■ **DIALOGUE 3**   A giant meteoroid (a rock in space) moves at high speed through the solar system. Being far from the sun and other large bodies, the meteoroid feels little gravitational pull. If we neglect this small gravitational pull, is any work being done on the meteoroid? What keeps it moving?

Anybody who has ever filled an automobile's gasoline tank knows that it is of practical importance to be quantitative about energy. Let's think about the quantity of work you do in various situations. If the word *work* is to agree roughly with common language, the work you do in lifting a load should be larger for larger loads. To see how much larger, compare lifting one book with lifting two identical stacked books (Figure 6.2). The effect is twice as big in the second case, so the work done should be twice as big. This means we want work to be proportional to force.

Now compare pushing a book across one table with pushing it all the way across two adjoining tables (Figure 6.3). Again the effect is twice as big in the second case. So work should be proportional to the distance moved.

Work should be proportional to both force and distance. Thus, we define the amount of **work** done by one object on a second object as the product of the force (exerted by the first object on the second) multiplied by the distance that the second object moves while experiencing that force:*

$$\text{work} = \text{force} \times \text{distance}$$
$$= Fd$$

For instance, if you push on a book with a 3-newton force while pushing it 2 meters, you have done $(3\text{ N}) \times (2\text{ m}) = 6$ newton-meters of work. Note that the unit of work is the newton-meter, a measurement that has been used so often that it has been renamed the **joule** (J) (rhymes with school), in honor of James Prescott Joule (Figure 6.4).

Although our definition of work is related to the way that the word is used in ordinary language, the scientific reason for this particular definition is that it is useful in connection with the upcoming idea of energy.

From the following dialogue you can see that the work done in lifting an object is the object's weight multiplied by the height lifted.

■ **DIALOGUE 4**   Suppose you slowly lift a book weighing 12 N from the floor to a shelf 2 m above the floor. While you are lifting it, (a) how large is the net force on the book? (b) how large is the force by your hand against the book? and (c) how much work did you do on the book?

---

3. No work, because there are no forces. Nothing is needed to keep it moving (law of inertia).

*This definition assumes that the motion is in a straight line in the direction of the force. If the motion and the force are not in the same direction, then in the formula we must use only that part or component of the force that is along the motion. We won't need this refinement here.

4. (a) Zero. (b) 12 N. (c) $12 \times 2 = 24$ J.

6.2 WORK AND ENERGY: A SIMPLE EXAMPLE

Do this two-step experiment, and observe carefully: First, place your book on your outstretched hand on the floor; slowly lift it to some height; hold it there a few seconds; and then slowly lower it back to the floor. Second, repeat the same lifting process to the same height, but this time suddenly remove your hand from the book so that it falls to the floor. We are going to look closely at this.

You do work on the book to lift it. But in lowering back to the floor, the book does work on you, because the book pushes downward against your hand all the way down. We could say that the raised book has an ability to do work and that the book actually does this work as it is lowered.

Let's look more closely at the *ability to do work*. We just saw that raised objects have this ability. And so do moving objects. For example, suppose that you throw your book, horizontally, at a wall. One way to get work out of the moving book would be to stick a thumbtack partly into the wall, directly in line with the book's motion, so that the book will hit the tack and drive it in farther (Figure 6.5). Any moving object has an ability to do work.

In the second step of our experiment, the book was again raised, and in the raised position it again had the ability to do work. But then you dropped it so that it simply fell, without doing work on your hand. As the falling book lost height, it gained speed, and so it retained an ability to do work. Just before hitting the floor, the book still had an ability to do work; only now this ability resulted from the book's speed rather than from its height. We could get the falling book to actually do this work by sticking a tack partly into the floor and letting the book drive in the tack farther (Figure 6.6).

Raised objects and moving objects have an ability to do work. When you lift or throw the book, you give it the ability to do work. The work you do is "stored" in the raised or moving book. In the case of the raised book you could get back this work at any time by letting the book push your hand back down slowly to the floor. Or you could just drop the book, in which case it would gain speed and so retain the "stored work" right up until it hits the floor.*

Physicists have a word for the ability to do work. It is called **energy**. We have seen that both raised objects and moving objects have energy. It is useful to distinguish these different forms. We say that a raised object has *gravitational energy*, because this energy is caused by the gravitational pull by Earth on the object, and that a moving object has *kinetic energy* (*kinetic* is related to the Greek word for motion). As we will see, there are several other energy forms.

The fact that you gave energy to the book when you raised it is one example of the **work–energy principle**, which states that work done on an object by outside agents such as your hand increases its energy.

As it falls, the book loses gravitational energy but gains kinetic energy and so retains its overall energy. This retention of overall energy when there is no outside agent (such as your hand) is one example of the **law of conservation of energy**.

FIGURE 6.5
One way to get work out of a moving book: Allow it to push a thumbtack into a wall.

FIGURE 6.6
One way to get work out of a falling book.

---

*For now, we'll worry about only what happens up until the book hits the floor. Later, we'll study the process of its actually hitting the floor.

These are the essential ideas about work and energy, presented in the context of a simple example. The rest of this chapter expands on these ideas.

6.3 A QUANTITATIVE LOOK AT ENERGY

Let's look quantitatively at the simple experiment described in the preceding section. As we know, the amount of work you do to lift an object is its weight multiplied by the height raised. Once the book is raised, it has gravitational energy because it can do work in pushing your hand back to the floor. How much gravitational energy does it have?

We define an object's energy quantitatively as the *amount of work it can do.** It is measured in joules (or newton-meters), just as work is. For the raised book, its energy is the amount of work it can do in slowly pushing your hand back down to the floor, which is just the book's weight multiplied by the distance to the floor. For this or any other raised object,

wt × ht at this point—

equals $(1/2)\, ms^2$ at this point

$$\text{gravitational energy} = \text{weight} \times \text{height}$$
$$\text{GravE} = \text{wt} \times \text{ht}$$

When you slowly lower the book, it uses up its energy in pushing your hand back to the floor. But when the book falls, its gravitational energy is transformed into kinetic energy. How much kinetic energy is possessed by a moving object such as the book? In other words, how much work can a moving object do because of its motion? It is possible to work out the answer to this question, starting from Newton's laws, although we won't work it out here. The answer turns out to be

$$\text{kinetic energy} = (\tfrac{1}{2}) \times (\text{object's mass}) \times (\text{square of object's speed})$$
$$\text{KinE} = (\tfrac{1}{2})\, ms^2$$

FIGURE 6.7
An amazing thing: The gravitational energy at the beginning precisely equals the kinetic energy at the end.

This formula tells us that a more massive object has more kinetic energy and that a faster object has more kinetic energy, as we might expect. It makes sense that the formula should involve the object's mass rather than its weight, because kinetic energy is possible even in the absence of gravity.

Now we note an amazing thing, also provable from Newton's laws: If we neglect air resistance (which we'll deal with later), the amount of gravitational energy the book has at the beginning of its fall will precisely equal the amount of kinetic energy it has at the end (Figure 6.7). The book's total ability to do work, its total energy, is quantitatively unchanged during the falling process. Its energy is simply changed in form—transformed—from gravitational to kinetic, but its total energy remains the same. Energy is precisely *conserved.*

Furthermore, since energy is conserved for any distance of fall, it must be conserved at halfway down, at three-quarters of the way down, and at every other point during the fall. The loss in gravitational energy during any portion of the fall precisely equals the gain in kinetic energy during that portion (Figure 6.8).

1

2

FIGURE 6.8
The total energy is conserved all the way down. The loss in gravitational energy between points 1 and 2 during the fall is precisely balanced by the gain in kinetic energy between these two points.

*But we'll find out in the next chapter that for the particular energy form known as *thermal energy,* there are limits on the efficiency of actually doing this work.

■ **DIALOGUE 5** (a) How much gravitational energy (relative to the water's surface) does a diver have when she stands 2 m above the water as compared with when she stands 4 m above the water? (b) One car moves at twice the speed of a second identical car. How much more kinetic energy does the faster car have?

■ **DIALOGUE 6** A book having a mass of 1.2 kg and a weight of 12 N is raised to a height of 2 m and then dropped. (a) How much gravitational energy does it have at 2 m high? How much total energy? (b) Neglecting air resistance, how much total energy does it have just before hitting the floor? How much gravitational energy? How much kinetic energy? (c) How much total energy does the book have at one-quarter of the way down from the top? How much gravitational energy? How much kinetic energy?

■ **DIALOGUE 7 MAKING ESTIMATES** Make an order-of-magnitude estimate (to within a factor of 10) of the energy in (a) this book (assume its weight is about 10 newtons) lifted to arm's length over your head, (b) you jogging, (c) you walking.

6.4 ENERGY: THE ABILITY TO DO WORK

Now we will expand on the preceding two sections, extending these ideas to a wide variety of situations.

Any system* having the ability to do work is said to have energy. A system's energy is quantitatively the amount of work it can do.[†]

Energy can be measured in joules, just as work can. But work and energy are not the same thing. A system does work, but it has energy. Work is a process, and energy is a property, an attribute, of a system. Think of a system's energy as the amount of work the system could do, regardless of whether it ever actually does this work: A raised boulder has energy, even though it might be tied up and left that way forever and never do any work.

If you think of energy as the ability to do work, it becomes apparent that there are many forms of energy. We'll discuss eight forms, beginning with the two we already know something about.

Kinetic energy is energy that is due to motion. It is the work that a system can do as a consequence of the system's motion, in other words, the work a system can do while coming to rest.

---

5. (a) Half as much, because gravitational energy is proportional to height. (b) Four times as much, because kinetic energy is proportional to the squared speed.

6. (a) GravE = weight × height = 12 N × 2 m = 24 J; TotalE = 24 J. (b) TotalE = 24 J because of the conservation of energy; KinE = 24 J, because there is no longer any gravitational energy. (c) TotalE = 24 J. But now the height is only 1.5 m, so GravE = wt × ht = 12 N × 1.5 m = 18 J (it's lost one-quarter of its gravitational energy). Because total energy is conserved, KinE = 24 J –18 J = 6 J.

7. (a) Suppose you can lift to about 2 m: GravE = wt × ht ≈ 10 N × 2 m = 20 J. (b) Suppose you can jog a kilometer in 6 minutes (a mile in 10 minutes). Your speed is then $s = d/t =$ 1000 m / 360 s ≈ 3 m/s. If your mass is about 70 kg (weight 150 pounds), your kinetic energy is $(1/2)ms^2 = (1/2) \times 70$ kg × $(3$ m/s$)^2 ≈ 300$ J. (c) You probably walk about half as fast as you jog, perhaps a kilometer in 12 minutes. Half the speed means one-fourth the kinetic energy, so the energy is about $(1/4) \times 300$ J = 75 J. *Note:* Your own estimates will probably differ from these.

*The words *system* and *process* come up a lot in science. A *system* means a part of the universe, for instance, a collection of objects, and a *process* means any sequence of events.

[†]In the case of thermal energy there are limits on the efficiency of actually doing this work.

Gravitational energy* is energy caused by gravitational forces. It is the work a raised system can do while Earth (or any other object that can pull gravitationally) pulls it back to its initial position. There is a quirk about gravitational energy: Its numerical value depends on the level chosen as the initial or *reference* level, simply because the amount of work you can get from a raised object depends on how far down it must go before you consider it no longer "raised." For instance, this book's gravitational energy is only a few joules relative to the floor of your room, but it may be thousands of joules relative to sea level. So we sometimes need to be explicit about the agreed-upon reference level when discussing gravitational energy.

If you stretch a rubber band or bend a ruler, it can snap back when released. There is energy in the stretched or bent system because it can do work while snapping back. For instance, a stretched rubber band can do work in slowly pulling your fingers together. This energy, resulting from the ability of a stretched or deformed system to snap back, is called **elastic energy**.

A pot of hot water has more energy than does a pot of cold water of the same size. How do we know? Well, if the hot pot is boiling, it can rattle its lid, and this requires work, so boiling water has the ability to do work. If the hot pot is below the boiling temperature, one way to get work from it would be to find another liquid that boils at a temperature below that of the hot pot's temperature. Then let the hot pot warm up the other liquid to boiling so that this other liquid rattles its lid.

This kind of energy that a system has due to its temperature is called **thermal energy**.†

It is enlightening also to consider thermal energy from a microscopic point of view. As discussed in Chapter 2, temperature is associated with random microscopic motions that are not visible macroscopically. For example, as the temperature of a pot of water rises, its molecules move around faster and faster. The molecules gain greater kinetic energies. Thermal energy is this microscopic kinetic energy that cannot be directly observed macroscopically.§

To prevent confusion, we will reserve the term *kinetic energy* for macroscopic kinetic energy and use *microscopic kinetic energy* when referring to thermal energy from a microscopic point of view.

Put one hand on a warm object and the other on a similar cooler object, for instance, on pots of warm and cool water. The only difference between the two pots is that the warm one has faster molecules. From a microscopic point of view, you are not really experiencing warmth and coolness, you are only experiencing rapidly and slowly moving molecules. Does that seem amazing to you? It does to me. How could the mere motion of microscopic particles, particles that I can't even see or feel individually, create the feeling of warmth in my hand? The idea of warmth has been replaced by, or "reduced to," motion. In a way, the notion of warmth has vanished, a classic example of science's reduction of a wide assortment of phenomena to a few basics.

---

*Also known as *gravitational potential energy. Potential energy* is energy resulting from a system's position or configuration. Except for kinetic, thermal, and radiant energies, all our listed energy forms are forms of potential energy. In the interest of brevity, we won't use the word *potential.*

†Thermal energy is sometimes called *internal energy* or *heat energy.*

§More precisely, thermal energy includes all of the microscopic forms of energy that are not directly visible at the macroscopic level. It includes *microscopic potential energy* that results from the forces between molecules, a point that is important to understanding melting and other *phase transitions.*

This reduction of sense impressions to the mechanical motion of atoms is precisely what Democritus (Section 2.6) was talking about when he proclaimed, "By convention hot is hot [and] cold is cold. . . . The objects of sense are supposed to be real—but in truth they are not. Only the atoms and the void are real." But Democritus's view does more than simply interpret sense impressions as mechanical motions; it proclaims sense impressions to be unreal, nonexistent, and claims that only the motions are real. As we discussed in Chapter 2, one can accept the scientific usefulness of the atomic viewpoint without necessarily accepting Democritus's philosophical view.

Historically, thermal energy was difficult to understand because it does not fit comfortably into the mechanical framework of Newtonian physics and because, as we will see in the next chapter, it is fundamentally different from the other energy forms. During the eighteenth and nineteenth centuries, Joule and others eventually demonstrated that thermal energy actually is energy. This was the key step in comprehending energy.

We will run quickly over the remaining four types of energy, returning to them in more depth in the appropriate later chapters.

The electromagnetic force (Chapter 8) is one of nature's fundamental forces. The energy that results from electromagnetic forces is called **electromagnetic energy**, or sometimes simply electric energy.

There is energy in a light beam, as you can tell from the fact that light (sunlight, for instance) can warm things, and you can get work out of warm things. The energy carried by a light beam is one form of **radiant energy**. There are other forms of radiant energy, some of them perhaps familiar to you: radio, microwave, infrared, ultraviolet, X ray, and gamma ray. All of these are similar to visible light. We will learn more about them in Chapter 9.

Chemical reactions can do work, as you can see from a wood fire used to boil water. This energy results from the molecular structure of the wood. This energy that results from a system's molecular structure is called **chemical energy**.

Whereas chemical energy results from molecular structure, **nuclear energy** results from nuclear structure, from the way protons and neutrons are arranged into nuclei. One obtains nuclear energy from nuclear reactions (Chapter 15), just as one obtains chemical energy from chemical reactions.

■ **DIALOGUE 8**   Name the type of energy possessed by each of the following: a raised book, gasoline, a stretched spring, sunlight, a speeding train, hot steam.

6.5 THE LAW OF ENERGY: ENERGY IS FOREVER

We have seen that the gravitational energy lost during any portion of an object's fall is exactly balanced by its kinetic energy gain, provided you neglect air resistance and observe the object only until it first touches the floor. The object's overall, or total, energy is conserved all the way down. Now we expand on this principle.

It is possible to prove, still based on Newton's laws, that any system that experiences only gravitational forces conserves its total energy, just as the falling

─────────────

8. Gravitational, chemical, elastic, radiant, kinetic, thermal.

object does. Any gains or losses in kinetic energy are exactly balanced by losses and gains in gravitational energy. Elastic forces also fit into this Newtonian picture. It is possible to prove that any system that obeys Newton's laws and that possesses only kinetic, gravitational, and elastic energy must conserve its total energy.

A perfectly elastic rubber ball that falls from some height, is squeezed elastically during its impact with the floor, and rebounds back to the initial height (this is what *perfectly elastic* means), is one illustration of such a system. Another example is the solar system: To a good approximation, only gravitational forces are important to determining the motion of the sun, planets, and moons of our solar system. In these cases, Newton's laws guarantee that the total energy remains unchanged despite the changes that occur within the system.

■ **DIALOGUE 9** What form or forms of energy does a falling elastic ball (Figure 6.9) have at each of the following times? (a) As it begins falling. (b) Halfway down. (c) As it first touches the floor. (d) After it touches the floor while it is still being further squeezed against the floor. (e) When it is at its most-squeezed position on the floor. (f) Just as it leaves the floor during its rebound. (g) When it is back at its high point.

This is as far as Newton's laws can take us. Even without going further, it is a remarkable result. There is this abstract quantitative idea, the ability to do work, that maintains the same quantitative value at every moment of some complicated process such as the rebound of an elastic ball or the motion of the solar system.

But even more remarkably, experiments show that the energy principle goes far beyond Newton's physics. Energy is conserved in every physical process yet observed, including processes lying far outside the range of validity of Newton's laws. We call this:

THE LAW OF CONSERVATION OF ENERGY
The total energy of all the participants in any process must remain unchanged throughout that process. That is, energy cannot be created or destroyed. Energy can be transformed (changed from one form to another), and it can be transferred (moved from one place to another), but the total amount always stays the same.

This statement is as true as any general idea ever gets in science: It is correct in every situation yet observed. It holds even in situations in which Newton's physics is not even remotely correct, such as near black holes, close to the speed of light, and for subatomic particles.

It's a useful principle, because once you have calculated or measured the total energy at one moment during some process, you will know it at any other moment without having to calculate or observe what happened between the two moments. In Dialogue 9, for instance, if you knew the rubber ball's initial height,

FIGURE 6.9
A bouncing ball. What forms of energy does a perfectly elastic ball have at each of these seven times during its fall to the floor and its rebound?

9. (a) Gravitational. (b) Gravitational and kinetic. (c) Kinetic and a little gravitational (measured relative to its lowest position "e" in Figure 6.9). (d) Elastic and kinetic and a very little gravitational (because it is still just a little above the floor). (e) Elastic. (f) Kinetic and a little gravitational. (g) Gravitational.

you would also know its initial energy; thus you would know that the ball has precisely this amount of energy at every moment during the bounce without having to calculate the forces and motions involved in the bouncing. For another example, the number of joules of chemical energy consumed from a car's fuel tank must equal the total number of joules appearing as the car's kinetic and gravitational energy, exhausted thermal energy, chemical energy of pollutants, and so forth, during that time. Energy conservation is like the motion of a bishop in chess. If a bishop is on a black square, then after any number of moves it will still be on a black square. You don't need to know the details of the moves to predict this, because a bishop's square color is conserved.

Some of the energy principle's scientific appeal lies in its elegance and its expression of timelessness. It is perhaps reassuring that amid nature's incessant change, energy is forever.

Physicists have discovered several other natural quantities that are conserved. Although we will not study the other conservation principles, it is interesting to list them. Throughout every physical process, the participants maintain or conserve each of the following:

- Total ability to do work, or *total energy*.
- Total motion through space, or *total linear momentum*.
- Total rotational motion, or *total angular momentum*.
- Total electric charge.
- Certain subatomic properties associated with microscopic interactions.

These are among the most far-reaching physical principles known.

There is a useful alternative way of stating conservation of energy. Whenever work is done, it is done by some system on some other system. The system doing the work must lose some of its ability to do work; in other words, it must lose energy. Since total energy is conserved, this energy cannot just vanish but must instead go into the system on which work is done. So work is an energy transfer from the system doing the work into the system having work done on it. We call this:

THE WORK–ENERGY PRINCIPLE
Work is an energy transfer. Work reduces the energy of the system doing the work and increases the energy of the system on which work is done, both by an amount equal to the work done.*

H O W   D O   W E   K N O W ?

Why do we believe energy is conserved? As stated above, in the realm where Newton's physics is correct, conservation of energy can be proved from Newton's laws for processes involving only kinetic, gravitational, and elastic energy forms.

---

*There is also another worklike process by which energy can be transferred, called *heating*. Heating is a thermal energy transfer due to a temperature difference, and it can be thought of as microscopic work. When expanded to include not only ordinary work but also microscopic work (heating), the work–energy principle is called the *first law of thermodynamics*. Thus conservation of energy, the work–energy principle, and the first law of thermodynamics all are equivalent ideas, stated in different forms. We won't need heating or the first law of thermodynamics in what follows.

The important question of including thermal energy in the energy principle will be discussed later. For now, we will just assume that thermal energy can be included in the energy principle, and we will describe how a thermal energy measurement process known as *calorimetry* can be used to demonstrate the conservation of the other non-Newtonian energy forms (electromagnetic, radiant, chemical, nuclear).

The idea of calorimetry is to measure a system's nonthermal energy by transforming it into thermal energy in some second system whose thermal energy can be measured. For example, the chemical energy in a gram of coal can be measured by combusting the coal and using the thermal energy to warm, say, a kilogram of water. The measured rise in temperature of the water tells us the amount of thermal energy deposited in the water, which is then the amount of chemical energy in the original coal (and in the oxygen used during combustion).

A historical example illustrates how calorimetry can test the conservation of such non-Newtonian energy forms as nuclear energy. Early in the twentieth century, nuclear physicists investigated *beta decay*, a process in which a nucleus spontaneously creates an electron and spits it out of the nucleus. This process alters the original nucleus. We represent this process as

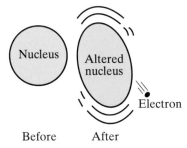

Before   After

FIGURE 6.10
The beta decay process, before and after. A nucleus spontaneously creates and ejects an electron. Physicists were once uncertain about whether energy was conserved in this nuclear process.

original nucleus → altered nucleus + ejected electron

(Figure 6.10). If energy is conserved, the nuclear energy of the original nucleus should equal the nuclear energy of the altered nucleus plus the energy of the ejected electron.

But nature had a surprise. Measurements showed that the energies did not balance! The energy was larger before than after. Being reluctant to conclude that energy was actually not conserved, physicists hypothesized that some undetected particle was also ejected along with the electron. It was thought that when this other particle's energy was included, the energies would balance.

Although the hypothesized particle had not been detected, it was thought that its energy could be directly measured by surrounding the nucleus with a large cylinder of lead. The lead cylinder was used as a *calorimeter*, or thermal energy measurer. The unseen particle would surely be slowed down and stopped inside a sufficiently large cylinder of lead and so deposit its energy in the lead. This energy would then show up as a small temperature rise in the lead, and by measuring this rise one could deduce the amount of energy deposited in the lead.

But nature sprang another surprise: There was no measurable temperature increase in the lead. Perhaps energy was *not* conserved in beta decay. This is where the matter stood from 1914 to 1930. Physicist Niels Bohr suggested in 1929 that perhaps energy conservation did not apply to the nucleus. But others did not accept this suggestion, and in 1930 physicist Wolfgang Pauli hypothesized that the new particle was so penetrating that it could pass right through the lead calorimeter without depositing any energy, and that energy would be found to be conserved once the elusive particle was found.

This set off a search for such a particle. Before long, physicists found other indirect evidence (other than beta decay), and they gave the hypothesized particle a name: the *neutrino*. It was finally detected directly in 1955, 25 years after Pauli's hypothesis. Experiments showed that, as Pauli had predicted, energy was conserved once the neutrino was included in the balance.

It takes about 8 solid light-years of lead to stop half the neutrinos emitted in a typical nuclear decay. They move like "greased lightning" through matter. . . . If you make a fist, there are thousands of neutrinos flying through it right now, because the entire universe is filled with neutrinos. . . . Another proposal, made tongue-in-cheek, is for a neutrino bomb, a pacifist's favorite weapon. Such a bomb would explode with a whimper and flood the target area with a high flux of neutrinos. . . . [T]he neutrinos would fly harmlessly through everything.

HEINZ PAGELS, IN *THE COSMIC CODE*

■ **DIALOGUE 10**   Farswell Slick invites your investment in a business venture to manufacture his remarkable "supertranspropulsionizer." His diagrams show a dazzling array of superconductors, lasers, liquid-helium coolants, and computer controls (Figure 6.11). Slick informs you that this ultimate propulsion system will accelerate spaceships to nearly lightspeed for interstellar travel. Amazingly, no fuel supply is needed, either on board or outside the spaceship. The principle involved, he explains, is "superconducting optical self-propulsion" (SOS). With SOS, the device operates in a continuous cycle that both accelerates the space-ship and "feeds back" some of its laser light to maintain, for as long as may be desired, the operation of the transpropulsionizer itself. Should you invest?

FIGURE 6.11
"A remarkable device," Farswell Slick remarks. Would you buy a supertranspropulsionizer from this man?

6.6 TRANSFORMATIONS OF ENERGY

Everything that happens can be described as an energy transformation. It's a useful way to look at things. This section uses conservation of energy to describe the energy transformations involved in some familiar processes.

FALLING   Once again, drop your book to the floor (it's coming in for a lot of rough treatment in this chapter!). We have studied this process up until its impact with the floor. Where is the energy after impact? It might appear to have vanished, but conservation of energy says that energy never vanishes. Let's proceed by elimination: The energy is not kinetic, gravitational, or elastic. There are no obvious electromagnetic, radiation, chemical, or nuclear effects. There is only one candidate left: thermal energy. So the impact must warm the book or the floor. This temperature rise is hard to detect, but you can demonstrate the same effect by driving a nail into a board with a hard hammer blow. Feel the nail before and after the blow. Try several blows.

It is useful to summarize the energy tranformations in the following way:

GravE (at the top) → KinE (just before impact) → ThermE (after impact)

Now let's add the effects of air resistance. The participants in the process are now the book, Earth (its gravitational pull), and the air. Since air resistance slows the book, the falling book has less kinetic energy than it did before. But this energy is not lost—you cannot lose energy. It must be transformed into thermal energy. The air and book must warm a little as the book falls.

Until the work of Joule and others around 1850, scientists had long believed that the work going into forces such as air resistance and friction, work that produces *heat*, was lost. The key to finally uncovering the principle of conservation of energy was discovering that these processes are simply another, more subtle, type of energy transformation and that the heat produced is a form of energy, a form we now call *thermal energy*.

---

10. Don't invest. And don't bother investigating his design. It violates conservation of energy. Work must be done by the transpropulsionizer to accelerate the spaceship, so the device transfers energy from itself. During a complete cycle of operation, energy must be supplied to the device in order to bring it back to its initial (at the beginning of the cycle) condition. So the device needs an external fuel supply. You can't get something for nothing.

H   O   W   D   O   W   E   K   N   O   W   ?

Joule's experiments

We predicted that air warms when you stir it up with a falling object. James Prescott Joule (Figure 6.4) did an experiment like this in the 1840s, using water instead of air. He placed a paddle wheel in a tub of water, stirred the water with the paddle wheel, and measured the temperature rise in the water.* He quantified the experiment by allowing a falling weight attached to a cord to turn the paddle wheel. The weight's energy loss is then its weight multiplied by the distance fallen. Joule found that the water's rise in temperature was precisely proportional to the work done, in other words, to the gravitational energy lost.

These ideas were murky in Joule's day because it was not understood that heat (thermal energy) is a form of energy. Joule clarified matters by showing that work is precisely, quantitatively convertible to thermal energy. This breakthrough extended the energy principle beyond the Newtonian energy forms to thermal energy.

Joule showed that a particular amount of work, about 4200 J, always produces a 1°C rise in the temperature of 1 kilogram of water. This amount of energy is the **dietitian's Calorie**.[†] Although the Calorie is often used to measure thermal energy, Joule's work showed that it is really a general energy unit, equivalent to 4200 joules.

Back to the falling book (isn't it surprising what we can learn just by thinking carefully about the fall of a book?): Just before impact, all the energy has converted to kinetic energy of the book and to the thermal energy of the air and book. Since air resistance has only a small effect on the motion, the thermal energy must form only a small fraction of the total. It is this small reduction in kinetic energy that shows up as thermal energy in the air. Finally, the impact converts the preimpact kinetic energy to the thermal energy of floor, book, and air.[§]

To summarize more complex energy transformations such as this one, an **energy flow diagram** of the sort shown in Figure 6.12 is useful. It shows energy transforming as though it were water flowing through pipes. The diagram shows energy beginning as gravitational energy, then transforming to kinetic energy and a small amount of thermal energy (of the air) during the book's fall, and finally transforming entirely to thermal energy when the book hits the floor. Since energy is conserved in each stage of the process, the pipe widths match up at each intersection.

ANIMAL ENERGY AND EFFICIENCY   The energy that enables you to do useful work is obtained from foods and is stored in your body as chemical energy

---

*When you stir hot water in the open air, the water *cools* because of evaporation. In Joule's experiment the stirring was done inside a closed container, thereby preventing any significant evaporation.

[†]The dietitian's Calorie is always spelled with a capital C. Physicists use the word *calorie* (lowercase c) to denote the energy needed to raise 1 gram of water by 1°C.

[§]And you can hear the impact. A small fraction of the energy is transformed into the energy of sound, which actually is a form of kinetic and elastic energy of the air.

FIGURE 6.12
Energy flow diagram for a falling book, with air resistance. A diagram such as this shows an energy transformation for some process. The widths of the "pipes" correspond to the amounts of energy involved in various parts of the process. Since energy is conserved in each stage of the process, the pipe widths match up at each intersection.

(Chapter 2). Dietitians measure this stored chemical energy in Calories (= 4200 J). For example, once you have eaten and respirated it, a 70-Calorie slice of bread gives you 70 Calories of stored chemical energy that can then provide 70 Calories of work and thermal energy.

When animal chemical energy is used to do work, for example when you lift a book, only a small fraction of this chemical energy is actually transformed into useful work. We say that the process is *inefficient* because so little of the input energy appears as useful work. A *highly efficient* process, on the other hand, is one in which most of the input energy is transformed into useful output energy and the wasted fraction is small.

It is useful to define this idea quantitatively: The **energy efficiency** of any energy transformation is the fraction of the input that appears as useful output:

$$\text{energy efficiency} = \frac{\text{useful energy output}}{\text{total energy input}}$$

It is usually expressed as a percentage.

The efficiency of transforming animal chemical energy into work done on the external environment is only about 10% for many human muscular activities. Most of the remaining 90% is transformed into thermal energy in your body. So when you slowly lift your book, the energy transformation is

ChemE → GravE (of book, 10%) + ThermE (of your body, 90%)

SLIDING Now give your book a quick hard push so that it slides across your tabletop, sliding to rest (Figure 6.13). Figure 6.14 shows the energy flow diagram for this process. Most of the initial chemical energy used to push the book turns into thermal energy in the body. The small amount that goes into the book then winds up as thermal energy created while the book slides to rest.

You will encounter many more energy transformations and energy flow diagrams throughout this book.

6.7 POWER: THE QUICKNESS OF ENERGY TRANSFORMATION

What is the difference between running and walking up a flight of stairs? Your gravitational energy increases by the same amount in both cases. The work you do is the same. And yet your body knows there is a difference between running

FIGURE 6.13
What energy transformations occur when you
briefly push a book and then let it slide?

and walking upstairs. The difference is that you do the work in less time, quicker, when you run.

We need a word for this notion of how quickly work is done. We call it **power**. Quantitatively, "how quickly a task is done" is the rate of doing that task, the amount done per second. So we define power as the rate of doing work, or the amount of work done divided by the time to do it:

$$\text{power} = \frac{\text{work done}}{\text{time to do it}}$$

Because work is an energy transformation, power can also be thought of as the rate of transforming energy.

Suppose you run up one flight of stairs and then walk up a second identical flight of stairs in twice the time it took to run up the first flight. You do the same work for each flight, but your power output during the first flight is double your power output during the second flight because you went up the first flight in half the time. The difference between the two processes is a power difference, not an energy difference.

The unit of power is the *joule per second* (J/s). It differs by an all-important *per second* from the unit of work or energy. Power is such a popular concept that its unit is given a special name. The joule per second is called the **watt** (W), in

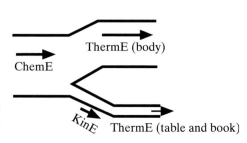

FIGURE 6.14
Energy flow diagram for a book that is given a
quick push and allowed to slide across a surface
while coming to rest. As the diagram indicates, the
pushing process is very inefficient, with most of the
initial chemical energy going into bodily heating
rather than into the book.

honor of the eighteenth-century developer of the steam engine, James Watt. The kilowatt (kW) is 1000 watts, and the megawatt (MW) is 1 million watts.

MAKING ESTIMATES

Running upstairs

What is your power output while running up a flight of stairs? If the energy efficiency of this process is 10%, what is your (chemical) power input, in watts and in Calories/second?

SOLUTION   Suppose that your weight is 500 newtons (110 pounds), the vertical height of one flight of stairs is 4 meters (measure it!), and you run up one flight in 5 seconds (try it!). The work done to lift yourself and the power output are

$$\text{work} = \text{weight} \times \text{height} = 500 \text{ N} \times 4 \text{ m} = 2000 \text{ J}$$
$$\text{power} = \text{work/time} = 2000 \text{ J/5 s} = 400 \text{ W}$$

This is a large power output for a human being, as you will discover if you do the experiment (see Home Project 2). To produce this output at 10% efficiency, you must convert chemical energy at a rate of 4000 W, or 4000 J/s. In Cal/s (1 Cal = 4200 J), this is $4000/4200 \approx 0.95$ Cal/s, your **metabolic rate** in this example. If you could maintain this rate for an hour (3600 s), you would "burn up" (transform) $3600 \times 0.95 = 3400$ Calories, the energy content of a sizable steak.

Think of several devices you use every day: automobile, lightbulb, electric blender, toaster, and so forth. These devices can be understood as energy transformers, that transform energy from one form to another form that you can use. An important feature is often the rate at which the energy is converted. For example, to get a certain lighting level from a lightbulb, the bulb must convert a certain number of joules per second to visible light. So lightbulbs and other devices are rated in power units (watts). In the United States, a popular power unit for automobile engines and other heat engines (next chapter) is the horsepower. One horsepower is about 750 watts.

Table 6.1 gives the power consumed (transformed into nonelectric forms) by typical household electrical appliances. The numbers give the power consumed only during the time that the appliance is turned on and consuming electric energy. Total electric energy consumption, perhaps over one day or over one year, often tells quite a different story. For example, one ordinarily uses a toaster for only a short time each day, so its daily energy consumption is low even though its 1200 W power consumption is high. And refrigerators, although their power consumption can be as low as 300 W, are leading household energy consumers because they operate for so many hours every day.

Although your total annual energy use determines the amount of fossil or other resources that your home electricity consumes, your impact on the need for new power plants is determined by your power use at so-called peak times. Each electric plant has a maximum power output, its *power rating*, usually several hundred megawatts. The plant's actual power output is largest at times such as hot afternoons when everybody is running an air conditioner. If the power

TABLE 6.1

Power consumption of household appliances while the appliance is turned on and consuming electric energy

Appliance	Power (W)
Cooking range	12,000
Clothes dryer	5,000
Water heater	4,500
Air conditioner, window	1,600
Microwave oven	1,400
Dishwasher (incl. hot water)	1,200
Toaster	1,200
Hair dryer	1,000
Refrigerator, frostless	600
Refrigerator, not frostless	300
TV, color	350
Radio-phonograph, stereo	100

peak approaches the plant's rating, the plant will cut its output by reducing the supply to all customers, causing lightbulbs and other appliances to dim in a *brownout*.

There are lots of opportunities today for electric power companies to save customers' money, enhance company profits, and protect the environment, all with no reduction in services, by finding ways to avoid building expensive new power plants. Pressures for new plants arise when existing plants can no longer provide the electricity needed during periods of peak demand. One way to reduce peak demand is with energy-efficient devices that provide the same services (the same amount of light, for example) with less energy. Because such efficiency measures are often far cheaper than the cost of building new plants, many power companies are actively seeking and providing new energy efficiency opportunities. For example, because it is often far cheaper to warm a house with additional insulation than with additional electrical heating, many power companies provide services and low-cost loans to encourage customers to insulate their homes. Everybody wins: the customer, who gets a warmer home cheaper; the power company, which profits from the fact that the insulation is cheaper than a new plant; and the natural environment, which benefits from reduced resource consumption and less pollution.

Energy storage is another way to reduce the need for new power plants. During off-peak times, a plant's excess capacity can be used to store energy for later use during peak times. Again, everybody wins because of the large avoided cost of a new power plant. One method, illustrated in Figure 6.15, is pumped storage, in which a plant's excess capacity during off-peak periods pumps water uphill into a reservoir. The water in the reservoir then provides hydroelectric power during peak periods.

Pricing is yet another way to reduce the need for new power plants. If electric companies charge higher rates at peak power times, balanced by reduced rates during off-peak periods, people have an incentive to switch their power use from peak to off-peak times, which in turn reduces the need for new power plants. Because of the large avoided cost of building a new power plant, this can actually reduce customers' average electrical bills while increasing company profits.

The most useful unit for measuring your home's electric energy consumption is the **kilowatt-hour** (kW·h), the amount of energy transformed when a power of 1 kilowatt operates for 1 hour. Since 1 kilowatt is 1000 joules/second and 1 hour is 3600 seconds,

$$1 \text{ kW·h} = 1000 \text{ J/s} \times 3600 \text{ s} = 3.6 \times 10^6 \text{ J}$$

If a known power in kilowatts operates for a known number of hours, it is easy to figure the number of kilowatt-hours of energy consumed: Just multiply the number of kilowatts by the number of hours.

■ **DIALOGUE 11**   (a) A clothes dryer is equivalent to how many 100 W lightbulbs? (b) If drying a load takes an hour and if you dry 15 loads in a month, how many kW·h of electric energy will you consume? (c) How much does this cost, at 10¢ per kW·h?

---

11. (a) Dryers operate at 5000 W (Table 6.1); 5000 W / 100 W per bulb = 50 bulbs! (b) 5 kW × 15 hours = 75 kW·h. (c) ($0.10 per kW·h) × 75 kW·h = $7.50.

FIGURE 6.15
The principle of pumped storage.
(a) During periods of low consumer
demand for electricity, the plant's
excess power capacity runs a pump
that lifts water uphill. (b) During peak
periods, this water flows back
downhill to produce electricity. What
energy transformations occur during
peak and off-peak periods? Pumped
storage moves some of the plant's
electrical production from peak
periods to off-peak periods, reducing
the need for new plants.

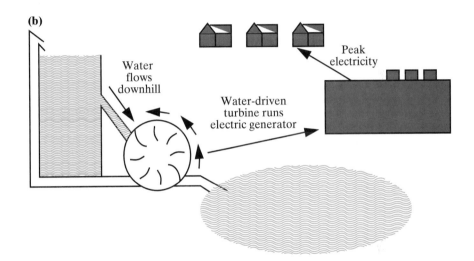

■ **DIALOGUE 12** How many kW·h of energy would you save in a month by switching from a frostless to a nonfrostless refrigerator? Assume that the refrigerator is drawing power 50% of the time. How much money do you save, at 10¢ per kW·h? By how much would this switch affect the local power company's peak power demand?

■ **DIALOGUE 13** An average U.S. household consumes electricity at an average rate of about 1000 W. How many kW·h of energy is this every month? How many joules? If this energy went instead into lifting, how high would a typical 1500 N (150 kg, 330 lb) family be lifted?

12. During one month, the refrigerator is drawing power for about 12 hr/day × 30 days = 360 hrs. According to Table 6.1, you use 300 W, or 0.3 kW, less power during this time. So your total energy consumption during the month is reduced by 0.3 kW × 360 hr = 108 kW·h. You save 108 kW·h × 10¢ per kW·h = $10.80 every month. The switch reduces the peak power demand by 300 W.

13. 1000 W = 1 kW. The number of hours in a month is about 30 days × 24 hr/day = 720 hr. So the amount of energy is 1 kW × 720 hr = 720 kW·h. In joules, this is 720 kW·h × 3.6 × 10^6 J/kW·h = 2.6 × 10^9 J. Using the formula GravE = wt × ht, the height to which this much energy can raise a 1500 N weight is ht = GravE / wt = 2.6 × 10^9 J / 1500 N = 1.7 × 10^6 m = 1700 km (about 1000 miles)!

"WAKE UP, DR. ERSKINE — YOU'RE BEING
TRANSFERRED TO LOW ENERGY PHYSICS."

SUMMARY OF IDEAS AND TERMS

Fossil fuel   The remains of ancient life, used as an energy resource, such as coal, oil, and natural gas.

Industrial Revolution   The beginning of the fossil-fueled industrial age, around 1750.

Work   Object A does work on object B if A exerts a force on B while B moves. Unit: newton-meter (N·m) = joule (J). Work = Fd.

Energy   The ability to do work. The energy of a system is the amount of work that system can do. Units: joule (J), Calorie, kilowatt-hour (kW·h).

Forms of energy:
- **Kinetic energy,** due to motion: KinE = $(1/2)ms^2$.
- **Gravitational energy,** due to gravitational forces: GravE = wt × ht.
- **Elastic energy** due to the ability of a deformed system to snap back.
- **Thermal energy**, due to temperature. Equivalently, thermal energy is microscopic energy, the kinetic (and other) energy of molecules that cannot be directly observed macroscopically.
- **Electromagnetic energy**, due to electromagnetic forces.
- **Radiant energy**, carried by a light beam, radio wave, microwave, infrared, ultraviolet, X ray, or gamma ray.
- **Chemical energy**, due to molecular structure.
- **Nuclear energy**, due to nuclear structure.

The law of conservation of energy   The total energy of all the participants in any process must remain unchanged throughout that process. There are no known exceptions.

The work–energy principle   Work is an energy transfer.

Energy efficiency   Useful energy output/total energy input.

Energy flow diagram   A diagram showing the energy transformations occurring during some process. Often, the various forms of energy are pictured as though energy were a fluid flowing through pipes.

Metabolic rate   The rate, usually measured in Cal/s, at which an animal transforms its bodily chemical energy into other thermal energy and other forms of energy.

Power   The rate of doing work. Units: joule/second, watt (W = J/s), horsepower. Power = work/time.

REVIEW QUESTIONS

INTRODUCTION
1. What type of culture was the first village culture?
2. What type of energy technology fueled the industrial revolution?
3. What is a fossil fuel? Name three kinds.

WORK
4. Is work done whenever a force is exerted? Explain.
5. Is work done whenever an object moves through a distance? Explain.
6. To what two quantities is work proportional?
7. You slowly lift a 3 N grapefruit, by 2 m. How much work did you do and on what object?
8. A 3 N grapefruit falls 2 m to the floor. Was work done during the fall? By what object on what other object?

ENERGY
9. Explain the difference between energy and work.
10. List the eight physical types of energy.
11. Explain thermal energy from a microscopic point of view.
12. Give one example of each of these energy forms: elastic, thermal, chemical, kinetic, radiant, gravitational.
13. If you double the speed, how is the kinetic energy affected? If you double the height, how is the gravitational energy affected?
14. Choose the correct answer(s): One joule is the same as 1 (watt-meter, newton-meter, meter per second squared, newton-second, kilowatt-hour).

CONSERVATION OF ENERGY AND ENERGY TRANSFORMATIONS
15. Has the law of conservation of energy been found to be correct in all situations observed so far? Is the same true of Newton's laws? Explain.

16. Choose the correct answer(s): For a system that returns to its initial state, during one complete cycle you can't get more (acceleration, force, energy, power, speed) out of the system than was put in.
17. What energy transformations occur when this book falls to the floor? When you lift this book?
18. Give an example of each of these energy transformations: kinetic energy → thermal energy, kinetic energy → elastic energy, elastic energy → kinetic energy.
19. What do we mean when we say that the energy efficiency of a lightbulb is 10%?
20. An apple in a tree has 90 J of gravitational energy (relative to the ground). It falls. If you neglect air resistance, what can you say about the amount of kinetic energy the apple has just before it hits the ground? What if you do not neglect air resistance?

POWER
21. Explain the difference between energy and power.
22. Choose the correct answer(s): The (watt, newton per second, joule, calorie, joule per second, meter per second, horsepower, kilowatt-hour) is a unit of power. Which are units of energy?
23. You lift a 2 N rock by 4 m in 3 s. What is your work output? Your power output?
24. Which do you pay for in your monthly electric bill, energy or power?

HOME PROJECTS

1. Rub one hand hard back and forth on a wooden tabletop for 30 seconds. Using your other hand, compare the temperature of the rubbed part with another part of the table. (This doesn't work on a metal surface because thermal energy quickly travels away from the rubbed region in a metal.) What happened at the microscopic level?
2. Time yourself as you run up a staircase. Measure the vertical height of the staircase, and find your total energy output (1 kg weighs 10 N on Earth). What energy conversion occurred? Find your power output in watts. How many 100 W bulbs could this light up if converted to electric energy?
3. Try a variation of Joule's experiment: Draw two glassfuls of room-temperature tap water; blend the water of one of

them at high speed for a few minutes; and pour it back into its glass. Dip your finger into both glasses. Notice how much work is needed to produce even a small temperature rise: There is a lot of stored work (energy) in hot liquids.
4. Measure a tennis ball's elasticity by dropping it and measuring the height of rebound as a fraction of the initial height. What fraction of the initial energy was transformed into thermal energy? What fraction was retrieved as gravitational energy? Measure this fraction for several different initial heights. Does it vary significantly? Now try it for several different balls: new tennis ball, old tennis ball, Ping-Pong ball, and so forth.
5. Suppose that you install a wind turbine (a windmill to provide electricity—see Chapter 17) to provide all your

home electrical needs. It provides, under the wind conditions in your area, a monthly energy output of about 90 kW·h, or about 3 kW·h per day. In order to keep the home going for as many as 4 consecutive windless days, the turbine's electricity is stored in storage batteries, which in turn provide electricity to your home as needed. The batteries can deliver a peak power of 2400 W. Use Table 6.1 to decide how you would allocate your electricity in order to comply with the 90 kW·h energy limit every month and the 2400 W peak power. Remember to include light bulbs, which are not listed in Table 6.1. (Thanks to the American Association of Physics Teachers' *Active Physics* educational program, and John Roeder, for suggesting this project.)

FOR DISCUSSION

1. Energy-awareness audit. Note the ways you use energy during one day, listing direct uses such as automobile travel, heating, and electric appliances. Make a second list of indirect uses such as energy-intensive foods (frozen foods, meats), packaging, and nonrecycled items. List as many items as you can think of. Describe the ways in which each item uses energy. (Example: Frozen foods use energy for growing, packaging, refrigeration, and transportation.)

2. Compare and discuss the energy audits of different class members. Are there ways to lower your energy consumption without lowering your quality of life? Are there ways that you or society can reduce energy consumption while enhancing the quality of life?

EXERCISES

WORK
1. Does Earth do gravitational work on you as you walk downstairs?
2. In order for you to get out of bed with the least amount of work, would it be better for your bed to be on the floor or about a meter high? Explain.
3. Describe some work you could do that would create elastic energy. Repeat for gravitational energy. For kinetic energy.
4. Describe some work you could do that would create thermal energy.

ENERGY
5. Which of the eight physical types of energy was the basis for the earliest human culture? Which was the basis for the Industrial Revolution?
6. Name the type of energy possessed by each of the following: Jill at rest at the top of a sliding board, Jill sliding off the bottom of the sliding board, sunlight, coal, hot air.
7. Name the main type of energy possessed by each of the following: dynamite, water at rest behind a high dam, a bow about to release an arrow, a wooden match, food.
8. Does your body contain any kinetic energy when you are sitting still? Explain.
9. Give an example of a system that has both kinetic and gravitational energy.
10. A rubber ball is thrown upward and bounced off a ceiling. What kinds of energy does it have at the highest point of the bounce?
11. Name two kinds of energy that are created as a result of a typical explosion, such as exploding dynamite. What kind of energy is consumed?

12. You lift a brick and put it on top of a wall. What quantities could you measure in order to determine how much work you did?
13. You throw a baseball. What quantities could you measure in order to determine how much work you did during the throw?
14. Explain, in microscopic terms, why the air pressure inside a tire increases on a hot day. If the air in a balloon is heated, will the balloon expand or contract? Why?
15. (a) Where would an apple have greater gravitational energy, at 100 km high or at 1000 km high? (b) Would the gravitational energy of an orbiting satellite be increased or decreased by moving it from an orbit that is 6000 km high up to an orbit that is 12,000 km high (see Figure 6.16)? (c) At which point, 6000 km high or 12,000 km high, does a satellite have the larger gravitational force on it?

6000 km	12,000 km
Before	**After**

FIGURE 6.16
What happens to the gravitational energy of the satellite when it is moved to a higher orbit?

CONSERVATION OF ENERGY AND ENERGY TRANSFORMATIONS

16. Give an example in which kinetic energy transforms into gravitational energy.
17. Give an example in which kinetic energy transforms into thermal energy.
18. Give an example in which chemical energy transforms into kinetic energy.
19. What is the main energy transformation (input and useful output) when an automobile speeds up? When a bicycle speeds up?
20. Neglecting air resistance, does the energy of a falling rock increase, decrease, or remain the same? What happens to its kinetic energy? Its gravitational energy?
21. Including air resistance, does the energy of a falling rock increase, decrease, or remain the same? What happens to its kinetic energy? Its thermal energy?
22. What is the main energy transformation (input and useful output) in the operation of an electric blender? A toaster? A lightbulb?
23. You squeeze an elastic spring and clamp it in the squeezed position. You then drop the clamped spring into acid, dissolving the spring. What happened to its elastic energy?
24. What energy transformation occurs when you climb a rope?
25. You throw a baseball horizontally, and Jill catches it. Neglect air resistance. Describe the energy transformation that occurs (a) during the throw (while the ball is in your hand) and (b) during the catch.
26. You throw a ball upward and then catch it at the same height. How does the ball's final speed compare with its initial speed, (a) neglecting air resistance and (b) including air resistance? Defend your answers.
27. Does an automobile use more gasoline when its lights are on? When the air conditioner is on? (*Note*: The battery does not run these devices while the engine is running.) Defend your answer.
28. Imagine a 100% efficient automobile. Would it emit any exhaust? Would its engine be hot?
29. Figure 6.17 is a graph of a roller coaster's height above the ground versus the length of track it covers. The coaster is powered up to its high point at 100 m from the

FIGURE 6.17
Elevation versus track distance for a roller coaster.

starting point. From the high point, the coaster coasts freely all the way to the end. Assume that the coaster starts from rest at the high point and encounters no friction or air resistance. Between 200 m and the finish, where is it moving slowest? Fastest?

30. Referring to the preceding exercise, is the roller coaster moving faster at 1000 m or at 1100 m? Describe how the coaster's speed changes during the last 300 m (900 m to the end)?

POWER

31. You start a bowling ball rolling by swinging it with your arm and releasing it. Then you start a second identical bowling ball rolling, at the same speed as the first, by hitting it sharply with a sledge hammer. Which process, the arm swing or the hammer blow, imparts more kinetic energy to the ball? Which process has the greater power output?
32. You lift bricks, one at a time, onto a table. After a while, you begin to lift bricks slower and slower. As you slow down, does the energy you put into lifting each brick increase, decrease, or remain the same? What about the power you put into lifting each brick?
33. What other unit(s) could automobile engines be rated in, instead of horsepower?
34. Why are electricity rates often higher in summer than in winter?

PROBLEMS

WORK

1. Your left hand lifts a 2-N apple by 1.5 m, and your right hand lifts a 4-N grapefruit by 0.5 m. Which hand did the most work? Which hand exerted the largest force?
2. Making estimates. About how much work would it take to lift the U.S. population by 1 km?

ENERGY

3. If you triple your altitude above the ground, how is your gravitational energy (relative to the ground) affected? What if you halve your altitude?

4. If you triple your speed, how is your kinetic energy affected? What if you halve your speed?
5. While driving down a highway you slam on the brakes, sliding a certain distance with locked brakes. How much further would you slide if you had been moving twice as fast? What if you had been moving half as fast?
6. Ned the skydiver weighs 600 N and is falling at *terminal speed*, that is, he has sped up until air resistance has built up so much that he has reached a final unchanging speed. How much work does Ned do on the air as he falls through 200 m? Into what type of energy

does this work go? From what type of energy did it come?

ENERGY CONSERVATION AND TRANSFORMATIONS

7. Referring to Exercise 29 (Figure 6.17), suppose the coaster has 160,000 J of gravitational energy at the high point. (a) How much kinetic energy does it have at the low point? (b) How much total (gravitational + kinetic) energy does it have at the low point? At the finish? (c) How much gravitational energy does it have at those points where it is 10 m high? How much kinetic energy does it have at these points?

8. Jack, who weighs 300 newtons, sits in a child's swing. You pull the swing back so that it is 2 m above its low point, and release it. (a) What form of energy, and how much energy, does Jack have when he is pulled back and held at rest? (b) What form of energy, and how much energy, does Jack have as he swings through the low point (neglect air resistance and friction in the moving parts)? (c) Still assuming no air resistance or friction, how high should Jack be when he has swung forward to his high point?

9. In a crash test, a 1000-kg automobile moving at 10 m/s crashes into a brick wall. (a) How much energy goes into demolishing and heating the wall and the auto? (b) What if the auto was moving twice this fast?

10. Referring to the previous problem, from how high a cliff would the automobile need to fall in order to sustain the same amount of damage upon hitting the ground? *Note*: A 1000-kg automobile weighs (on Earth) about 10,000 N.

POWER

11. Which process has the largest power output: 2 J of work performed in 0.1 s, or 1000 J of work performed in an hour? Which has the largest energy output?

12. An automobile travels 60 km in 50 minutes, doing 30×10^6 J of work against outside forces (air resistance and rolling resistance) in the process. What is the automobile's average power output in watts? (The actual power output varies above and below this average value during the 50 minutes.) If the auto's energy efficiency is 10%, what is its power *input* (its rate of converting the gasoline's chemical energy into other forms) in watts? How many 100-W light bulbs could this light up?

13. A cyclist delivers power to her bicycle at a rate of 150 W, while her metabolic rate is 1000 W. What is her body's bicycling energy efficiency?

14. How much does it cost to run a nonfrostless refrigerator for a month? Assume the refrigerator consumes power for 8 hours each day, and the electricity costs 10¢ per kW·h. What if it is frostless?

15. Making estimates. Use Table 6.1 to estimate the number of kilowatt-hours of electric energy a typical single-family home consumes in one month. Don't forget light bulbs (they aren't in Table 6.1).

7

Second Law of Thermodynamics

—and you can't even break even

If you glance back at Figures 6.12 and 6.14, you get the impression that the energy that goes into many processes eventually turns into thermal energy. This tendency of nonthermal forms of energy to end up as thermal energy is an important general feature of the universe, known as the **second law of thermodynamics.** It is our focus in this chapter.

As you saw in Chapter 6, the big breakthrough in understanding energy was the discovery that heat (thermal energy) is actually a form of energy, in other words that heat can do work and can be created by doing work, just as can other forms of energy. This breakthrough established the validity of the law of conservation of energy even in processes involving thermal energy, and, in fact, in every known natural process. Because of the central role of thermal energy in understanding the general principles of energy, the study of energy is called **thermodynamics**, and the law of conservation of energy is often called the **first law of thermodynamics.*** These laws of thermodynamics have no known exceptions and are among the most general scientific principles known.

There are three different ways of stating the second law. In its most straightforward form, it is a familiar observation about thermal energy flow (Section 7.1). Like many everyday observations, this one has profound consequences. Section 7.2 discusses one of these, namely, another form of the second law that highlights the special nature of thermal energy. Unlike other energy forms, there is a restriction on the transformations of thermal energy: It can be transformed into other forms only with limited efficiency. This leads to discussion of a device that is central to society's use of energy: the heat engine (Sections 7.2 and 7.3). Section 7.4 presents yet another way of stating the second law, known as the law of increasing entropy. This law provides insight into some fundamental philosophical and cultural issues, including the direction of time, the ultimate fate of the universe, and why there is a second law of thermodynamics. Sections 7.5 and 7.6 apply the laws of thermodynamics to our continuing study of the physics and social implications of a significant heat engine: the automobile. Section 7.7 looks at another significant heat engine: the steam-electric power plant. Because these topics bring up the question of the use and depletion of energy resources, it is

---

*Textbooks usually state the first law of thermodynamics in a form that looks different from the law of conservation of energy. Nevertheless, the two ideas are logically equivalent to each other.

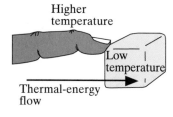

FIGURE 7.1
When you touch a piece of ice, thermal energy flows from your higher-temperature hand to the lower-temperature ice, so the ice gets warmer and your finger gets colder.

FIGURE 7.2
When you touch a hot cup of coffee, thermal energy flows from the high-temperature cup to your lower-temperature hand, so your finger gets warmer and the coffee cools.

natural at this point to discuss exponential growth and its implications for resource depletion (Section 7.8).

7.1 HEATING

Touch a piece of ice (Figure 7.1). Energywise, what happens? Your hand gets cooler, and the ice begins to melt. So thermal energy must have flowed from your hand to the ice. Now touch a hot cup of coffee (Figure 7.2). Your hand gets warmer while the coffee cools, so thermal energy must have flowed from the cup to your hand. Notice that in each case, thermal energy flowed from the high-temperature object to the low-temperature object. There is a general principle operating here, a principle that you experience whenever you touch an object that feels hot or cold: Thermal energy flows spontaneously from hot to cold. Any such flow of thermal energy from a high to a low temperature is called **heating**.

Heating has a one-way nature: It flows "downhill" from a high to a low temperature. It won't flow spontaneously (without external help) "uphill" from a low to a high temperature. Like a lot of simple ideas, this one-wayness of heating has profound consequences. It is one way of stating the second law of thermodynamics:

> **THE SECOND LAW OF THERMODYNAMICS, STATED AS THE LAW OF HEATING**
> Thermal energy flows spontaneously (without external assistance) from a higher-temperature object to a lower-temperature object. It will not spontaneously flow the other way.

Now we need to be more quantitative about temperature. **Temperature** is a quantitative measure of warmth. We can measure temperature by using any material that expands as it warms and contracts as it cools. Most materials do this. Such an expansion and contraction can be used as the basis for a temperature-measuring device, or **thermometer**. One choice is liquid mercury, which expands inside a glass tube. The standard (metric) temperature unit is the **degree Celsius** (°C). The Celsius scale assigns 0°C and 100°C to the freezing and boiling points of water. The United States still uses the Fahrenheit scale, related to the Celsius scale as shown in Figure 7.3.

Temperature and thermal energy are related but different. For instance, a cool lake contains much more thermal energy than does a hot cup of coffee, even though the lake has a lower temperature, because the lake is so much larger.

■ **DIALOGUE 1**   To develop a feel for Celsius temperatures, answer the following questions in °C. What is the approximate temperature of a nice day? Of a hot day? What temperatures are below freezing? Above boiling?

---

1. 20–25°C, 35–40°C, below 0°C, above 100°C.

°C   °F

FIGURE 7.3
The Celsius and Fahrenheit scales, compared.

7.2 HEAT ENGINES: USING THERMAL ENERGY TO DO WORK

Drop a book on the floor. Slide it across the table. Smack it with your hand. Imagine tearing out a page and burning it up. (Books come in for a certain amount of rough treatment in physics courses.) Consider the energy transformations during these processes: Thermal energy is created during each one! It's hard to think of a physical process that doesn't create at least a little thermal energy. It is easy—almost inevitable—to create thermal energy.

But processes that consume thermal energy as the input and then convert it to other forms are less common. Can you think of any?

—Time out, for thinking.

One example is the automobile engine, which uses thermal energy to do work. Any device that uses thermal energy to do work is called a **heat engine**.

One significant feature of an automobile engine is that, in addition to doing work, it ejects a lot of thermal energy. The automobile's radiator ejects excess thermal energy, and its tailpipe exhausts residual hot gases. So *not all the thermal energy input is actually used to do work*. This turns out to be a general feature of every heat engine. A heat engine's thermal energy output is called its **exhaust**.

So the energy transformation for any heat engine is

ThermE (input) → Work (which could then produce any form of energy)
+ ThermE (exhaust)

See Figure 7.4.

Because of the social significance of heat engines, we should quantify their performance. Recall from Chapter 6 that the energy efficiency of any device is its useful energy output divided by the total energy input, often expressed as a percentage. Thus the **energy efficiency** of any heat engine is its work output divided by the thermal energy put into it:

$$\text{energy efficiency} = \frac{\text{work output}}{\text{thermal energy input}}$$

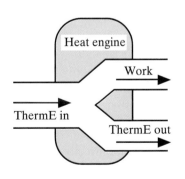

FIGURE 7.4
Energy flow for a heat engine. Heat engines consume thermal energy and turn part of it into work, which could then produce any form of energy.

The notion that heat engines always have an exhaust can then be stated very simply: Heat engines are always less than 100% efficient.

This special feature of thermal energy—that you can entirely create it but you can't entirely consume it—has been found to be true every time we have checked. It is, in other words, a fundamental principle of nature. But as we will see, it turns out not to be another *new* principle of nature. It has a one-way quality about it that is reminiscent of the law of heating. Perhaps it is not surprising then that it turns out to be the second law of thermodynamics, only put into words that are different from our previous statement. We call it:

Many times I have been present at gatherings of people who . . . are thought highly educated and who have with considerable gusto been expressing their incredulity at the illiteracy of scientists. Once or twice I have been provoked and have asked the company how many of them could describe the Second Law of Thermodynamics. The response was cold: it was also negative. Yet I was asking something which is about the scientific equivalent of: "Have you read a work of Shakespeare's?"

C. P. SNOW, BRITISH
SCIENTIST AND AUTHOR

THE SECOND LAW OF THERMODYNAMICS, STATED AS THE LAW OF HEAT ENGINES

Any process that uses thermal energy as the input to do work must also have a thermal energy output or exhaust. In other words, heat engines are always less than 100% efficient.

H O W   D O   W E   K N O W ?

Why do we believe the law of heat engines? Rather than appealing directly to experiment, we offer a theoretical argument based on the law of heating, which is in turn based on experiment. It is a type of argument that logicians call an argument by contradiction or a reduction to an absurdity.

Let's temporarily suppose that (in violation of the law of heat engines) there is a heat engine that can convert thermal energy entirely to work. We could then use that heat engine to extract thermal energy from, say, a pot of warm water and convert this energy entirely to work. This work could produce thermal energy (by friction, for example) in some hotter system, such as a pot of hotter water. The net result would be to transfer thermal energy "uphill," from a lower to a higher temperature, without any other change taking place. But this is absurd—it's exactly what the law of heating says we cannot do. In other words, any violation of the law of heat engines implies that the law of heating is also violated. But we know directly from many experimental observations that the law of heating is not violated. So it follows that the law of heat engines cannot be violated either.

FIGURE 7.5
Heat engines use a portion of the thermal energy that flows naturally from a high to a low temperature and convert it to work.

Following the above argument, it should come as no suprise that the flow of thermal energy from a high to a low temperature is essential to heat engines. In fact, a heat engine may be described as a device that uses the natural hot-to-cold flow of thermal energy, by shunting aside some of the flowing thermal energy to do work (Figure 7.5). Keep in mind that Figure 7.5 is only a graphical way of showing the energy flow and is not a picture of an actual heat engine.

Since heat engines are driven by the flow of thermal energy from hot to cold, you must have a difference in temperatures before you can have a heat engine. You cannot turn thermal energy into work if you have only a single temperature to work with. The ocean, for example, contains a lot of thermal energy, but you cannot use it to do work unless you have a colder system for the ocean's thermal energy to flow into.* Heat engines always operate between two systems with different temperatures, as Figure 7.5 shows.

How efficient can the very best heat engine be? It's an important question, because most of the world's primary energy (energy that comes directly from nature) passes through heat engines. In the United States, fully 60% of the primary energy consumed each year passes through either a transportation vehicle or a steam-electric power plant. Both of these devices are heat engines, subject to the second law.

---

*This means that temperature differences between different depths of ocean water could be used to run a heat engine.

Since a heat engine operates because of the ability of thermal energy to flow from hot to cold, we expect its efficiency to be influenced by both its hot **input temperature** (the temperature at which thermal energy is put into the engine) and its cooler **exhaust temperature**. Since temperature differences drive heat engines, we expect a higher efficiency for larger temperature differences between the input and the exhaust. Nineteenth-century physicists found a quantitative formula that predicts the best or ideal efficiency of a heat engine operating at any predetermined input and exhaust temperatures.* As examples, Table 7.1 lists the predicted ideal efficiencies for several specific types of heat engines, along with the actual efficiencies obtained by these heat engines in practice.

Table 7.1 shows how important the second law is to society, as even the ideal efficiencies of these heat engines are 60% or less. Friction, air resistance, and other nonideal processes reduce this even further. Less than half the natural energy resources used in these heat engines are employed to do work. This doesn't necessarily mean, however, that all of the remainder is wasted, because the exhausted thermal energy might be warm enough to be used directly for heating. Indeed, many countries' steam-electric power plants make use of such **cogeneration** of both electricity and useful thermal energy.

Table 7.1 shows the importance of "burning hot" and "exhausting cool." For example, the fossil, nuclear, and solar generating plants have progressively lower input temperatures, but they all exhaust to cooling water that is at about 40°C. As you can see, the efficiency declines as the difference between T_{in} and T_{ex} declines.

Ocean-thermal generation of electric power is a plan to use some of the ocean's thermal energy by exploiting temperature differences at different ocean depths. In the tropics, the ocean's temperature drops rapidly from 25°C at the surface to 5°C at 300 m beneath the surface. This small temperature difference could be used to run a heat engine with an efficiency of less than 7%. Because

TABLE 7.1

Heat engine efficiencies.   Typical input and exhaust temperatures, ideal (best possible) efficiencies, and actual efficiencies.

			Efficiency (%)	
Engine type	**T_{in} (°C)**	**T_{ex} (°C)**	**Ideal**	**Actual**
Transportation				
Gasoline automobile/truck	400	25	55	10–15
Diesel auto/truck/locomotive	500	25	60	15–20
Steam locomotive	180	100	18	10
Steam-electric power plants				
Fossil fuel	550	40	60	40
Nuclear powered	350	40	50	35
Solar powered	225	40	38	30
Ocean-thermal	25	5	7	???

*The formula is efficiency = $(T_{in} - T_{ex})/T_{in}$. In this formula, the temperatures must be measured in degrees *Kelvin* (K), a new temperature scale. The temperature in K is found by adding 273 to the temperature in °C. 0 K (or –273°C) is known as absolute zero, because it is the lowest possible temperature—the temperature at which all microscopic motion is at its absolute minimum.

the primary energy resource would be solar energy falling on the ocean, the low efficiency would be of little concern. But because the intake point for cool water must be 300 m below the ocean's surface, the engineering problems would be formidable.

■ **DIALOGUE 2**   (a) Assume that the energy flows in Figure 7.5 are proportional to the width of each pipe. Estimate this engine's efficiency. (b) Suppose this engine consumes 400 J of thermal energy and converts 300 J of this to exhaust. Find the engine's efficiency.

■ **DIALOGUE 3**   A typical large coal-fired electric generating plant burns about 1 tonne (1000 kg) of coal every 10 seconds. According to Table 7.1, how much of the tonne actually goes into producing electric energy? How much goes into exhaust?

■ **DIALOGUE 4**   A coal-burning steam locomotive heats steam to 180°C and exhausts it at 100°C. During 1 s of operation, it consumes 500 million J of energy from the burning coal. According to Table 7.1, how much work can be obtained from this locomotive during 1 s of operation under ideal conditions (no friction or other imperfections)? How much work can be obtained under actual conditions?

7.3 ENERGY QUALITY: THINGS RUN DOWN

The second law tells us that thermal energy is less useful than are other types of energy, because unlike the other forms, not all its energy can be used to do work. In this sense, thermal energy is lower-quality energy. A moving bullet or a raised rock can convert its kinetic or gravitational energy entirely to work, but thermal energy can be only partially used to do work. So whenever you transform other energy forms into thermal energy—say by friction or combustion— you reduce the energy's quality.

So there is an irreversibility about any process that creates thermal energy. Once a system creates thermal energy, that system will never by itself (spontaneously) be able to return to its previous condition. To return, it would have to convert spontaneously all the created thermal energy to its original form, and the second law prohibits this. The system can return to its initial state only with outside help.

A good example is a rock swinging back and forth on a string tied to a hook (Figure 7.6). Air resistance and friction (between the string and the hook) gradually bring the rock to rest. Although the complete system (rock, string, hook, and surrounding air) loses no energy, it can't return to its initial condition because thermal energy is created, and this cannot be entirely reconverted to kinetic or gravitational energy. Something is permanently lost when systems run down like this, but it cannot be energy because energy is conserved. Instead, en-

FIGURE 7.6
A rock swinging on a string tied to some fixed point overhead. The rock "dies down," illustrating the irreversibility of natural processes implied by the second law.

---

2. (a) Something like 30%, judging from the widths of the "pipes." (b) Work = 400 − 300 = 100 J. Efficiency = 100/400 = 0.25 = 25%.

3. 40% of 1000 kg is 400 kg; the remaining 600 kg goes into exhaust.

4. Ideal efficiency = 0.18, $0.18 \times 500 \times 10^6$ J = 90×10^6 J. Actual efficiency = 0.10, $0.10 \times 500 \times 10^6$ J = 50×10^6 J.

ergy *quality* is lost. The system's energy gradually decreases in quality until all of it has turned into thermal energy at a temperature only slightly warmer than the surroundings. Now the original energy is almost entirely useless.

When we use Earth's energy resources, we don't use up energy. Instead, we degrade energy from highly useful forms such as the chemical energy of oil to less useful forms such as thermal energy.

One of the two great laws of energy says that the quantity of energy is conserved, and the other says that the quality of energy runs down. You can't get ahead, and you can't even break even.

7.4 THE LAW OF ENTROPY: WHY YOU CAN'T BREAK EVEN

You have seen the second law stated in two different ways. Now we'll look at yet a third way, which puts it in terms of nature's microscopic behavior.

Suppose that a box of hot gas and a box of cold gas are put into contact. The second law says that the hot box will heat the cold one and will continue heating it until there is no longer a temperature difference between the boxes.

Figure 7.7 shows this process from the microscopic point of view. Microscopically, the difference between the hot gas and the cold gas is that the hot box's molecules are moving at higher speeds on the average (note that they aren't all moving at the *same* speed—they have various speeds), while the cold box's molecules are moving at lower speeds on the average; Figure 7.7a shows just a few of the molecules in the hot gas on the left and in the cold gas on the right. The exchange of thermal energy from the hot gas to the cold gas causes the molecules of the hot gas to slow down and the molecules of the cold gas to speed up (Figure 7.7b) until both gases come to some intermediate temperature. Once the two boxes have reached the same temperature, we no longer have the separation of speeds with which we started. We can no longer say, with any

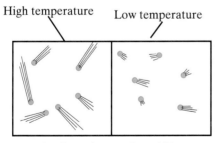

(a) *A hot box of gas and a cold box of gas, at the instant they are put into contact: Most of the molecules in the hot box move rapidly, while most of the molecules in the cold box move slowly.*

(b) *After there has been time for the hot box to warm up the cold box, the molecules in the left-hand box have slowed down while those in the right-hand box have sped up.*

FIGURE 7.7
Microscopic view showing just a few of the molecules in a box of hot gas and a box of cold gas that are put into contact. The final situation shows greater disorganization than the initial situation, because the faster molecules are no longer separated from the slower molecules.

confidence, that we can find a fast molecule by reaching into the left-hand box or a slow molecule by reaching into the right-hand box. Instead, the molecules in both boxes have come to an intermediate average speed, with slower and faster molecules mixed together in both boxes.* From the microscopic point of view, the system is less organized. *Microscopic disorganization has increased.*

This turns out to be the general situation, no matter whether the materials are gases or anything else. When thermal energy flows from hot to cold, microscopic disorganization always increases. In fact, this idea—that microscopic disorganization must increase—is equivalent to the fact that thermal energy flows from a high to a low temperature. In other words, this idea is another way of stating the second law.

Physicists have found a quantitative measure of the disorganization at the microscopic level of any system. It is called **entropy**. We don't need to delve into the precise definition of entropy here. Suffice it to say that it can be specified by measurements of temperature and thermal energy.

So we have yet a third way of stating the second law:

> ## THE SECOND LAW OF THERMODYNAMICS, STATED AS THE LAW OF INCREASING ENTROPY
> The total entropy of all the participants in any physical process must either increase or remain unchanged; it cannot decrease.

The law of increasing entropy is similar to the law of conservation of energy. Both place restrictions on natural processes: The total energy of all the participants in any process must remain unchanged, and the total entropy must not decrease.

But unlike the law of energy, the law of entropy predicts that most processes are **irreversible**—they cannot proceed in the opposite direction. Processes must go in the direction of increasing, not decreasing, entropy. In fact, except for a very subtle effect at the subatomic level,[†] the second law is the only principle of physics that distinguishes between a forward and a backward direction of time. If it weren't for the second law, everything could just as well run backward. For example, a book resting on a table could spontaneously leap into the air by converting some its thermal energy into kinetic and gravitational energy. This is the reverse of dropping a book onto a table. It might appear to violate such principles as Newton's law of motion, but if viewed at the microscopic level there is no violation: It is possible, although highly improbable, for the randomly moving molecules in the book to all just happen to be moving upward at the same instant, with sufficient speed to cause the book to leap from the table as though

For some reason, the universe at one time had a very low entropy for its energy content, and since then the entropy has increased. So that is the way toward the future. That is the origin of all irreversibility, that is what makes the processes of growth and decay, that makes us remember the past and not the future, remember the things which are closer to that moment in the history of the universe when the order was higher than now, and why we are not able to remember things where the disorder is higher than now, which we call the future.

RICHARD FEYNMAN, IN
*THE FEYNMAN LECTURES
ON PHYSICS*

---

*If the boxes contain different kinds of gases, O_2 in one box and N_2 in the other, for example, then the average kinetic energies, rather than the average speeds, of the individual molecules of the two gases become equal when a common intermediate temperature is attained.

†The effect involves the decay (or disintegration) of a particle known as the K-meson. Physicists James Cronin and Val Fitch discovered in 1964 that K-mesons distinguish between forward and backward in time. It is not known whether this subatomic physics discovery is in any way related to the second law or to our sense of a forward direction in time, although it does appear to be responsible for the fact that the big bang origin of the universe created mostly matter rather than antimatter. One member of the 1980 committee that awarded Cronin and Fitch their Nobel Prize remarked, "It would take a new Einstein to say what it means."

the table had thrown the book upward. Similarly, water could run uphill. Thermal energy could flow from cold to hot. And people could grow younger instead of older. Perhaps you have seen a movie run backwards. Despite the odd appearance of the events in such a movie, the only law of physics that would be violated if these events actually occurred in real time is the second law of thermodynamics!

The law of entropy shows us why there is a limit on the possible efficiency of even the most perfectly constructed heat engine. When a heat engine uses some of its thermal energy to do work, it turns disorganized energy into organized energy. If there were no exhaust, this use of thermal energy to do work would violate the law of increasing entropy. The exhaust prevents this violation, because we saw in Figure 7.7 that disorganization increases when thermal energy flows from hot to cold. In a heat engine, the increase in disorganization caused by the thermal energy flow from hot to cold must at least balance the increase in *organization* caused by the conversion of some of the thermal energy into work. Overall, there must be a net increase (or, in the case of a perfectly constructed heat engine, zero increase) in disorganization. So there must be an exhaust.

The law of increasing entropy suggests a deeper reason behind the second law. Increased disorganization is common in everyday life. For example, suppose that you have a partially organized deck of cards, perhaps with all the spades collected together. If you now shuffle the deck, you will almost certainly disorganize the deck further. It is possible, but improbable, that the shuffling will further organize the cards, perhaps by leaving the spades together and, luckily, collecting the hearts together also. But the deck of cards is much more likely to evolve toward a more disorganized state, simply because there are so many more ways to disorganize the deck than there are ways to organize it. As you know whenever you clean your house, it is easy to disorganize things, but it takes effort—or a lot of luck—to organize them.

Thus the second law arises for simple *statistical* reasons. Like a deck of cards, molecular systems are much more likely to evolve toward greater disorganization than toward greater organization, simply because there are many more ways for a molecular system to become disorganized than to become organized. It is highly likely (in fact, overwhelmingly likely) that such systems will become more disorganized.

The second law leads to some interesting speculation about the possible long-term fate of the universe. Unless there are other universes of which we are unaware, our universe would appear to be an isolated system—a system to which one can apply the laws of thermodynamics. If one applies the second law to the universe as a whole, one concludes that the natural evolution of the universe must be toward states of greater and greater disorganization. The long-term fate of the universe would then have to be a state of maximum disorganization, in which no further macroscopic developments would be possible (more precisely, such developments would be highly unlikely).

Such a maximum entropy state would be pretty boring. Among other things, all the stars would have burned out and no new ones could form because all possible nuclear reactions would have run their course. Life could not exist because of the lack of sunlike stars. A more fundamental reason why life could not exist is that living organisms represent a highly organized state of matter—just the kind of low-entropy state that could not exist once the universe had evolved to its maximum-entropy state. This long-term fate of the universe has been termed

*Humpty Dumpty sat on a wall.
Humpty Dumpty had a great fall.
All the king's horses,
And all the king's men,
Couldn't put Humpty together
  again.*
MOTHER GOOSE RHYMES

It's in the cards.
OLD SAYING

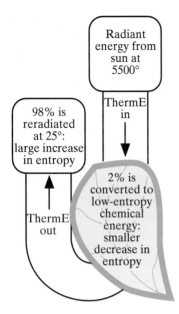

FIGURE 7.8
Energy flow through a leaf. The leaf is similar to a heat engine. A growing leaf illustrates how Earth has become more organized, despite the universe's trend toward increased entropy.

A living organism . . . feeds upon negative entropy. Thus the device by which an organism maintains itself at a fairly high level of orderliness (= fairly low level of entropy) really consists in continually sucking orderliness from its environment.
    ERWIN SCHROEDINGER, PHYSICIST, IN *WHAT IS LIFE?*

Until evolutionists can not only speculate, but demonstrate, that there does exist in nature some vast program to direct the growth toward higher complexity of the [biosphere] . . . the whole evolutionary idea is negated by the Second Law. . . . There seems to be no way of modifying the basic evolutionary model to accommodate this Second Law.
    HENRY M. MORRIS, DIRECTOR OF THE INSTITUTE FOR CREATION SCIENCE, IN *SCIENTIFIC CREATIONISM*, CREATION-LIFE PUBLISHERS, SAN DIEGO, 1974

the *heat death* of the universe. It's not exactly the world's most pressing issue, since it is hundreds of billions of years in the future, but it raises interesting and perhaps troubling philosophical questions. I hasten to add that there is a large speculative element in the notion of a universal heat death, because we are not necessarily justified in applying the laws of thermodynamics to the entire universe or to very long times. It is always risky to assume that the principles of physics as we now understand them are so precisely correct that they can be applied to the universe as a whole, or to all future time.

Biological systems provide interesting examples. For example, a growing leaf manufactures complex glucose molecules out of simple CO_2 and H_2O molecules. Glucose is highly organized when compared with the randomly moving CO_2 and H_2O that goes into its formation. The leaf must create this organization. How does it manage to produce this decrease in entropy, in apparent violation of the second law?

The answer is that the leaf had help. The second law says that the total entropy of all the participants in any process cannot decrease. In the growth of a leaf, the other vital participant is the sun. Solar radiation has a temperature, the 5500°C surface temperature of the sun. When this radiation is absorbed by a leaf, only about 2% of the energy is converted to chemical energy. The remaining solar energy is reradiated out into space, at the 25°C temperature of the leaf. So most of the solar energy flows from 5500°C to 25°C, and the large entropy increase of this thermal energy flow allows the remaining solar energy to be organized into low-entropy chemical energy (Figure 7.8). Solar radiation organizes the leaf despite (but not in violation of) the second law. Solar radiation is both the energizer and the organizer of all life on Earth.

Creationism, the belief that the various biological species (and especially humans) originated from specific acts of instantaneous creation rather than by the natural process of biological evolution, argues that biological evolution is impossible because it violates the second law. It is an idea that seems plausible at first glance. After all, it does seem paradoxical that life on Earth could have evolved on its own from simple single-celled structures into the much more organized plants and animals of today. It seems overly lucky, and it seems to violate the second law's demand that disorganization should always increase.

But like a leaf, biological evolution had help from the sun. The increase in entropy that occurs when sunlight is absorbed and reradiated by Earth is much more than adequate to compensate for the decrease that occurs when plants evolve into new organisms. And animals, which do not use solar energy directly, reduce their entropy by eating highly organized food—another form of outside help. Biological evolution is fully consistent with the second law. Over billions of years, Earth has evolved toward greater organization at the cost of a much greater disorganization of the solar radiation that has passed through biological systems.

Your brain is one result of this long evolution toward greater organization. As an information-storage device, the human brain is the most highly organized form of matter on Earth.* It is surely the most organized form of matter within some 4×10^{13} km from Earth—the distance to the nearest star beyond the sun—and perhaps within the entire Milky Way galaxy (see Chapter 12). It is remarkable that, in the human brain, nature has finally created a self-aware collection

---

*Carl Sagan makes this point in *The Dragons of Eden* (Random House, New York, 1977), in which he estimates the information content of the brain's neurons and compares it with computers, with other animal brains, and with the information content of genetic material.

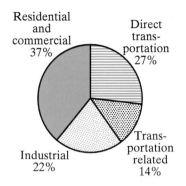

FIGURE 7.9
The fraction of total U.S. primary energy consumed by each economic sector.

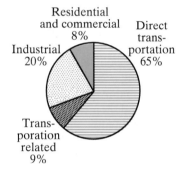

FIGURE 7.10
The fraction of U.S. petroleum consumed by each economic sector.

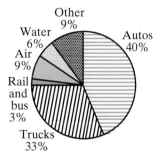

FIGURE 7.11
The fraction of U.S. direct transportation energy consumed by each transportation mode.

of molecules, molecules so well organized that they are capable of knowing that they are a collection of molecules! Nature has spent billions of years of evolution getting to this point. So take good care of yourself and of all of us.

7.5 THE AUTOMOBILE

Few technologies shape our culture as strongly as the automobile. It brings great freedom while affecting our quality of life, family structure, self-perceptions, physical environment, health, work, community structure, resource use, economy, and even issues of war and peace. We cannot go into all these far-flung issues, but as part of our science-and-society theme, we will look at the automobile's role in energy-related matters.

Transportation consumes much of the energy used by the United States (Figure 7.9) and most of the nation's oil (Figure 7.10). Most of this is used by cars and trucks (Figure 7.11).

MAKING ESTIMATES

Calorimetry experiments show that upon combustion, 1 gallon (3.8 liters) of gasoline releases (converts to thermal energy) about 130 million joules of energy. Use this figure to estimate the rate, in watts, at which a typical car consumes chemical energy while moving at highway speeds without acceleration. *Hints*: Typical gasoline consumption is 25 miles/gallon (10 kilometers/liter), and highway speeds are about 50 mi/hr (80 km/hr).

SOLUTION During 1 hour at 50 mi/hr, a car travels 50 miles and so consumes 2 gallons. Since 1 gallon contains 1.3×10^8 J, the chemical energy in 2 gallons is 2.6×10^8 J. This is consumed in 1 hour. Since there are 3600 (3.6×10^3) seconds in 1 hour, the chemical energy consumed per second is

$$2.6 \times 10^8 \text{ J}/3.6 \times 10^3 \text{ s} \approx 0.7 \times 10^5 \text{ J/s} = 70,000 \text{ J/s}$$

or 70,000 watts or 70 kW.

The 70 kW consumed by a typical car is equivalent to the electric power going into 700 100-W bulbs burning the entire time the car is moving. As another comparison, the average rate at which a typical U.S. household consumes electric energy is about 1 kW, so while a car is moving, its energy use is equivalent to that of 70 households! If the car is accelerating, all of these figures should be multiplied by about 5.

Most transportation energy goes into heat engines, where it first burns to produce thermal energy and then is partially transformed into useful work. Cars and trucks are powered by **internal combustion** engines. Heat engines of this type burn a fuel–air mixture. The mixture's high combustion temperature gives it a high pressure, which enables the hot gases to push strongly on a **piston**, a movable metal plate connected to a rod (Figure 7.12). The piston does the work

FIGURE 7.12
The conversion of thermal energy to useful work using a piston.

FIGURE 7.13
General Motors' experimental Sunraycer. It weighs 390 pounds unloaded, carries 68 silzer-zinc storage batteries, and is energized by 8000 photovoltaic cells arrayed over its exterior.

FIGURE 7.14
Toyota's battery-powered RAV4-EV has a range of 200 km (125 mi) on one battery charge, a recharge time of 6.5 hours, and a top speed of 125 km/hr (78 mph).

FIGURE 7.15
The Honda Civic GX is the first fully mass-produced natural gas passenger car. It has a range of 320 km (200 miles). Because natural gas burns cleaner and more efficiently than gasoline, this car produces lower hydrocarbon, carbon dioxide, carbon monoxide, and nitrogen oxide emissions than other internal-combustion cars.

that turns the drive wheels. Combustion is *internal* because the combustion occurs directly inside the gases that do the work, in contrast with **external combustion**, which occurs in a fuel that then provides thermal energy to a second substance, such as steam, that does the actual work.

The most convenient and abundant automobile fuel is *gasoline,* a form of petroleum. But there are many other possibilities. As gasoline's pollution problems loom larger and as petroleum's availability declines, other, less-polluting fuels may become more popular. Table 7.2 (see also Figures 7.13 through 7.16) lists and describes possible automotive fuels.

Figure 7.17 shows the energy transformations in a typical automobile. On the average, 1 kW is lost to evaporation. It isn't really lost, as energy is never lost. But 1 kW of hydrocarbons goes into the atmosphere where its chemical energy essentially cannot be recovered and where it contributes to chemical pollution. The remaining 69 kW go to the engine, which produces about 17 kW of work and exhausts the remaining 52 kW as thermal energy and unused chemical energy.

About half of the exhaust energy is removed by the car's radiator, and the other half goes out through the exhaust pipe as polluting gases. Gasoline is a hydrocarbon, made of hydrogen (H) and carbon (C) atoms. Both the hydrogen and the carbon combust with oxygen from the atmosphere, so the exhaust gases are mostly CO_2 and H_2O. Although neither one is toxic (poisonous to humans), CO_2 contributes to environmentally harmful global warming (Chapter 9).

The tailpipe exhaust carries various other molecules, mainly CO, NO, NO_2, and unburned hydrocarbons. These are toxic pollutants, and their unused chemical energy also represents an energy inefficiency. The carbon monoxide and unburned hydrocarbons are the result of incomplete combustion (partial combustion) of the fuel. The two oxides of nitrogen—collectively called NO_x—form from atmospheric oxygen and nitrogen, which combine under the influence of the engine's high temperatures. Cars and trucks produce about two-thirds of the nation's CO pollution, one-third of its hydrocarbon pollution, and half of its NO_x pollution.

The automobile's main loss of useful energy occurs in the engine, mainly as a consequence of the second law. The engine's best possible, or ideal, efficiency is 30%, and its actual efficiency is 17 kW/69 kW = 25%. Several losses combine to

TABLE 7.2

Fuels for automobiles and trucks

Fuel	Source of fuel	Description
For internal combustion		
Gasoline	petroleum	liquid, widely used
Diesel fuel	petroleum	liquid, widely used
Compressed natural gas	natural gas	high-pressure gas
Liquefied natural gas	natural gas	low-temperature liquid
Liquefied petroleum gas (propane)	petroleum, natural gas	high-pressure liquid
Methanol (wood alcohol)	wood, natural gas, coal	liquid
Ethanol (grain alcohol)	grain, sugar, trash	liquid
Hydrogen[a]	any source of electricity	hydrogen produced from water
For noncombustion		
Storage batteries[a]	any source of electricity	recharge electrically
Hydrogen fuel cell[a]	any source of electricity	hydrogen produced from water

[a]Since electricity for hydrogen production or for storage batteries can come from any source, the ultimate energy resource can be wind, hydroelectric, nuclear, coal, photovoltaics, and the like.

FIGURE 7.16
A high-pressure cylinder in the rear of the Honda Civic GX is the only indication that the car is powered by compressed natural gas. The cylinder is made of high strength, fiber composite material.

bring the engine's efficiency from the maximum 30% allowable by the second law down to 25%: incomplete combustion, the formation of NO_x, friction in the engine, and thermal losses through the engine's wall.

Of the 17 kW of work produced by the engine, 5 kW go to internal devices like the water pump and air conditioner. The remaining 12 kW go to the transmission and drive train that couples the engine to the drive wheels. This coupling is about 75% efficient, so 9 kW finally arrive at the drive wheels.

The 9 kW carry the car down the road. About half of it goes into overcoming air resistance. Air resistance increases with speed and dominates at higher speeds. The other half goes into rolling resistance (Section 4.6), which dominates at lower speeds. The overall energy efficiency of the entire automobile (not just the engine) is 9/70 =13%, or about one-eighth.

7.6 TRANSPORTATION EFFICIENCY

Our definition of energy efficiency—useful energy output divided by total energy input—doesn't really capture the automobile's purpose. Although its purpose is to move people, its energy goes mostly into moving the car itself rather than people. Gasoline mileage, the common measure of automobile efficiency, suffers from the same defect. Neither of these measures really captures people-moving efficiency, in other words, transportation efficiency. On the other hand, gasoline mileage is useful for comparing different cars with one another (Table 7.3, Figure 7.18).

The efficiency of using energy to move people can be directly measured by using as the output the number of *passenger-kilometers* delivered. For example, if a bus moves 20 passengers a distance of 3 km, it has delivered 20 passengers × 3 km = 60 passenger-km. Similarly, the efficiency of using energy to move goods can be directly measured by using as the useful output the number of *tonne-kilometers* delivered. For example, if a truck moves 5 tonnes a distance of 80 km, it has delivered 5 tonnes × 80 km = 400 tonne-km.

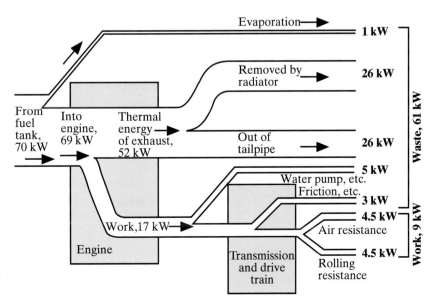

FIGURE 7.17
Typical energy flow in an unaccelerated car at highway speed.

FIGURE 7.18
General Motors' Ultralite experimental automobile has a fuel efficiency of 40 km/l (100 mi/gal).

TABLE 7.3

Fuel efficiencies of four-passenger automobiles

	mi/gal	km/l
Averages, all cars		
U.S. fleet average	20	8
World fleet average	25	10
Averages, new cars		
U.S.	26	11
European, Japanese	30	13
Cars produced today		
Ford Escort	53	22
Honda Civic	57	24
Suzuki Sprint	57	24
Prototypes		
Volvo LCP 2000	71	30
Volkswagen E80	85	36
Toyota AXV	98	41
General Motors Ultralite	100	42
Renault VESTA	124	52

SOURCE: D. Bleviss, *The New Oil Crisis* (New York: Quorum Books, 1988).

Table 7.4 compares the passenger-moving and freight-moving efficiencies of several transportation modes. Walking and bicycling come out far ahead because no energy is put into moving a heavy vehicle and because no heat engine is employed; thus there is no loss due to the second law of thermodynamics. Bicycling is more efficient than walking because wheels keep rolling—wheels take advantage of the law of inertia. With each step, walking requires you to start and stop your legs—an acceleration, requiring a force, requiring you to do work. Trains are far more efficient than other vehicles because trains can more efficiently overcome both of the resistive forces acting on rolling vehicles (compare Figure 7.17). Because a train presents a small frontal area relative to its large load, its air resistance per kilogram of load is far below that of cars and trucks. And because a train rolls on steel wheels on steel tracks, there is little of the tire-squeezing that causes rolling resistance (see Section 4.6).

TABLE 7.4

Passenger-moving and freight-moving efficiencies of several transportation modes, per gallon of gasoline or of gasoline's energy equivalent in food or other fuels

Passenger-moving efficiency	Pass-mi per gal	Pass-km per l	Pass-km per MJ
Auto (average occupancy =1.2)	24	10	0.29
Carpool auto (occupancy =4)	80	32	1.0
Commercial airline	50	21	0.62
Transit bus	50	21	0.62
Intercity rail	250	100	3.0
Person walking	400	170	5.0
Person on bicycle	1200	500	15.0

Freight-moving efficiency	Ton-mi per gal	Tonne-km per l
Airplane	13	5
Truck	65	25
Rail	260	100
Waterway	290	110

SOURCE: *Transportation Energy Data Book*, Oak Ridge National Laboratory.

TABLE 7.5

Mass-moving efficiencies of animals and machines, in gram-kilometers per joule

	g-km/J
Human on bicycle	1.5
Salmon	0.6
Horse	0.5
Jet transport	0.4
Human walking	0.3
Automobile	0.3
Typical bird	0.2
Hummingbird	.05
Fly, bee	.02
Jet fighter	.01
Mouse	.005

SOURCE: S. Wilson, "Bicycle Technology," *Scientific American*, March 1973.

For fun, let's look at transportation efficiencies throughout the animal kingdom. Which animal or machine is most efficient at moving itself and any passengers that might be along? Fruit flies? Horses? Jet planes? To compare fruit flies and horses fairly, we must incorporate the fact that the horse's energy goes into moving a much larger mass. So the useful output should be measured as total body (or vehicle) mass times distance moved. Table 7.5 gives several such mass-moving efficiencies, in gram-kilometers per joule of energy. For vehicles such as the bicycle and the automobile, the total mass of the occupant(s) plus the vehicle is included as the output. Again, bicycling comes out far ahead, because animals don't have wheels, and among the vehicles used for transportation, the bicycle is the only one that is not a heat engine.

Animals with wheels would have a big energy advantage. They have not evolved on Earth's rough surface, but tumbleweeds take advantage of rolling in order to spread their seeds. One could speculate that on another planet with surfaces created by smooth lava flows, wheeled animals might evolve and be abundant!

■ **DIALOGUE 5** You wish to move 130 tons of freight a distance of 100 miles. About how many gallons of gasoline (or gasoline equivalent) will you need to move it by truck? By air? By rail?

■ **DIALOGUE 6 MAKING ESTIMATES** About how much gasoline would a 55-mph speed limit save every year, for an average American car and for all American cars? Use the following data: There are some 150 million cars in the United States; a typical car travels 15,000 miles in a year, about half of this being highway driving; American cars average 20 miles per gallon; and fuel mileage is about 15% worse at 65 mph (the approximate average speed without a 55 mph limit) than it is at 55 mph.

7.7 THE STEAM-ELECTRIC POWER PLANT

We turn now from the automobile to society's other major heat engine, the steam-electric power plant.

Figure 7.19 is a schematic diagram of the operation of a typical coal-burning electric power plant. Coal is the most widely used primary energy source for electricity. Other plants that use oil, natural gas, nuclear energy, or solar energy to turn water into steam in a boiler operate much as a coal-burning plant does in producing energy from steam.

The coal combusts externally in a furnace, and its thermal energy is transferred to water inside a boiler. Most combustion products escape through the

---

5. 130 tons × 100 miles = 13,000 ton-mi. Trucks use 1 gallon of gasoline for each 65 ton-mi (Table 7.4), so 13,000 ton-mi requires (13,000 ton-mi)/(65 ton-mi per gal) = 200 gallons. For the other two questions, it's easiest to use proportions: Table 7.4 shows that air transport is one-fifth as efficient as truck transport, and so it requires five times more gasoline, or 1000 gallons. Rail transport is four times more efficient than truck transport, and so it requires one-fourth as much gasoline, or 50 gallons.

6. An individual car drives 7500 highway miles in 1 year. Compare driving this at 55 mph (getting about 20 miles per gallon) with driving it at 65 miles per hour (where fuel efficiency is 15% worse, or only 17 miles per gallon): 7500/20 = 375 gallons, 7500/17 = 441 gallons, so an extra 441 − 375 = 66 gallons is needed at the higher speed. For the entire U.S. auto fleet, $150 \times 10^6 \times 66$ gallons = 10×10^9 gallons. This figure represents some 12% of U.S. oil imports.

stack but some pollutants are removed first. The boiler produces high-pressure steam at over 500°C, far above the normal boiling temperature. The steam moves through pipes to a large rotating device called a **steam turbine** that is designed to turn when it feels a higher pressure on the front (upstream) side than on the back (downstream) side. Like the piston in a car, the steam turbine is the key device that transforms thermal energy into work. The turbine turns an **electric generator**, a device containing wires and magnets that creates electricity when the turbine causes it to rotate.

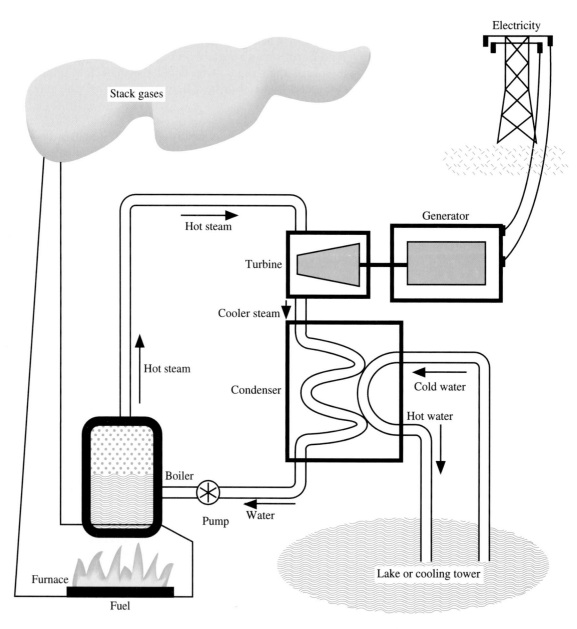

FIGURE 7.19
A schematic diagram showing the operation of a coal-fueled steam-electric generating plant.

The rotating turbine extracts some of the hot steam's thermal energy and converts it to work. The second law tells us that this is possible only if the remaining thermal energy flows to a cooler temperature. To maintain the required temperature difference, the exhaust side of the turbine is cooled by an external stream or lake, or by evaporative cooling in the atmosphere.

To obtain the greatest efficiency, the exhaust is cooled sufficiently to turn the steam back into liquid water, because this greatly reduces the pressure against the back side of the turbine.* The steam is then sucked forcefully through the turbine, from very high pressure on one side to a near vacuum on the other. Because the transition from steam to water is called *condensation*, the cooling device is called a *condenser*. Once condensed, pumps move the water back around to the boiler, where the cycle begins again.

As you can see by inspecting Figure 7.19, the plant is a heat engine. It converts thermal energy to work. Thermal energy flows in at the boiler and out at the condenser, and work is done by the turbine (compare Figure 7.5).

Figure 7.20 shows the energy flow. A large plant generates about 1000 MW of electric power, enough for a large city. Because a typical plant has an efficiency of about 40% (Table 7.1), this electrical output requires about 2500 MW, or 2500 million joules in every second, of energy input from coal or another fuel. Calorimetry experiments show that the combustion of 1 kilogram of coal produces about 25 million joules of thermal energy, so the plant requires 100 kilograms of coal every second.

Of the 2500 MW input, 300 MW go out through the stack, accompanied by gases, such as sulfur oxides (which cause acid rain) and CO_2 (which causes global warming), and small incombustible solid particles called *ash*. Modern plants remove about 90% of the sulfur oxides and 99% of the ash, which then presents a significant solid-waste disposal problem. The turbine converts the thermal en-

FIGURE 7.20
Energy flow in a typical 1000 MW coal-fueled electric generating plant.

---

*In practice, not all of the steam is condensed; some of it is removed from the turbine's exhaust and reused without condensation to heat the water on its way to the boiler. Steam also is reused by being reheated and passed through lower-pressure turbines several times before finally going to the condenser.

FIGURE 7.21
Cooling towers and stacks at a coal-fueled generating plant. Cool air is sucked into the bottom of the cooling towers, where it cools hot water from the plant. The tower's shape promotes a rapidly rising hot-air column.

The world's most important arithmetic is the arithmetic of the exponential function.
ALBERT BARTLETT, PHYSICIST

ergy of steam to useful work at a rate of 1000 MW of power, which is in turn used to drive a generator that creates 1000 megawatts of electric power. The plant's biggest loss in useful energy, the 1200 MW exhausted, is primarily an unavoidable consequence of the second law. This exhaust goes into the condenser's cooling water. If the cooling water comes from a lake or river, the exhaust warms the water, an effect called *thermal pollution*. Many plants use the atmosphere as the coolant by employing large evaporative coolers known as *cooling towers* (Figure 7.21). Finally, some 100 MW of the generated electricity is transformed into thermal energy during transmission over electric power lines that might extend hundreds of miles, and 900 MW gets to the user.

■ **DIALOGUE 7** How efficient is the boiler? The electric generator?
■ **DIALOGUE 8 MAKING ESTIMATES** (a) Estimate the amount of coal (in tonnes) that this plant uses in one day. How many large highway truckloads of coal is this, at about 50 tonnes per truck? (b) Estimate the amount of carbon dioxide that this plant puts into the atmosphere every day. *Hint*: The oxygen atom's mass is 33% larger than the carbon atom's mass; for your estimate, treat their masses as roughly equal, and assume that coal is 100% carbon.

7.8 RESOURCE USE AND EXPONENTIAL GROWTH

Consideration of the social implications of energy raises many growth-related issues. As our growing numbers and environmental impact begin to affect the entire natural world, it is important that we analyze where all of this growth might be taking us.*

We begin with a familiar example. Suppose you invest $100 at a rate of return or **growth rate** of 10% per year. When will you double your money? You earn $10 during the first year, so you have $110. In the second year you earn $11, one dollar more than you earned the first year. Now you have $121, so you earn $12.10 during the next year. Each year you earn more than you did the previous year, because your *percentage* increase is the same each year.

Figure 7.22 graphs your account and compares it with a second graph that also increases by $10 during the first year but then follows a straight line. The second graph illustrates **linear** (straight line) **growth**, which increases by a fixed amount each year. As you can see, there is a big difference between $10 per year and 10% per year. When a quantity grows by a fixed percentage in each unit of time, its growth is said to be **exponential**.†

---

7. 2200/2500 = 88%. For the generator, 1000 MW of input power (from the turbine) produces 1000 MW of electric power (a little less, really), so the generator is essentially 100% efficient.
8. (a) The number of seconds in a day is 60 s/min × 60 min/hr × 24 hr/day ≈ 10^5 s. One hundred kilograms enter the plant every second, so the amount entering in a day is about $10^5 \times 100 = 10^7$ kg, or 10,000 tonnes. This is 10,000/50 = 200 large truckloads every day! (b) The CO_2 molecule is about three times as massive as the C atom. The mass of CO_2 created is thus three times the mass of coal consumed: 30,000 tonnes!
*This section draws on the work of University of Colorado physicist Albert A. Bartlett.
†It is called *exponential* because the account is worth 100×1.1 after 1 year, 100×1.1^2 after 2 years, 100×1.1^3 after 3 years, and so forth. The number of years is in the *exponent*.

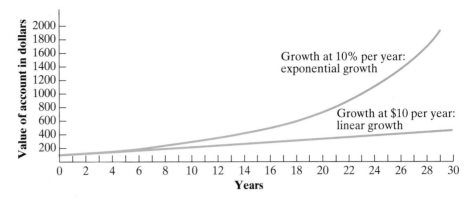

FIGURE 7.22
The 10% investment account. The initial investment is $100. Would you rather have exponential growth at 10% per year or linear growth at $10 per year?

The arithmetic of growth is the forgotten fundamental of the energy crisis.
    ALBERT BARTLETT

If you continue adding on 10% each year, your investment account will reach about $195 after 7 years. In 7 years, you will roughly double your money.

What will happen if you keep your money deposited for another 7 years? Since the same arithmetic applies, it will double again, to nearly $400.* In the next 7 years it will double again, to nearly $800. And so forth. Exponential growth has a fixed, unchanging, **doubling time**.

Pressures resulting from unrestrained population growth put demands on the natural world that can overwhelm any efforts to achieve a sustainable future. If we are to halt the destruction of our environment, we must accept limits to that growth. . . . The United Nations concludes that the eventual total [world population] could reach 14 billion. . . . But, even at this moment, one person in five lives in absolute poverty without enough to eat, and one in ten suffers serious malnutrition. . . . We must stabilize population.

FROM *WORLD SCIENTISTS'
WARNING TO HUMANITY*,
A DECLARATION SIGNED BY
NEARLY 1700 LEADING
SCIENTISTS FROM 71
COUNTRIES, INCLUDING 104
NOBEL LAUREATES—A MAJORITY
OF THE LIVING RECIPIENTS
OF THE NOBEL PRIZE IN THE
SCIENCES. THE DECLARATION
WAS SPONSORED BY THE
UNION OF CONCERNED
SCIENTISTS AND PUBLISHED
IN 1993.

MAKING ESTIMATES

The exponential salary

Suppose you take a job requiring you to work every day for 30 days and your employer offers you just 1¢ for the first day and then a doubled salary every day for the 30 days. About how much will you earn on day 30?

SOLUTION   Starting with 1¢ on day 1, you will earn $5.12 on day 10 (1, 2, 4, 8, 16, 32, 64, 128, 256, 512¢). Poverty wages! To simplify matters, round off the $5.12 to just $5. Now continue doubling. On day 20, your earnings will be $5120. Now you are getting rich. Round this off to $5000, and continue. On day 30 alone, your earnings are more than $5 million!

Exponential growth can be surprising, because our intuitions tend to be linear.

In a population of animals, the number of newborns each year is roughly proportional to the number of potential parents in the population that year. This

*If you aren't convinced of this, think of it in this way: You begin the second 7-year period with about $200. Think of this as *two* $100 accounts. During the second 7-year period, each of these accounts must grow to about $200, so your total is $400.

means that the percentage increase—the number of newborns divided by the total population—should be roughly the same from year to year. At least, this should be true so long as there are no offsetting deaths due to starvation or other consequences of large population size. This unchanging percentage increase means that the growth is exponential.

The world is projected to add at least 960 million people during this decade, up from 840 million in the eighties and 750 million in the seventies.

WORLD POPULATION DATA SHEET, 1990

■ **DIALOGUE 9 POPULATION GROWTH IN A FINITE ENVIRONMENT** Bacteria reproduce themselves by means of division. If you start with 1 bacterium, it will divide into 2; they will divide into 4, then into 8, and so forth. Since each population doubling occurs in the same time interval, it is an exponential process. Suppose that some strain of bacteria has a dividing time of 1 minute. You put 1 bacterium into a bottle at 11 A.M., and at noon you note that the bottle is full of bacteria. When was the bottle half full? If you were one of the bacteria, when might you have noticed that you were running out of space?

When you consume a finite resource exponentially, it is easy to use nearly all of it before you realize there is a problem (Figure 7.23). Continuing with Dialogue 9, suppose that at 11:55 A.M. some visionary bacteria, realizing they have a problem, launch an all-out search for new bottles. By 11:58 A.M., this program has been successful in discovering a huge, new reserve: three new bottles! It took the bacteria an entire hour to fill the first bottle. When will the new bottles be full? The answer is at 2 minutes past noon. Continued exponential growth eventually overwhelms all attempts to expand the resource base.

There is a simple and useful quantitative relation for exponential growth. Any increase in the growth rate must decrease the doubling time, so we might expect to find a relation between these two. It turns out that they are inversely proportional. The relation is, approximately,

$$\text{doubling time} \approx \frac{70}{\text{growth rate}}$$

$$T \approx \frac{70}{P}$$

where T stands for the doubling time and P is the growth rate (the percentage growth per unit time, expressed in percent). This relation can also be turned around to read

$$P \approx \frac{70}{T}$$

Either quantity, the doubling time or the growth rate, is equal to 70 divided by the other quantity. For instance, the 10% savings account has a doubling time of $T = 70/P = 70/10 = 7$ years.

---

9. Work backward from noon: The bottle must have been half full at 1 minute before noon. There is no single answer to the second question, but Figure 7.23, obtained by working backward from noon, gives a rough idea. A few forward-looking bacteria might begin to be concerned when the bottle got to be a few percent full, at a few minutes before noon.

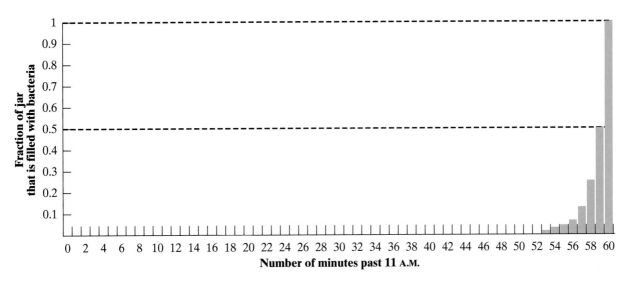

FIGURE 7.23
Bacterial growth in a jar. A finite resource, consumed exponentially, runs out surprisingly rapidly toward the end. On the scale of this graph, growth is imperceptible until 11:53 A.M. when the bottle is 1% full.

For a significant historical example involving energy, we consider the growth of electric power in the United States. As you can see from examination of Figure 7.24, electric power production grew exponentially between 1935 and 1975, as it roughly doubled every 10 years during this period. The percentage growth rate between 1935 and 1975 was then $P \approx 70/T = 70/10 = 7\%$ per year. What would the consequences have been if this growth rate had continued for very many years past 1975? In 1975, all of the electric energy in the United States could have been provided by about 400 large plants. If the 10-year doubling time had continued, 800 plants would have been needed in 1985, 1600 in 1995, 3200 in 2005, and 6400 in 2015. Sixty-four hundred power plants would mean an average of more than one hundred twenty-five in every state, with everybody living within a few miles of a large power plant!

Obviously, expansion at a fixed growth rate cannot be sustained forever. In fact, electric power production increased by 50% during the 15-year period between 1973 and 1988, for an average growth rate of $50/15 \approx 3\%$ per year. Although growth continued, the growth rate declined from 7% to 3%. The dramatic shift in growth rate was caused by the Mideast oil embargo, which raised energy prices, which depressed energy consumption.

U.S. oil production illustrates what can eventually happen when a finite resource is consumed exponentially. Like many industries, oil production grew exponentially during its early years, maintaining a growth rate of 8% per year between 1870 and 1930. But this could not be maintained, because the United States' recoverable oil resources (the oil that is economically feasible to recover) would be gone by now. The growth rate declined, and then around 1970 U.S. oil production in the forty-eight contiguous states began to drop.

Figure 7.25 shows the bell-shaped graph that is typical for the yearly production of a **nonrenewable resource**—a resource that cannot be readily replaced

And in each of these decades (the 1950s and 1960s), more oil was consumed than in all of man's previous history combined.

PRESIDENT JIMMY CARTER, 1977

FIGURE 7.24
The history of electric power production. The annual electric energy produced, in billions of kW·h, is graphed from 1900 through 1990. The growth was roughly exponential between 1935 and 1975.

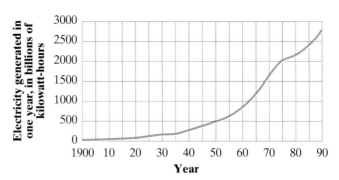

FIGURE 7.24
The history of electric power production. The annual electric energy produced, in billions of kW·h, is graphed from 1900 through 1990. The growth was roughly exponential between 1935 and 1975.

within a reasonable time span, such as a human lifetime. U.S. oil production is following this pattern and is in the early stage of region C on the graph. World oil production is not as far along on this curve and might be in the early stages of region B. In any case, resource depletion is inevitably driving the world toward the end of the oil age, although other problems such as global warming or toxic pollution might end the oil age even sooner.

Renewable resources such as wood or solar energy can follow a different kind of history (Figure 7.26). In their early stages, renewable and nonrenewable resource use follow the same pattern. But for a renewable resource, consumption can be sustained indefinitely, provided that it proceeds at less than the replacement rate.

Finally, let's consider world population growth. The number of humans in 6000 B.C. has been estimated at 30 million—less than the combined populations

FIGURE 7.25
A typical bell-shaped curve showing the life history of consumption of a nonrenewable resource. Exponential growth must slow as the resource is depleted. Consumption eventually levels off and declines as the resource nears exhaustion.

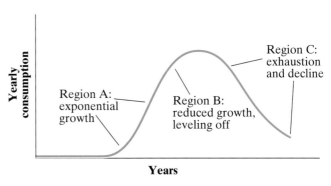

FIGURE 7.26
A typical life history of a renewable resource such as hydroelectric power. Exponential growth slows as consumption reaches its natural limits. Consumption eventually levels off at some sustainable value.

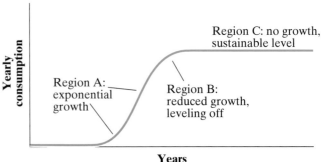

of New York City and Tokyo today. By A.D. 200 the number had reached 200 million. Then it doubled during A.D. 200 to 1300, doubled again during 1300 to 1650, again during 1650 to 1800, and once more during 1800 to 1900. Because the doubling times decreased, the growth was faster than exponential.

Figure 7.27 charts this growth. It took about 4 million years for the human population to grow to its first billion in 1825. It reached its next billion in 1930, its third billion in 1960, its fourth in 1975, and its fifth in 1987. The sixth billion is expected in 2000. We can find the approximate current rate of growth from the population doubling during 1960 to 2000: $P = 70/T = 70/40 = 1.75\%$, or a little under 2%. Although this might not sound like a large growth rate, you can see from the graph that the results over time can be explosive.

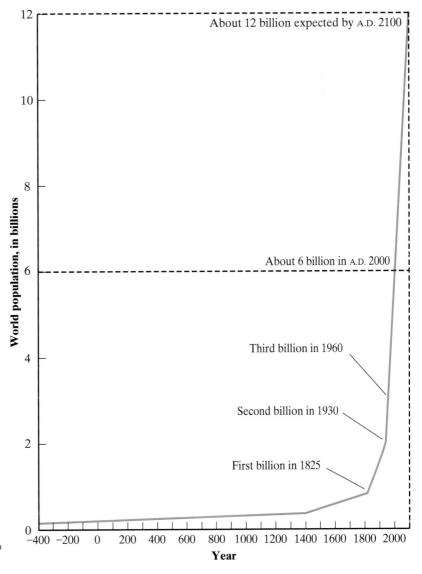

FIGURE 7.27
The population explosion: faster than exponential. Note the resemblance to Figure 7.23.

When you ask the question of an ecologist, "What is the most important environmental issue," they will be unanimous in saying it is exponential population growth.

GAYLORD NELSON, FORMER U.S. SENATOR AND FOUNDER OF EARTH DAY IN 1970

■ **DIALOGUE 10** Between 1870 and 1930, U.S. oil production had an 8% growth rate. What was its doubling time? It reached 1 billion barrels (1 barrel = 42 gallons) per year by 1930. What would the production be this year if it had maintained this growth rate?

■ **DIALOGUE 11** During the 1980s, U.S. car and truck miles traveled increased by 4% per year, but the length of highway increased by only 0.1% per year. Find the doubling time for vehicle miles traveled and for miles of highway. Suppose that these rates are maintained in the future. While vehicle miles double (a 100% increase), by roughly what percentage will the amount of highway increase? Roughly, how much worse will traffic congestion be at that time?

10. $T = 70/P = 70/8 \approx 9$ years. At this rate, production in 1995 alone (65 years after 1930—about seven doubling times) would be 128 billion barrels, nearly the entire amount of recoverable U.S. oil in the ground as of 1992.

11. For vehicle miles, $T = 70/P = 70/4 = 18$-year doubling time. For miles of highway, $T = 70/0.1 = 700$ yrs. During 18 years, the number of miles of highway will increase by about $(0.1\%/\text{yr}) \times (18 \text{ yrs}) = 1.8\%$. Congestion will be about twice as bad.

SUMMARY OF IDEAS AND TERMS

Thermodynamics   The study of the general properties of energy. Thermal energy plays a central role in understanding these properties.

Second law of thermodynamics   This law describes the general tendency of nonthermal forms of energy to end up as thermal energy. This law can be stated in three logically equivalent forms: The **law of heating** states that thermal energy flows spontaneously from higher to lower temperatures. The **law of heat engines** states that any cyclic process that uses thermal energy to do work must have a thermal energy exhaust. The **law of increasing entropy** states that the total entropy of all the participants in any physical process must either increase or remain unchanged.

First law of thermodynamics   An alternative way (not stated or used in this book) of stating the law of conservation of energy.

Heating   The spontaneous flow of thermal energy from a higher temperature object to a lower temperature object.

Temperature   A quantitative measure of warmth. The units normally used to measure temperature are called **degrees Celsius.** Any device that measures temperature is called a **thermometer.**

Heat engine   Any device that uses thermal energy to do work. Its **energy efficiency,** the fraction of its input thermal energy that is converted to work, will be higher if the **input temperature** is higher and the **exhaust temperature** is lower. The portion of the input energy that is not converted to work is called the **exhaust.**

Cogeneration   The simultaneous production of both electricity and useful thermal energy.

Energy quality   Thermal energy is considered to be of lower quality than other energy forms, because of the second law's restrictions on converting it to other forms of energy.

Entropy   The amount of microscopic disorganization a system possesses.

Irreversibility   The second law says that most processes are **irreversible,** i.e. that physical systems proceed spontaneously toward states of higher entropy, and will not proceed spontaneously in the reverse direction. This principle is responsible for the difference between forward and backward in time.

Microscopic interpretation of the second law   Entropy increase results from the fact that microscopic disorganization is overwhelmingly likely to increase.

Biology and the second law   The entropy of a growing plant decreases, at the expense of a far greater entropy increase in the absorbed and re-radiated solar energy that passes through the plant. A similar situation exists for all biological processes, including the evolution of species. Growth and biological evolution are organized by solar radiation and are in full accordance with the second law.

Creationism   The belief that the various biological species, and especially humans, originated from specific acts of instantaneous creation rather than by the natural process of biological evolution.

Internal combustion engine   A heat engine in which burning occurs within the hot gases that push directly on mechanical parts, such as a **piston,** to provide useful work. Typically, its energy efficiency is low, largely because of the second law.

Transportation efficiency   Useful output (such as distance traveled, passengers moved, freight moved, or mass moved) per unit of fuel input.

Steam-electric power plant   Uses thermal energy from an external source such as burning coal, to turn water into steam that pushes on a **steam turbine** that provides work to generate electricity. The device that converts the rotational motion of the turbine into electricity is called an **electric generator.** Since the steam is heated by fuel that burns outside of the boiler, the plant is an **external combustion engine.** Typically, its energy efficiency is less than 50%, mostly because of the second law.

Exponential growth   Growth by a fixed percentage in each unit of time. It has a fixed **doubling time,** related to its fixed percentage **growth rate** by $T \approx 70/P$.

Linear growth   Straight-line growth; it increases by the same amount (rather than the same percentage, as in exponential growth) in each unit of time.

Nonrenewable resource   A natural resource that can be used up. Its use follows a curve that begins exponentially, levels off, and declines.

Renewable resource   A natural resource that is continually replaced. Its use typically begins exponentially, then levels off at some sustainable level (provided that it is not overconsumed).

REVIEW QUESTIONS

HEATING
1. What is heating?
2. In your own words, state the law of heating.
3. Give an example showing that thermal energy and temperature are really two different things.

HEAT ENGINES AND ENERGY QUALITY
4. Give an example of each of these energy transformations: kinetic to thermal, gravitational to thermal, thermal to kinetic.
5. In your own words, state the law of heat engines.

6. What properties of the input and the exhaust are the most important to determining a heat engine's efficiency? Describe the manner in which the efficiency depends on these properties.
7. Is the actual overall energy efficiency of an automobile closest to 98%, 90%, 40%, 10%, or 2%? What about a steam-electric power plant?
8. If a heat engine operated entirely without friction, would it then be 100% efficient? Explain.
9. In terms of energy, what happens when the motion of a rock swinging from a string "dies down"? In what sense is this behavior irreversible?

THE AUTOMOBILE AND TRANSPORTATION

10. Name two general types of heat engines of major social importance.
11. Why do we call it an *internal combustion* engine? Describe in your own words how it works.
12. Name two alternative fuels (not gasoline or diesel fuel) for the automobile.
13. What is the source of the largest inefficiency in an automobile's operation?
14. Describe two different ways of measuring a transportation mode's efficiency, and give an appropriate measurement unit for each.
15. Why are trains more efficient than other transportation vehicles?

THE STEAM-ELECTRIC POWER PLANT

16. In a steam-electric power plant, what is the purpose of the turbine? Generator? Condenser? Cooling tower? Stack?
17. What is thermal pollution?
18. What is the source of the most important inefficiency in a steam-electric power plant's operation?
19. What part of a steam-electric power plant is analogous to the piston in an automobile?

EXPONENTIAL GROWTH

20. What is the difference between exponential and linear growth?
21. Your savings account grows at 7% per year. What is its doubling time?
22. Draw a typical life-history graph for a nonrenewable resource. Is any part approximately exponential?
23. Repeat the previous question, but for a renewable resource.

THE LAW OF ENTROPY

24. What is entropy?
25. In your own words, state the entropy form of the second law.
26. Which law or laws of physics distinguish between forward and backward in time?
27. When a growing leaf increases its organization, does it violate the second law? Explain.

HOME PROJECTS

1. Line up three containers filled, respectively, with hot (not scalding) water, warm water, and ice water. For a few moments, put one finger into the hot water and one finger into the cold water. Now plunge both fingers into the warm water. Can you always trust your sense of hot and cold?
2. How much of your neighborhood is devoted to the automobile? Choose a single city block and measure the area devoted to the automobile, using a tape measure or pacing. Include garages, parking lots, gas stations, car dealers, and the like. Include only half the total area of streets bordering your block; the other half belongs to the adjoining block. Find the fractional area devoted to the car; it is nearly 50% in most U.S. cities. Your class could divide into groups and measure several blocks and then pool its results for an overall figure. Your local newspaper might be interested in your results.
3. Study your home electric bill. How many kW·h of electric energy were used during the month? Use this figure to calculate your home's average power consumption in kilowatts. Use Table 6.1 to estimate the electric energy, in kW·h, that your home appliances consume during one month, and compare your estimates with the total shown on the bill. Compare this with your classmates' results.
4. Is electricity a good buy? Look at your electric bill to see how much you pay for 1 kW·h. Calculate the height to which 1 kW·h of energy could lift you. *Hint*: Recall that GravE = weight × height; find your weight in newtons from the fact that 1 pound = 4.5 newtons and 1 kilogram weighs about 10 newtons.
5. Play the population explosion game. Put 100 pennies into a box. Bring a penny into "penny world" by taking one penny out of the box and putting it on a table. Flip it. *Heads* means it had a baby that year, so you should bring another penny into penny world, and *tails* means it didn't have a baby. Now penny world's population is 1 or 2. Flip all the pennies in penny world, getting 0 or 1 or 2 new babies during the second year. Continue in this fashion, and draw a graph showing the population of penny world versus the number of years beginning at 0 years with a population of 1. If 100 represents overpopulation, how many years does it take until overpopulation is reached? Play the game several times. Is it the same number of years each time? Is this growth approximately exponential? From your graph, measure the doubling time and yearly growth rate.

FOR DISCUSSION

1. Assess the transportation possibilities for your region: (a) List your region's major transportation needs such as campus to shopping, suburbs to downtown, campus to evening entertainment, travel to nearest large city. (b) List all the possible ways to fill each need, assuming that the proper facilities (roads, bikeways, sidewalks, and so on) exist. Include city-planning options such as relocation of workplaces and reduction in the distance to outlying suburbs. (c) Underline the modes that actually do fill each need. (d) Discuss the pros and cons of the different modes listed for each need. Include energy consumption as one consideration. Other considerations might include convenience, pollution, safety, cost, traffic congestion, community quality of life, and so forth.

2. List 10 globally significant inventions or technologies. Which two or three of these have had the greatest impact on our way of life?

3. List several quantities whose growth is exponential. Which ones are likely to continue growing exponentially, which are likely to follow a bell-shaped growth-and-decline curve, and which are likely to level off at a sustainable level?

4. Should we seek alternatives to the car? If so, which alternatives, and under what circumstances?

EXERCISES

HEATING

1. How would you describe the weather on a day when the temperature was –3°C? +3°C? 22°C? 35°C?

2. Give the approximate temperature, in °C, of each of the following: Your body; water boiling in an open pot; ice water; a nice day.

3. Which is larger, a Celsius degree or a Fahrenheit degree?

4. How does the flow of thermal energy through a closed window illustrate the second law? Which direction is this flow when it is cold outside? Hot outside?

5. Try to think of at least one technological device that causes thermal energy to flow "uphill," from colder to hotter. Does this device violate the law of heating? Explain.

6. In the operation of a refrigerator, does thermal energy flow from hot to cold, or is it from cold to hot? Does this happen spontaneously, or is outside assistance required?

HEAT ENGINES

7. Is it possible to convert a given quantity of kinetic energy entirely into thermal energy? Is it possible to convert a given quantity of thermal energy entirely into kinetic energy? In each case, either give an example or explain why it is impossible.

8. Is it possible to convert a given quantity of chemical energy entirely into thermal energy? Is it possible to convert a given quantity of thermal energy entirely into chemical energy? In each case, either give an example or explain why it is impossible.

9. Which are not heat engines: natural gas-burning power plant, hydroelectric power plant, ethanol-fueled automobile, bicycle, solar-thermal electric power plant, steam locomotive?

10. Which of the following are heat engines: nuclear power plant, diesel locomotive, electric locomotive, geothermal power plant, wind turbine (windmill for generating electricity), solar hot water heater?

11. What does the second law tell us about the efficiency of heat engines?

12. Can you think of any way to drive a ship across the ocean by using the ocean's thermal energy, without violating the second law?

13. Farswell Slick (see Dialogue 10 of Chapter 6) approaches you with plans for a revolutionary transportation system. He has noticed that when he drives an automobile without accelerating, all the input energy eventually shows up as thermal energy. Slick proposes to use this thermal energy to drive the car at a constant speed. The car will still need fuel, but only for accelerating. It will be possible to travel cross-country on only a few gallons of gasoline. He describes his scheme as a "computerized advanced-technology exhaust feedback afterburner." Should you invest in Slick's scheme? Explain.

ENERGY QUALITY AND THE LAW OF ENTROPY

14. When your book falls to the floor is this a thermodynamically irreversible process? Is energy conserved? Does entropy increase?

15. When we say that the motion of a rock swinging on a string is irreversible, do we really mean that it is impossible to get the rock back to its starting condition? Explain.

16. When a block of wood slides down a sliding board, is this a thermodynamically irreversible process? Does this mean that it is impossible to make a block of wood slide up a sliding board? Explain.

17. As an egg develops into a chicken, its contents become more ordered. Does this violate the law of increasing entropy? Explain.

18. A pan of liquid water freezes when you place it outside on a cold day. Liquid water has greater molecular disor-

der than ice does. Is the freezing process then an exception to the law of entropy? Explain.
19. When orange juice and grapefruit juice are mixed, does entropy increase?

THE AUTOMOBILE AND TRANSPORTATION
20. Describe the energy input for walking and bicycling. How do walking and bicycling illustrate the second law?
21. Suppose an automobile's fuel could be made to burn hotter without harming the engine's operation (for instance, without cracking the engine). Would you still get the same amount of useful work from each gallon of gasoline?
22. Suppose an automobile could run on hard wheels that were not squeezed by the weight of the car on the road. Would this alter the car's efficiency? How might this affect the gas mileage? What kind of wheels and road might you suggest?
23. According to Figures 7.9, 7.10, and 7.11, which of the three main sectors of the U.S. economy (industry, residential–commercial, transportation) consumes the most oil?
24. One car has twice the overall energy efficiency (in km/l) of a second car. Compare the amounts of pollution they produce when they both travel the same distance.

THE STEAM-ELECTRIC POWER PLANT
25. Which type of generating plant would you expect to be more energy efficient, steam-electric or hydroelectric? Defend your answer.
26. Would it be more energy efficient to heat your home electrically or to heat it directly using a natural gas heater, assuming that the electricity comes from a steam-electric plant?
27. Which method of fueling your car is likely to be more energy efficient, and why: gasoline used in a standard car engine or electricity taken from a coal-fueled generating plant and stored in lightweight car batteries? Assume that the batteries convert electricity to work at 100% efficiency.

EXPONENTIAL GROWTH AND RESOURCE USE
28. A lily pond doubles its number of lilies every month. One day, you notice that 2% of the pond is covered by lilies. About how long will it be before the pond is entirely covered?
29. On June 1 there are a few water lilies in a pond, and they then double daily. By June 30 they cover the entire pond. On what day was the pond still 50% uncovered?
30. Company X increases its profits every year by $50 million. Is its growth in profits exponential? Company Y increases its profits by 1% every year. Is its growth in profits exponential?
31. According to Figure 7.24, did electric power production grow exponentially between 1910 and 1935? Estimate the number of kW·h produced in 1935, 1945, 1955, 1965, and 1975, and verify that production grew approximately exponentially between 1935 and 1975.
32. Which of the following are renewable energy resources: coal, firewood, nuclear power, wind, water behind a dam?
33. What is the original source of energy in each of the following energy resources: oil, firewood, wind, water behind a dam, geothermal, ocean-thermal electricity? Which of these are renewable resources?
34. The most recent world population doubling, to a total population of about 6 billion, has occurred in about 40 years. Making the (unrealistic!) assumption that this rate of population will continue for two centuries, what would the world's population be two centuries from now? Such unrealistic assumptions are often useful in projecting future trends, because they give us a sense of what is likely or unlikely. For example, this exercise shows us that it is very unlikely that our present population growth will continue for two more centuries.
35. The most recent world population doubling has occurred in about 40 years. Suppose that the next doubling occurs also in 40 years, but that a new agricultural "green revolution" manages to also double food production. Then how many people will be starving 40 years from now, as compared to the number starving now?

PROBLEMS

HEAT ENGINES
1. In one cycle of its operation, a heat engine does 100 J of work while exhausting 400 J of thermal energy. What is its energy input? Its efficiency?
2. In one cycle of its operation, a heat engine consumes 1500 J of thermal energy while performing 300 J of work. What is its efficiency? How much energy is exhausted in each cycle?
3. The steam entering the turbine in a certain coal-burning power plant is heated to 500°C. The steam is cooled and condensed to water at 80°C. Calculate the ideal efficiency of the power plant. Remember that you must convert Celsius temperatures to degrees Kelvin before using the formula.
4. A solar-heated steam-electric generating plant heats steam to 250°C. After passing through the turbine, cool-

ing towers cool the steam to 30°C. Calculate the ideal efficiency of this power plant. Suppose that, for every 1000 J of thermal energy going into this plant, the cooling towers remove 750 J as exhaust. What is the actual efficiency of this power plant?

THE AUTOMOBILE AND TRANSPORTATION

5. Out of every 100 barrels of gasoline, about how many actually go into driving a typical car down the road?
6. A bus carries 30 people 200 km using 300 liters of gasoline. Find its passenger-moving efficiency.
7. You travel alone from New York to Los Angeles, about 2800 miles. Working from Tables 7.3 and 7.4, how many gallons of gasoline will you use if you travel by car (assuming your car gets average gasoline mileage for U.S. automobiles)? How many gallons if you travel by air? By bus? By train?
8. A 100-car freight train hauls 16,000 tons of freight from New York to Los Angeles, about 2800 miles. How many trucks would be needed for this load, assuming that each truck carries 32 tons of freight? Working from Table 7.4, how many gallons of gasoline are saved if this load is carried by train rather than by truck? How many 42-gallon barrels of gasoline is this?

THE STEAM-ELECTRIC POWER PLANT

9. Out of every 100 tons of coal fed into an electric generating plant, roughly how many tons produce the electricity you can use at your home, and how many go into waste energy? Use the energy flows given in Figure 7.20.
10. For every 100 kilograms of coal entering a generating plant (recall that this much enters every second), about 15 kilograms of sulfur oxides and ash are removed, producing a significant solid-waste disposal problem. For a typical 1000 MW plant, how much of this solid waste is produced every day? Express your answer in tonnes (1 tonne = 1000 kg).

11. How would the pollution from two coal plants compare, if the first plant is twice as energy efficient as the second? Assume that they both produce the same amount of electric power.
12. Making estimates. In the U.S., solar energy strikes a single square meter of ground at an average rate (averaged over day and night and over the different seasons) of 200 watts. At what average rate does solar energy strike a football field (about 100 m by 30 m)? A typical home in the U.S. consumes electricity at an average rate of 1 kW. How much surface area would be needed to provide this electric power, assuming a 10% conversion efficiency? What dimensions would a square-shaped photovoltaic collector need to cover this area?

EXPONENTIAL GROWTH

13. How much electric energy would have been produced in 1985, if the exponential growth of 1935–75 had continued for another 10 years beyond 1975? If this growth had continued, roughly how many power plants would have been needed in 1985, as compared with 1975?
14. During 1985–1990, annual U.S. population growth was 0.8% per year, for Mexico it was 2.2%, and for Kenya (the highest) it was 4.2%. At these rates, how long will it take for the populations of each of these countries to double?
15. World population is now about 6 billion. The growth rate has been roughly 2% per year since the end of World War II (1945). If a 2% per year growth rate continued, when would world population be 12 billion?
16. Centerville, with a growth rate of 7% annually, is using its only sewage treatment plant at maximum capacity. If it continues its present growth rate, how many sewage treatment plants will it need 40 years from now?

CHAPTER 8

Light and Electromagnetism

Light is all around us, yet it is not easy to say what light is. How are you able to see this page? Are invisible rays emitted by your eyes that move from your eyes toward the page, as the ancient Greek thinkers Plato and Euclid thought? Or does the page send out or reflect a stream of particles that is received by your eyes, as the Pythagoreans (Section 1.3) and Isaac Newton thought? Or is light, as the Greek thinker Empedocles and Newton's contemporary Christian Huygens thought, a wave?

What is light? It is one of science's oldest questions, and it is the focus of this and the following chapter. Pursuit of this question led, historically, to both of the great post-Newtonian theories, relativity theory and quantum theory. Light and other *electromagnetic radiations* are also important to understanding societal topics like solar energy, ozone depletion, and global warming.

This chapter presents two topics that are essential to understanding the nature of light: waves (Sections 8.1 through 8.3) and the electromagnetic force (Sections 8.4 and 8.5). Sections 8.6 and 8.7 discuss the planetary atom, an atomic model that incorporates electromagnetic forces and helps explain light. Chapter 9 then presents the electromagnetic wave theory of light and explores the human impacts of these ideas.

8.1 WAVES: SOMETHING ELSE THAT TRAVELS

You are probably familiar with some kinds of waves (Figure 8.1). Stretch a few meters of flexible rope along the floor, fix one end (perhaps under a friend's foot), and give the free end a single shake. Something travels down the rope. Figure 8.2 is a series of pictures taken with a movie camera, showing a similar "something" traveling down a long spring that has been given a single up-and-down shake at the right-hand end. As another example, imagine (better yet, try it!) stretching a Slinky toy between your two hands along a tabletop. Quickly move the left end a short distance toward the right and then back to the left, holding the right end fixed. Something travels along the Slinky from your left to your right hand (Figure 8.3).

This "something" that travels across the water, along the rope and along the Slinky, is called a *wave*. As another example, the continued shaking of one end

FIGURE 8.1
Water waves.

FIGURE 8.2
A series of pictures taken with a movie camera, showing a wave moving along a spring. A ribbon is tied to the spring at the point marked by the arrow. The ribbon moves up and down as the wave goes by but does not move in the direction of the wave.

of a rope causes a long, continuous wave to travel along the rope (Figure 8.4). But *wave* is just a word that names the behavior without telling us what it really is. What actually happens here?

Observe the motion carefully. In Figure 8.2, how do the individual parts of the spring actually move? As you can see from the motion of the small ribbon tied to the spring, each loop just moves up and then back down.

How does a particular part of the water move in Figure 8.1? Fill a bowl with water, float a small cork in it, and drop a small pebble in, several centimeters from the cork, to create ripples. Observe the cork as ripples pass by. If the ripples are small, the cork will move up and down, not outward along with the ripples. Each portion of the water just shakes or *vibrates* up and down, but does not travel along the water surface. The Slinky wave is similar, except that the vibrations are parallel instead of perpendicular to the Slinky.

One thing that is traveling along with each of these waves is energy. You can verify this for yourself by holding the fixed end of a rope while a friend shakes the other end. Your hand vibrates as the pulse arrives. It takes work to force your hand back and forth this way, and we know that work requires energy. So waves transfer energy.

On the other hand, no material substance is transferred by waves: No water is transferred outward in Figure 8.1; no part of the spring is transferred from right to left in Figure 8.2; and no part of the Slinky is transferred from left to right in Figure 8.3. This type of motion is unlike any motion we have examined before. Previously we studied balls, books, molecules, and other material objects actually moving from one place to another. We call such cases **projectile motion**, because a material object is actually projected (thrown or transferred) from one place to another.

In discussing waves, it is important to distinguish between the small vibrational motions of the substance through which the wave travels and the nonvibrational motion of the wave itself.

What do we see traveling down the spring in Figure 8.2? Very simply, we see a bump in the otherwise straight spring. In Figure 8.3, we see a compression, a squeezed region traveling down the otherwise evenly stretched Slinky. We could describe both as *disturbances* that travel down the otherwise undisturbed rope or Slinky. The situation is similar for water waves. The material through which the disturbance travels—the spring or Slinky or water—is called the **medium** for the wave.

So we have a useful definition: A **wave** is a disturbance that travels through a medium, and that transfers energy without transferring matter. Such **wave motion** is different from projectile motion.

If you have ever attended a sporting event in a large stadium filled with people, you have probably observed a sports wave. It is an instructive example of

FIGURE 8.3
With the right-hand end of the Slinky held fixed, a quick motion of the left-hand end to the right and back again to the left creates a pulse that travels down the Slinky.

Compression

Compression moves toward the right

FIGURE 8.4
Continued shaking of the end of a rope creates a continuous wave that travels down the rope.

Direction of wave motion

the wave concept. A sports wave begins when all the people in one portion of the stadium, often at one end, stand up briefly with their hands in the air. As they sit down, the people sitting in the adjoining part of the stadium stand up briefly with their hands in the air, and so forth all around the arena. To the viewing spectator, a sports wave gives the appearance of something traveling all around the stadium. But this "something" that travels is obviously not any single thing. Each individual part (each person) of the "medium" (the crowd) simply stands up with hands in the air, and sits back down. The people who are momentarily standing constitute a disturbance of the otherwise-seated crowd, and it is this disturbance, rather than any individual person, that travels through the crowd. This is precisely the sort of situation that we have in mind when we use the word *wave*.

We need some quantitative terms. The **wavelength** of a continuous, repeated wave is the distance from any point along the wave to the next similar point, for example, from crest to crest or from trough to trough in Figure 8.5.

A wave's **frequency** is the number of vibrations that any particular part of the medium completes in each second. Waves are usually sent out by a vibrating source of some kind, in which case the wave's frequency must be the same as the source's frequency. The frequency could also be defined as the number of waves that the source sends out during each second. The unit for measuring frequency is the *vibration per second*, also called a **hertz** (Hz).

A wave's **amplitude** refers to its width of vibration, the distance that each part moves back and forth during that part's vibrations. Quantitatively, the amplitude is measured from the midpoint of vibration to the farthest point, so it is just half of the overall vibration width.

The **wavespeed** is the speed at which the disturbance moves through the medium, for example, the speed at which a compression moves along a Slinky or the speed at which a crest moves along the rope in Figure 8.5.*

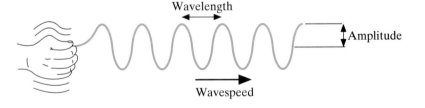

Wavelength

Amplitude

Wavespeed

FIGURE 8.5
The meaning of wavelength and amplitude. The wavespeed is the speed at which a crest or a trough moves down the rope.

---

*Quantitatively, a wave's wavelength λ, frequency f, and wavespeed s are related by $s = f\lambda$. For example, if three waves are sent out by the wave source every second ($f = 3$ vib/s) and each wave has a length of 2 meters ($\lambda = 2$ m), then it seems reasonable that the wavespeed should be $3 \times 2 = 6$ m/s. Extending this argument, you can see that the general relation is $s = f\lambda$.

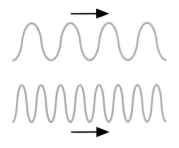

FIGURE 8.6
Which wave has the higher (larger) frequency, assuming that both have the same wavespeed?

Disturbances are able to travel through a medium because of the connections between the parts of the medium. For example, when you shake one end of a rope, this disturbance is transferred down the rope because the different parts of the rope are connected, so that when one part is lifted, its neighbor soon is lifted also. So it is reasonable to suppose that the wavespeed is determined mainly by the medium and is roughly the same for different disturbances within the same medium. Experiments confirm this notion that different disturbances travel through any particular medium at roughly the same wavespeed.

■ **DIALOGUE 1**   Surfers are able to ride breaking water waves coming into a beach. Are these really waves as we defined them? What about a row of falling dominoes, in which each domino knocks over its neighbor as it falls—is this a wave? If either answer is yes, then what is the medium?

■ **DIALOGUE 2**   Which wave has the larger or "higher" frequency: a long-wavelength wave moving along a rope or a short-wavelength wave moving along the same rope (Figure 8.6), assuming that both have the same wavespeed? Which carries more energy, assuming their amplitudes are the same? Note these useful rules: Short wavelength means high frequency, and high frequency means high energy.

FIGURE 8.7
Two waves travel in opposite directions along a rope. What happens when they meet?

8.2 INTERFERENCE: A BEHAVIOR UNIQUE TO WAVES

How do different waves traveling through the same medium interact with one another? For example, what happens when the large upward wave and the small downward wave shown in Figure 8.7 meet? If we do an experiment like this, we will find that the two waves just pass through each other without distortion (Figure 8.8). This is what we might have expected, because each wave just lifts or lowers the rope as it travels along the rope, so when the two waves meet, the rope is raised a lot by the large wave at the same time that it is lowered a little by the small wave. Experiments like this show that when different waves move through the same medium, they pass through each other without disturbing each other and continue on their way as though nothing had happened.

The effects that occur when two waves are present at the same time and place are called **wave interference** (or just interference) effects.

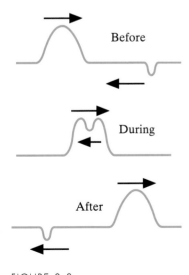

FIGURE 8.8
Two waves meeting: interference.

■ **DIALOGUE 3**   What would be the interference effect if the two waves of Figure 8.7 had identical heights, one upward and one downward? What if both were upward?

Two equal waves of opposite orientation interfere by canceling each other (Figure 8.9). Two equal waves of the same orientation interfere by adding up to

---

1. Breaking waves carry water into the beach (and then back out again as an undercurrent). This is projectile motion, not wave motion. A row of falling dominoes is a true wave, because the dominoes don't go anywhere. The medium is the row of dominoes.

2. The short-wavelength wave had to be sent out more frequently; the rope must vibrate more frequently as this wave passes through it; and the short-wavelength wave must carry the most energy. For instance, it could vibrate any "receiver" more often, which would take more work.

3. See Figures 8.9 and 8.10.

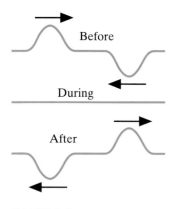

FIGURE 8.9
Two waves meet and interfere destructively.

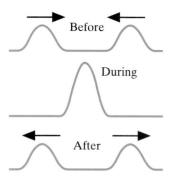

FIGURE 8.10
Two waves meet and interfere constructively.

a disturbance that is twice the size of either wave (Figure 8.10). These two cases, cancellation and reinforcement, are called **destructive interference** and **constructive interference**.

Wave interference shows, once again, the stark difference between projectile motion and wave motion. Two moving material objects—say two freight trains moving toward each other on the same track—do not interfere in the way that waves do. Two trains certainly don't pass through each other undisturbed! This distinction between projectile motion and wave motion—that projectiles do not exhibit wave interference but waves do—has proved crucial at several points in the history of physics, because it offers a clear-cut experimental method for distinguishing between particles and waves. It will turn out to be significant at several points later in this book.

Interference effects become more interesting when they happen in two or three dimensions. An undisturbed rope has only one significant *dimension*, length. The surface of a lake is *two dimensional* because it has length and width. And the space in a room is *three dimensional* because it has length, width, and height.

In a two- or three-dimensional medium, the waves created by a small source spread out into circles or spheres. If you send out a continuous water wave from a small vibrating source such as your finger tapping on the water surface, waves will spread out into circles, as in Figure 8.1, and eventually cover the entire surface.

Suppose you fill a rectangular pan with water and tap your fingers at the same steady rate on the surface at two points along one side of the pan. Continuous waves will soon cover the water surface, spreading out from each of the two sources. The wave crests spreading out from each individual source form circles (Figure 8.1), as do the wave troughs (the low points). What will the interference effects look like?

With the help of Figure 8.11, we can predict these interference effects. The figure shows the wave crests as viewed from above, looking down on the surface. The two sources are marked A and B. The solid circles represent the crests from source A acting alone, while the dashed circles represent the crests from source B acting alone. The troughs, not drawn, lie midway between the crests. To predict the interference pattern, work through Dialogue 4.

■ **DIALOGUE 4**   Draw an *x* (a color enhances the effect) at every point where a crest meets a crest. These are points of constructive interference. Also draw an *x* at every point where a trough meets a trough. These are also points of constructive interference, because the overall displacement of the water (a large trough) is larger than either individual effect (smaller troughs). Can you see a pattern? Next, draw a small o (in a different color) at every point of destructive interference. Now can you see a pattern?

Figure 8.12 is a photograph of this experiment, looking down onto the water's surface. The photographic technique causes crests to appear bright and troughs to appear dark. As you can see, the interference pattern has lines of undisturbed water radiating outward as though they came from a point somewhere between the two sources. The interference is destructive along these lines. Between these undisturbed lines are other lines of constructive interference, with large crests and troughs. This is just what Dialogue 4 predicted.

---

4. See the small x's and o's in Figure 8.14.

FIGURE 8.11
Continuous surface waves spreading out from two sources. What will the interference effects look like?

FIGURE 8.12
Interference between continuous surface waves spreading out from two sources: experimental results. Crests are bright, and troughs are dark.

FIGURE 8.13
Lines of constructive and destructive interference remain fixed in place, and constructive (large) waves move outward within the constructive regions in the directions indicated by the arrows.

All of our analysis so far has been at one instant in time. Now "turn on time" by imagining a moving picture that begins with the snapshot in Figure 8.12. Since the individual circular waves move outward from A and B, the entire interference pattern must move outward also. The rays of destructive and constructive interference remain fixed in place, and the large waves within the constructive rays move outward, as shown by the arrows in Figure 8.13.

Finally, think about how all this would appear to an observer who could only see these waves roll in to the far right border of Figure 8.13. Imagine that the body of water is a rectangular swimming pool. How does water-wave interference appear to an observer who examines the waves arriving along the right-hand wall?

We can predict the answer using Figure 8.13. The observer should find some points where large waves pound against the wall and other points where no waves roll in. These are the points of constructive and destructive interference along the wall, as diagrammed in Figure 8.14. An observer examining waves arriving at the right-hand wall would find alternating points (marked **X**) where large waves roll in and points (**O**) where no waves roll in.

The difference here between the pattern from a single source of water waves and the pattern from two sources is especially striking. Waves from a single source, say source A, spread out in circles that cover the entire surface. These waves roll into all parts of the bordering wall (Figure 8.15). If we now also turn on the second source, B (Figure 8.14), the pattern along the wall shifts to an interference pattern. The most dramatic change is that now no waves come into the points marked **O**, even though they did come into these points when only one source was operating. It seems paradoxical: We added a second source and got a reduced (in fact, zero) effect at the points marked **O**. Interference effects such as these give us a good way to identify wave behavior.

Now let's look at light.

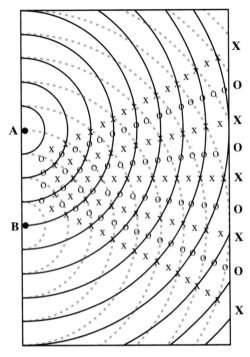

X
O
X
O
X
O
X
O
X

Observer stands on this side and looks down at pool, observing large waves rolling into points marked **X** at side of pool, and no waves rolling into points marked **O**. Small x's and o's are places on surface of pool where interference is constructive (x's) and destructive (o's).

FIGURE 8.14
An observer scanning a wall at the far border of the pool finds points where large waves come into the wall, interspersed with points where no waves come in.

FIGURE 8.15
If one of the two wave sources shuts off, an observer scanning the wall will find that waves arrive at all points. Since waves now spread out from only a single source, there is no longer any interference.

8.3 LIGHT: PARTICLES OR WAVE?

When you turn off the light at night, it gets dark. So the light in your room must have come from the lightbulb. It couldn't, for example, have come from your eyes, because even though your eyes were still open after you turned off the lightbulb, you could not see. So light is something that enters your eyes from the outside. When you look at a luminous (light-emitting) object like a lightbulb, light goes from the bulb to your eyes. In order for you to see a nonluminous object, such as the wall of your room, light from the lightbulb must bounce (*reflect* is the official word) off the wall and then into your eyes. The light reflecting from the wall does not give you a nice mirror reflection, however, as light does when it reflects from a glass mirror, because the rough surface of most walls scatters the incoming light in many different directions.

But what enters your eyes when you see light? The question has been debated for centuries. Most of the suggested answers fall into one of two different categories: particles and waves. Perhaps luminous objects send out tiny particles that enter our eyes. Or perhaps light is a wave emitted by some kind of vibrations within the luminous object. Experiment is the ultimate judge.

We need an experimental test that distinguishes between the projectile and wave models of light. The preceding section of this chapter suggests a good candidate: wave interference. When waves meet, they exhibit wave interference. Particles might interact in other ways, but they do not interfere in the way that waves do. What does light do? To answer this, we need an experiment like the water-wave interference experiment, but with light.

The first experiment you might think of is just to shine two flashlights on a movie screen. This gives no observable interference effects, as you can see by trying it yourself. So maybe light is a stream of particles.

But recall that in order to observe water-wave interference, we assumed that the two wave sources had identical and synchronized vibrations. For example, if the two sources in Figure 8.12 were changing their frequency all the time and in different ways, we would not expect to see any recognizable interference pattern because the wavelengths coming from the two sources would be continually changing in completely different ways. It is possible that light is a wave but that the waves from the two flashlights are not synchronized. A flashlight bulb's light is produced by heating up the bulb's thin wire, or *filament*, until it glows. The microscopic thermal motions that make the filament hot enough to glow are mixed up and random, so we would not expect two different bulbs to have synchronized vibrations.

H O W   D O   W E   K N O W ?

The double-slit experiment with light

In 1801, Thomas Young solved the problem of finding two synchronized light sources. Young's trick was to use a single light source but to split its light into two parts that should then have identical vibrations. He then recombined these parts to see whether they interfered.

Figure 8.16 shows one way to do this. A light source sends light to two very small and narrow parallel slits (typically each slit is much less than 1 mm wide, and the two slits are less than 1 mm apart) in a partition that blocks all the light except that going through the slits. The two slits then act as the two sources of light, like the two sources in Figure 8.12. With this arrangement, if light is a wave, these two sources should have synchronized vibrations, because the light from each slit originated in the same filament.

Using this arrangement, Young found an experimental result like that shown in Figure 8.17. This photograph was made by placing photographic film at the position of the receiving screen in Figure 8.16. Figure 8.18 depicts the experimental arrangement and its outcome. As we will soon see, this outcome is excellent evidence that light is a wave.

In order to interpret Figure 8.17, let us return to water-wave interference. Figure 8.14 shows the interference pattern observed along a wall placed in the path of water waves from two synchronized sources. The receiving screen of Figure 8.16 is just like this wall: The screen is placed in the path of the light from the two sources. But water waves occur on the two-dimensional surface of water,

FIGURE 8.16
The double-slit experiment with light. What will we see on the screen?

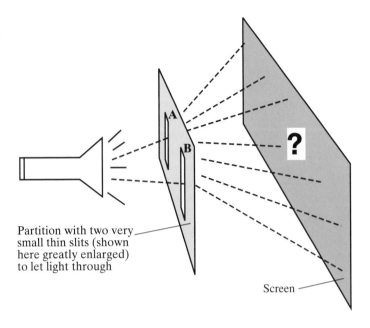

Partition with two very small thin slits (shown here greatly enlarged) to let light through

Screen

FIGURE 8.17
The double-slit experiment with light: experimental results.

whereas light fills up three-dimensional space. The two sources of light are not tiny points like A and B in Figure 8.14 but instead are slits that extend into the third dimension. If these sources send out light waves, we would expect the interference pattern seen on the receiving screen to be alternating lines of constructive and destructive interference running parallel to the slits, not small points like the points marked **X** and **O** along the right-hand border in Figure 8.14. In other words, we would expect alternating bright (lit) and dark lines. We would, in fact, expect precisely the outcome seen in Figure 8.17.

What would we expect if we closed one of the slits, leaving only one slit open? If light is a wave, we would expect that waves would spread out from the open slit, without interference, just as water waves spread out from a single source without interference in Figure 8.15. A broad band of light should then cover a large area of the receiving screen (Figure 8.19). This is, in fact, what happens.

The clearest evidence that the flashlight is sending out waves and not particles can be found at the positions of the dark lines in the double-slit experiment, the points in Figures 8.17 and 8.18 where no light arrives. With only one slit open, light spreads out over the entire receiving screen (Figure 8.19). How is it, then, that if we simply open a second slit, no light will arrive at these particular positions? It is difficult to see how particles coming through the two slits could cancel one another out in this way, but it is just what we expect of waves. We conclude that light is a wave.

By measuring the distance from one bright constructive-interference line to the next such line in a pattern such as Figure 8.17 and using a little geometry, it is possible to calculate the wavelength of the light that created the pattern. Measurements of interference patterns like this are the usual method of measuring the wavelength of light. This wavelength turns out to be very small. Light sources have wavelengths ranging from about 0.4×10^{-6} m to 0.7×10^{-6} m (0.4 to 0.7 millionths of a meter).

FIGURE 8.18
The double-slit experiment with light: the experimental setup and results.

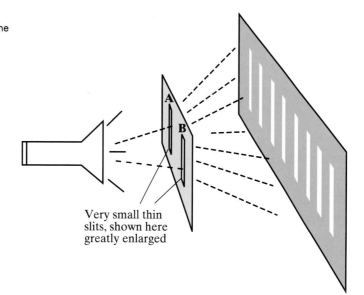

Very small thin
slits, shown here
greatly enlarged

H O W   D O   W E   K N O W ?

You can demonstrate light-wave interference yourself, using a single-slit wave-interference effect that occurs when the slit is much wider than the wavelength of the light. In order to get the spread-out, noninterfering result shown in Figure 8.19, the single slit must be very narrow—comparable in width to the wavelength of the light being used. In the following demonstration, we use a slit whose width is hundreds of times larger than the wavelength of the light. With such a wide

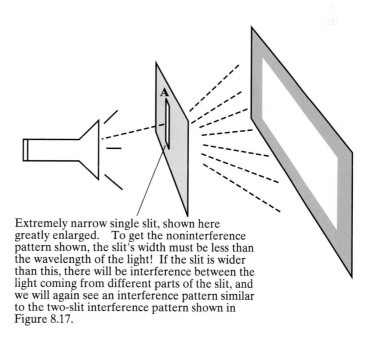

FIGURE 8.19
If one of the two slits is closed, light will arrive at all points along the receiving screen. Compare Figures 8.18 and 8.19 with Figures 8.14 and 8.15.

Extremely narrow single slit, shown here greatly enlarged.   To get the noninterference pattern shown, the slit's width must be less than the wavelength of the light! If the slit is wider than this, there will be interference between the light coming from different parts of the slit, and we will again see an interference pattern similar to the two-slit interference pattern shown in Figure 8.17.

slit, the light coming through the slit does not behave as though it came from a single tiny source. Instead, the slit acts like hundreds of tiny sources. If light is a wave, then all the individual waves from these hundreds of sources should interfere with one another to form an interference pattern.

Here's the experiment: Focus your eyes on a well-lit wall or other surface. Make a slit by holding your thumb and forefinger close together but not touching, about a millimeter apart and several centimeters in front of your eye. Focus on the light source (the wall), and not on your fingers, because you want to learn about the light from this source. Your fingers should look blurred, and where the blurs overlap, you should be able to see narrow bright and dark lines running parallel to your fingers.

These light and dark lines are constructive and destructive interference regions, formed at the position of your eye, created by the light coming through the single slit. A half-millimeter-wide slit between your thumb and forefinger is about 1000 times larger than the wavelength of light, so as explained earlier, the observed pattern is just what we would expect if light is a wave.

FIGURE 8.20
Light beams cannot be seen from the side. What invisible medium is carrying the light waves?

FIGURE 8.21
You can see a light beam by allowing it to reflect off dust particles in the air.

Since light is a wave, what kind of wave is it? What is it a wave in? Water waves, rope waves, and Slinky waves are waves in water, ropes, and Slinkies. What medium vibrates when light waves travel? It's not an easy question. We don't see light beams directly the way we see water waves (Figure 8.20). It is as though we could see the impact of water waves against a bordering wall, but without being able to see the water. We can see light beams in dusty air (Figure 8.21), but only because the light is reflected off the dust particles. The medium for light waves is itself invisible.

Could the medium be air? This sounds plausible because you cannot see air, and air does fill up space, at least near Earth. But what about light traveling far from Earth? In any appreciable amounts, air extends only a few miles above Earth's surface, yet light arrives here from distant stars, across great reaches of nearly empty space. So air cannot be the medium for light waves.*

The odd thing about light is that it moves through empty space where there is essentially no matter at all. But something must be out there, in so-called empty space, because you can't have a wave without having a medium to do the waving. After all, we began our study of waves with the idea that a wave is a disturbance in a medium. You can't have a disturbance without having something to disturb. The medium for light, then, must be nonmaterial, not made of atoms or other forms of matter.

Nineteenth-century scientists devoted a lot of time and effort to learning what kind of wave light is. It turned out that the answer is bound up with a new phenomenon that was beginning to be understood during that century: electricity. The answer, briefly, to our questions about light waves is that light is an *electromagnetic wave* and that the medium through which these waves travel is an *electromagnetic field*. In order to understand these ideas, we will devote the rest of this chapter to electricity. Then we will see, in the following chapter, what electricity has to do with light.

---

*Air is the medium for sound waves rather than for light waves. Since sound does not bear directly on the major purposes of this book, we won't discuss it further here.

■ **DIALOGUE 5**   If we compared two water-wave interference experiments, one using a short wavelength and the other using a long wavelength, how would the interference patterns observed along the right-hand wall in Figure 8.14 differ? How would the use of longer-wavelength light waves affect the interference pattern in Figure 8.17?

■ **DIALOGUE 6 MAKING ESTIMATES**   Roughly, how does a typical light wavelength compare with the thickness of a piece of paper?

8.4 ELECTRIC FORCE: PART OF THE ELECTROMAGNETIC FORCE

Find a couple of plastic transparencies, the kind that teachers use on overhead projectors. Rub them vigorously with tissue paper and hold them from one edge in separate hands (Figure 8.22). If you hold them parallel and just a few centimeters apart but not touching, they should repel each other. At short separations, they repel each other strongly enough to stand noticeably apart. The force weakens at larger separations. Now spread out the tissue on a level surface, and hold one transparency directly above it. The tissue is pulled upward and clings to the transparency. You may experience similar forces when you take a synthetic shirt from a clothes dryer or when you brush your hair. Figure 8.23 shows an extreme example.

What is the source of these forces? This force can act across a distance such as that between the transparencies. The only force acting at a distance that we have encountered so far in this book is gravity. This force has other similarities with gravity. For example, the force acting between the two transparencies weakens as the separation widens, just as gravity does.*

Can this force actually be gravity? There are several reasons that it cannot be. First, the transparencies repel each other, whereas gravity attracts. Second, this new force is far stronger than gravity could possibly be between relatively low-mass objects such as transparencies and tissues. The force by the transparency on the tissue is so strong that it easily lifts the tissue against the downward gravitational pull on the tissue by the entire Earth. Third, the existence and size of this force depend on whether the transparencies are rubbed, and it is hard to see how rubbing could affect the gravitational force.

We call this force the **electric force**. An "electrified" object such as the transparency or the tissue (after rubbing) is said to be **electrically charged**, or *charged*, and any process that produces this state is called **charging**.

The experiments show that when we electrically charge two identical objects in identical ways, they repel each other. But the charged transparency and the

FIGURE 8.22
Two rubbed transparencies exert electric forces on each other.

FIGURE 8.23
An extremely bad hair day: electric hair. The source of this effect is the "charged" metal sphere.

---

5. Longer-wavelength water waves would create a more spread-out pattern, with more distance between the **X**'s and **O**'s in Figure 8.14. Longer-wavelength light would produce broader, more spread-out lines in Figure 8.17.

6. Choose a typical wavelength of light, say 5×10^{-7} m (in making estimates, choose simple but reasonable numbers). To estimate the thickness of a sheet of paper, estimate the thickness of, say, 500 sheets (about 5 cm) and divide by 500 (5 cm / 500 = .01 cm = 10^{-4} m). The number of wavelengths in this thickness is $10^{-4}/5 \times 10^{-7} = 10^{-4+7}/5 = 10^3/5 = 1000/5 = 200$.

*In fact, this force even turns out to be inversely proportional to the square of the separation distance between the two objects experiencing this force, precisely like gravity. Nature seems to favor "inverse square" force laws.

charged tissue attract each other. So the transparency must be charged differently from the tissue. After experimenting with all sorts of other charged objects, one finds that every charged object falls into just two categories: those that repel the transparency but attract the tissue, and those that attract the transparency but repel the tissue. We name these two categories of charged objects **positive** and **negative**. Do not attach much significance to the names—we could just as well call them red and blue, or charming and revolting.

Our experiments demonstrate:

THE ELECTRIC FORCE LAW
Electrically charged objects exert forces on each other at a distance. Objects may be charged in either of two ways, known as positive and negative. Two objects possessing like charges repel each other, and two objects possessing unlike charges attract each other.*

8.5 MAGNETIC FORCE: THE OTHER PART

Have you ever played with magnets? Everybody should have the opportunity to experience the intriguing forces of attraction and repulsion that magnets exert on each other. You can buy inexpensive magnets at a toy store.

If you bring two bar magnets near each other, you will discover that the ends of two magnets either attract or repel each other even when they are separated by a distance. The ends are called **magnetic poles**. The two ends are called the *north* and *south* pole of the magnet and might be indicated by N and S printed on the ends. Experiment shows that similar poles repel each other and dissimilar poles (north and south) attract each other. This reminds us of the forces between electric charges: Likes repel, and unlikes attract.

It seems plausible to hypothesize that the force acting between the magnets actually is the electric force. This hypothesis is easy to check, for it predicts that magnets should exert forces on electrically charged objects such as a rubbed transparency or tissue. If you try this, you will find that the magnets do not exert forces on a rubbed transparency or tissue.† So our hypothesis is wrong. Despite the similarity to electricity, the force between bar magnets is not the electric force, and the two ends of a bar magnet are not electrically charged.

There are other important differences between magnetism and electricity. First, a bar magnet's magnetism is permanent and has nothing to do with rubbing. Second, every magnet has both a north and a south pole—magnetic poles always come in pairs. Nobody has ever found an object that possessed either kind of pole without the other kind, although many experiments have looked for

---

*The electric force is quantitively proportional to the product of the *strengths* (or magnitudes) of the charge of each of the two objects and inversely proportional to the square of the separation distance: $F \propto q_1 q_2 / d^2$. Note the similarity to the gravitational force law: $F \propto m_1 m_2 / d^2$.

†A small attractive force is sometimes obtained with the transparency, an electric effect called *electrical polarization*. This force, if it does occur, is attractive at both the north and south poles of the magnet, and it occurs equally strongly even if an unmagnetized piece of metal is used in place of the magnet, so it is not caused by the presence of the magnetic poles. Rather, it is caused by the redistribution of electric charge that occurs in a metal when a charged object like the transparency is brought near it. A similar electric polarization occurs quite dramatically when a charged transparency is brought near an empty aluminum can that is free to roll. Try this!

evidence of such "monopoles." On the other hand, it is easy to find objects that are entirely positively charged or entirely negatively charged.

This new kind of force is called the *magnetic force*. Despite their differences, the similarities between the electric force and the magnetic force suggest that they might be related. One of the great triumphs of nineteenth-century physics was the demonstration that electricity and magnetism are in fact related. The most concrete evidence was an experiment, first conducted in 1820, in which electrically charged particles that were in motion exerted a measurable force on a magnet.

Note that it is only *moving* charged objects that can exert forces on magnets. As we have seen, *stationary* charged objects do not exert forces on magnets.

Further experiments during the nineteenth century showed that *all* magnetic forces can be traced to the motion of charged objects. Moving charged objects exert and feel magnetic forces over and above whatever purely electric forces they would feel if they were at rest. This additional force, due solely to the motion of charged objects, is the magnetic force.

For example, the forces observed with bar magnets are due to subatomic charged particles moving inside each magnet (Section 8.7). For another example, Earth's magnetic effects are due to electrically charged material flowing inside Earth.

We summarize this idea as:

THE MAGNETIC FORCE LAW
Charged objects that are moving exert and feel an additional force beyond the electric force that exists when they are at rest. This additional force is called the **magnetic force**. All magnetic forces are caused by the motion of charged objects.

This means that the separate concept of magnetic poles is not needed, that we can drop the idea of magnetic poles and just think of moving charges instead.

A goal of science is to find connections between apparently different phenomena. For example, Newton united the heavens and Earth by finding similarities between a falling apple and the moon. Similarly, nineteenth-century scientists found that both electricity and magnetism are due to the existence of electrically charged objects and that these two forces can be united into a single **electromagnetic force** between charged objects.

■ **DIALOGUE 7**   Could the force between two bar magnets be due to gravity? Defend your answer.

■ **DIALOGUE 8**   If you charge two transparencies by rubbing and then hold them at rest several meters apart, will they exert an electrical force on each other? Will they also exert a magnetic force? What if you shake both of them back and forth?

■ **DIALOGUE 9**   What will you have if you saw off one end of a magnet?

One day, Sir, you may tax it.
MICHAEL FARADAY, CODISCOVERER OF ELECTROMAGNETISM, WHEN ASKED BY THE CHANCELLOR OF THE EXCHEQUER ABOUT THE PRACTICAL WORTH OF ELECTRICITY

---

7. Although the magnetic force can be repulsive, gravity cannot. Also, the force between bar magnets is far too strong to be caused by gravity acting between the small bars.

8. With both at rest, you will get only an electric force. With both in motion, you will get not only an electric force but also a magnetic force. However, if you try an experiment like this, the magnetic part of the overall electromagnetic force will be immeasurably small.

9. Two magnets. Each piece will have its own north and south poles. It is impossible to isolate a north or south pole.

8.6 THE ELECTRIC ATOM: THE PLANETARY MODEL

We saw in Chapter 2 that we can learn a lot by picturing phenomena in terms of what is going on at the microscopic level. This and the next section explain what happens microscopically when electric and magnetic effects occur macroscopically. We will find that such explanations require a new, electrical model (officially known as the *planetary model*) of the atom. Because nineteenth-century studies of electricity and magnetism lead directly to this new view of the atom, and because we need this new atomic model in order to understand light, it is appropriate at this point to study the electric atom.

Physicists have always sought microscopic explanations of macroscopic events. So far in this book, we have discussed, from a microscopic point of view, chemical reactions, thermal energy, and much more. All of these phenomena are comprehensible on the basis of the **Greek model of the atom** (Chapter 2). According to this theory, all matter is made of small unchangeable particles, *atoms* (Greek for *not divisible*), which come in many varieties or "elements" and which move and combine in various ways. This model of the microscopic world, conceived by the ancient Greeks, formed the background for Newtonian physics. During the nineteenth century, this model was developed into the theory of chemical elements, compounds, and reactions.

It is difficult to fit electromagnetic phenomena into this indivisible-particle picture of the atom. Where does electric charge come from? How can rubbing produce it? Why are there two kinds of charge? These and other questions led, early in the twentieth century, to the **planetary model of the atom.*** According to this theory, the atom is not a solid, indivisible object. To the contrary: The planetary atom is continually changing, is almost entirely empty, is divisible, and is made of many parts.

As its name implies, the planetary atom resembles a miniature solar system. Figure 8.24 portrays single atoms of two different elements, helium and carbon. The defining feature of the planetary atom is the tiny **nucleus** at the center, surrounded by a number of even tinier **electrons** (*electrified ones*) that orbit the nucleus at a relatively great distance, much greater than the size of the nucleus itself. The overall size of an atom, the distance across its electron orbits, is typically about 10^{-10} m. This is roughly the distance between neighboring atoms in solid materials. But a typical nucleus is 10,000 times smaller, on the order of 10^{-14} m. To put this into perspective, if we built a scaled-up model of an atom in which the nucleus were represented by a soccer ball, the orbiting electrons would be dust specks several kilometers away! Atoms are nearly totally empty. But despite the fact that the nucleus is far smaller than the atom, more than 99.9% of an atom's mass resides in its nucleus.

You cannot remove the charge from an electron—it is permanently negatively charged. Also the electron has a mass that is about 2000 times smaller than the mass of even the least massive atom. Nobody has any idea of why either fact is so.

The nucleus is itself made of several subatomic particles, of two different types: protons and neutrons. A **proton** (*positive one*) is another permanently

---

*The term *planetary atom* is self-contradictory. *A-tom* means *indivisible*, and *planetary* refers to the parts into which the atom can be divided! The old Greek name, *atom*, has stuck, but not its essence.

FIGURE 8.24
Two examples of the planetary model of the atom. Protons are blue, neutrons are white, and electrons are black. The diagrams are not drawn to scale! If they were drawn to scale on the page, the nuclei would be too small to be seen.

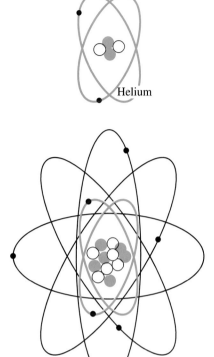

Helium

Carbon

charged particle, like an electron. It is charged precisely as strongly as the electron, but positively instead of negatively. When we say that electrons and protons are charged "equally strongly," we mean that when they are placed near some other charged object, both exert the same amount of force (but in opposite directions) at the same distance away. Although the electron and proton have equally strong charges, they do not have equal masses. The proton is about 2000 times more massive. The **neutron** (*neutral one*) is an uncharged, or neutral, particle whose mass is nearly the same as the proton's mass. Between one and a few hundred protons and neutrons form the nucleus of any atom.

The "glue" that holds electrons in their orbits around the nucleus is the electric attraction between the electrons and the protons in the nucleus. The glue that holds the nucleus together, however, must be some nonelectric force, because the electric force between the positively charged protons is repulsive and neutrons do not exert an electric force. Chapter 15 discusses this nuclear glue.

Since an atom's electrons are relatively distant (compared with the size of the nucleus) from the nucleus, it is not surprising to learn that the forces binding them to atoms are rather weak, so that it is not difficult to remove electrons from atoms. On Earth, unattached atoms (atoms that are not combined into molecules) usually have just as many electrons as protons. The reason is that any atom having fewer electrons than protons carries a net positive charge and so tends to attract electrons from its environment, while any atom having more electrons than protons carries a net negative charge and tends to lose its outer-

FIGURE 8.25
J. J. Thomson, discoverer of the
electron. He was the first to show that
atoms are made of smaller,
electrically charged, parts.

most electrons to its surroundings. This is why it took so long to discover many electrical phenomena: Most individual atoms are normally uncharged and exhibit no obvious electrical effects. Any atom having an excess or deficiency of electrons is called an **ion**.

HOW DO WE KNOW?

The discovery of the electron

One key experiment leading to the planetary atom was by the English physicist J. J. Thomson (Figure 8.25). In 1897, Thomson was investigating a type of invisible beam known as a *cathode ray*. Cathode rays were produced in a nearly evacuated (emptied of air and other gases) glass tube whose two ends were attached by metal wires to a source of electric power. When the power was switched on, rays of unknown composition streamed along the length of the tube, as could be observed by the flashes of light where they hit one end of the tube.

Suspecting that these rays were electrically charged, Thomson placed electric charges and magnets around them. The flashes of light shifted in position. Because the charges and magnets deflected the rays, the rays themselves had to be electrically charged. The only charged microscopic objects then known were ions, observed in certain chemical experiments. Thomson hypothesized that the electrically charged cathode rays were streams of such ions.

He then measured the deflections of the rays. Using the known electric and magnetic force laws, he deduced from these measurements that these rays were streams of charged particles whose charge was the same as the charge of typical ions but whose mass was some 2000 times smaller.* These, then, were not ions.

This was revolutionary. It established that atoms had parts. According to Thomson, "At first there were very few who believed in the existence of these bodies smaller than atoms. . . . It was only after I was convinced that the experiments left no escape from it that I published my belief in the existence of bodies smaller than atoms."

Thomson had discovered the electron. Cathode rays are now also called *electron beams*. Today, electron-beam tubes are in wide use as TV tubes, fluorescent bulbs, computer screens, and many other devices.

FIGURE 8.26
Ernest Rutherford, codiscoverer of the
nucleus and one of the greatest
experimental physicists of the
twentieth century, talks with a
colleague. Because Rutherford's
resounding voice could upset delicate
experimental apparatus, the sign
overhead was playfully aimed at him.
His research on radioactivity and
nuclear physics influenced a
generation of experimental physicists
early in this century.

HOW DO WE KNOW?

The discovery of the nucleus

Like others around 1910, New Zealander Ernest Rutherford (Figure 8.26) was trying to determine the atom's internal structure. He knew that atoms contained electrons and that a positive charge must be present too. It was known that

*More precisely, he found that the *ratio* of the mass to the charge was 2000 times smaller than it was for any known ion.

atoms are pressed right up against one another in solid materials and that huge forces are required to compress solids into smaller volumes. This means that it is difficult to squeeze atoms into other atoms. So Rutherford and others hypothesized that atoms were filled with matter throughout most of their volume.

To probe the atom's structure and test this hypothesis, Rutherford and his coworkers used what has become a traditional physics technique: He threw tiny things at other tiny things in order to see what would happen. He threw a recently discovered ray known as an alpha ray at the atoms in pieces of thin metal foil, similar to aluminum foil. The alpha ray was a high-energy stream of positively charged and fairly massive (about four times the proton's mass) "alpha particles" that emerged from certain substances (Chapter 15).

The idea was to observe how far the foil deflected the fast-moving alpha particles from their original directions and, from this, to deduce how matter must be distributed within the foil's atoms (Figure 8.27). The deflection was measured by observing flashes of light where the alpha particles hit a screen placed partially around the foil. Similar experiments had been done before, and it had been found that the foil had surprisingly little effect on the motion of the alpha particles. Most deflections were less than one angular degree. Since even the thinnest foils were about 500 atoms thick, alpha particles apparently passed straight through most atoms without deflection. Apparently, atoms were fairly porous, open structures.

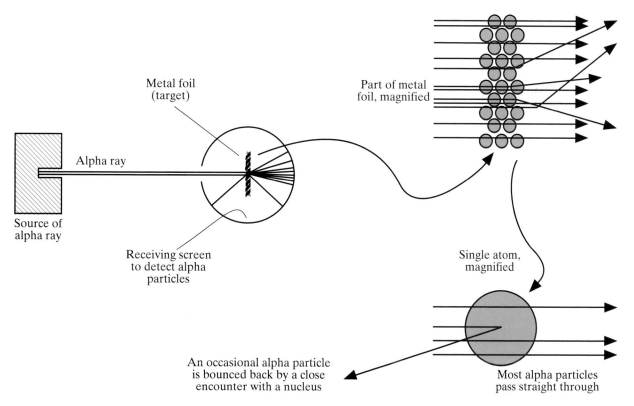

FIGURE 8.27
Rutherford's alpha-scattering experiment.

The energy produced by the breaking down of the atom is a very poor kind of thing. Anyone who expects a source of power from the transformation of these atoms is talking moonshine.

ERNEST RUTHERFORD, DISCOVERER OF THE ATOMIC NUCLEUS, MADE THIS FAMOUS WRONG PREDICTION IN 1933, WHILE ADDRESSING THE BRITISH ASSOCIATION FOR THE ADVANCEMENT OF SCIENCE IN THE SAME HALL WHERE PHYSICIST LORD KELVIN HAD ASSERTED, IN 1907, THAT THE ATOM WAS INDESTRUCTIBLE. THE *NEW YORK HERALD TRIBUNE* ARTICLE, DATED SEPTEMBER 12, 1933, CARRIED THE HEADLINE "ATOM-POWERED WORLD ABSURD, SCIENTISTS TOLD. LORD RUTHERFORD SCOFFS AT THEORY OF HARNESSING ENERGY IN LABORATORIES."

Then in 1911, Rutherford decided to see whether any alpha particles were deflected through very large angles, perhaps greater than 90 degrees. His coworkers studied this by completely surrounding the foil with the detection screen (with a gap to allow the alpha ray to enter). Rutherford expected to see no large deflections, because a fast-moving and massive alpha particle was thought to pass through an atom somewhat like a high-velocity cannonball through pudding. An alpha particle would have to experience an enormous force to be deflected by very much.

His coworkers came to Rutherford a few days later with the news that a few alpha particles had been deflected backward. In Rutherford's words, "It was quite the most incredible event that has ever happened to me in my life. It was almost as incredible as if you fired a 15-inch shell at a piece of tissue paper and it came back and hit you."

The reason was, apparently, that nearly all of the matter and all of the positive charge of the atom were concentrated in a tiny, dense region in the center. The cannonball had struck an even more massive cannonball and bounced back. Rutherford and his colleagues had discovered the atomic nucleus.

8.7 THE PLANETARY ATOM: A USEFUL THEORY

Like all good theories, the planetary atom explains many things.

It explains our experiments with charged objects. When you rub a transparency with tissue, some of the loosely attached outermost electrons in the transparency's atoms are rubbed off and transferred to the tissue's atoms, charging the transparency positively and the tissue negatively. That is why the two attract each other after rubbing. The two rubbed transparencies repel each other because both are positively charged.

Not exist—not exist! Why I can see the little beggars there in front of me as plainly as I can see that spoon!

RUTHERFORD, WHEN ASKED OVER A DINNER TABLE WHETHER HE BELIEVED THAT ATOMIC NUCLEI REALLY EXISTED

When the ends of a copper wire are attached to the positive and negative attachments (called *electrodes*) of a battery, electric forces are suddenly exerted on every charged particle in the wire (Figure 8.28). Chemical processes in the battery cause its electrodes to be permanently charged, positively and negatively, and these charged electrodes exert electric forces on every charged particle in the wire. These forces produce practically no disturbance of the positively charged copper nuclei, which are fixed in position. But in copper or any other metal, each atom's outermost electrons are nearly free of their parent atom. As soon as these electrons feel the electric forces established by a battery, they all begin to move along the wire. Such a flow of charged particles is called an **electric current**.

If this *electric circuit*, from one electrode around to the other electrode and then through the battery back to the first electrode, also includes a lightbulb filament (Figure 8.28), electrons will flow through the thin filament and heat it by simply bumping into lots of atoms. The filament becomes so hot that it creates light. All of the standard electrical appliances—toasters, lightbulbs, electric motors, and so forth—are based on the flow of the unseen, microscopic electrons that ordinarily orbit in atoms. Figure 8.28 shows how the process works when a battery is the source of the electric force. Electrical outlets in your home work in a similar way (Figure 8.29).

FIGURE 8.28
A battery produces electrical forces that cause electrons to move through the metal wire. Electrons flow around the wire circuit from one electrode (terminal) to the other electrode and are then pushed, by chemical forces, through the battery from the second electrode back to the first.

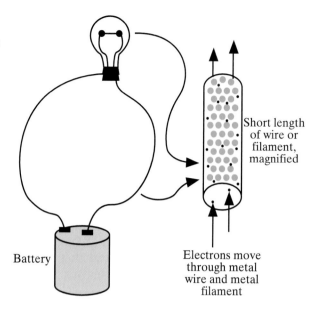

Short length of wire or filament, magnified

Battery

Electrons move through metal wire and metal filament

The magnetic force law says that all magnetic effects come from electric charges in motion, in other words, from electric currents. Where, then, are the electric currents responsible for permanent magnets? The answer is that they are found at the subatomic level, in the motions of electrons in atoms.* Each moving electron in a bar magnet causes its own tiny magnetic force on each moving electron in another bar magnet. In most materials, these tiny magnetic forces cancel one another because all the electrons have different orientations.

FIGURE 8.29
Electrical appliances are based on the motion of unseen, microscopic, charged particles called electrons that move through appliances. A wall socket (more precisely, a generating station connected by wires to a wall socket) is the electrical equivalent of a water pump. It "pumps" (energizes) the electrons so that they can flow around the circuit.

Wall socket

Plug

When the plug is plugged in, electrons flow in the direction of the arrows. Normally, the two wires shown stretching from the plug to the bulb are both placed inside a single electrical cord.

---

*More precisely, in orbiting and spinning electrons. Electrons not only orbit around a nucleus, but they also spin on their own axis. This spin is a kind of electric current also, with magnetic effects. Electron spin, rather than orbital motion, causes the large effects seen in ordinary magnets.

But the treatment of the iron when a magnet is made locks many electron orbits into similar orientations. This causes the many atomic-level magnetic forces to add up to a large macroscopic effect, creating the forces you see between permanent magnets.

Permanent magnets can temporarily magnetize objects such as nails by forcing many of the nail's electrons to orient themselves in similar directions. This is why nails are attracted to permanent magnets.

Earth's magnetism is thought to be caused by molten iron and electric currents flowing in Earth's hot interior. This magnetism is, in turn, responsible for the behavior of compass needles, which are small permanent magnets, and for Earth's other magnetic effects.

The planetary model explains many chemical phenomena. Chemical reactions and other properties of the elements occur because of the behavior of the orbital electrons of the atoms of that element. Table salt, NaCl, makes a good example. Sodium (Na) combines readily with chlorine (Cl) because the Cl atom has a stronger attraction for electrons than does Na, so one electron from Na is attracted to Cl, leaving an overall positive charge on the Na and a negative charge on the Cl. The electric force between opposite charges then attracts and holds the Na and the Cl atoms together.

The property that really defines an **element** is the structure of the electron orbits in the atoms of that element. Two neutral atoms with the same number of electrons have the same chemical properties because their electron orbits have the same structure. The different elements are in fact numbered according to their number of electrons or, what is the same thing, their number of protons. This number is called the element's **atomic number**.

Since an atom is mostly empty space, what keeps solid matter solid? Why don't you fall through the floor, for instance? The planetary model's answer is that the repulsion of electrons by electrons keeps atoms from penetrating one another. Although an atom itself is mostly empty space, it occupies a considerable amount of space because of the forces that its electrons exert on surrounding objects. All contact forces can be interpreted at the microscopic level as forces by the orbital electrons in atoms on the orbital electrons in other atoms. Every time you touch or feel something, you are experiencing the electric force between orbiting electrons! In fact, all the forces in your daily environment come down to just two kinds of forces. The gravitational force explains weight, and the electromagnetic force explains all the contact forces and also the forces between charged or magnetized objects. Just two fundamental forces underlying all ordinary phenomena. This is quite a unification!

Despite the many successes of the planetary model, more recent experiments require it to be replaced with yet another model, the *quantum atom*. Today, the quantum atom is regarded as the correct theory, and the planetary model and the Greek model are regarded as approximations that are useful for many purposes.

■ **DIALOGUE 10**   The magnetic north pole of a compass needle points roughly toward Earth's geographic North Pole. What type of magnetic pole must be at Earth's northern end?

---

10. A magnetic *south* pole.

"THESE DAYS EVERYTHING IS HIGHER."

SUMMARY OF IDEAS AND TERMS

Wave   A disturbance that travels through a **medium** and that transfers energy without transferring matter.

Projectile motion or **particle motion**   An actual transfer of matter. The motion of a material object.

Wavelength   The distance from any point to the next similar point along a continuous wave.

Wave frequency   The number of vibrations that any part of the medium completes in each second. Also the number of complete wavelengths sent out in each second. Higher-frequency waves have shorter wavelengths and (assuming equal amplitudes) higher energies. Unit: vib/s = **hertz** (Hz).

Amplitude   The half-width of vibration of each part of a wave's medium.

Wavespeed   The speed at which a disturbance (a wave) moves through a medium.

Wave interference   The effects that occur when two waves of the same type are present at the same time and place. Interference can be either **constructive** or **destructive**, depending on whether the two waves reinforce or cancel each other.

Double-slit experiment with light   The interference of light coming from two synchronized sources is observed on a screen, demonstrating that light is a wave.

Electrically charged object   Any object that can exert or feel the electric force. There are two types of charge, **positive** and **negative.**

Electric force law   Electrically charged objects exert forces on each other at a distance. Like charges repel each other, and unlike charges attract each other.

Electric current   A flow or motion of charged particles. Electric currents in wires are due to electrons moving along the wire.

Magnetic poles   The ends of a permanent magnet. There are two types, north and south. Like poles repel, and unlike poles attract.

Magnetic force law   Charged objects that are moving exert and feel an additional force, called the **magnetic force**, beyond the electric force that exists when they are at rest.

Electromagnetic force   The total (electric and magnetic) force between charges.

Greek model of the atom   This model pictures the atom as a tiny indestructible object, like a small and rigid pea.

Planetary model of the atom   Atoms are made of negative **electrons**, positive **protons**, and uncharged **neutrons**. From one to a few hundred protons and neutrons form a tiny central **nucleus**, which the electrons orbit.

Ion   Any atom having an excess or deficiency of electrons.

Electric current   A flow of electrically charged particles.

Atomic number   The number of protons in an atom. Also the number of electrons in a neutral atom. An atom's atomic number determines its chemical properties and the **element** to which it belongs.

REVIEW QUESTIONS

WAVES

1. Describe the motion of the parts of the medium when a wave travels along a rope, along a Slinky, and across the water.
2. What is the difference between wave and projectile motions?
3. What does *hertz* mean?
4. Waves A and B have the same wavespeed, but A's wavelength is longer. Which has the larger frequency, or are they the same?
5. Choose the correct answer(s): Two different continuous waves on the surface of a pond would be expected to have roughly the same (wavespeeds, wavelengths, frequencies, energies, amplitudes, none of these).

INTERFERENCE

6. Two continuous water waves each have a 2-cm amplitude. What is the water surface's displacement (a) when a crest meets a trough? (b) When crest meets crest? (c) When trough meets trough?
7. Give an example of (a) a one-dimensional medium. (b) A two-dimensional medium. (c) A three-dimensional medium.
8. You tap one finger on the water's surface, on one side of a rectangular pan of water. Describe the waves rolling into the far side. How will the wave pattern at the far side change if you now begin tapping at the same rate, at two different points, using one tapping finger at each point?

LIGHT

9. Describe the experimental evidence supporting the claim that light is a wave.

10. How would the pattern on the screen in Figure 8.18 change if one of the slits were closed?
11. How do we know that the medium for light waves is not air?

ELECTRIC FORCE, MAGNETIC FORCE

12. When you rub two transparencies with tissue and hold them close together, they stand apart. Give two reasons that the force causing this cannot be gravity.
13. Cite the evidence supporting the claim that there are two, and only two, types of electric charge.
14. Give two reasons that the force between bar magnets cannot be the electric force.
15. What is meant by an *electric current*?
16. Magnetic forces are always caused by what types of objects?

THE PLANETARY ATOM

17. List several phenomena that require the planetary atom, rather than the Greek atom, for their explanation.
18. Name and briefly describe the three kinds of subatomic particles found in atoms.
19. Give evidence supporting the claim that most of an atom's mass is concentrated in a tiny nucleus at the center.
20. Explain what happens at the microscopic level in a wire when a battery creates an electric current in the wire.
21. What is an *atomic number*, and how is it related to the chemical elements?

HOME PROJECTS

1. Float a cork at one end of a tub of water. Drop a pebble into the other end to make a wave that reaches the cork. Observe the cork's motion. Does it vibrate back and forth? Up and down? Try it with larger pebbles. Is there some point at which the water's motion is no longer a true wave?

2. In a tub of still water, drop two small pebbles in at different places. Observe the interference effects as the two spreading circular waves cross. Do they appear to pass through each other without distortion? Try a tiny pebble and a larger pebble. Try tapping one finger at the water's edge to produce a continuous wave.

3. Use a razor blade to cut a thin slit in an index card, a millimeter or less in width and a few centimeters long. Look at a well-lit wall or other surface through the slit, focusing on the wall. Hold the card at an angle to narrow the width through which light can come. You should see rather striking light and dark lines running parallel to the long dimension of the slit. What causes these lines?

4. Ask a friend to stand at the far end of a carpeted room. Scuff your way across the rug until your noses are close together. What happens? Do you get a stronger effect when either of you are barefoot or when wearing shoes? Would the weather make any difference? Try touching wood, metal, and plaster. Explain any differences.

EXERCISES

WAVES
1. Is a mountain stream rushing downhill an example of wave motion? Defend your answer.
2. A gust of wind moves across a wheat field, causing a noticeable ripple that crosses the field. Is this ripple a wave? If not, why not? If so, then what is the medium?
3. When you send a brief wave down a rope or Slinky, it eventually dies out. What has become of the energy?
4. A cork floats on the water as a water wave passes by. What happens to the cork? Will the cork's vibrational frequency be related to the water wave's frequency, and if so, how?
5. Most waves have fixed wavespeeds that are determined by the properties of the medium through which the wave travels. What about a sports wave—does it have a fixed, predetermined wavespeed? Why or why not?

INTERFERENCE
6. The two waves shown in Figure 8.30 are moving toward each other along a string. Sketch the string during the interference of the two waves.

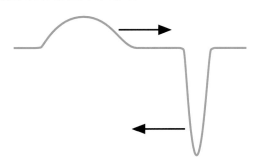

FIGURE 8.30
How will the two waves interfere?

7. Suppose the frequency of the waves in Figure 8.12 is 1 Hz. Describe the appearance of the water surface one-half second later (one-half second after the snapshot was taken). One second later.
8. What happens to the energy of the two waves in Figure 8.9 when they interfere destructively, as shown in the second of the three sketches? Did the energy vanish? (*Hint*: What are the parts of the rope doing as the wave moves? Which type of energy is the wave carrying?)
9. Two small sources of water waves send out circular waves. Each source produces waves whose amplitudes are 1.5 cm. What is the water level (relative to the undisturbed level) at the point where crest meets crest? Where trough meets trough? Where trough meets crest?

LIGHT
10. Do waves of any sort travel to Earth from the moon? Explain.
11. How do you know that light is not something that comes out of your eyes?
12. You shine two flashlights on a wall. Why don't you see an interference pattern?
13. Which of these objects are luminous (light sources), during their normal operation: camera, polished chrome, firefly, electric stove heating element, camera flashbulb, mirror, diamond, sun, moon?
14. Can light travel through a vacuum? Cite evidence for your answer.
15. If you were on the moon and you exploded a firecracker there, would you hear the explosion? Would you see the flash of the explosion? (*Note*: The moon has no atmosphere).

ELECTRIC FORCE, MAGNETIC FORCE
16. Since matter is made of electrically charged particles, why don't we and the objects around us feel electric forces all the time?
17. When you remove a wool dress from a garment bag, the sides of the bag might stick together. Explain.
18. Figure 8.31 shows an *electroscope*. The leaves (made of metal foil) normally hang down, but they spread apart

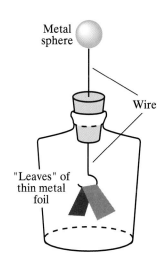

Metal sphere

Wire

"Leaves" of thin metal foil

FIGURE 8.31
Why do the leaves stand apart?

when the metal sphere on top touches a charged object. Explain.

19. How does the operation of the electroscope (previous exercise) demonstrate electric current?

20. You have three iron bars, only two of which are permanent magnets. Because of temporary magnetization (Section 8.7), all three bars at first appear to be magnetized. How can you determine which one is not magnetized, without using any other objects?

21. Suppose you have two iron bars (see the previous exercise), one magnetized and one not magnetized. Can you then determine which one is magnetized, without using any other objects?

22. When tiny scraps of paper are placed between two flat metal plates that have been oppositely charged (one plate charged positively and the other charged negatively), they bounce back and forth between the plates. Explain this phenomenon.

23. Highway trucks can become electrically charged as they travel. How can this happen? This can be dangerous, especially for gasoline tank trucks. How can it be prevented?

THE PLANETARY ATOM

24. While brushing your hair, you find that the hairs tend to stand apart from one another and that they are attracted toward the brush. Explain this in microscopic terms.

25. A covered mystery shoebox is placed on a table. What are a few ways that you could learn something about its contents without directly touching it or having it lifted?

26. After you walk across a rug and scuff electrons off the rug, are you positively or negatively charged?

27. According to Figure 8.24, what are the atomic numbers of carbon and helium? Roughly how much more massive is the carbon atom than the helium atom?

28. Some science fiction stories portray atoms as true miniature solar systems populated by tiny creatures. What are some differences, other than size, between our solar system and the planetary model of an atom?

29. An atom loses its two outermost electrons. How does the resulting ion behave when it is near a positively charged transparency? A negatively charged tissue? Would anything be different if it lost only one electron?

PROBLEMS

WAVES

1. If you doubled the wavelength of a wave without changing its wavespeed, what would happen to the frequency? What if you halved the wavelength?

2. Typical AM radio wavelengths are about as long as a football field, while typical FM radio wavelengths are about a meter long. Which one then has the highest (largest) frequency? If an AM wavelength is 100 times longer than an FM radio wavelength, then how do the frequencies compare quantitatively, assuming that the 2 wavespeeds are the same?

3. Mary attaches one end of a long rope to a wall, and sends waves down the rope by shaking the other end at a steady rate of 4 times every second (Figure 8.5). If the wavelength is 1.5 m, find the wavespeed. Suppose that Mary slows down to 2 times every second; assuming that the wavespeed is unchanged, find the new wavelength.

4. You note that waves roll past your stationary boat at a rate of one wave every 3 seconds, and you estimate the distance between waves to be 15 meters. What is the frequency of these waves? How fast are these waves moving?

5. By tapping his finger on the surface of a pond, Tom sends out waves at a frequency of 3 Hz. If the wavespeed

is 2 m/s, find the wavelength of these waves. Suppose that Tom speeds up his tapping to 4 Hz. Find the new wavelength, assuming that the wavespeed is unchanged.

6. Radio waves travel at lightspeed, 3×10^8 m/s. Find the wavelength of an FM radio wave whose frequency is 100 MHz (1 MHz = 1 million Hertz). Find the wavelength of an AM radio wave whose frequency is 1000 kHz (1 kHz = 1 kilohertz).

LIGHT

7. Red light of wavelength 0.6 micrometer (*micro* means *millionth*) shines from a stoplight to your eye, a distance of 10 m. How many wavelengths are contained in this distance?

8. Making estimates. Estimate the number of wavelengths of light in one millimeter.

THE PLANETARY ATOM

9. Making estimates. About how many atoms thick is a sheet of paper?

10. Making estimates. Which is bigger, an atom or a wavelength of light? Roughly how much bigger?

CHAPTER

Electromagnetic Waves

This chapter continues our quest to understand light. We found in the preceding chapter that light must be a wave because it shows interference. Then we looked at a new force: electromagnetism. Now we will see what electromagnetism has to do with light.

In this chapter we meet a radically new concept: fields. "Radically" is not too strong a term, for our studies so far have centered on the Newtonian worldview of a universe made of particles moving mechanically in empty space, and fields are essentially the opposite of particles. Particles are discrete and isolated in particular places, whereas fields are continuous and spread out over a region of space. Fields were first introduced into physics during the nineteenth century, as a way of visualizing electric and magnetic forces. Fields are characteristic of post-Newtonian physics, cropping up as gravitational fields, electromagnetic fields (this chapter), psi fields (Chapters 13 and 14), and quantized force fields (Chapter 18).

Section 9.1 explains the field concept, and Section 9.2 explains light in terms of electromagnetic fields. Section 9.3 asks whether fields are actually real physical entities or simply imaginary figments of our theories, and looks at the implications for the Newtonian worldview. The electromagnetic field theory of light leads to an understanding of a wide assortment of other lightlike "radiations," explained in Section 9.4. Section 9.5 studies the radiations coming to us from the sun.

With our knowledge of atoms and molecules (Chapter 2), energy (Chapters 6 and 7), and electromagnetic radiation (Chapters 8 and 9), we have the tools needed to discuss two global environmental problems: ozone depletion (Section 9.6) and global warming (Section 9.7). These topics continue one of this book's themes, namely the social context of physics. They are examples of a new kind of social issue: environmental effects that are truly global because they cannot be contained locally. If Earth is to pull through its current experiment with powerful technologies, we all must think seriously about such issues.

9.1 FORCE FIELDS: A DISTURBANCE OF SPACE

In order to visualize electric and magnetic forces, nineteenth-century scientists invented a new idea called a **force field**. Broadly speaking, a force field is the effect that the source of a force (a charged particle, for instance) has on the surrounding space: not on the things in the space, but on the space itself. Think of a field as a stress in space or a disturbance of space. When this idea was invented, force fields were generally considered to be only mental images—imaginary concepts that are made up because they are useful. At that time, forces were thought to affect only material objects and not space itself. It seemed impossible, for example, that a charged object could have real effects on distant regions of empty space. Nevertheless, scientists invented the notion that force fields could exist even in empty space.

The force field idea applies to every type of force. Let's begin with gravity. A **gravitational field** is said to exist throughout any region of space where an object would feel a gravitational force if it were placed there. A gravitational field is "the possibility of a gravitational force." For example, Earth's gravitational field exists wherever another object would feel a gravitational force from Earth if another object were placed there. This means that Earth's gravitational field exists even outside Earth in empty space. For example, Earth's gravitational field exists halfway between Earth and the moon, despite the fact that there is no material object there. Since a gravitational field can exist even in empty space, we must think of it as a property of space itself—as a stress or alteration of space.

Fields help us visualize forces by one object on another object some distance away. Consider the force by Earth on the moon. Rather than thinking of this force as something that happens directly between Earth and the moon, we can picture Earth's gravitational field filling the surrounding space in the way that smoke from a fireplace can fill a room. The moon then feels the force exerted by this field, in the same way that a person can smell the smoke.

Every material object can exert gravitational forces and so is surrounded by its own gravitational field. We can speak of the gravitational field of (or created by) the sun, the moon, a rock, your body, and so forth. Strictly speaking, each object's gravitational field fills all space, but from a practical point of view, most fields can be neglected at large distances from the objects that create them.

Just as gravitational fields surround massive objects, an **electromagnetic field** surrounds every charged object; this field exists wherever any other charged object would feel electromagnetic forces exerted by the first charged object. An electromagnetic field is the possibility of an electromagnetic (electric or magnetic or both) force. An electromagnetic field with no magnetic component is referred to as an *electric field*, and the term *magnetic field* is used in a similar way.

For instance, an electrically charged transparency creates an electric field around itself, and this field exists even in the absence of other material objects around the transparency. Even if the transparency is isolated in outer space, it still creates an electric field throughout the space around it. Magnetic fields surround electric currents and permanent magnets. Earth's magnetic field is created by electric currents within Earth and can be detected by a compass needle on Earth or in space.

We often draw "field lines" to represent fields and to help make the field concept more concrete. For example, Figure 9.1 shows the magnetic field lines in

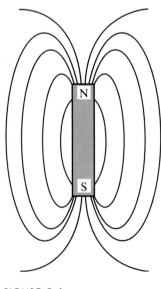

FIGURE 9.1
A representation of the magnetic field of a bar magnet.

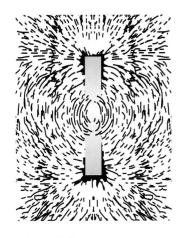

FIGURE 9.2
An experimental demonstration, using iron filings, that the field of a bar magnet actually does have the shape drawn in Figure 9.1.

the vicinity of a bar magnet. The lines are drawn parallel to the direction of the magnetic force that would be exerted on any other magnetic pole that happened to be placed in the vicinity of this magnet. Figure 9.2 is an experimental demonstration that a bar magnet's field actually does have the shape drawn in Figure 9.1. Figure 9.2 is made by sprinkling small iron filings in the vicinity of a bar magnet. The long slender filings are temporarily magnetized by the magnetic field (in the way that an iron nail is temporarily magnetized), and the magnetic forces on their north and south poles then cause the filings to line up parallel to the bar magnet's field.

All of these ideas can be made quantitative and mathematically precise, although there is no need to do that here. For example, Earth's gravitational field has been precisely measured and tabulated out to large distances. Such a mapping can be useful for satellite missions and other projects.

■ **DIALOGUE 1**   A proton is placed at rest in the middle of a vacuum chamber, an enclosure that has been emptied of all matter. Consider some point X near a particular corner of the chamber. Neglecting all influences other than the proton, is there an electric field at X? Is there an electric force at X? Is there a magnetic field at X?

9.2 THE ELECTROMAGNETIC WAVE THEORY OF LIGHT

Suppose you pull a rubber comb through your hair, scuffing electrons from your hair onto the comb. The charged comb then creates an electric field in the surrounding space, a field that can be detected by a rubbed transparency held near the comb.

If you quickly shake the comb once, up and back down, the electric field in the surrounding space will shake too. This can be detected by the transparency, which will shake in response to the comb's motion. The comb also creates a magnetic field during the brief time that it is moving. This temporary magnetic field could be detected as a brief shake of a magnet placed near the comb. In summary, the motion of the comb causes an alteration in the electromagnetic field around the comb, an alteration that can be detected by other charged objects and magnets (Figure 9.3).

There is an interesting question about this experiment, a question that many scientists asked during the nineteenth century: *When* will a distant detector feel the altered force? Is the effect instantaneous? Suppose you shake the comb precisely at noon. Does the transparency shake precisely at noon, too, or a little later? At that time, it was not easy to perform an experiment to answer this question.

During the 1860s, British physicist James Clerk Maxwell (Figure 9.4) did some hard thinking about electromagnetism. He gathered all that was then known about the subject and put it together in a single consistent quantitative theory. Maxwell's theory emphasized the field concept and described how electrically charged objects create electromagnetic fields.

FIGURE 9.3
If you shake a charged object such as a charged comb, other charged objects will shake in response.

---

1. Yes, no, no (because the proton is not moving).

FIGURE 9.4
James Clerk Maxwell, the "Isaac Newton of electromagnetism." He cast the principles of electricity and magnetism into the form of four equations involving the electric and magnetic fields created by electric charges and electric currents. Maxwell's equations contained a term that he inserted in order to enhance the symmetry—the beauty—of the theory. Experiment confirmed Maxwell's invention. The theory led to a unification of the electric with the magnetic force, and to the understanding of electromagnetic radiation which underlies much modern technology including radio, television, and lasers.

Three basic principles of the electromagnetic force were known at that time. The first was the electric force law, and the second was the magnetic force law. Stated in terms of fields, these two laws say that charged objects create electric fields and that electric currents create magnetic fields.

Maxwell's third principle was one we have not yet mentioned. It is called *Faraday's law*. This idea can be expressed entirely in terms of fields, without direct reference to electric charge or current. It states that any change in a magnetic field must create an electric field. This means that electric fields can be created in two ways: by electric charge (this is the electric force law) and by a change in a magnetic field (this is Faraday's law).

To these three known principles, Maxwell added a fourth idea that he invented for reasons that were as much aesthetic as physical. Maxwell was impressed with the symmetries, or similarities in form, between electric and magnetic effects. Because of these symmetries, it seemed to Maxwell that, since any change in a magnetic field creates an electric field (this is Faraday's law), then any change in an electric field ought to create a magnetic field (this is Maxwell's new principle). It turned out that Maxwell's invention was experimentally correct. Once again (compare Chapter 1) we see the importance of beauty in science.

All of this worked remarkably well. Maxwell's theory, which he formulated in precise mathematical language, described all known electromagnetic phenomena and it predicted an answer to our question about the time delay for electromagnetic forces. The key ingredient was the symmetry between electric and magnetic effects. Since a changing magnetic field creates an electric field and also since a changing electric field creates a magnetic field, these two fields can create and recreate each other. Once the fields at one point in space are altered—for instance, by giving a charged comb a single shake—Maxwell's theory implied that the alteration is transmitted outward as an alteration in the nearby fields a short time later, and these changing fields in turn transmit the change farther outward, and so forth. It follows from this that electric and magnetic forces are not transmitted instantaneously. There must be a time delay between the motion of a charged comb and the resulting force felt by a charged transparency.

Maxwell's theory showed that an electromagnetic field behaves like a continuous elastic substance such as foam rubber. The different places in the field are connected in the way the different places in a piece of foam rubber are connected. Just as with elastic foam, if you disturb the field at one point, these connections will cause the disturbance to move outward through the field. Any alteration of an electromagnetic field, as might be caused by shaking a charged comb, is transmitted as a pulse, a wave pulse, that spreads outward through the field. But the medium through which such a pulse travels is not foam rubber and in fact is not a material (made of matter) medium at all. Rather, the medium is the electromagnetic field itself. Any such disturbance that moves through an electromagnetic field is called an **electromagnetic wave**.

You can't directly see electromagnetic waves in the way that you can see water waves, because the medium for electromagnetic waves is an invisible electromagnetic field instead of a material substance such as water. Nevertheless, an electromagnetic wave can be detected by other charged or magnetized objects at some distance from the source of the wave and at some later time after the wave was sent out. Figure 9.5 pictures these invisible waves.

FIGURE 9.5
When you shake a charged object
back and forth, it sends out an
electromagnetic wave in all directions.

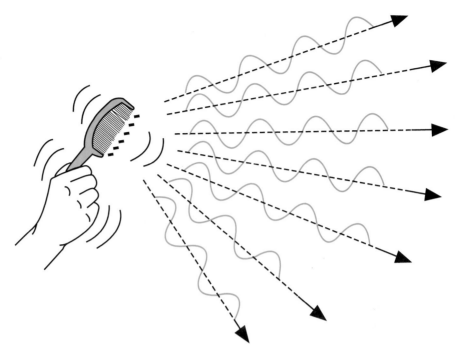

FIGURE 9.5
When you shake a charged object
back and forth, it sends out an
electromagnetic wave in all directions.

This all was worked out quantitatively in Maxwell's theory. The theory predicted not only that there must be a delay in the transmission of electromagnetic forces but also how long that delay must be for a given separation between the source and the recipient of any such force. In other words, the theory predicted the speed of transmission of electromagnetic waves and forces. The predicted speed was 300,000 km/s.

This particular speed had come up before, in experiments performed nearly two centuries before Maxwell invented his theory. But these previous experiments seemed entirely unrelated to the electromagnetic effects that Maxwell was studying. This speed, 300,000 km/s, was the known speed at which light travels!

H O W   D O   W E   K N O W ?

Measurement of lightspeed

People once thought that light requires no travel time—that its speed is infinite. Light certainly travels much faster than sound, as you can verify whenever you see lightning before you hear thunder.

Galileo was one of the first to try to measure lightspeed, by measuring the total round-trip time for light to travel to a distant mountain and back again. His experiment didn't work because the time turned out to be far too short to measure using Galileo's timing methods. Either greater timing accuracy or a greater travel distance was needed.

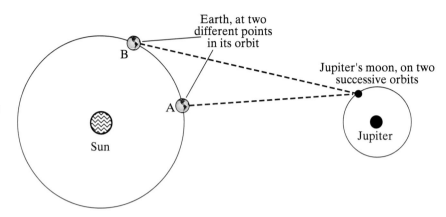

FIGURE 9.6
If Earth moves from point A to point B while Jupiter's moon is orbiting Jupiter one time, the orbital time as measured on Earth will be longer than the true orbital time, because it takes longer for light to get from the moon to point B than it takes for light to get from the moon to point A. This effect creates a variation in the measured orbital time that can be explained by assuming that light has a finite, not infinite, speed.

The first evidence for a finite, and not infinite, speed of light came from astronomical observations several decades later. Pointing telescopes at a moon of Jupiter, astronomers found that the time they measured for this moon to orbit Jupiter did not remain constant, as they had expected. Why should a moon take longer for some orbits than for others? Earth's moon takes 27.3 days, every time. The astronomers found that these variations were not caused by Jupiter's moon at all but were instead related to Earth's motion around the sun. They found that the variations were just what would be expected if the light from Jupiter's moon traveled at a finite, and not infinite, speed. The light from Jupiter's moon would then arrive at Earth sooner or later, depending on whether Earth was closer to or farther from Jupiter in Earth's orbit around the sun (Figure 9.6). From the measured variations in orbital time, lightspeed could be estimated.

So electromagnetic waves were predicted to spread through electromagnetic fields at the speed of light. Was this because light actually was an example of Maxwell's predicted electromagnetic waves, or was it just a coincidence? Perhaps light had nothing to do with electromagnetism. Scientists knew of no way to verify the tantalizing suggestion that light was an electromagnetic wave. So Maxwell's work didn't attract much attention until two decades later, when it became possible to check Maxwell's theory directly, by causing charged objects to oscillate and observing the effects some distance away.

HOW DO WE KNOW?

Verification of the theory of electromagnetic waves

One way to verify Maxwell's theory would be to shake a charged object at the frequency of visible light, about 10^{15} hertz, and to see whether the shaking created light. But such high frequencies of oscillation are hard to achieve, even today.

German physicist Heinrich Hertz (the hertz is named for him) figured out how to do an experiment of this sort, but at a frequency of only about 10^9 hertz, far below that of visible light. He constructed an electric circuit that contained a small open space or air gap. Ordinarily, such a gap stops the flow of electric charge, but Hertz built the endpoints of the gap in such a way that large amounts of charge could build up on them. After enough buildup, the charge was forced to jump across the gap. We observe such charge jumping as a spark. Lightning is a spark of this sort.

In Hertz's circuit, the first jump of charge triggered a brief series of such jumps back and forth across the gap, at a rate of about a billion per second. If Maxwell was correct, these oscillations should create electromagnetic waves with a frequency of a billion hertz.

At some distance from this predicted source of electromagnetic waves, Hertz placed a second circuit. This circuit was entirely passive, with no battery or other internal source of electric current. If Maxwell's theory was correct, electromagnetic waves from the first circuit should cause an electric current to oscillate in the second circuit, also at a billion hertz. The transmission from one circuit to the other should occur at lightspeed.

Hertz performed several experiments during 1887/1888. The results entirely confirmed Maxwell's predictions. Hertz found oscillations in the second circuit at the expected frequency. He showed that the transmission occurred at a speed that was at least the same order of magnitude as lightspeed (as you might expect, this was a difficult measurement). And he showed that these waves had many other properties that light waves have, such as reflection.

Although Hertz's waves were not light waves, his work convinced scientists that electromagnetic waves really existed and that light is actually an electromagnetic wave. As a by-product, Hertz's basic scientific work came to the attention of an ingenious Italian inventor named Guglielmo Marconi, launching the radio and television revolution. Today, we know Hertz's waves as radio waves.

Maxwell and Hertz had uncovered a vast new range of physical phenomena, invisible waves similar to light. In Hertz's words, "The connection between light and electricity, of which there were hints and suspicions and even predictions in the theory, is now established. . . . Optics [the study of visible light] appears merely as a small appendage of the great domain of electricity. We see that this latter has become a mighty kingdom."

It is a stunning unification: Maxwell's theory correctly describes electricity, magnetism, light, and radio. All of these are different aspects of one underlying reality: electric charge.

We summarize these ideas as follows:

THE ELECTROMAGNETIC WAVE THEORY OF LIGHT
Every vibrating charged object creates a disturbance in its own electromagnetic field, which spreads outward through the field at 300,000 km/s.*
Light is just such an electromagnetic wave.

---

*More precisely, the speed of light, and all electromagnetic waves, is 299,792.458 km/s in a vacuum. When traveling through matter, however, light is slower than this because it is continually absorbed and reemitted by atoms.

■ **DIALOGUE 2** If you electrically charge a comb by running it through your hair and then shake it back and forth at a frequency of 1 Hz, will this produce an electromagnetic wave? What will be the wave speed? The frequency? How long is the wavelength?

■ **DIALOGUE 3** Hertz's waves had a frequency of 10^9 hertz. How many megahertz is this? How many kilohertz? Is this on your radio dial?

■ **DIALOGUE 4 MAKING ESTIMATES** About how long does it take light to get to your eyes from a lightbulb in your room?

9.3 THE DECLINE OF THE NEWTONIAN UNIVERSE

Since ancient Greek times, scientists have generally thought that the universe was made of tiny material particles, atoms. As Democritus put it, there are only atoms and empty space (Chapter 2). It is a view that is remarkably compatible with Newton's physics, which can be thought of as the rules according to which atoms move. This general worldview—atomic materialism coupled with Newtonian physics—dominated science at least during the eighteenth and nineteenth centuries (Section 5.5).

When the electromagnetic field was first proposed around 1830, it was thought of as only a useful way to picture electromagnetic forces and not as a real physical entity. Then Maxwell and Hertz showed that waves can travel through electromagnetic fields and that light is one example of these electromagnetic waves. This meant that electromagnetic fields were not just a useful fiction. They were physically real, as real as light.

The most convincing argument for the reality of electromagnetic fields comes from the conservation of energy. Suppose that a radio transmitter sends a message (an electromagnetic wave) to a receiver on the planet Mars and that the message's travel time is 20 minutes. Energy must travel from the sender to the receiver because it takes energy to cause the receiver to respond. Where is this energy during the 20 minutes between sending and receiving the message? Not in the sender. Not in the receiver. And energy never just vanishes. So it must be in the space between sender and receiver, in the electromagnetic field.

So *electromagnetic fields contain energy*. And to a physicist, nothing is more real than energy.

"Empty space" is a contradiction in terms. Even when devoid of matter, space cannot be empty. The universe is filled with a sea of nonmaterial and mostly invisible fields.

Physicists resisted giving up the Newtonian clockwork universe. So ingrained was the Newtonian worldview that scientists could not contemplate that energy

---

2. Yes, you produce an electromagnetic wave whose wave speed is 300,000 km/s and whose frequency is 1 Hz. Since you produce one complete wave every second and the wave travels outward at 300,000 km/s, the wavelength must be 300,000 km.

3. One thousand megahertz; 10^6 kilohertz; this is on neither your AM nor your FM radio dial (please verify this by looking at your radio dial).

4. The travel time for light is the distance to the lightbulb divided by lightspeed, 3×10^8 m/s. To make the arithmetic easy (remember that in estimates you want to choose approximate numbers that make the arithmetic easy), suppose the distance is 3 m: $3 \text{ m}/(3 \times 10^8 \text{ m/s}) = 10^{-8}$ s, or 0.00000001 seconds, a hundredth of a millionth of a second.

might exist apart from tiny material particles. They developed the idea that a gaslike material substance called **ether** filled all space. Ether was assumed to be a form of matter but made of some unknown substance rather than of the atoms that are familiar to us. Electromagnetic forces and other forces that act at a distance were assumed to be transmitted by ether. Light and other electromagnetic waves could then be explained mechanically, in terms of the motions of material ether particles and in keeping with the Newtonian tradition. Maxwell, Hertz, and others accepted this ether theory of the electromagnetic force. After two centuries of the Newtonian worldview, it was difficult to think in any other way.

Early in the twentieth century, Albert Einstein was one of the first to break out of this mold. His work showed that the ether theory had to be rejected. But surprisingly, this had no effect on Maxwell's theory or on the interpretation of light and radio as electromagnetic waves. It only affected the underlying mechanistic interpretation of the electromagnetic field. After Einstein's work, electromagnetic fields could no longer be interpreted as properties of a material substance.

So the electromagnetic field turned out to be philosophically revolutionary, the first of many post-Newtonian physical ideas. Although not made of particles and not made of any material substance at all, electromagnetic fields are real nonetheless. Apparently the universe is not made like a mechanical clock.

Although the Newtonian worldview still dominates much popular culture and even lies behind many scientists' intuitive view of nature, the mechanical universe is seriously out of tune with twentieth-century physics. The two major post-Newtonian theories, relativity theory and quantum theory, contradict both the specific predictions and the conceptual underpinnings of Newtonian physics. Physics is still in the middle of the post-Newtonian revolution, and it is not clear what new scientific worldview will emerge. But it is clear that the Newtonian assumptions have broken down.

9.4 THE COMPLETE SPECTRUM

Light is only a small portion of the "mighty kingdom" (as Hertz puts it) of electromagnetic waves. The entire kingdom is called the **electromagnetic spectrum** (Figure 9.7). *Spectrum* simply means range, or types.

As you can see from the figure, scientists have observed electromagnetic waves over an enormous range of wavelengths. Wavelengths are arranged from the longest at the bottom to the shortest at the top, with typical objects having the size of these wavelengths listed for comparison. Frequencies are shown from the lowest at the bottom to the highest at the top, with typical sources of these frequencies listed.

In order to display a large range, Figure 9.7 shows the wavelengths and frequencies on a *logarithmic scale* in which each increment is a factor of 10: 1, 10, 100, 1000, and so forth. A *linear scale*—for example, 10, 20, and 30—could not do justice to the entire range. Logarithmic scales are useful to graph quantities that span a large range.

All these waves are basically the same phenomenon, namely, an electromagnetic field disturbance that is created by a vibrating charged object and that travels outward through the field in the form of an electromagnetic wave moving at

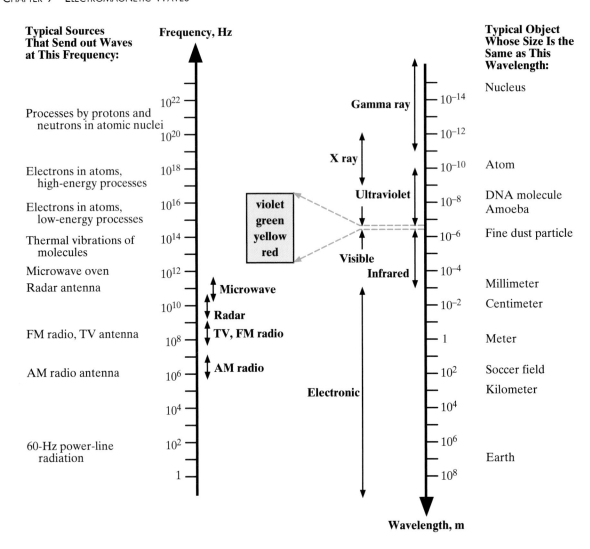

FIGURE 9.7
The electromagnetic spectrum. There are no definite ends to the spectrum and no sharp boundaries between the regions.

lightspeed. All of these waves can be received by other charged objects that are caused to move by the wave as it passes by. All can travel through empty space.

All these waves carry energy. The energy of an electromagnetic wave is called **radiant energy**. For example, the energy in transit from the sun to Earth is radiant energy. Electromagnetic waves are often called **electromagnetic radiation** because they "radiate" out from charged objects. Since higher frequency means higher energy (provided that other things such as amplitude are not changed), energies increase as we move up the scale from bottom to top.

It is useful to arrange the electromagnetic spectrum into the six regions shown in Figure 9.7. These regions correspond to different ways of either sending or receiving electromagnetic radiation. The longest wavelengths, down to about a millimeter, form the **electronic waves**, or radio waves. Humans can create and control these waves electronically by causing electrons to vibrate in human-made electric circuits. Hertz's waves fall into this category, and so does a lot of modern technology. AM radio waves at around 1000 kilohertz (10^6 Hz) and FM radio and TV waves at around 100 megahertz (10^8 Hz) are created by electrons vibrating along a metal antenna that is part of an electric circuit. Radar and microwaves, with frequencies up to a trillion (10^{12}) hertz, also are created electronically.

Many natural processes also create radio waves. Radio astronomers use radio receivers pointed at stars or other astronomical objects to learn about the universe. In fact, astronomical objects produce electromagnetic radiation in all parts of the spectrum. During the past few decades, many new sorts of receivers or "telescopes," stationed on or above Earth, have produced an explosion of astronomical knowledge.

Infrared radiation has wavelengths ranging from 1 millimeter to below 1/1000 of a millimeter, which is the size of a fine particle of baby powder. Infrared is typically created by the chaotic thermal motion of molecules due to their thermal energy. Since all objects have thermal energy, all objects produce infrared. Hotter objects produce more than cooler objects do. Infrared detectors, such as certain infrared-sensitive chemicals, can detect warmer objects against a cooler background, which is the basis for night-vision devices and infrared photography.

Although you cannot see infrared radiation, you can feel it. Since it is created by thermal motion, it is not surprising that it has the proper frequency to shake molecules into thermal motion. So it can warm the objects it hits. When you feel the warmth of a fire or a hot plate at some distance away, you are using your skin as an infrared detector, "seeing" with your skin. Through competitive evolution, many animals have highly developed infrared sensors for nocturnal vision.

Most higher animals have sensors that detect a narrow range of frequencies just above infrared. This range of **visible radiation** or *light* has wavelengths spanning less than a single factor of 10, centering on 5×10^{-7} m. This is smaller than the finest dust particles and some 5000 times larger than an atom. Light is created typically by electrons moving within individual atoms.

This region's defining characteristic is simply that the human eye is sensitive to it. Light waves entering the pupil of the eye strike the retina at the back (Figure 9.8). The retina is covered with light-sensitive cells that act like tiny antennae to receive electromagnetic waves in the visible range. Some cells respond differently to different wavelengths, and the brain interprets these as different colors. All of these cells contain light-sensitive chemicals, within which the light wave's vibrating field causes electrons to move, which in turn initiates chemical reactions in the cell. The signal received by the retina is transmitted to the optic nerve, which transmits the signal along the optic nerve to the brain.*

Suppose that you have a variable-frequency source of electromagnetic waves and that you set it to produce green light with a frequency around 6×10^{14} Hz. If you gradually decrease the frequency, the green light will change from green to yellow, then orange, and finally red. As you continue decreasing the frequency, the red becomes deeper and darker until, at about 4×10^{14} Hz, the frequency is so low that your retina can no longer respond to it. The source no longer emits

Light beam
focused on
retina

Light beam
entering eye

Lens

Retina

Optic
nerve

FIGURE 9.8
The human eye.

---

*Nerve cells transmit signals not as electromagnetic waves but rather as waves of chemical reactions moving at only about 30 m/s along the cell.

Heater

X rays

Electrons

High voltage

Metal plate creates X rays

FIGURE 9.9
The operation of an X-ray tube.

visible light. The waves have crossed the boundary into *infra-* (below) red. The source still radiates, but your eye cannot detect it.

Now go in the other direction. Beginning with green, increase the frequency. The color changes from green to blue to violet. The violet light darkens until, around 8×10^{14} Hz, your eye can no longer detect it. The waves have crossed into the *ultra-* (above) violet region. **Ultraviolet radiation** is created in the same way that light is created, by electrons moving within individual atoms. Although similar to light, ultraviolet's higher energy has important consequences.

Ultraviolet radiation has the proper frequency to shake many biological molecules, so it is readily absorbed by living things. And it has enough energy to split molecules, which can disrupt or kill living cells. When absorbed by DNA, ultraviolet radiation can cause genetically significant changes, or *mutations*. This can produce a cancerous growth if the mutated cell reproduces many copies of itself, or it can lead to a mutant offspring if the cellular mutation occurs in a sperm or egg cell.

X-ray radiation also comes from electrons in individual atoms, but only from the highest-energy electron activities within atoms. X-ray wavelengths span a range around 10^{-10} m, which is about the size of an individual atom. Humans make X rays in high-energy X-ray tubes (Figure 9.9). Electrons are boiled off a thin heated wire filament and are accelerated toward a metal plate at the other end of the vacuum tube. When the electrons smash into this plate, they cause the plate's atoms to create X rays.

X rays have important interactions with biological matter. They have enough energy to ionize molecules within biological cells, that is, to knock the electrons right out of some molecules. The evicted electrons can then move on to cause still more damage. Like ultraviolet, this radiation can cause mutations and cancers. Radiation energetic enough to ionize biological matter is called **ionizing radiation**. X rays and gamma rays are ionizing radiations, but ultraviolet has insufficient energy to be an ionizing radiation. Because X rays are able to penetrate deeply into biological matter, they can be put to the useful cause of examining the interior of the human body without surgery.

There is a certain logic to our tour through the electromagnetic spectrum. As we move to smaller wavelengths, we are moving toward higher frequencies, which imply higher-energy radiation, which in turn implies higher-energy processes to create the radiation. We also have progressed toward processes that occur in smaller and smaller regions of space: Electronic waves are created in macroscopic electric circuits; infrared is created in molecules; and the next three (visible, ultraviolet, and X ray) are created in atoms.

It should come as no surprise then that the shortest-wavelength radiation, **gamma radiation**, carries the highest frequency and highest energy and comes from the highest-energy processes and the smallest regions of space. Gamma rays are created within atomic nuclei by high-energy nuclear processes caused by the strong forces that hold the nucleus together. Gamma rays are created in radioactive materials (see Chapter 15) and in the nuclear reactions known as *fission* and *fusion* (see Chapter 16).

Like X rays, gamma rays are a form of ionizing radiation and can damage biological matter. But this very feature is often put to use to destroy diseased cells and so cure some cancers. Since gamma-ray wavelengths are much smaller than individual atoms, atoms cannot readily respond to them, and so they can penetrate rather deeply into matter.

Consider the electromagnetic waves all around you. The room you are in is full of them. Hundreds of television and radio broadcasts, radio pulsations from neutron stars, radio noise from millions of normal stars, the faint cosmic background radiation, possibly communications from extraterrestrial life, radiations from the sun and from the center of our galaxy, and much more are passing through your room right now. Your body is equipped to receive directly only the tiny visible portion of the complete spectrum of these waves. With the proper receiver, you could tune in to any of the other frequencies. The universe would appear far different in other wavelength ranges and would appear complex indeed if we could directly receive the entire spectrum. The reality that meets the eye is only a tiny fraction of nature's mighty kingdom.

■ **DIALOGUE 5**   Your radio is tuned to 98 on the FM dial. What electromagnetic wave frequency is your radio receiving? How fast is this wave traveling? Roughly how big is the wavelength? How about the sound that comes from this radio—is it moving at the same speed as the radio wave? Is sound an electromagnetic wave?

■ **DIALOGUE 6 MAKING ESTIMATES**   In the following list, which have wavelengths much bigger than your room (a few meters), which have wavelengths between a millimeter and a few meters, and which have wavelengths of less than a millimeter? AM radio, light, electromagnetic waves from the alternating current that oscillates 60 times each second in your house circuits, warming rays from a fire, rays from a microwave oven, radar, electromagnetic radiation from shaking an electrically charged blouse that you remove from the dryer.

9.5 SOLAR RADIATION: THE LIGHT FROM OUR STAR

The sun, "Sol," transmits electromagnetic waves in every region of the spectrum. Most solar radiation is visible, infrared, and ultraviolet and is created at the sun's visible surface. Other solar radiation is created in a rarefied, very hot gas that surrounds the sun in the same way that Earth's atmosphere surrounds Earth. Processes within the sun's atmosphere create high-energy X rays and some gamma rays, along with radio waves. The intense radiation created by high-energy processes deep within the sun is absorbed and altered within the sun, and little of it escapes directly.

Figure 9.10 graphs the relative amounts of radiant energy created by the sun at different wavelengths. When you sit in the sunlight, your eyes detect the sun's visible radiation, and your skin detects its infrared as warmth. Your skin also detects ultraviolet, but you don't notice it until a little later, as the cellular damage that we call *sunburning*.

The solar radiation graph (Figure 9.10) reaches its highest point around a yellow-green wavelength of 5.5×10^{-7} m and remains high throughout the visi-

---

5. 98 megahertz = 98×10^6 hertz = 98,000,000 hertz. 3×10^8 m/s. Roughly 5 m (Figure 9.7). Sound moves much more slowly; sound is not an electromagnetic wave.

6. Use Figure 9.7. Much bigger than a few meters: AM radio, waves from alternating current, waves from the blouse. One millimeter up to a few meters: microwaves, radar. Less than 1 mm: warming rays, light.

FIGURE 9.10
The relative amounts of energy at different wavelengths in the solar spectrum. Most of the sun's radiant energy is in the ultraviolet, visible, and infrared portions of the electromagnetic spectrum.

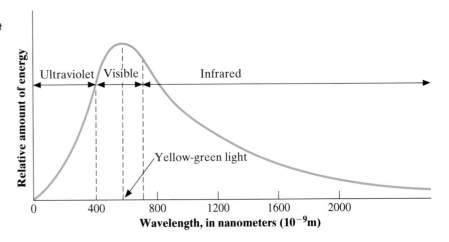

ble region. Solar energy is most concentrated throughout the region that our eyes can detect.* Even more strikingly, our eyes have evolved in such a way that their maximum sensitivity is to yellow-green light, precisely where the solar spectrum peaks. Some municipalities take advantage of this by painting their fire trucks yellow-green. What a strange coincidence that all this radiation happens to occur precisely in the region that our main sensors, our eyes, can detect. How lucky for us! But of course it was not just luck. Through biological evolution, animals developed detectors of the dominant radiation reaching Earth. An animal whose sensitivity peaked in, say, the microwave region would be at a biological disadvantage, because there isn't much naturally occurring microwave energy by which such an animal could see.

The amount of solar energy reaching Earth is different at different locations, in different seasons, in different weather conditions, and at different times of the 24-hour day. In the United States, an average 200 watts (200 joules every second) strikes every square meter. This figure, 200 watts per square meter, is an average that takes into account all of the mentioned variations.

■ **DIALOGUE 7 MAKING ESTIMATES** Photovoltaic cells are devices that transform solar energy into electric current. If such devices were 100% efficient, about how much area would need to be covered by these cells in order to provide the average of 1 kilowatt of electric power that a typical family home uses? Actual photovoltaic cells are only about 15% (one-seventh) efficient. At this efficiency, how much area must be covered? Could you put this on your roof?

■ **DIALOGUE 8** Once radiant energy from the sun is absorbed by your skin, is it still radiant energy? What has it become?

---

*More precisely, the solar energy per unit wavelength range is largest in this region.

7. One kilowatt (1000 watts) of solar energy falls on 5 square meters (1000/200) of surface. At an efficiency of about one-seventh, it would take seven times this much area: 35 square meters. If square shaped, this would be about 6 m on a side. You could probably fit it on your roof.

8. No. It has been transformed into thermal energy, plus some chemical energy from chemical reactions caused by ultraviolet radiation.

9.6 GLOBAL OZONE DEPLETION

The concept is not obvious: a perfume spray in Paris helps to destroy an invisible gas in the stratosphere and thereby contributes to skin cancer deaths and species extinction half a world away and several generations in the future. Neither traditional environmental law nor traditional diplomacy offers guidelines for confronting such situations.

RICHARD BENEDICK, CHIEF U.S. NEGOTIATOR OF THE MONTREAL OZONE TREATY

For the first time in my life I saw the horizon as a curved line. It was accentuated by a thin seam of dark blue light—our atmosphere. Obviously this was not the ocean of air I had been told it was so many times in my life. I was terrified by its fragile appearance.

ULF MERBOLD, GERMAN ASTRONAUT

Ozone depletion is a classic science and society problem, involving technology, politics, business, scientific understanding, public understanding, and the daily decisions you and I must make. Because local actions have worldwide effects on ozone, it is perhaps the first truly global environmental problem. Public understanding of science is essential if democracies are to cope with such problems. Ozone is a good historical case study because its societal history is mostly completed: Humankind has damaged Earth's ozone, belatedly recognized that damage, and done nearly all that can be done to rectify it. The verdict, as to how bad the damage is and what its effects will be, is now largely back in nature's hands. That verdict will not be in until the mid–twenty-first century.

We can best capture the social lessons of the ozone story by treating it historically. A good place to begin is in 1930 when the General Motors Corporation first synthesized **chlorofluorocarbons (CFCs)**, molecules made from atoms of chlorine, fluorine, and carbon, for their Frigidaire refrigerators. CFC molecules are chemically *inert* or *stable*, meaning that they do not readily react with other substances. They normally form a gas but become liquid when put under high pressure. Being inert, they are nontoxic in the human body, noncorrosive in mechanical devices, and nonflammable in the atmosphere. And they are inexpensive to make.

Such a chemical can have many uses, one of which is as a coolant. Refrigerators and air conditioners operate in the same way that heat engines do, only in reverse (Figures 9.11 and 9.12). Just as heat engines use hot gases as the "work-

FIGURE 9.11
Energy-flow diagram for a heat engine. The input thermal energy flows naturally "downhill" to a lower temperature, while some of the energy is used to do work.

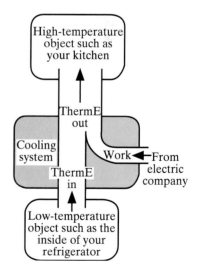

FIGURE 9.12
Refrigerators operate like heat engines in reverse. An outside energy source does work to push thermal energy "uphill," from a low to a high temperature.

TABLE 9.1

Composition of the atmosphere

Molecule	Concentration
Major constituents	
N_2	78.084 %
O_2	20.946 %
Ar	00.934 %
total	99.964 %
Trace gases (in ppm[a])	
CO_2	350
H_2O	20–20,000
Ne	12
He	5
NO_2	2
CH_4	2
Xe	2
Kr	1
O_3	<1
H_2	<1
NO	<1
CO	<1
N_2O	<1
SO_2	<1
NH_3	<1

[a]Parts per million.

ing fluid" that does the work, so refrigerators and air conditioners use a "coolant" that has work done on it while extracting thermal energy from a refrigerator or from a house.

CFCs had just the right properties and soon became a universal coolant, marketed as *Freon*. Production soared after 1931. In the 1940s, CFCs were found to be useful as pressurized gases to propel aerosol sprays. Then they began to be used as blowing agents to form lightweight, closed bubbles in plastics that could then be used as "thermal insulators" against heat and cold and to make soft foam rubber. Beginning in the 1950s these wonderfully inert chemicals created the air-conditioning revolution that facilitated America's shopping malls and summer automobiling odysseys and population shifts to Houston, Los Angeles, and other Southwestern cities. Later, as useful solvents to clean small electronic parts, they assisted the computer revolution.

CFCs created a lot of business and little fuss until 1973 when scientists began to ask where all these inert and hence supposedly harmless gas molecules might be drifting. After all, being inert they were indestructible by chemical reactions, and so essentially all the CFCs manufactured since 1930 should still be in the atmosphere. But where? And what became of them there? During the previous four decades of production, nobody had thought to ask such questions.

In 1973, two university scientists were studying chemical emissions from rocket launches into the upper atmosphere or **stratosphere**, 10 to 50 kilometers overhead. They were concerned about how the emissions reacted with O_3, or **ozone**. Ozone is one of a long list of **trace gases** in the atmosphere—gases that, even when they are all added together, make up far less than 1% of the atmosphere. Table 9.1 shows only a few of these trace gases, namely those whose concentrations are around one part per million (that is, one molecule per million atmospheric molecules) or larger. The list would be far longer if it were extended down to one part per billion!

Ozone is produced naturally in the stratosphere from O_2 with the help of ultraviolet radiation from the sun at that altitude. The radiation breaks up the O_2 molecule, and the resulting oxygen atoms then combine with O_2 to create O_3. But ozone is not a particularly stable molecule, and it can be broken down by various chemical reactions including, as the scientists discovered, reactions with the chlorine released by rockets. The ozone-destroying reaction is

$$Cl + O_3 \rightarrow ClO + O_2$$

This produces normal molecular oxygen, O_2. Under the bombardment by ultraviolet radiation at that altitude, a second reaction then occurs:

$$ClO + ClO + sunlight \rightarrow Cl + Cl + O_2$$

The second reaction frees the chlorine, which is then able to destroy more ozone. Scientists found that a single Cl atom broke down about 100,000 ozone molecules.

This didn't seem too important, because not much chlorine is released from space rockets. But the next year, in 1974, other university chemists suggested a more alarming possibility. Mario Molina and Sherwood Rowland (Figure 9.13) discovered that because CFC molecules are inert and gaseous, they are not

FIGURE 9.13
Paul Crutzen, Mario Molina, F. Sherwood Rowland.
The theory of stratospheric ozone depletion by means of human-made chlorine compounds was recognized by the awarding of the 1995 Nobel Prize in chemistry—an award given only to those whose research has been widely accepted by the mainstream scientific community—to the three scientists who discovered it. During the 1960s, Crutzen was among the first to study ozone destruction. His studies on the effects of nitrogen compounds in the atmosphere continue to motivate concerns about the use of fertilizers and about emissions from high-flying jet planes, especially supersonic passenger jets. In 1974 Mario Molina, then a postdoctoral student in Rowland's lab at the University of California at Irvine, and F. Sherwood Rowland uncovered the threat posed by adding CFCs to the atmosphere. In conferring the Prize on Rowland and Molina, the Nobel Committee commended them for having "contributed to our salvation from a global environmental problem that could have catastrophic consequences."

chemically broken down or rained out in the lower atmosphere. Instead, they migrate slowly up to the stratosphere, where they may remain intact for decades to even centuries. Molina and Rowland theorized that solar ultraviolet radiation should eventually break down CFCs, releasing large quantities of ozone-destroying chlorine into the stratosphere.

Sratospheric ozone destruction poses a problem for humankind. The problem is that the ozone molecule vibrates naturally at ultraviolet frequencies and so absorbs much of the sun's radiation at these frequencies. Because this protects us from the sun's biologically harmful ultraviolet radiation, stratospheric ozone is important to life on Earth.* Furthermore, being extremely dilute in the atmosphere, ozone is easily altered by human activities. Out of every 10 million atmospheric molecules, only 3 are ozone! If compressed to atmospheric pressure, ozone would form a layer only about 2 millimeters thick. We see in ozone the potential importance of each of the many atmospheric trace gases: This dilute gas absorbs a significant fraction of the sun's ultraviolet radiation.

In 1974, nobody had dreamed that CFCs, seemingly safe and thoroughly tested, could pose dangers many kilometers above Earth's surface. In fact, many

---

Ground-level ozone, on the other hand, is a toxic pollutant. It is a part of automobile exhaust and is the primary component of urban smog.

FIGURE 9.14
Atmospheric chemist Susan Solomon in Antarctica. She organized the first National Ozone Expedition in 1986 and devised the theory that correctly explained the destruction of ozone by chlorine compounds in the stratosphere over Antarctica.

From time to time we are all a bit taken aback by the fact that this thing that fascinates us from an intellectual point of view has serious consequences. I know a number of others feel as I do, that it would be nice to be doing medical research . . . and say "Wow, I have a cure for some terrible disease." But we don't have that kind of message. Still, intellectually it is incredibly exciting. It's so complicated and so beautiful.

SUSAN SOLOMON, ATMOSPHERIC OZONE SCIENTIST

people regarded the suggestion that coolants and spray cans could cause such a catastrophe to be extreme and absurd. It certainly violated the popular opinion that human activities were far too puny to alter the global environment. But with CFC annual production running at 800,000 tonnes in 1974, and with CFC residence times in the atmosphere running at a century or more, some people became alarmed.

A debate, of a type that has now become familiar, ensued, continuing from 1974 to 1978. The chemical industry argued that the theory was speculative and that there should be no economically damaging CFC restrictions until there was clear evidence of harm. Environmentalists countered that the theory was plausible enough that we should not risk further delays. Consumer pressures for non-CFC goods played a role. In 1978, the United States and a few other governments initiated a ban on the least-essential CFC technology, spray propellants. It was the first time that a substance suspected of causing global harm had been regulated before the effects had been fully demonstrated.

This partial concession to environmental concerns ended the debate. By 1980, the problem had faded from public view. But CFCs continued building up in the atmosphere, because U.S. spray propellants formed only 25% of the world's CFCs and other production was on the increase.

It was fortunate that a British meteorological survey team had been performing routine operations in Antarctica since 1956. The team observed a new trend beginning in 1977. Ozone concentrations dropped every spring and then returned to normal in a few months. The decline was greater each year, plummeting 40% by 1984. The area involved was larger than the United States. So unbelievable at first were these measurements that the scientists delayed publication for several years while rechecking their data and instruments. Their report, in 1985, was greeted skeptically. Perhaps the effect was coincidental or part of a natural cycle and would soon vanish.

In early 1986, U.S. atmospheric chemist Susan Solomon (Figure 9.14) organized an expedition to Antarctica to study the possible causes of ozone depletion there. With the arrival of the Antarctic spring in October, she found not only that ozone was now 50% below normal but also that high levels of chlorine were present. There was indeed an "ozone hole," and it was getting bigger. It was entirely unpredicted. Molina and Rowland and others had thought that the effects would occur only gradually and worldwide. Although their predictions had at first been greeted with disbelief and charges of extremism, reality was proving to be even more extreme.

But was this sudden localized depletion at the South Pole actually caused by chlorine from CFCs? The Solomon expedition provided the key to understanding the origin of the ozone hole. Solomon offered a hypothesis involving the ice clouds that form high above Antarctica during the long dark polar winter. When the sun reappeared in the Antarctic springtime, solar energy caused chemical reactions on the surfaces of the tiny ice particles forming these clouds. The reactions released ClO into the stratosphere from chlorine compounds that had been mixed in the ice particles, and sunlight then split the ClO molecule to free the chlorine. It is a complex mechanism, not the sort of thing that scientists were likely to have predicted in advance. The stratosphere turned out to be surprisingly subtle. A second ozone-observing expedition, in 1987, confirmed this hypothesis.

One should never make the mistake of underestimating the power of consumers when they are educated and motivated. . . . What consumers were reading about and seeing on television persuaded them to buy something different. And this provided a very powerful signal to industry that they should come up with substitutes, which they soon did.

RICHARD BENEDICK,
ON CFC-POWERED SPRAYS

When we build a bridge, we build it to withstand much stronger pressures than it is ever likely to confront. And yet, when it comes to protecting the global atmosphere, where the stakes are so much higher, the attitude of some people seems to be equivalent to demanding certainty that the bridge will collapse as a justification for strengthening it. If we are to err in designing measures to protect the ozone layer, then let us, conscious of our responsibility to future generations, err on the side of caution.

RICHARD BENEDICK, U.S.
CHIEF NEGOTIATOR OF THE
OZONE TREATY

HOW DO WE KNOW?

Figures 9.15 and 9.16 show some of the evidence that implicated chlorine as the culprit. The two graphs show data recorded by aircraft flights into the Antarctic stratosphere on August 23, 1987, as the sun was reappearing, and three weeks later on September 16. The ClO and ozone concentrations are graphed in parts per billion (ppb) for ClO and parts per million (ppm) for ozone, at different southern latitudes from 63 degrees (just south of South America) to 70 degrees (inland over Antarctica). On August 23, ClO appeared in the stratosphere over Antarctica, south of 65 degrees, but had not yet begun to destroy ozone. By September 16, ozone had been destroyed in the polar regions, and at precisely the latitudes where ClO was most concentrated. The correlation between the presence of ClO and the depletion of ozone in Figure 9.16 is exquisite: Ozone zigs precisely where ClO zags. This detailed correlation leaves little doubt of a causal relationship between ClO buildup and ozone depletion.

Ozone history offers an instructive but unfortunate example of the dangers of mixing political preconceptions with science. Despite the overwhelming evidence supporting the theory of ozone destruction by human-made chlorine compounds, there remains a handful of politically-motivated skeptics who find it difficult to believe that human actions could cause so great a global change. They argue that so much chlorine is getting into the stratosphere from sea salt, volcanoes, and biomass burning that CFCs couldn't possibly make much difference.

Measurements have checked the sea salt, volcano, and biomass hypotheses in detail. Because chlorine is soluble in water, nearly all the chlorine from these natural sources gets rained out of the lower atmosphere. CFCs can make it into

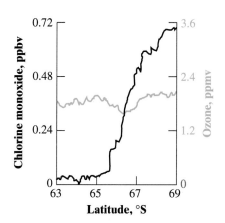

FIGURE 9.15
ClO concentrations (black) and ozone concentrations (blue) measured on August 23, 1987, before the ozone had begun vanishing near the South Pole for that year.

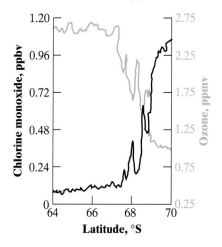

FIGURE 9.16
Concentrations measured on September 16, 1987, after the ozone was depleted.

During the last 2 or 3 years, a new generic question appears regularly which runs more or less as follows: "Don't volcanoes cause the Antarctic ozone hole?" A very simple answer of course is that the very large losses of ozone over Antarctica . . . first became noticeable around 1980 in measurements conducted continually since 1956. . . . Because volcanoes have been around for geologic time, with no unusually violent activity during the past decade, they can't be the direct cause of these . . . losses.

F. SHERWOOD ROWLAND

[The Antarctic volcano] Mount Erebus pumps out 50 times more chlorine annually than an entire year's production of CFCs.

DIXY LEE RAY AND LOU GUZZO, IN *ENVIRONMENTAL OVERKILL: WHATEVER HAPPENED TO COMMON SENSE?*

Conclusion: mankind can't possibly equal the output of even one eruption from Pinatubo, much less a billion years' worth, so how can we destroy ozone?

RUSH LIMBAUGH, RADIO TALK SHOW HOST, IN *THE WAY THINGS OUGHT TO BE*

In one eruption, Mount Pinatubo spewed forth more than a thousand times the amount of ozone-depleting chemicals . . . than all the chlorofluorocarbons manufactured by [all of the] corporations in history. . . . [Proponents of the ozone depletion theory] have deliberately obfuscated the facts about ozone research [and are now] in top posts with command power over scientific journals and associations and, ultimately, public opinion.

ROBERT MODURO, IN *THE HOLES IN THE OZONE SCARE: THE SCIENTIFIC EVIDENCE THAT THE SKY ISN'T FALLING*

the stratosphere only because they are inert and insoluble. If sea salt or biomass products were getting into the stratosphere, there should be chemical evidence of these substances in both the lower and upper atmosphere. Measurements by satellites and by airborne instruments at both levels have failed to find such evidence. Satellite measurements also reveal a total stratospheric chlorine "budget" that is exactly what is expected if CFCs are the dominant factor. Skeptics reply that tall volcanoes can directly loft chlorine into the atmosphere explosively at great heights. Although a few eruptions, such as Mount Pinatubo in 1991, are high enough and energetic enough to accomplish this, most are not. The Mount Pinatubo eruption did show up temporarily in the Antarctic data, as slightly larger ozone holes in 1992 and 1993. By 1994, this small effect had subsided while the hole continued to broaden and deepen, but at a slower rate now that the debris from Pinatubo had been removed from the atmosphere, further confirming synthetic chemicals as the primary cause.

Now that such dissenting hypotheses as volcanoes and sea salt have been tested, with negative results, these dissenting hypotheses should be withdrawn. Many ozone skeptics have still not done this. To continue to accept hypotheses that have already been disproven by the evidence is to operate in a netherworld of ideology and irrationality that flies in the face of the methods of science.

The Antarctic ozone hole has continued deepening, with springtime depletions reaching 70% and spreading across 10% of the Southern Hemisphere. Australian monitoring stations have noted sharp yearly depletions since 1987, and shifting patches of ozone-depleted atmosphere have occasionally pushed ultraviolet radiation in Australia 20% above normal. Due to the longevity of CFCs in the atmosphere, such deep Antarctic ozone holes may be the norm for the twenty-first century, no matter what we do now to stop further CFC production.

Ozone depletion is now showing up around the globe, and a smaller hole has opened over the Arctic. The worldwide trend of declining ozone levels began in the 1980s; by 1992, ozone levels over the United States had dropped by 8%.

In 1987, in Montreal, Canada, 47 nations forged a treaty that was unique to world history.* The **Ozone Treaty**, developed independently of and mostly before the Solomon Antarctic expedition, was not based on clear experimental evidence of ozone depletion and was not a response to a disaster that had already occurred. Instead, the treaty was preventive action on a global scale, based mostly on scientific theories about what might occur. The wisdom of this treaty, in which the United States played a leading role, has now been amply confirmed. The treaty is important not only for its effect on the atmosphere, but also for its inspiring effect on our perceptions of what humans can do to solve global environmental problems. It provides what mathematicians call an "existence proof"—a demonstration that something that might at first glance appear impossible can actually be accomplished. In recognition of the scientific work that led the way to this treaty, the 1995 Nobel Prize in Chemistry was conferred on Rowland and Molina (Figure 9.13)—whose names deserve to become known to every schoolchild—for having "contributed to our salvation from a global environmental problem that could have catastrophic consequences."

---

*I recommend U.S. chief negotiator Richard Benedick's account, *Ozone Diplomacy* (Cambridge, MA: Harvard University Press, 1991).

The Ozone Treaty calls for CFCs and other ozone-destroying chemicals (Table 9.2) to be essentially phased out by the year 2000. The restrictions are extraordinarily broad and rapid. The entire treaty is unprecedented. The treaty will definitely make a difference (Figure 9.17). Since the 1987 treaty, the chemical industry, sensing an expanding new market for ozone-friendly products, has moved rapidly to develop replacements. In fact, industries have been adopting these so quickly that the phaseout has moved faster than expected.

The verdict is now mainly up to nature. Because of the long atmospheric lifetimes of ozone-destroying chemicals (Table 9.2), that verdict will take a while. As the lower graph of Figure 9.17 shows, chlorine concentrations reached what is expected to be a maximum in 1998, and are then expected to fall slowly to below 2 ppb by 2050. Because the effects in the Antarctic were first observed when global chlorine concentrations had reached 2 ppb, this might be a safe level. The maximum ozone depletion over the entire globe, reached in 1998, ranged from 7% to 13% below the levels of the 1960s, depending on the season and latitude. These declines let in 8% to 15% more biologically harmful ultraviolet radiation than before. For comparison, Figure 9.17 also shows the soaring chlorine concentrations that would have occurred if controls had been delayed until 2010; as you can see, stratospheric chlorine would have climbed to three times the peak level reached in 1998.

What effects are likely? The U.S. Environmental Protection Agency estimates 12 million additional U.S. skin cancer cases, including 200,000 additional deaths, during the next 50 years. This is a 50% increase above current U.S. skin cancer rates. Other expected human effects include eye cataracts, suppression of immune system responses, and premature aging of the skin. Some plant species are sensitive to ultraviolet, and crop production is expected to decrease. Microscopic plants and animals living near the ocean's surface may be depleted, along with shellfish and other fish. Scientists are divided on the threat to life in the most affected region, Antarctica. Most of the organisms there live underwater, where they are somewhat protected. But microscopic plants living near the surface are fundamental to the Antarctic food chain and may be vulnerable. By 1996, measurements showed a striking decline—as much as 25%—in the populations of these plants in the ocean around Antarctica. Possible scenarios run the gamut from a temporary drop in these microorganisms to a collapse of the entire Antarctic ecosystem.

TABLE 9.2

Ozone-destroying chemicals in the atmosphere

Chemical	Uses	Average lifetime in atmosphere (yr)	Share of problem (%)	U.S. share of use (%)
CFC-11	coolant, aerosol, foam	60	28	22
CFC-12	coolant, aerosol, foam	130	47	30
CFC-113	solvent	90	5	45
Carbon tetrachloride	solvent	50	15	27
Methyl chloroform	solvent	7	2	50
Halon 1211	coolant, foam	25	1	25
Halon 1301	fire extinguisher	110	2	50

SOURCE: U.S. Office of Technology Assessment, 1991

FIGURE 9.17
The effect of the Ozone Treaty on atmospheric chlorine concentrations. The Antarctic ozone hole first became obvious when levels reached 2 ppb. If controls had been delayed until 2010, chlorine levels would have soared. Concentrations are in parts per billion. (Source: *Science* 5 January 1996, p 32.)

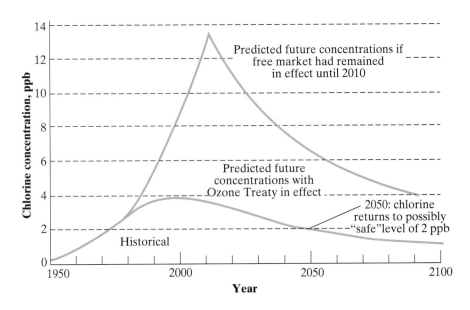

Ozone loss will be our legacy for much of the next century.

MICHAEL KURYLO, UPPER ATMOSPHERIC PROGRAM MANAGER, NASA

Human alteration of Earth is substantial and growing. Between one-third and one-half of the land surface has been transformed by human action; the carbon dioxide concentration in the atmosphere has increased by nearly 30% since the beginning of the Industrial Revolution; more atmospheric nitrogen is fixed by humanity than by all natural terrestrial sources combined; more than half of all accessible surface fresh water is put to use by humanity; and about one-quarter of the bird species on Earth have been driven to extinction. By these and other standards, it is clear that we live on a human-dominated planet.

P.M. VITOUSEK, H.A. MOONEY, JANE LUBCHENCO, J.M. MELILLO, INTRODUCING A SPECIAL SEGMENT ON "HUMAN DOMINATION OF EARTH'S ECOSYSTEMS," IN THE JOURNAL *SCIENCE*, 25 JULY 1997

■ **DIALOGUE 9 MAKING ESTIMATES: THE MEANING OF PPM** Although Earth's atmosphere extends upward for about 250 km, if it were squeezed down in such a way as to bring the entire atmosphere to the density and pressure that it normally has at sea level, it would extend only 8 km. From this information, estimate the thickness of atmospheric ozone (concentration of 0.3 ppm) if it all were compressed to atmospheric density and pressure.

■ **DIALOGUE 10** Refrigerators (Figure 9.12) cool the objects inside. How do they affect your kitchen?

9.7 GLOBAL WARMING

If you took Earth's temperature from space, you would find it to be a chilly −19°C (−2°F). The way you could take Earth's temperature, if you were in space, would be by studying its electromagnetic radiation output. This would be like taking the approximate temperature of an electric hot plate by looking at it: If it glowed white, it would be very hot, bright red means hot, and a dimmer red means less hot. A refinement of this method, using devices that measure the precise spectrum of the radiation, would tell you the precise temperature. Earth is not hot enough to glow visibly in the way a hot plate or the much hotter sun glows. Instead, Earth "glows infrared."

Fortunately for us, Earth's average surface temperature is +14°C, far warmer than the −19°C detected from space. The extra 33 degrees of warming is due to Earth's surrounding "blanket" of atmospheric gases. This warming is called the **greenhouse effect.**

---

9. Since ozone makes up only three-millionths of the atmosphere, the thickness would be three-millionths of 8 km. $(0.3 \times 10^{-6}) \times 8$ km $= 2.4 \times 10^{-6}$ km $= 2.4 \times 10^{-3}$ m $= 2.4$ millimeters.

10. The thermal energy removed from the refrigerator, and the energy from the electric company, must be exhausted from the refrigerator into your kitchen. This warms your kitchen.

Surprisingly, the gases that create the greenhouse effect are not the nitrogen and oxygen that form the bulk of Earth's atmosphere. Rather, the effect is due to certain trace gases, mainly water vapor and carbon dioxide (Table 9.1). The reason that these gases, called **greenhouse gases**, are so important to causing the greenhouse effect is that they are strong absorbers of infrared radiation.* We might expect any atmospheric gas that absorbs infrared to be important to Earth's energy balance, since Earth radiates in the infrared. Again, we see the importance of trace gases.

Like many of nature's ways, the greenhouse effect is best understood in terms of energy flows. For starters, let's study the energy flow near the surface of an imaginary Earth that has no greenhouse gases (Figure 9.18). We further imagine that the sun has suddenly turned on for the first time and begun warming a cold Earth.[†] Some of the solar radiation hitting Earth is reflected back into space, and the rest is absorbed by the atmosphere and the surface. The absorbed energy warms the atmosphere and the surface. As it warms, Earth itself begins radiating energy out to space, just as any object radiates energy because of the thermal motion of its molecules. As Earth warms, its radiated energy increases. Unlike the sun, which is hot enough to glow in the visible region of the spectrum, Earth's radiated energy is invisible because Earth is only warm enough to "glow" in the infrared region. Eventually, Earth reaches a temperature at which its radiated energy just balances the energy it absorbs from the sun. This balance temperature is −19°C. Earth then ceases getting warmer and maintains this −19°C temperature, because the radiated energy just balances the absorbed energy. The energy flow is then as shown in Figure 9.18.

When we add the greenhouse gases, something new happens (Figure 9.19): Greenhouse gases absorb and reradiate the infrared energy that Earth radiates. When these gases reradiate, they do so in all directions, downward as well as upward. This makes Earth's surface warmer than it would have been because there is now additional, "recycled," energy striking the surface. Notice that the top of the atmosphere is still at −19°C, even though the temperature of Earth's surface has been raised to +14°C. The greenhouse gases act like a thermally insulating blanket that warms you by retaining some of your radiated warmth, but without warming the outside of the blanket.

The glass on a greenhouse operates in a roughly analogous way. Sunlight passing inward through the glass is absorbed inside the greenhouse and is reemitted as infrared, which is absorbed and reradiated both outward and inward by the glass.[§] The inside of your car also warms this way when you park it in sunlight.

Combustion, the source of most of the energy for our industrial society, produces carbon dioxide, a greenhouse gas. This is beginning to create a problem. For hundreds of years before the Industrial Revolution, which began around

Responses to prediction of a greenhouse effect fall into three categories:

- It is too soon to panic. We do not yet know enough to make rational plans.
- It is too late to panic. The world is irrevocably committed to a rise in greenhouse gases and we should learn to adjust.
- We should promptly begin steps to minimize production of greenhouse gases.

While scientists do not customarily call for panic, there are widespread calls to Pursue Actions Now Inhibiting CO_2. The third of these responses appears to be the appropriate one.

DAVID BODANSKY, PHYSICIST AND ENERGY POLICY ANALYST

---

*Three-atom molecules such as H_2O and CO_2 are efficient infrared absorbers because they have many different ways of vibrating in response to radiation. The dominant atmospheric molecules, N_2 and O_2, have only two atoms, and so each molecule has only one way it can vibrate (atoms moving closer together and then farther apart).

[†]Something like this did happen once, about 5 billion years ago.

[§]But the analogy isn't perfect because a greenhouse also traps the air inside so that it cannot cool by mixing with outside air. Without this trapping, a greenhouse wouldn't get very warm.

FIGURE 9.18
Energy flow near the surface of an imaginary Earth that has no greenhouse gases but that has an otherwise normal atmosphere. The numbers represent percentages, relative to the radiant energy received from the sun.

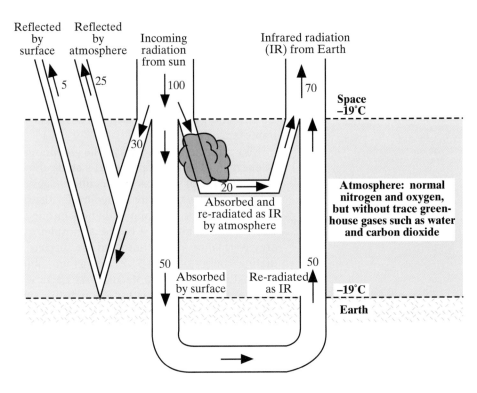

The insurance business is first in line to be affected by climate change . . . it could bankrupt the industry.
FRANKLIN NUTTER, PRESIDENT OF THE REINSURANCE ASSOCIATION OF AMERICA

[Even if global warming occurs,] any necessary adjustments would be small relative to the adjustments that we make during the year to temperature differences where we reside and as we travel.
JULIAN SIMON, SCARCITY OR ABUNDANCE: A DEBATE ON THE ENVIRONMENT

The only appropriate response [to global warming] would be to be sure that more and better air conditioning were available.
GEORGE REISMAN FROM HIS ESSAY "THE TOXICITY OF ENVIRONMENTALISM" IN RATIONAL READINGS ON ENVIRONMENTAL CONCERNS

1750, carbon dioxide maintained a natural atmospheric concentration of about 280 ppm (Figure 9.20). The history of the fossil-fueled industrial age is reflected in the CO_2 accumulation since that time.

In 1995, the atmospheric CO_2 concentration passed the 360 ppm mark. This represents an increase of nearly 30% in the concentration of this important contributor to the greenhouse effect, an effect which, as you can see by comparing Figures 9.18 and 9.19, plays a large role in the relationship between the sun and Earth.

This increase in greenhouse gases has warmed Earth slightly during the past century, and is expected to warm it further in the future. Any such excess warming due to human-created greenhouse gases is known as **global warming**. This problem has been predictable ever since the beginning of the industrial age, and was in fact predicted during the nineteenth century, but it was not taken seriously as long as it was still thought to be far in the future. But the problem no longer lies only in the future. In 1995, 2500 scientists from 60 countries evaluated some 20,000 scientific papers from such fields as climatology, ecology, chemistry, biology, economics, and medicine, and came to a broad consensus that global warming is occurring now. This effort, known as the Intergovernmental Panel on Climate Change, represents the current state of human understanding on climate change. The group's summary states:

Viewed as a whole, these results indicate that the observed trend in global mean temperature over the past 100 years is unlikely to be entirely natural in origin. Most importantly, there is evidence of an emerging pattern of climate response to forcings by greenhouse gases . . . in the observed climate record. . . . The probability

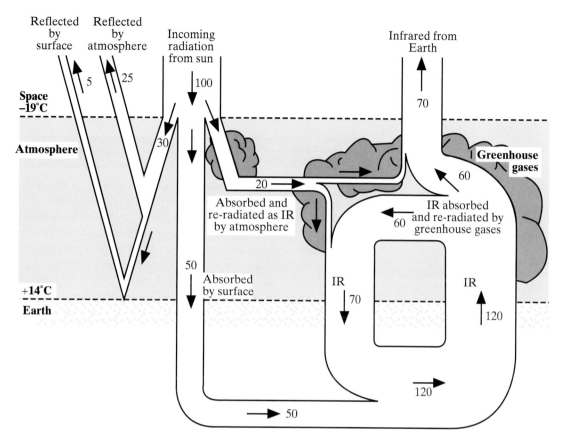

FIGURE 9.19
Energy flow near Earth's surface. The large difference between this and Figure 9.18 is due to trace amounts of greenhouse gases, gases that have significant leverage over Earth's energy flows.

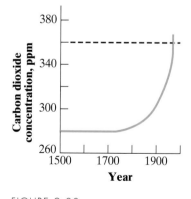

FIGURE 9.20
Measured or estimated past atmospheric carbon-dioxide concentrations. The concentration reached 360 ppm in 1995, an increase of 29% above the pre-industrial level.

is very low that these correspondences could occur by chance as a result of natural internal variability.

It isn't easy to predict the weather. There is considerable uncertainty about how severe global warming is today and what effects it will create in the future. Scientists approach these questions as they approach other questions, with theories and observations.

One important approach is computerized models of the climate. Modelers predict atmospheric quantities throughout the atmosphere, at each point of a three-dimensional grid of points separated by several hundred kilometers. Sophisticated climate models, using closely spaced grid points and using lots of data, are complex and involve many uncertainties.

One uncertainty in all the predictions is future CO_2 levels, which will be determined mainly by fossil-fuel use. To separate this societal part of the problem from the science part, modelers simply assume some CO_2 level and ask, "What will happen if we reach this level?" A common level to choose is an effective CO_2 doubling, an increase of all human-created greenhouse gases to concentra-

If postponing action [on global warming] for ten years means that in the long run average global temperature would be 0.2C [sic] higher than it otherwise would be, well that gives us plenty of time to change into a lighter shirt.

WILFRED BECKERMAN, IN HIS OPINION PIECE "WHY WORRY ABOUT THE WEATHER" IN *THE WALL STREET JOURNAL*

First, no evidence for the existence of the greenhouse effect can be found in the temperature records of the last 100 years; and second, current forecasts of global warming for the 21st century are so inaccurate and fraught with uncertainty as to be useless to policy-makers. . . . It is difficult to argue with the contention that "the only question concerning the warming is how much, and by when." However . . . the answer to the question "how much" may turn out to be very little.
Our second recommendation [is] that major policy actions not be undertaken until the implications are better understood.

RICHARD LINDZEN, METEOROLOGIST, AND WILLIAM NIERENBERG, OCEANOGRAPHER

tions having the same effect as 560 ppm of CO_2, twice the preindustrial CO_2 concentration.

Putting together the uncertainties, the models predict that an effective CO_2 doubling will cause the global average surface temperature to increase by 1.0–3.5°C, over and above the 0.3–0.6°C increase that has already occurred during the past century.

H O W   D O   W E   K N O W ?

How valid are the computer models? One way to check is to see what they retroactively "predict" about past events. For example, a climate model was used to predict the fluctuations of global temperature caused during 1991–1994 by the 1991 Mount Pinatubo eruption, and the results agreed closely with observations. For another example, when the known emissions of greenhouse gases and other industrial gases were fed into the models for the period 1860–1990, the models gave predictions in good agreement with the observed 0.3 to 0.6°C temperature increase (Figure 9.21) and other observed climate changes over the same period.

Long-term geological records also provide useful checks. Figure 9.22 shows temperatures and CO_2 concentrations during the past 160,000 years. Air bubbles

FIGURE 9.21
Global average surface temperatures since 1860, measured relative to the average temperature between 1961 and 1990. The bar graph shows the temperature measured each year, and the smooth curve is a 21-year "running average." The data indicates a 0.3 to 0.6°C temperature increase since 1860. As you can see, 1995 was the warmest year in this record, and 9 of the 10 warmest years have occurred since 1980. (Source: British Meteorological Office)

FIGURE 9.22

CO₂ concentrations (left-hand scale, upper graph) and temperatures (right-hand scale, lower graph) during the past 160,000 years. The period of low temperatures, from 120,000 to 15,000 years ago, was the last ice age. CO₂ levels have been about 280 ppm for the past few thousand years, up until the beginning of the industrial age. During only the past 250 years, the level has risen to over 360 ppm—far above the highest level reached during the past 160,000 years. A doubled CO₂ level, 560 ppm, could occur by 2030.

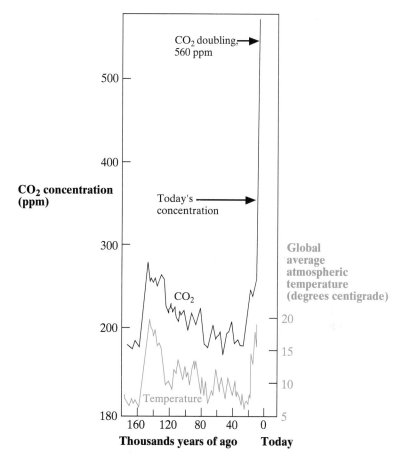

Tapping the limited resources of our world—particularly those of the developing world—simply to fuel consumerism, is disastrous. If it continues unchecked, eventually we all will suffer. We must respect the delicate matrix of life and allow it to replenish itself.

THE DALAI LAMA, SPEAKING AT THE 1992 EARTH SUMMIT IN RIO DE JANEIRO

in ancient Antarctic ice sheets were analyzed to obtain the changing atmospheric CO_2 content over this period. The air temperature at the time the ice was deposited can be inferred by studying the oxygen in the snow.* The resulting temperatures and CO_2 concentrations are correlated (they track each other), which may indicate a causal relationship between the two. For comparison, Figure 9.22 also shows today's CO_2 concentration and the doubled 560 ppm concentration that might occur by 2030.

The U.S. Environmental Protection Agency predicts that present trends will cause CO_2 levels to double, relative to their pre-industrial level of 280 ppm, by about 2030. Simply to hold CO_2 concentrations at their present high levels, there would have to be a net 50 to 80% reduction in emissions during the next few decades. Figure 9.23 shows the history of CO_2 emissions since 1950. Emissions increased at 4% per year between 1950 and 1973, dropped following the 1973 and 1979 world oil price increases, and rose at 2 to 3% per year between 1983 and 1990.

---

*The ratio of the oxygen isotope $^{18}_{8}O$ to ordinary oxygen $^{16}_{8}O$ is dependent on the temperature.

FIGURE 9.23
The history of carbon-dioxide emissions from fossil fuels. Annual emissions in billions of tons, from three groups of countries and total. (Source: Oak Ridge National Laboratory)

FIGURE 9.24
The main human-related greenhouse gases and the contribution of each to global warming. (Source: U.S. Office of Technology Assessment)

Figure 9.22 puts this in long-term historical perspective. The highest CO_2 concentration of the past 160,000 years was slightly above 280 ppm and occurred 130,000 years ago at the warmest part of Earth's last interglacial (between the ice ages) period. Today's 360 ppm is already far above that previous maximum. The last ice age began 120,000 years ago and ended 15,000 years ago. Earth warmed by 5 degrees as it emerged from that ice age.* Because enormous changes in world climate accompanied that 5-degree increase, we should expect similarly enormous changes if anything like a 5°C warming occurs.

Carbon dioxide is not the only problem. As Figure 9.24 shows, three other gases contribute, including the CFCs that are also causing ozone depletion. Deforestation is another major contributor to global warming, because trees store CO_2. Destroyed trees release their CO_2 to the atmosphere either rapidly (when burned) or gradually (when used for wood products). Permanent deforestation in the tropics causes between 7 and 30% of worldwide CO_2 emissions. However, the long-term effects are reduced, perhaps to zero, if the destroyed trees are replaced.

What other consequences of global warming are predicted, besides a temperature increase? A 10 to 25 cm rise in ocean levels has already occurred. Assuming a doubled CO_2 level, oceans will rise by another 15 to 95 cm during the next century because of partial melting of the polar ice caps and *thermal expansion* of the heated water (water, like most substances, expands when heated). Computer models predict more extreme weather: droughts, floods, heavy precipitation, and hurricanes, with more extremely hot days and fewer extremely cold days. Climate zones will shift some 500 km northward, resulting in extinctions and displacements of tree species and other plant and animal species. Snow and ice will retreat. Researchers already report a decrease in the extent of Arctic ice cover and enormous rifts in Antarctic ice sheets since 1978. In 1995, a 4200 square kilometer piece of the Larsen Ice Shelf broke away into the Antarctic Ocean. Mountain glaciers have been in worldwide retreat for the past century. Coastal land will be flooded and low-lying islands such as the Maldives and the Marshall Islands could vanish. The effects on agriculture will be mixed. If temperature increases are in the 1–2°C range rather than the 2–3.5°C range, then yields could actually increase in temperate regions, while decreasing in equatorial regions. Mosquitoes and other disease vectors will thrive, and tropical diseases such as malaria, dengue, yellow fever, and sleeping sickness are likely to spread northward.

Many consequences will have effects that further influence their original cause. Such **feedback effects** can be unpredictable and dramatic. As one example, warming should increase the evaporation of water, which might increase the cloud cover, cooling Earth and reducing global warming. Such a reduction would be a negative feedback effect. On the other hand, increased evaporation might create additional atmospheric water vapor, which would enhance the warming, because water vapor itself is a greenhouse gas. This would be a positive feedback or amplifying effect. Positive feedbacks can "run away" unpre-

---

*Figure 9.22 exaggerates this warming because temperatures in the Antarctic, where the data for the figure were gathered, vary more than do global temperatures.

Analysis of the CO_2 budget illuminates the diversity in processes and feedbacks that control and respond to atmospheric CO_2. . . . [B]oth land and sea will respond dynamically, not passively, to anthropogenic [human-caused] increases in atmospheric CO_2. Several important feedbacks will extend to millennia and longer. . . . Thus, fundamental uncertainties . . . extend to time scales well beyond the period of foreseeable management policies.

ERIC SUNDQUIST, U.S. GEOLOGICAL SURVEY

We travel together, passengers in a little space-ship, dependent on its vulnerable supplies of air and soil.

ADLAI STEVENSON, DEMOCRATIC PARTY PRESIDENTIAL CANDIDATE IN 1952 AND 1956

TABLE 9.3

Carbon dioxide emissions from fossil fuels, per person, in selected countries in 1993

Country	CO_2 per person (tonnes)
United States	19.8
Russia	12.4
Germany	11.3
Japan	8.4
World average	**3.9**
Mexico	3.7
China	2.6
Brazil	1.5
Nigeria	1.1
India	0.7
Zaire	0.1

SOURCE: Worldwatch Institute

dictably. For example, warming could cause increased evaporation, causing increased water vapor, causing even further warming, causing even further evaporation, and so forth. Scientists do not know which of these effects will actually dominate.

What can be done? To hold greenhouse gases to current levels, there would have to be a net 50 to 80% reduction in the annual emissions of all greenhouse gases during the next few decades. This could be accomplished only with large reductions in fossil-fuel use.

Are large reductions in fossil-fuel use possible? Economic growth went hand in hand with fossil-fuel growth up until the 1973 Mideast oil crisis. Since then, oil price increases have given industrialized countries incentives to achieve growth without more fossil fuel (Figure 9.23), primarily by using energy more efficiently and also by switching to nuclear and solar energy. Emissions from developing countries have increased recently (Figure 9.23), for two reasons: Their fossil-fuel use per person is increasing rapidly as they begin to industrialize, and their population is increasing rapidly as well. Their per capita fossil-fuel use is still quite low today (Table 9.3) but could triple or quadruple. If this happens, world emissions will rise greatly no matter what happens elsewhere.

Action on global warming has been much weaker than on ozone depletion, for several reasons: Solutions will have a far larger economic impact; the required lifestyle changes are difficult; and the world's leading economic power, the United States, has been lagging on this issue instead of leading as it did on ozone depletion. In a hopeful note, the European Union, United States, and Japan agreed in 1997 to reduce their greenhouse emissions by 6 to 8 percent, by the year 2012. Twenty-one other industrialized countries agreed to meet similar binding targets. Some analysts have suggested that future large reductions of greenhouse gasses, of up to 80%, are both feasible and necessary for all industrialized nations.

Recommendations typically include installing better building insulation, manufacturing more efficient automobiles, using nuclear or solar energy instead of coal to generate electricity, reducing deforestation, and reforesting. Figure 9.25 graphs the effects of several energy-efficiency measures. We will discuss some of these options in Chapter 17. All these measures taken together would reduce the total annual U.S. CO_2 emissions by over 50%. Studies indicate that the United States could make significant (25%) reductions in total greenhouse gas emissions if it were willing to invest about $10 per ton of CO_2 emissions. In more familiar terms, this is equivalent to an 11¢ rise in the price of a gallon of gasoline.

In fact, many studies suggest rebates and taxes as incentives for energy conservation. A tax on fossil fuels would encourage switching to noncarbon fuels. Revenue-neutral taxes and rebates on, respectively, gas-guzzling and gas-sipping cars would promote fuel efficiency. An underlying view, sometimes called the "polluter pays" principle, is that environmental costs that are not charged to the polluter really amount to hidden subsidies that encourage the destruction of the environment. Ultimately, this harms not only the environment but also the economy. In order to maintain both a free market and a healthy environment, environmental costs such as global warming must be incorporated into the economic system.

FIGURE 9.25
The effects on CO_2 emissions of several possible energy-efficiency measures. Each measure could, if vigorously pursued, reduce the total U.S. emissions by the percentage indicated. (Source: U.S. Office of Technology Assessment)

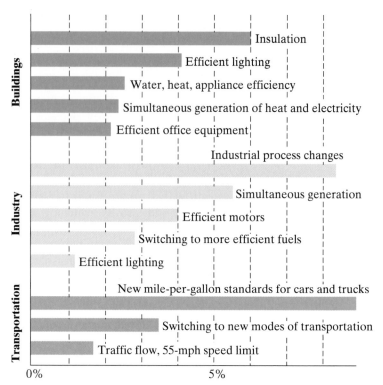

> Of course, the overarching issue now is global warming, which is admittedly much more complex than the ozone problem.
>
> RICHARD BENEDICK

■ **DIALOGUE 11** How do you suppose we know that the visible surface of the sun has a temperature of about 5400°C?

■ **DIALOGUE 12** By inspecting Figure 9.19, verify that energy is conserved at each of the energy-flow "intersections" (points where the flow separates or joins together).

■ **DIALOGUE 13** According to Figure 9.22, what were the approximate dates of the last interglacial (or warm) period before the present interglacial period? Approximately when did the coldest global average temperature occur during the last glacial period, and what was this coldest temperature?

---

11. By measuring the radiation emitted at different frequencies.

12. No answer.

13. The last interglacial period lasted from roughly 150,000 years ago to 120,000 years ago. The coldest time during the last glacial period was about 25,000 years ago, when the temperature was about 6°C.

EINSTEIN SIMPLIFIED

s.harris

SUMMARY OF IDEAS AND TERMS

Force field   The effect that the source of a force has on the surrounding space. Examples include gravitational fields and electromagnetic fields.

Gravitational or electromagnetic field   Exists wherever an object can feel a gravitational or electromagnetic force. Every material object creates a gravitational field throughout the space around it, and every charged object creates an electromagnetic field throughout the space around it.

Electromagnetic wave theory of light   Every vibrating charged object creates a disturbance in its own electromagnetic field, which spreads outward through the field at 300,000 km/s. Light is just such an **electromagnetic wave**.

Electromagnetic radiation   Any electromagnetic wave.

Lightspeed   300,000 km/s, or 3×10^8 m/s.

Ether   The gaslike material substance that nineteenth-century scientists believed filled the universe. Ether's mechanical motions were thought to be responsible for light waves and for the transmission of forces, such as the electromagnetic force, that act at a distance. Einstein's work led to the rejection of the ether theory and the acceptance of the electromagnetic field as a nonmaterial, but nonetheless physically real, medium.

Electromagnetic spectrum   The complete range of electromagnetic waves. Divided into the following regions, listed from longest to shortest wavelength (lowest to highest frequency):

- **Electronic**, or **radio waves**   Created by humans in such forms as AM and FM radio, TV, radar, and microwave.
- **Infrared**   Created by the thermal motion of molecules.
- **Visible**   Detectable by the human eye. Created by lower-energy motions of electrons in atoms. Colors are due to different wavelengths, ranging from red (longest) to violet (shortest).
- **Ultraviolet**   Created by higher-energy motions of electrons in atoms. Has sufficient energy to damage biological molecules but not enough to ionize them. Can cause mutations and cancers.
- **X rays**   Created by the highest-energy motions of electrons in atoms. X rays are one type of **ionizing radiation** because they can ionize biological molecules.
- **Gamma rays**   Created in atomic nuclei. Very high energy and highly penetrating.

Radiant energy   The energy of an electromagnetic wave.

Solar radiation   Electromagnetic radiation from the sun. Concentrated mainly in the infrared, visible, and ultraviolet.

Trace gases   Gases that form only a minute fraction of the atmosphere. Examples: ozone, carbon dioxide, and water vapor.

Ozone   The O_3 molecule. A dilute layer of ozone fills the upper atmosphere or **stratosphere**. Ozone absorbs and shields biological life from most ultraviolet radiation.

Chlorofluorocarbons or **CFCs**   Chemicals whose molecules are made of chlorine, fluorine, and carbon. Manufactured as coolants, spray propellants, foaming agents, and solvents. They destroy stratospheric ozone.

Ozone Treaty   Calls for a nearly total phaseout by A.D. 2000 of CFCs and most other ozone-destroying chemicals.

Greenhouse effect   The warming created by Earth's surrounding blanket of atmospheric gases.

Greenhouse gas   The atmospheric gases, mostly water vapor and carbon dioxide, that cause the greenhouse effect.

Global warming   Additional greenhouse-effect warming of Earth, caused by fossil-fuel use, deforestation, and other human activities.

Feedback effect   Any effect that further influences the phenomenon that caused it. **Negative feedback** diminishes its cause, and **positive feedback** enhances its cause.

REVIEW QUESTIONS

FORCE FIELDS
1. Name two kinds of force fields.
2. Which force fields would be felt by a stationary charged object: gravitational, electric, or magnetic? By a moving uncharged object? By a moving charged object?
3. What does it mean to say that there is an electric field throughout this room?

ELECTROMAGNETIC WAVES
4. Make a list of phenomena that are caused by electric charge.
5. What is an electromagnetic wave, and how fast does it go?
6. Are electromagnetic fields physically real? Defend your answer.
7. Describe the ether theory of light.
8. In what way do electromagnetic waves violate the philosophical underpinnings of Newtonian physics?

THE COMPLETE SPECTRUM
9. List the six main regions of the spectrum, from longest to shortest wavelength.
10. Describe a typical source of each: X rays, infrared, gamma rays, electronic waves, and ultraviolet.
11. Describe physically how humans see.
12. List several practical applications of electronic waves.
13. Which parts of the spectrum are ionizing radiations?

14. Which three regions of the spectrum dominate solar radiation? Which single region dominates?

GLOBAL OZONE DEPLETION
15. Where does ozone occur naturally on Earth, and what is its human significance?
16. What are CFCs? How have they have been useful to humans?
17. Outline the process by which CFCs destroy ozone.
18. List three atmospheric trace gases.
19. What is the Ozone Treaty?
20. List some expected effects of ozone depletion.

GLOBAL WARMING
21. What are the two major greenhouse gases, and what property makes them greenhouse gases?
22. What would happen if we removed all the CO_2 from the atmosphere? All the ozone?
23. Explain the greenhouse effect.
24. Which would increase global warming, forestation or deforestation? Why? Would this also increase ozone depletion?
25. What is a feedback effect? Describe one feedback effect related to global warming.
26. List three recommendations that have been made to combat global warming.

HOME PROJECTS

1. Count seconds during the next thunderstorm, to estimate the distance to several lightning strokes. Sound travels at 340 m/s. If a lightning stroke is 1 km away, about how many seconds will elapse before you hear it? Verify that the travel time for light is negligible, by calculating light's travel time for a few of the lightning strokes.
2. Turn on the heating coil of an electric stove, heater, or toaster. Before it begins to glow, hold your hand at some distance from the coil and feel its warmth. You are feeling infrared. When you sense a warm object near you, you are using your body as an infrared detector.
3. Find a black sheet of paper (you could paint it black with a marker pen) and a white sheet of paper of similar composition. When illuminated with white light, which one absorbs the visible radiation? Where does this energy go? Try putting the two sheets out in strong sunlight, sheltered from wind. Which one warms up more?

FOR DISCUSSION

1. What scientific topics are important to understanding ozone depletion and global warming? Should we be educating people about such problems? How should we go about it? Are we doing this now? At what age (if any) should students learn about these problems?
2. At what points in the history of ozone depletion were good decisions or poor decisions made, and what were the consequences? What would have been the consequences of different decisions? In dealing with global warming, what (if any) lessons can we learn from the history of the ozone problem?
3. Dealing with uncertainty. There often are large uncertainties about complex systems such as the atmosphere. How should society deal with these uncertainties when they involve possible human or environmental harm? What principles should be applied when making deci-

sions about products such as CFCs? "Presumed innocence" is an accepted principle of criminal law: Defendants are considered innocent until legally proven guilty. Should new products and technologies be considered environmentally harmless until proven harmful? Some people support the "precautionary principle" that says that in case of uncertainty, society should act as though new technologies are harmful, in order to ensure that harm does not occur. What is the proper balance between caution and innovation?
4. List all of your personal activities during the last 24 hours that contributed to global warming. Which of these activities might you alter in order to reduce global warming?
5. List several policies that the United States might pursue, as a nation, to combat global warming. Discuss the pros and cons of each such policy.

EXERCISES

FORCE FIELDS
1. Do the electric circuits in your home produce magnetic fields? Suggest a measurement that might check your answer.
2. Is an electric field a form of matter? Explain. What about a gravitational field?
3. If you place a proton at some point in an electric field and then release it, what will happen? How would the proton's motion differ from the motion of an electron placed at the same point in the same electric field? What if the field were gravitational instead of electric?
4. A proton is placed, at rest, at some point A within a room that is otherwise devoid of all matter. At some other point B within the room is there an electric field? An electric force? A magnetic field? A magnetic force? Is there energy at point B?
5. Suppose that, in the preceding exercise, the proton is oscillating back and forth. Is there an electric field at point B? An electric force? A magnetic field? A magnetic force? Energy?

ELECTROMAGNETIC WAVES
6. What evidence is there that the medium that carries light waves is not air?
7. Imagine a region of space that had no atoms and no subatomic particles. Would it contain matter? Might it contain energy?
8. Which travels faster, light or radio waves? Which has a longer wavelength? Which has a higher frequency?
9. It takes light 20 minutes to travel from Mars to Earth (depending on where Mars and Earth are in their orbits). How might this affect a conversation between an astronaut on Mars and people on Earth? Suppose an ultra-

powerful telescope were used for visual communication. Would this speed things up?
10. Are radio waves filling your room right now? Describe a simple experiment that could demonstrate your answer.
11. If your ears actually detected radio waves instead of sound waves, what would your ears be "hearing" right now?
12. List three characteristics that all electromagnetic waves have in common.

THE COMPLETE SPECTRUM
13. Is there any physical difference between light and a radio wave? Explain.
14. You can get a sunburn even on a cloudy day. Why?
15. You do not get a sunburn, even on a sunny day, if you are behind glass. Why?
16. What kinds of animals might evolve on a planet orbiting a star whose radiation is mostly in the infrared?
17. Suppose you viewed Earth's dark side from space. Could you "photograph" Earth using ultraviolet-sensitive film? Light-sensitive film? Infrared sensitive film? If so, what would you expect to see in the photo?
18. What energy transformation occurs (from what type to what type?) when solar radiation warms your skin? In a flashlight? When microwaves warm your food?
19. Your radio is tuned to 92 on the FM dial. What frequency of electromagnetic wave is your radio receiving? How fast is this wave traveling? About how long is one wavelength? Does an electromagnetic wave actually reach your ear? What does reach your ear?

GLOBAL OZONE DEPLETION
20. Which of these worsens the ozone problem: coal-fired power plants, gasoline engines, CFC coolants, nuclear power plants, hydroelectric power plants, solar heating of homes, deforestation?

21. Why is it important that spray-can aerosols be propelled by inert gases? Why does the inertness of CFCs lead to problems with stratospheric ozone?

22. According to Table 9.2, which will be the first ozone-destroying chemical to decrease to harmless levels, and how long will it take? Which will be the last, and how long will it take?

23. What was the chlorine concentration in 1950? Under the Ozone Treaty, what is the highest predicted level? When is the level predicted to return to the 1975 level? What would the level be at that time if there were no Ozone Treaty?

24. Which of the ozone-destroying chemicals has the longest atmospheric lifetime? Today, which one forms the largest share of the problem?

25. Ozone, O_3, typically "lives" for only 100 to 200 seconds in the stratosphere before reacting with an O atom to form two molecules of O_2. Since it is so short-lived, how can it offer any protection from ultraviolet radiation?

26. The total quantity of energy that reaches Earth in the infrared region is greater than that in the ultraviolet region. Why then are we more concerned about the direct health effects of UV than of IR?

GLOBAL WARMING

27. Which of these automobile fuels would not contribute to global warming: natural gas, synthetic gasoline made from coal, electricity from a solar power plant, electricity from a coal power plant, electricity from a nuclear power plant, ethanol (grain alcohol) made from corn?

28. What are some alternatives to burning fossil fuels? Consider electric power generation, transportation, home heating.

29. Will the Ozone Treaty help in solving global warming? How?

30. Which of these contributes to global warming: coal-fired power plants, gasoline engines, CFC coolants, nuclear power plants, hydroelectric power plants, solar heating of homes, deforestation?

31. It has been suggested that global warming could affect the CO_2 concentration in the oceans. What evidence is there in your everyday environment to show that CO_2 can be dissolved in water? As the temperature is raised, would you expect the amount of dissolved CO_2 to increase, decrease, or remain the same? If this effect happens in ocean water, will it represent a positive or negative feedback on global warming?

PROBLEMS

ELECTROMAGNETIC WAVES AND LIGHTSPEED

1. How long does it take a radio signal to travel from New York to San Francisco, about 5000 km?

2. A radar transmitter pointed at the moon receives a reflection 3 seconds after the signal is sent. How far is it to the moon?

3. Find the radius of Earth's orbit, from the fact that light gets here from the sun in 8.3 minutes.

4. Light from the sun takes roughly 8 minutes to reach Earth, while light from Alpha Centauri, the nearest star beyond the sun, takes about 4 years. How many times farther is it to Alpha Centauri than it is to the sun?

5. Using the data in the preceding problem, about how far is it to Alpha Centauri in kilometers?

6. What is the time delay for a television signal that is sent via satellite? Communications satellites orbit above the equator in circles of radius 36,000 km (6 times larger than Earth's radius!). To simplify the problem, assume that the signal goes straight up and straight down.

7. Use the formula $s = f\lambda$ (Chapter 8) to find the wavelength of 60-Hz power-line radiation.

8. Use the formula $s = f\lambda$ (Chapter 8) to find the wavelength of microwaves (see Figure 9.7).

OZONE DEPLETION AND GLOBAL WARMING

9. One cubic centimeter of normal air at Earth's surface contains about 2.7×10^{19} molecules. Working from Table 9.1, find the number of each of the following kinds of molecules in 1 cm^3 of air: O_2, CO_2, O_3.

10. To better understand the meaning of 1 ppm, find the number of Krypton atoms in a cubic centimeter of normal air. According to Table 9.1, the concentration of Krypton (a single-atom molecule) in normal air is 1 ppm. Work from the data given in the preceding problem.

11. According to Figure 9.22, how far back before the present century must one go to find a time when the CO_2 concentration was above 280 ppm? How large was the CO_2 concentration at that time? How did the temperature at that time compare with present temperatures?

12. Assuming that differences in CO_2 emissions listed in Table 9.3 are due only to differences in fossil fuel use, how many joules of fossil energy are consumed by a typical American for each joule consumed by a typical Japanese? How many joules are consumed by a typical Japanese for each joule consumed by a typical Indian?

13. According to Figure 9.25, by what total percentage would U.S. CO_2 emissions be reduced if cogeneration of heat and electricity were vigorously pursued? What if all three listed transportation options were vigorously pursued?

14. Making estimates. The mass of one liter of gasoline is about 0.7 kg, and the mass of one gallon is about 2.5 kg. A CO_2 molecule is about 3 times as massive as a carbon atom. Assuming that gasoline is pure carbon (actually only about 85% of its mass is carbon), estimate the mass of CO_2 a typical automobile puts into the atmosphere each year. Compare with Table 9.3.

PART FOUR

THE POST-
NEWTONIAN
UNIVERSE
▼
the observer
intrudes

10

Relativity Theory

Nature and Nature's laws lay hid in night:
God said, "Let Newton be" and all was light.
ALEXANDER POPE

It did not last: the Devil, shouting "Ho
Let Einstein be" restored the status quo.
SIR JOHN COLLINGS SQUIRE

This chapter enters the domain still referred to as "modern" physics, even though it is now a century old. *Post-Newtonian* physics is a more accurate term. The introduction of fields during the nineteenth century represented a break from the materialistic, mechanical tradition of Newtonian physics. The theory of relativity steps further away from the Newtonian tradition. Time, space, and mass now become to some extent "observer dependent" or *relative* rather than *absolute*. This emphasis on the observer is even more characteristic of the other great twentieth-century physical theory, quantum theory.

From the scientific point of view, the transition from Newtonian to post-Newtonian physics is less radical than the transition from Aristotelian to Newtonian physics (Chapter 3). Newtonian physics rejected Aristotelian physics, but post-Newtonian physics only limits the validity of Newtonian physics. As discussed in Section 5.6, the predictions of Newtonian physics agree with observation in "normal" human situations, defined as those for which speeds are very small compared with lightspeed; distances are far larger than molecules but far smaller than galaxies; and gravitational forces are not many times stronger than on Earth. But outside this "normal" or Newtonian range, Newtonian physics differs significantly from observation—in other words, Newtonian physics is wrong outside this range.

From the philosophical point of view, the transition from Newtonian to post-Newtonian physics is revolutionary, and the revolution continues today. The field concept, which contradicts the atomistic materialism that is at the heart of the Newtonian worldview, has already given us a taste of this revolution (Chapter 9).

Although it has the reputation of being difficult, Einstein's theory of relativity is actually fairly simple. It is based on two very simple ideas, and all of its major principles are offshoots of these two. Rather, the theory's reputation for difficulty comes from its strangeness. Its conclusions seem to violate common sense. To understand this theory one needs the mental flexibility to follow an argument wherever it might lead.

This chapter focuses on the two key ideas behind the theory of relativity, and the theory's most basic new prediction. Chapter 11 looks at several other predictions and presents Einstein's extension of these ideas to form a new theory of gravity. Section 10.1 provides historical context, and Section 10.2 discusses the older "Galilean" (Newtonian) way of viewing the phenomena with which Einstein was concerned. Sections 10.3 and 10.4 cover the principle of relativity and the principle of the constancy of lightspeed, the two key ideas behind the new theory. In line with two of the themes of this book, Section 10.5 discusses why we believe Einstein's theory and compares the new view with Newtonian views. Sections 10.6 and 10.7 explain one of Einstein's most fascinating predictions: the relativity of time.

10.1 EINSTEIN: REBEL WITH A CAUSE

> Common sense is nothing more than a deposit of prejudices laid down by the mind before you reach eighteen.
>
> EINSTEIN

Albert Michelson, whose highly accurate measurements of lightspeed earned him America's first Nobel Prize in physics, was one of the most highly esteemed physicists of his time. In 1894, in a speech dedicating a new physics lab at the University of Chicago, Michelson proclaimed that the era of new discovery in fundamental physics was probably finished:

> While it is never safe to affirm that the future of Physical Science has no marvels in store even more astonishing than those of the past, it seems probable that most of the grand underlying principles have been firmly established and that further advances are to be sought chiefly in the rigorous application of these principles to all the phenomena which come under our notice. . . . The future truths of Physical Science are to be looked for in the sixth place of decimals [that is, in improvements in the accuracy of known results].

FIGURE 10.1
Never one to take himself too seriously, Einstein stuck his tongue out when asked to smile on his seventy-second birthday.

It was a powerful acknowledgment of the triumph of Newtonian physics. Many scientists of that day shared Michelson's confidence that the known "grand unifying principles"—Newton's laws and the laws of thermodynamics and electromagnetism—were complete and permanent.

But the world changed only 6 years later. In 1900, the physicist Max Planck introduced a revolutionary idea, the quantum of energy (Chapters 13 and 14). The new idea was little noticed. The post-Newtonian revolution began quietly.

It is remarkable that in a span of only 5 years at the outset of the twentieth century, the seeds of both of the great post-Newtonian theories were planted. In 1905, 5 years after Planck's introduction of the quantum, a quite different but equally revolutionary idea was hatched in the brain of an obscure patent clerk in Bern, Switzerland. His name was Albert Einstein (Figures 10.1 and 10.2).

> To punish me for my contempt for authority, Fate made me an authority myself.
>
> EINSTEIN

Einstein was a rebel in more ways than one. In his midteens, he got fed up with high school and dropped out. This surprised no one, for he had been a mediocre student and a daydreamer since beginning elementary school. Before that he had been a slow child, learning to speak only at three years of age. His high school teachers were glad to see him go, one of them informing Einstein that he would "never amount to anything" and another suggesting that he leave

FIGURE 10.2
Albert Einstein during his student days
in Zurich, a few years before he
created his special theory of relativity.

My intellectual development was
retarded, as a result of which I
began to wonder about space and
time (things which a normal adult
has thought of as a child) only
when I had grown up.

EINSTEIN

If I were a young man again and
had to decide how to make a
living, I would not try to become a
scientist or scholar or teacher.
I would rather choose to be a
plumber or a peddler, in the hope
of finding that modest degree of
independence still available under
present circumstances.

EINSTEIN, A REMARK MADE
NEAR THE END OF HIS LIFE

Were I wrong, one professor
would have been quite enough.

EINSTEIN, WHEN ASKED
ABOUT A BOOK IN WHICH 100
NAZI PROFESSORS CHARGED
HIM WITH SCIENTIFIC ERROR

school because his presence destroyed student discipline. Einstein was only too
happy to comply. He spent the next few months as a model dropout, hiking and
loafing around the Italian Alps.

After deciding to study engineering, he applied for admission to the Swiss
Federal Polytechnic University in Zurich, but unfortunately he failed his entrance exams. It seems he had problems with biology and French. To prepare for
another try, he spent a year at a Swiss high school where he flourished in this
particular school's progressive and democratic atmosphere. He recalled later
that it was here that he had his first ideas leading to his theory of relativity.

The university now admitted Einstein on the basis of his high school diploma.
He was known as a charming but indifferent university student who attended
cafes regularly and lectures sporadically. He managed to pass the necessary
exams and eventually graduate with the help of friends who shared their systematic class notes with the nonconforming Einstein.

Einstein recalled that his college experience

had such a deterring effect upon me that, after I had passed the final examination, I
found the consideration of any scientific problems distasteful to me for an entire
year. . . . It is little short of a miracle that modern methods of instruction have not
already completely strangled the holy curiosity of inquiry, because what this delicate little plant needs most, apart from initial stimulation, is freedom; without that
it is surely destroyed. . . . I believe that one could even deprive a healthy beast of
prey of its voraciousness, if one could force it with a whip to eat continuously
whether it were hungry or not.

Following his graduation in 1900, Einstein applied for an assistantship to do
graduate study, but it went to someone else. After looking unsuccessfully for a
teaching position, in 1902 a friend helped him land a job as a patent examiner.
Einstein often referred to his seven years at this job as "a kind of salvation" that
paid the rent and occupied only 8 hours a day, leaving him the rest of the day to
ponder nature.

The two great theories of twentieth-century physics are the theory of relativity and quantum theory. Quantum theory is the creation of many physicists, including Einstein and several others whom we will meet in Chapters 13 and 14.
One remarkable aspect of the theory of relativity is that it was invented nearly
single-handedly.

10.2 GALILEAN RELATIVITY: RELATIVITY ACCORDING TO NEWTONIAN PHYSICS

Here is a typical relativity question: Suppose that a train passenger, call her
Velma, throws a baseball toward the front of the train. Both she and Mortimer,
who is standing on the ground watching the passing train, measure the speed of
the baseball (Figure 10.3). Will they get the same answer? If not, how will their
answers differ?

This question concerns two observers who are moving differently. We say
that Velma and Mort are in **relative motion** whenever they are moving differently. A **theory of relativity** is any theory that works out answers to questions
concerning observers who are in relative motion. We can think of the train as

FIGURE 10.3
Velma throwing a ball, observed by Mort.

being Velma's laboratory, or her **reference frame**, within which Velma measures things like the speed of the ball. We can think of the ground beside the tracks as a second reference frame, Mort's reference frame, for his measurements. The standard question that any theory of relativity asks is how measurements made in one reference frame compare with those made in another. Scientists have thought about questions like this since at least the time of Galileo.

To be more specific, suppose that the train moves at 70 meters per second (150 miles per hour, a typical modern train speed). Suppose that Velma throws the baseball toward the front of the train at 20 m/s *relative to Velma* (as measured on the train, using meter sticks and clocks that are on the train). How fast does the baseball move *relative to Mort* (as measured on the ground)?

The answer: Mort observes the baseball to move at 90 m/s. The reason is that during each second, the baseball moves 20 meters toward the front of the train as measured by Velma within the train and (as by observed by Mort) the baseball moves an additional 70 meters, because the train itself moves 70 meters during that second.

Centuries ago, Galileo would have given the same answer, 90 m/s. This straightforward, fairly intuitive form of relativity is called **Galilean relativity**. Here are a few more questions along this line:

■ **DIALOGUE 1**   Alphonse moves eastward in a car at 30 m/s. Velma, moving eastward in a train at 70 m/s, passes him. Are Velma and Al in relative motion? What is Velma's velocity (including direction) relative to Al? What is Al's velocity relative to Velma?

■ **DIALOGUE 2**   While the train moves eastward at 70 m/s, Velma throws a ball toward the rear of the train, throwing it just as hard as she did earlier when she threw it at 20 m/s toward the front of the train. (a) Find the velocity (including direction) of the ball relative to Velma. (b) Find the velocity of the ball relative to Mort, who is standing beside the tracks. (c) Since the two answers differ, whose answer is correct, Velma's or Mort's?

---

1. Yes; 40 m/s eastward; 40 m/s westward because Velma observes that Al is 40 meters farther away on her west side every second.

2. (a) 20 m/s westward. (b) 50 m/s eastward. (c) Both answers are correct—they differ because they refer to different observers.

■ **DIALOGUE 3**   The train moves eastward at only 20 m/s, and Velma throws a ball at 20 m/s toward the rear of the train. Find the ball's velocity relative to Mort.

■ **DIALOGUE 4**   Let's give the ball to Mort. (a) He throws it at 20 m/s eastward as measured by him. What is the ball's velocity relative to Velma, who is in the train moving eastward at 70 m/s? (b) What if Mort had instead thrown the ball westward at 20 m/s?

Let's turn to a similar example involving light beams instead of baseballs. We learned in Chapter 9 that light is an electromagnetic wave moving at 300,000 km/s, a speed that we will symbolize by the letter c. It is difficult to imagine such a high speed. A light beam travels from New York to Los Angeles in a hundredth of a second and travels a distance equal to 7.5 times around Earth in 1 second. Trains, jet planes, and even Earth satellites moving at 8 km/s are slowpokes by comparison.

Imagine that Velma pilots a futuristic rocket ship that is much faster than today's fastest vehicles. Suppose Velma moves past Earth at 75,000 km/s, or $0.25c$ (25% of lightspeed). Mort stands on Earth holding a source of light such as a flashlight or a laser. Suppose that Velma flies eastward past Mort and that Mort directs the light beam eastward. What would be the speed of the light beam—the moving tip of the beam—relative to Mort and relative to Velma? Experiments with light beams from sources on Earth have repeatedly shown that a light beam moves away from its source at speed c. So Mort's light beam moves eastward at speed c relative to Mort. What speed would Velma measure for the same light beam (Figure 10.4)? The intuitively sensible answer is $0.75c$.

This is the answer Galileo would have given, the answer given by Galilean relativity. It is the answer that all scientists would have given up through the end of the nineteenth century. It is certainly a sensible answer. Nevertheless, it is experimentally wrong. Nature does not comply with our intuitions. To see why there might be something wrong with this answer and to see what nature's answer is, we turn in the following section to some thoughts that Albert Einstein had around the beginning of the twentieth century.

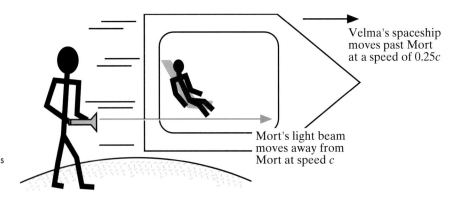

FIGURE 10.4
How fast is Mort's light beam moving, as observed by Velma relative to her own reference frame?

Velma's spaceship moves past Mort at a speed of $0.25c$

Mort's light beam moves away from Mort at speed c

---

3. Zero velocity relative to Mort—the ball falls straight down.

4. (a) 50 m/s westward. (b) The ball's velocity relative to Velma would be 90 m/s westward.

10.3 THE PRINCIPLE OF RELATIVITY: RELATIVITY ACCORDING TO EINSTEIN

Imagine riding in a smoothly moving unaccelerated vehicle such as a jet airplane in level flight. Suppose that the flight attendant pours you a cup of coffee. Where should you hold your cup: a little behind the spout, to take account of the fact that your airplane is moving forward, or under the spout? In other words, does the coffee pour straight downward relative to the airplane? Try it sometime and see. Or try dropping a coin from one hand to the other in a moving vehicle (but not when you are driving): Is the catching hand directly beneath the dropping hand?

The result is that the coffee pours straight downward relative to the plane.

You could carry out other, more sophisticated experiments along this line, all on a smoothly moving reference frame such as a jet airplane. You might take multiple-flash photos of a falling ball (Figure 3.12) or perform experiments with frictionless air gliders (Figure 3.6) or experiments designed to check the validity of Newton's laws. You could perform electromagnetic experiments involving electric charges, electric currents, or magnets. Just as for the poured coffee, you would find in all these cases that the experimental result is exactly the same as it is when the experiment is performed in a reference frame that is at rest on the ground: A falling ball falls straight downward with an acceleration of 9.8 m/s^2; an object's acceleration equals the force exerted on it divided by its mass; and so forth—all as measured relative to the moving reference frame (the interior of the jet airplane). In other words, all physics experiments performed entirely within the jet airplane come out precisely as they would if performed at rest on Earth.

Suppose you were the only passenger on an airplane with no windows in the passenger compartment and that you went to sleep and woke up sometime later to find yourself alone in the compartment. Could you tell, without receiving information from the outside world,* whether your airplane was flying at a constant velocity or parked on the ground? The answer is no. You could throw a ball, do handstands, pick up nails with magnets, and the like, and everything would be the same, regardless of whether your plane was in flight or parked.

As long as you are in a nonaccelerating reference frame (no bumps, no turns, no speed changes), everything is normal. All physics experiments, and consequently all the laws of physics, are normal. We summarize this idea as:

THE PRINCIPLE OF RELATIVITY
Every nonaccelerated observer observes the same laws of nature. In other words, no experiment performed within a sealed room moving at an unchanging velocity can tell you whether you are standing still or moving.

---

*Information from the pilot would be from the outside world, because the pilot's information enters through the cockpit window and through radio receivers. Any actions, such as changes in the air pressure, taken by the crew upon landing would also be based on information from the outside world.

Unless you look outside, you can't tell how fast you're going. This is a fairly plausible idea and was the key to Einstein's thinking about relativity. It is called the *principle of relativity* because it says that as far as the laws of physics are concerned, all motion is just relative motion. When we say, for example, "the car is moving at 25 km/hr westward," we really mean that "the car is moving at 25 km/hr westward relative to the ground" or that "the car and the ground are in relative motion at a speed of 25 km/hr." In regard to the basics (the laws of physics), we could just as well say that the car is standing still and that the ground is moving eastward at 25 km/hr. We could even say that the ground is moving eastward at 1600 km/hr (which it is, relative to Earth's center, due to Earth's spin) and that the car is moving eastward at only 1575 km/hr. It is only the relative velocity, the 25 km/hr, that really counts.

■ **DIALOGUE 5** If you drop a coin inside a car that is speeding up, where will the coin land? What if the car is slowing down? Turning a corner to the right? Try it sometime (when you are not driving).

■ **DIALOGUE 6** The principle of relativity states that unless you look outside, you cannot detect the velocity of your own reference frame. What about acceleration of your own reference frame—can this be detected without looking outside?

10.4 THE CONSTANCY OF LIGHTSPEED: STRANGE BUT TRUE

One of the many questions that Einstein enjoyed pondering in high school was "What if I could catch up with a light beam? How would a light beam appear if I were moving along with it?"

To Einstein, the possibility of moving along with a light beam seemed paradoxical, contradictory. The reason is that to an observer moving along with a light beam, the light beam itself would be at rest. To this observer, the light beam would appear as an electromagnetic "wave" that was standing still! To Einstein, that seemed impossible.

Our understanding of electromagnetic waves, such as light, is based on Maxwell's theory of electromagnetic fields (Chapter 9). Recall that Maxwell's theory predicts that any disturbance in an electromagnetic field, such as a disturbance caused by the motion of an electrically charged object, must propagate as a wave moving outward through the field *at speed c*. This particular speed, 300,000 km/s, is built into Maxwell's theory.

Einstein believed that Maxwell's theory should, like all other laws of nature, obey the principle of relativity. Maxwell's predictions should be correct even within any moving reference frame. Since speed c is built into Maxwell's theory, Einstein concluded that *every observer ought to observe every light beam to move at speed c*, regardless of the observer's motion. No matter how fast you move, a light beam always passes you at speed c, relative to you. This is why the idea of catching up with a light beam seemed contradictory to Einstein. If every

---

5. To the rear of the drop point. Toward the front. To the left of the drop point.

6. Yes, you can detect accelerations by doing experiments like those of Dialogue 5.

observer sees every light beam move at speed c, then nobody can even begin to catch up with a light beam, much less catch all the way up with one and observe it at rest.

This idea is simple but odd, which is why it took an original genius like Einstein to think of it. We will discuss a few of its oddities later. This idea is so odd that other turn-of-the-century physicists who might have discovered it did not. It is the second important principle underlying Einstein's theory. We summarize it as:

> **THE PRINCIPLE OF THE CONSTANCY OF LIGHTSPEED**
> The speed of light (and of other electromagnetic radiation) in empty space is the same for all nonaccelerated observers, regardless of the motion of the light source or of the observer.

Like the principle of relativity, this principle is valid only for nonaccelerated observers. The reason is that Maxwell's theory, like most laws of physics, is valid only for nonaccelerated observers.

To get a feel for Einstein's idea, we will apply the principle of the constancy of lightspeed to several "thought experiments," experiments that could in principle be performed but that are impractical for one reason or another. Each experiment involves a light beam, which we take to be a laser beam but which could just as well be a flashlight beam or any other electromagnetic radiation such as a radio wave.

If both our observers, Velma and Mort, are on Earth and Mort switches on the laser beam, they will observe the beam moving at speed c. No surprises here.

If Velma and Mort are together in any other reference frame, such as a spaceship moving rapidly past Earth or on a distant galaxy that is moving away from Earth at 90% of lightspeed, the result will be the same: When Mort switches on the laser beam, Einstein's principle predicts that both observers will see the laser beam move at speed c. This is not very surprising, perhaps, since both observers are in the same reference frame.

Things get more interesting when Velma and Mort are in relative motion. Let's give the laser to Mort and imagine that Velma is in a spaceship approaching Mort at 75,000 km/s, or $0.25c$ (Figure 10.5). You can imagine Mort to be on

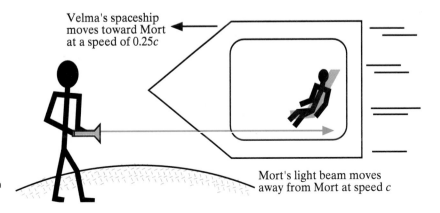

Velma's spaceship moves toward Mort at a speed of $0.25c$

Mort's light beam moves away from Mort at speed c

FIGURE 10.5
What is the speed of Mort's light beam relative to Velma?

Earth, but Mort could as well be on a distant galaxy moving at a high speed relative to Earth or on a spaceship moving rapidly past Earth. The important point is that Velma moves at 0.25c relative to Mort. Mort switches on the laser and observes the beam to move away from him at speed c. What speed does Velma observe for the laser beam as it passes her spaceship? Galilean relativity and our intuitions answer 1.25c, or 375,000 km/s. But Einstein's relativity predicts that the answer is c, or 300,000 km/s.

At this point, you might be inclined to ask, "But why? Why should it be c rather than 1.25c?" We'll save that question for the next section.

Now let's give a laser beam to Velma, who is still approaching Mort at a speed of 0.25c (Figure 10.6). She switches on her laser and observes that the beam moves away from her at speed c relative to her reference frame. What speed does Mort measure for the same beam, relative to his reference frame? Galileo and common sense again predict 1.25c, but Einstein predicts c.

Let's return to the example posed toward the end of Section 10.2 (Figure 10.4). Mort has the laser, and Velma moves away from Mort at a speed of 0.25c. Mort switches on the laser, observing the beam to move away from him at speed c. Galilean relativity predicts that this beam will pass Velma at a speed of 0.75c as measured by her, but Einstein predicts that she will measure this beam to move at speed c.

To dramatize the oddness of this, let's assume that Velma is moving away from Mort at a speed of 0.999,999c, just a little slower than lightspeed (Figure 10.7). Mort switches on his laser and sees the light beam move away from him at speed c. As observed by Mort, Velma is moving only 0.000,001c slower than this light beam—he says that she is nearly keeping up with the light beam. Galilean relativity predicts that Velma will observe the light beam passing her very slowly, at only 0.000,001c. This is one-millionth (10^{-6}) of lightspeed, or 0.3 km/s, or 300 m/s—about the speed of sound and of fast jet airplanes. If the Galilean prediction is correct, she should be able to watch the tip of the beam passing her at about the speed of a jet airplane! The Galilean prediction says that from Velma's point of view, she is nearly keeping up with the light beam.

But Einstein's relativity says that from her point of view Velma isn't even beginning to catch up with the light beam. She sees it move past her at precisely c, 300,000 km/s, despite the fact that she is moving away from the light source at nearly lightspeed.

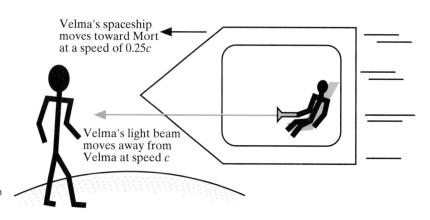

Velma's spaceship moves toward Mort at a speed of 0.25c

Velma's light beam moves away from Velma at speed c

FIGURE 10.6
What is the speed of Velma's light beam relative to Mort?

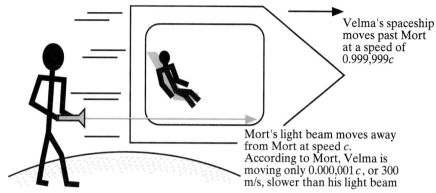

FIGURE 10.7
How fast is Mort's light beam moving, as observed by Velma?

Velma's spaceship moves past Mort at a speed of $0.999,999c$

Mort's light beam moves away from Mort at speed c. According to Mort, Velma is moving only $0.000,001\,c$, or 300 m/s, slower than his light beam

You might have noticed that we didn't allow Velma to have precisely speed c in these examples. We did allow her to move at nearly, but not quite, speed c relative to Mort. If we imagine that she moves at precisely speed c, then we get into the difficulty that Einstein noted about moving along with a light beam: An observer moving at speed c would observe a light beam to be at rest. So, according to Einstein's relativity, an observer can move at nearly, but not precisely, speed c relative to another observer. We will find in the next chapter that there are additional reasons that no observer can move at lightspeed relative to any other observer.

H O W   D O   W E   K N O W ?

Strange as it may seem, these ideas are experimentally verified every day in laboratories around the world. However, the experiments do not usually involve macroscopic objects like spaceships moving at speeds near lightspeed, because it isn't easy to get large objects moving so fast. Instead, most experiments involve fast-moving microscopic particles.

For example, in 1964 an experiment was performed in which a subatomic particle moving at nearly lightspeed emitted electromagnetic radiation in both the forward and backward directions. Galilean relativity predicts that the forward-moving radiation should move at a speed much faster than c and that the backward-moving radiation should move much slower than c, as measured in the laboratory. But the actual measurement showed that both radiation beams move at speed c relative to the laboratory, in agreement with the principle of the constancy of lightspeed.

10.5 EINSTEIN'S LOGIC, MATERIALISM, AND THE LOGIC OF SCIENCE

Nobody thought of the principle of the constancy of lightspeed before Einstein did, because it is so odd. Maxwell and other nineteenth-century scientists had a more conventional view of light beams. As we explained in Section 9.3, they believed that light waves were caused by vibrations of a material medium, just as, for example, water waves were caused by vibrations of water. They called this medium **ether**. Nobody had ever observed ether. It was known not to be made of ordinary chemical atoms, because light waves travel through regions of space where there are essentially no atoms. Instead, ether was thought to be a gaslike substance filling the entire universe and made of some unknown, nonatomic form of matter.

The ether theory predicts that the "natural" speed of light, 300,000 km/s, is light's speed relative to the ether. Observers moving through the ether should then measure other speeds for light beams, speeds that should depend on the observer's speed through the ether and on the light beam's direction (Figure 10.8).

The **Michelson–Morley experiment**, performed by Albert Michelson (whom we met at the outset of this chapter) and Edward Morley in 1887, 18 years before Einstein's theory, provided the earliest evidence that light has the same speed for all observers, in disagreement with the ether theory. Michelson and Morley made accurate comparisons of the speeds of light beams that were moving in different directions. Since their laboratory (like all of Earth) was presumed to be moving through the ether, they predicted that different light beams would have slightly different speeds relative to the lab. But they found that all light beams had the same speed relative to their lab. The physicists of that time, including Michelson and Morley, failed to appreciate this experiment's fundamental implications, preferring instead to assume that light beams merely *appeared* to have the same speed in that particular experiment and to seek explanations for that appearance. Michelson and Morley themselves were disappointed at their "failure" to detect the expected difference in the speeds of different light beams. Actually, their experiment was a stunning breakthrough to an exciting new experimental fact. It's just that nobody could see it that way.

Einstein, in 1905, was the first to state that light has the same speed for all observers. Furthermore, and surprisingly, Einstein paid little attention to the Michelson–Morley experiment in developing his theory. To him, it seemed obvious that light should have the same speed relative to all observers, for the reasons given at the beginning of Section 10.4. Here was a truly independent and original mind: Other physicists rejected the constancy of lightspeed even despite the Michelson–Morley experiment's evidence that lightspeed is constant, but Einstein accepted the constancy of lightspeed even without paying attention to the evidence of the Michelson–Morley experiment!

By contradicting the ether theory, Einstein's theory undermines the notion that light waves are caused by vibrations of a material medium. So light waves and other electromagnetic waves must be nonmaterial. Since Einstein, electro-

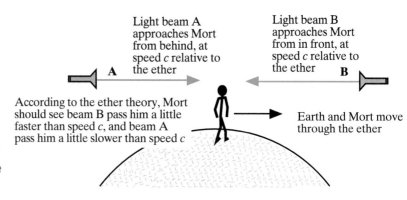

Light beam A approaches Mort from behind, at speed c relative to the ether

Light beam B approaches Mort from in front, at speed c relative to the ether

According to the ether theory, Mort should see beam B pass him a little faster than speed c, and beam A pass him a little slower than speed c

Earth and Mort move through the ether

FIGURE 10.8
The ether theory of light predicts that an observer moving through the ether will measure different light beams moving at different speeds.

The Lord is subtle, but He is not malicious.

EINSTEIN

magnetic waves have been viewed as caused by the vibrations of an electromagnetic field, which itself is not made of any material substance. As we discussed in Section 9.3, this represents an important deviation from the materialist worldview associated with Newtonian physics.

The principle of the constancy of lightspeed is the key new idea in the theory of relativity, and gives the theory its odd, nonintuitive quality. It is natural that we should question this principle. Why is it true? How can it be true?

The answer is very simple and goes to the heart of science. The principle of the constancy of lightspeed is true because nature says so. Observation of nature, not our preconceived beliefs, determines truth in science. And numerous experiments show that every light beam in nature moves at speed c, regardless of the motion of the observer. It is a simple answer, but profound. The acceptance of nature—acceptance of the universe—is the key to science.

The two key ideas of Einstein's theory are the principle of relativity and the principle of the constancy of lightspeed. These should be regarded as the first and second laws of Einstein's relativity. They play a role in the theory of relativity that is identical to the role of Newton's laws in Newton's theory of force and motion: They form the logical basis of the theory, from which everything else in the theory is derived, and which are themselves justified directly by experience.

Physicists call the theory founded on these two principles the **special theory of relativity**. The word *special* distinguishes this theory from the Galilean theory of relativity and also from another, related theory of Einstein's called the **general theory of relativity**. The distinguishing feature of the general theory of relativity is that it does not assume an unaccelerated observer. That is, the general theory of relativity applies to all observers, both accelerated and unaccelerated, and so it is more general, broader, than the special theory. We will look at Einstein's general theory of relativity in the latter part of Chapter 11.

Strictly speaking, Earth itself is an accelerated reference frame, both because it spins on its axis and because it rotates around the sun. But these accelerations are small compared with the accelerations that are significant in the general theory of relativity. So we can assume, as a good approximation, that the predictions of the special theory of relativity are correct for any Earth-based observer.

Beginning in the next section and continuing through most of the next chapter, we explore five of special relativity's most important predictions: the relativity of time, the relativity of space, the relativity of mass, c as the limiting speed, and $E = mc^2$. These results are predictions that follow from the two basic principles of the theory; in other words, if you believe the two basic principles, then you must logically believe these predictions. Not surprisingly, these predictions seem fairly odd to our intuitions. The source of this oddness is the principle of the constancy of lightspeed.

■ **DIALOGUE 7** Velma is moving away from Mort at a speed of 0.75 c. She turns on two lasers, one pointed forward and the other pointed backward. (a) What speed does Velma observe for each of the two laser beams? (b) According to Galilean relativity, what speed does Mort observe for each beam? (c) According to Einstein's relativity, what speed does Mort observe for each beam? (d) What speed does Mort actually observe for each beam (according to nature)?

---

7. (a) c. (b) $1.75c$ for the forward beam, $0.25c$ for the backward beam. (c) c. (d) c.

10.6 THE RELATIVITY OF TIME: TIME IS WHAT HUMANS MEASURE ON CLOCKS

If you do not ask me what is time, I know it; when you ask me, I cannot tell it.

ST. AUGUSTINE, A.D. 400

Absolute, true, and mathematical time, of itself, and from its own nature, flows equably without relation to anything external, and by another name is called duration. Relative, apparent, and common time, is some sensible and external measure of duration by means of motion, which is commonly used instead of true time; such as an hour, a day, a month, a year.

NEWTON, JUSTIFYING THE IDEA THAT TIME IS ABSOLUTE

We feel that we understand what "time" is, but the meaning of time fades from our grasp whenever we begin to think about it. The problem is that, being enmeshed in time, we have no way to study it from a distance. So we have deeply embedded intuitions about time, unconscious intuitions that might turn out to be false in nature.

Einstein's basic insight into time was that it was physical, part of the physical universe, rather than apart from or beyond the physical universe. Just as one can measure the properties of a stone or of a light beam, one can measure the properties of time itself. And how should we measure the properties of time? With clocks, obviously. This reply is more profound than it appears. The only way we can measure time is with real, physical clocks (including all time-keeping phenomena such as a heartbeat and Earth's rotation around the sun). Physically, the concept of a clock really defines time. In other words, time is what humans measure on clocks.

So if we want to investigate the properties of time, we must investigate the properties of clocks. How do clocks really behave? Einstein's goal in studying time was to predict the properties of clocks using as his starting point only the two principles of the special theory of relativity.

An ordinary spring-wound or battery-driven clock would be hard to study using only Einstein's two simple principles because these clocks are too complex. They involve springs, electric current, vibrational motions, gears, and so forth. So Einstein invented a simple kind of clock, a simple thought experiment, really. His clock involves no mechanically moving parts; its only motion is the motion of a light beam.

Figure 10.9 shows this **light clock**. Two parallel mirrors face each other, one above the other, and a light beam bounces up and down between them. It is convenient for our discussion (but not too practical for the clock maker) to imagine that the two mirrors are separated by 150,000 kilometers, because then the time for one complete round-trip of the light beam is just 1 second. We know that this time is 1 second because the principle of the constancy of lightspeed tells us that all light beams travel 300,000 kilometers in 1 second.* In fact, this device defines what we mean by "1 second": It is the interval during which the tip of the light clock's light beam makes one round-trip. We will assume that this light clock ticks at the end of every round-trip, in other words, that it ticks once every second.

Now that we have a simple physical definition of time, we can investigate the properties of time. We begin by installing one light clock in Velma's spaceship and another in Mort's laboratory. We imagine that Mort is on Earth and Velma is moving eastward past Earth in a spaceship. Let's begin by thinking about Velma's light clock. As seen by her, her light beam bounces straight up and down, covering 150,000 km upward and then 150,000 km downward, ticking when the light beam gets back to the bottom mirror (Figure 10.10(a)). Simple enough.

How does Velma's light clock appear to Mort? From his point of view, the tip of Velma's light beam is not only moving up and down, it is also moving east-

---

*We are assuming that the clock is carried by an unaccelerated observer.

FIGURE 10.9
A light clock. A light beam bounces up and down between two mirrors. If the distance between mirrors is 150,000 km, then 1 second will elapse during one complete round-trip up and back down.

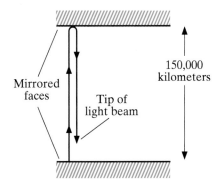

Mirrored faces

Tip of light beam

150,000 kilometers

ward because of the eastward motion of Velma's spaceship. This means that the tip of Velma's light beam, as seen by Mort, moves along a diagonal path as it moves up to the top mirror and then back to the bottom mirror. Figure 10.10(b) shows this effect by depicting Mort's observations of Velma's spaceship at three different instants: when the tip of her light beam is at the bottom mirror, when her light beam has moved up to the top mirror, and when her light beam is back at the bottom mirror.

Recall that the top-to-bottom distance between the two mirrors is 150,000 km. As you can see from the figure, the distance along one of the two diagonals must then be greater than 150,000 km. This means that the total round-trip distance traveled by Velma's light beam, as measured by Mort, must be greater than 300,000 km. There is nothing surprising or subtle about this; Galileo would have said the same thing.

Now comes the part that Galileo (and our intuitions) wouldn't agree with: The principle of the constancy of lightspeed predicts that Mort observes Velma's light beam to move at just 300,000 km/s. Since Mort observes the round-trip distance to be greater than 300,000 km, it follows that according to Mort it takes

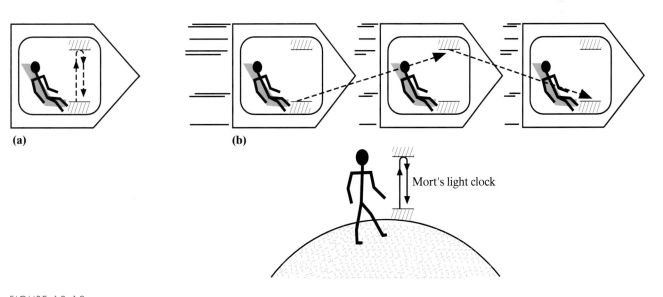

(a)   **(b)**

Mort's light clock

FIGURE 10.10
(a) Velma in her spaceship, observing her light clock. (b) Velma's spaceship and the light beam on Velma's light clock as observed by Mort using his own light clock. According to Mort's observations, the tip of Velma's light beam moves along a diagonal path shown by the dashed arrows.

more than 1 second for Velma's light beam to make the round-trip! So, as measured by Mort using his clock, more than 1 second elapses between Velma's ticks. According to Mort, Velma's clock runs slow.

Velma's second is different from Mort's second. The two observers measure different time intervals for the same event (one round-trip of Velma's light beam). That is, *time is relative to the observer.*

Let's turn things around. How does Mort's clock appear to the two observers? To Mort, his own clock's light beam travels 300,000 km in one round-trip and requires 1 second to do so. But from Velma's viewpoint, Mort's clock is moving westward, so the tip of Mort's light beam is moving along a diagonal, so the total round-trip distance traveled by Mort's light beam as observed by Velma is greater than 300,000 km. But because Velma observes Mort's light beam to move at 300,000 km/s, she must observe that more than 1 second elapses between Mort's ticks. According to Velma, Mort's clock runs slow.

Mort observes that his own clock is accurate and that Velma's runs slow. Velma says that her clock is accurate and that Mort's runs slow. This is not your normal situation caused by an inaccurate clock, in which if my clock runs slow according to your clock, then your clock must run fast according to my clock. Instead, if you and I are in relative motion, then my clock runs slow according to you, and your clock runs slow according to me.

This raises an interesting question: Who is right? Whose clock is really running slow, and whose clock is really accurate?

The answer is that both Velma and Mort are right. Velma observes that Mort's clock is slow, and Mort observes that Velma's clock is slow, and both observations are accurate. This situation is not caused by inaccurate observations or inaccurate clocks; it is instead a property of time itself. There is no single "real" time in the universe, no "universal time"; there is only Mort's time and Velma's time and all the other possible observed times.

The general rule here is that moving clocks run slow.

This effect—the slowness of moving clocks—occurs whenever the two observers are in relative motion. The effect is small if the relative speed is small compared with lightspeed, because for such small speeds the diagonal path length (Figure 10.10) is only slightly longer than 300,000 km. At familiar human speeds, the effect is so small that it wasn't noticed before Einstein's work. Even at a speed as high as 3000 km/s (1% of lightspeed, which would carry you across the United States in about a second), the effect is very small.

As you might expect, there is a formula that quantitatively describes the relativity of time.* Table 10.1 gives a few of the numerical results that can be calculated from this formula, and Figure 10.11 is a graph based on the same formula. As you can see from the table, the effect is negligible at jet airplane speeds (0.3 km/s) and at orbiting satellite speeds (10 to 20 km/s). It is not until speeds of $0.1c$—a speed that would get you around the world in about 1 second—are reached that the effect amounts to even a half of 1%. But at large fractions of lightspeed, the effect becomes quite large: If Velma passes Mort at 99.9% of lightspeed, their seconds will be more than 22 seconds long as measured by the other observer. The relativity of time is also called **time dilation**, because a time

I do not define time, space, place and motion, as being well known to all.
NEWTON

It requires a very unusual mind to undertake the analysis of the obvious.
ALFRED NORTH WHITEHEAD, TWENTIETH-CENTURY PHILOSOPHER

---

*This formula can be derived by studying Figure 10.10 and using the *Pythagorean theorem*, which states that a right triangle's short side lengths *a* and *b* are related to its diagonal length *c* by $c^2 = a^2 + b^2$. The formula that results from this is $T = T_o/\sqrt{(1 - s^2/c^2)}$, where *s* is the relative speed, T_o is the time between two of Velma's ticks as observed by Velma ($T_o = 1$ second for a normal clock), and *T* is the time between two of Velma's ticks as observed by Mort.

TABLE 10.1

The relativity of time: some quantitative predictions.   To give you a feel for these speeds: 0.3 km/s is a typical subsonic jet plane speed. 3 km/s is twice the speed of a high-powered rifle bullet. 30 km/s is about the speed of Earth in its orbit around the sun. At 300 km/s you could cross the United States in 10 seconds, at 3000 km/s you could cross the United States in 1 second, and at 30,000 km/s you could circle the globe in about 1 second. Clearly, relativistic effects are small until the speed becomes very large!

Relative speed (km/s)	Relative speed as a fraction of lightspeed (c)	Duration of one "tick" on a moving clock, as measured by an observer past whom the clock is moving (s)
0.3	10^{-6}	1.000,000,000,000,5
3	10^{-5}	1.000,000,000,05
30	10^{-4}	1.000,000,005
300	0.001	1.000,000,5
3000	0.01	1.000,05
30,000	0.1	1.005
75,000	0.25	1.03
150,000	0.5	1.15
225,000	0.75	1.5
270,000	0.9	2.3
297,000	0.99	7.1
299,700	0.999	22.4

interval of 1 second on a moving clock is expanded or *dilated* to more than 1 second as measured by an observer past whom the clock is moving.

Although we investigated the relativity of time by studying light clocks, the conclusion holds for every type of clock. We are investigating time, not just light clocks. All clocks must behave the way that light clocks behave: spring-driven clocks, pendulum clocks, and so forth, because time behaves this way.

Furthermore, everything that happens in time, every phenomenon that occurs during an interval of time, must also behave in this way. Think, for example, of an ice-cream cone melting. Suppose that an ice-cream parlor makes ice-cream cones that melt in exactly 10 minutes and that both Velma and Mort have one of these cones. These cones are a kind of clock, a clock that ticks in 10 minutes.

> The testimony of our common sense is suspect at high velocities.
>
> CARL SAGAN, ASTRONOMER AND AUTHOR

■ **DIALOGUE 8** Velma passes Mort at 75% of lightspeed. Use Table 10.1 to find (a) how long Velma's 10-minute cone will take to melt as measured by Mort; (b) how long Velma's cone will take to melt as measured by Velma; (c) how long Mort's cone will take to melt as measured by Velma; and (d) how long Mort's cone will take to melt as measured by Mort.

■ **DIALOGUE 9** Answer the questions of Dialogue 8, assuming that Velma moves past Mort at 99% of lightspeed.

■ **DIALOGUE 10** How fast must Velma move, relative to Mort, in order for Velma's ice-cream cone to melt in 20 minutes as measured by Mort (use Figure 10.11)? How fast must she move in order for her cone to melt in 30 minutes as measured by Mort?

---

8. (a) Since a 1-second tick on Velma's clock is observed by Mort to take 1.5 s (1.5 times as long), Velma's 10-minute ice-cream cone must be observed by Mort to take 15 minutes to melt; (b) 10 minutes; (c) 15 minutes; (d) 10 minutes.

9. (a) 71 minutes; (b) 10 minutes; (c) 71 minutes; (d) 10 minutes.

10. About 0.87c. About 0.94c.

FIGURE 10.11
Relativistic time dilation. The graph shows the duration of one clock tick (representing 1 second in the clock's reference frame) on a moving clock, for various speeds of the clock relative to the observer.

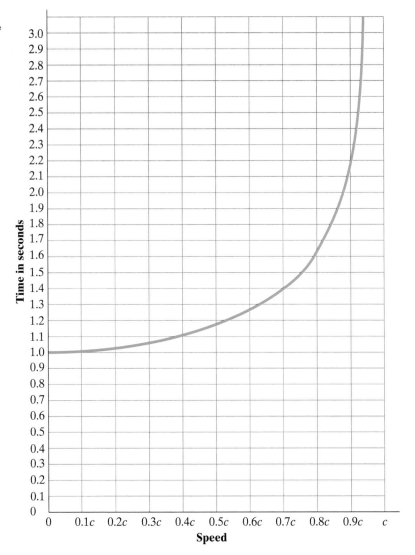

Instead of ice-cream cones, it could be frogs. Suppose your local biology department hatches guaranteed 10-day frogs, having a 10-day lifetime. Biological life occurs in time, too, just as clocks do, so these frogs can be thought of as a kind of clock. So if Velma passes Mort at 75% of lightspeed, he says that her frog lives 15 days but that his frog lives only 10 days (see Dialogue 8). And she says that his frog lives 15 days but that her frog lives only 10 days.

This raises the interesting question of which frog dies first. According to Mort, his frog dies first, and Velma's frog "ages slowly." According to Velma, her frog dies first, and it is Mort's frog that hangs on 5 days longer. Whose frog really dies first? The answer is that both observations are right! Mort observes his frog to die first, Velma observes her frog to die first, and both observations are accurate.

Again, the unspoken belief in the back of our mind when we ask "whose frog really dies first?" is that there is one universal real time. But there is not any single, universal time. Instead there is Mort's time, and Velma's time, and every other individual observer's time.

H O W   D O   W E   K N O W ?

The relativity of time has been verified again and again in laboratories, by observing fast-moving subatomic particles. One experiment, similar to the frog example, involved a type of subatomic particle known as a *muon*. Muons, unlike most ordinary matter, are not permanent objects. Instead, they have a "lifetime" after which they disintegrate spontaneously into other particles. The lifetime of a muon is only 2.2 microseconds (a microsecond is a millionth of a second). If you observe a muon that is at rest relative to you, its lifetime is 2.2 microseconds as measured by you. But a muon moving rapidly past you lives much longer as measured by you, because of time dilation. For example, if the muon is moving past you at 99% of lightspeed (muons often move this fast in high-energy physics labs), then Table 10.1 says that its lifetime will be lengthened by a factor of 7.1 and that it will not disintegrate until $7.1 \times 2.2 = 15.6$ microseconds have passed. This experiment has been done, and the moving muons were observed to have lifetimes that were lengthened by just the predicted amount.

10.7 TIME TRAVEL: YOU CAN'T GO HOME AGAIN

As you might have suspected, the next step is to investigate the life span of Velma and Mort themselves. Suppose they are born at the same time* and that they have lifetimes of 80 years. In other words, Velma measures her lifetime to be 80 years, and Mort measures his lifetime to be 80 years. If Velma and Mort spend their lives moving at 75% of lightspeed relative to each other, then Table 10.1 informs us that Mort's descendants (his grandson perhaps) will observe that Velma lives for 120 years, as measured by Mort's clocks. And Velma's granddaughter will observe that Mort lives for 120 years, as measured by Velma's clocks. According to Mort and his grandson, Velma ages slowly, by just a year during each of Mort's 1.5 years; he dies after just 80 of his years, and she dies after 120 of Mort's years, but having the physical appearance of a person who is only 80. According to Velma, all of this is reversed: He ages slowly, and she dies first. And both of them are correct.

■ **DIALOGUE 11** (a) When Mort observes himself to be 60 years old, how old will he observe Velma to be? (b) When Velma observes herself to be 45 years old, how old will she observe Mort to be?

This suggests a fascinating question. Suppose that Velma and Mort are born at the same time on Earth, as twins perhaps, and that then Velma boards a spaceship, takes a fast trip to a far star, and returns to Earth. This scenario is different from the scenario in the preceding paragraph, because now Velma and

You could see that Einstein was motivated not by logic in the narrow sense of the word but by a sense of beauty. He was always looking for beauty in his work. Equally he was moved by a profound religious sense fulfilled in finding wonderful laws, simple laws in the universe. It was really a religious experience for him, of the most profound sort, even though he did not believe in a personal god.

BANESH HOFFMANN, MATHEMATICIAN AND AUTHOR, IN *SOME STRANGENESS IN THE PROPORTION*

*You might wonder what "at the same time" means, since we are assuming that Mort and Velma are in different reference frames. To simplify matters, suppose that Mort and Velma are just passing each other. Then "at the same time" means that as either one comes into the world, he or she observes that the other is coming into the world too.

11. (a) 40, because he observes that 1.5 of his years pass for every 1 of hers. (b) 30.

Mort begin and end in only one reference frame. Once they are back together they must agree on who is older, because there is only a single time in any single reference frame. Which twin will be older, or will they be the same age?

Recall that the special theory of relativity applies only to nonaccelerated observers. But in the scenario for the two twins, Velma had to leave Earth, speed up enormously, reverse directions at least once in order to get back to Earth, and then slow down and come to rest on Earth. Since this trip necessarily involves enormous accelerations, the special theory does not apply to Velma's observations.

The special theory does, however, apply to Mort's observations, since he did not accelerate. As we have seen, the theory predicts that he will observe Velma to age slowly during her entire trip, because she is moving relative to him. For example, if she moves at an average speed of 75% of lightspeed, then he should observe that 1.5 of his years elapse for every 1 of her years (Table 10.1). Suppose that Velma's trip takes 60 years as measured by Mort. He observes that during these 60 of his years, only 40 of her years elapse. So he observes that when they get back together on Earth, he is 60 and she is 40! Her observations must agree with this, since the two are now in the same reference frame. This is how you can get to be 20 years younger than your twin brother.

HOW DO WE KNOW?

This conclusion has been experimentally verified, but in a less dramatic fashion. Atomic clocks were flown around the world on commercial jet plane flights. Although the predicted time difference was only a fraction of a second, it was measurable using high-accuracy clocks. As predicted, the clock that went on the trip came back younger (it hadn't ticked as many times) than the clock that stayed home. And the quantitative difference in elapsed time was precisely as predicted.

This suggests some strange and wonderful possibilities. Suppose your mother leaves Earth for the star Vega, a sunlike star lying relatively close to our sun and a possible candidate for a planetary system. The distance to Vega is 26 light-years, by which we mean that it takes light 26 years to reach Vega from here. A **light-year** is the distance light travels in one year.

Suppose that your mother's spaceship averages a colossal 99.9% of lightspeed. She spends 3 years on a planet that is orbiting Vega and returns home. Since she travels at nearly lightspeed, each one-way trip takes slightly more than 26 years, as measured on Earth. So she is gone for about 26 + 3 + 26 = 55 years, as measured on Earth. If you were 5 and your mother was 30 when she departed, you would be 60 when she returned. But your mother would no longer be 25 years older than you! Table 10.1 informs us that during the 52 "Earth-years" of space travel at 99.9% of lightspeed, she aged by only 1 year for every 22.4 years of "Earth time." So she aged by only 52/22.4 = 2.3 years during the 52 Earth-years. Including the 3 years spent on Vega, she aged by only 5.3 years during the entire trip. So she is 35.3 years old when she returns, and you are 60! This is how you can get to be older than your mother.

Your mother took a trip to the future. She could travel much further into the future, hundreds or thousands of years into the future, by moving much faster during a longer Earth time. But she couldn't go home again, in time. You can go as far forward in time as you like, if you only move fast enough, but you can never go backward in time.

Time dilation suggests that humans can travel to distant stars within a single human lifetime. Suppose you travel to a star 200 light-years away, at an average speed of 0.999 c relative to Earth. Even though the trip takes a little over 200 years as measured on Earth clocks, it takes you only 200/22.4 = 9 years as measured in your spaceship. You can get there within your lifetime. But you can't go home again in time, to your own time on Earth. By the time you arrive at the star, two centuries will have elapsed on Earth. Even if you immediately hurry back to Earth, you will time-travel four Earth centuries into the future during the round-trip but will age by only 18 years. On Earth, you will be a relic from four centuries earlier.

Although there are many practical difficulties in accelerating humans to speeds like 0.999c, it might be possible someday (Section 12.5).

■ **DIALOGUE 12**   Could you leave Earth and travel fast enough to return before you were born? Could your mother leave Earth after you were born and return before you were born? Explain.

12. No to both questions. The relativity of time allows time travel into the future but not into the past.

SUMMARY OF IDEAS AND TERMS

Relative motion   Two objects are in relative motion whenever they have different velocities.

Reference frame   The laboratory or other surroundings within which an observer makes measurements. Measurements made in a particular reference frame are said to be *relative to* that frame.

Theory of relativity   Any theory that provides answers to questions about observers in relative motion.

Galilean relativity   The intuitive theory of relativity, in which time and space are absolute (in other words, different observers measure the same time intervals and the same distances) and light has different speeds relative to different reference frames.

Special theory of relativity   Einstein's theory based on the principle of relativity and the principle of the constancy of lightspeed. In this theory, time and space are not absolute, and light has the same speed in all nonaccelerated reference frames.

General theory of relativity   Describes the observations of accelerated observers, while the special theory of relativity applies only to nonaccelerated observers.

Light clock   A clock whose time-keeping is based on the motion of a light beam.

Principle of relativity   Every nonaccelerated observer observes the same laws of nature. "Unless you look outside, you can't tell how fast you're moving."

Principle of the constancy of lightspeed   Light (and other electromagnetic radiation) has the same speed for all nonaccelerated observers, regardless of the motion of the light source or of the observer.

Ether theory   The idea that a gaslike, nonatomic, material medium, the ether, fills the entire universe and that light waves are waves traveling through this medium. This the-ory was rejected after 1905 because it contradicted Einstein's theory. Philosophically, this amounts to a rejection of the idea that every physical thing is made of a material substance. Light waves are physical, but they are not made of matter.

Michelson–Morley experiment   An experiment in 1887 to compare the speeds of light beams moving in different directions. The ether theory predicted that the light beams should have slightly different speeds, but the experiment detected no such difference.

Time   Time is defined by clocks, in other words, by the operations we perform to measure time. The light clock, based on the motion of light beams, is a simple instrument to define time.

The relativity of time or **time dilation**   The elapsed time (the number of seconds) between two particular events, such as two ticks on a particular clock or the birth and death of a person, is different for two observers who are in relative motion. The duration of one clock tick is longer for observers who are moving relative to the clock than it is for observers for whom the clock is at rest: Moving clocks run slowly.

Light-year   The distance that light travels in one year.

Time travel   An observer who accelerates to a high speed and then returns to the initial reference frame experiences a shorter elapsed time than does an observer who remains in the initial frame. So objects in the initial reference frame have aged more than the traveler has. This effect makes it possible to travel to stars that are many light-years distant in only a few years' travel time. It also makes travel to the future possible, by going on a fast trip and returning.

REVIEW QUESTIONS

INTRODUCTION

1. Name the two revolutionary physical theories initiated at the beginning of the twentieth century.
2. Which inaccuracies of Newtonian physics are corrected by relativity theory? By quantum theory?
3. Describe the new element found in both relativity theory and quantum theory that is not found in Newtonian physics.

GALILEAN RELATIVITY

4. What is meant by relative motion, reference frame, a theory of relativity?
5. A train moves at 70 m/s. A ball is thrown toward the front of the train at 20 m/s relative to the train. How fast does the ball move relative to the tracks? What if the ball had instead been thrown toward the rear of the train?

6. A spaceship moves at 0.25c relative to Earth. A light beam passes the spaceship, in the forward direction, at speed c relative to Earth. According to Galilean relativity, how fast does the light beam move relative to the spaceship? Is this answer experimentally correct? If not, then what answer is correct?

THE PRINCIPLES OF RELATIVITY
AND CONSTANCY OF LIGHTSPEED

7. How does travel in a jet airplane illustrate the principle of relativity? How must the airplane be moving in order to illustrate this principle?
8. State the principle of relativity in your own words. Does it apply to every observer? Explain.
9. State the principle of the constancy of lightspeed in your own words. Does it apply to every observer? Explain.

10. Why can't you keep up with a light beam—what law of physics prevents it?
11. Use the principle of the constancy of lightspeed to explain why no observer can move at precisely speed *c* relative to any other observer.

EINSTEIN'S LOGIC
12. Describe the ether theory. What prediction does this theory make regarding the speeds of light beams? Is this prediction correct?
13. What did the Michelson–Morley experiment measure, and what was the result?
14. Describe the philosophical implications of the rejection of the ether theory.
15. In Galilean relativity, space and time are absolute, and lightspeed is relative. What is the situation in Einstein's relativity?
16. What distinguishes the special from the general theory of relativity?
17. List the basic "laws" of the special theory of relativity.

THE RELATIVITY OF TIME
18. How is time defined in physics?
19. Describe the light clock.

20. Use the principle of the constancy of lightspeed to show that moving clocks must run slowly.
21. Velma passes Mort at a high speed. Both observers have clocks. What does each observer say about Velma's clock? What do they each say about Mort's clock?
22. Velma passes Mort at a high speed. Mort and Velma have identical ice-cream cones, frogs, and life spans. What does each observer say about the melting of both ice-cream cones, the aging of both frogs, and the aging of Velma and Mort?
23. One twin goes on a fast trip and returns. Does the special theory of relativity apply to the observations of both twins? Why, or why not?
24. One twin goes on a fast trip and returns. Have the two twins aged differently during the trip? If so, how do their ages differ?
25. Could you be older than your mother? If so, how?
26. A certain star is 200 light-years distant from Earth. If you traveled there at nearly lightspeed, how many years would the trip take as measured by observers on Earth? Would the trip take this many of your years? Explain.
27. Explain how you can travel to the future.

HOME PROJECT

Do some experiments while traveling on a smoothly moving, unaccelerated jet plane, train, or car. Also try to do the same experiments while your vehicle is speeding up, slowing down, or turning. Suggested experiments: Drop a ball with one hand held directly beneath the drop point; drop a piece of tissue paper (close all windows and vents to avoid air currents); roll a ball on a smooth tabletop. Don't try this while you are driving!

FOR DISCUSSION

1. Does it seem to you that Einstein's personality might have had much to do with his creation of the theory of relativity? Discuss.
2. Does Einstein's comment on his college experience (Section 10.1, including the quotations in the margins) seem to you to be a valid criticism of education? Is this criticism valid today, at the grade school, high school, or college level?

EXERCISES

GALILEAN RELATIVITY
1. Two bicyclers, on different streets in the same city, are both moving directly north at 15 km/hr. Are they in relative motion?
2. Two bicyclers, both moving at 15 km/hr, pass each other in opposite directions. Are they in relative motion? What if one is traveling northward at 15 km/hr and the other is traveling eastward at 15 km/hr?

3. According to the Galilean theory of relativity, does every observer measure the same speed for a lightbeam?
4. Velma moves away from Mort at half of lightspeed. Mort shines a searchlight toward Velma. What does Galilean relativity predict about the speed of the searchlight beam as observed by Velma?
5. Velma moves toward Mort at half of lightspeed. Mort shines a searchlight toward Velma. What does Galilean

relativity predict about the speed of the searchlight beam as observed by Velma?

THE PRINCIPLE OF RELATIVITY

6. Does the principle of relativity require that *every* observer observes the same laws of physics? Explain.

7. If you were riding on a train moving at constant speed along a straight track and you dropped a ball directly over a white dot on the floor, where would the ball land relative to the dot?

8. Suppose that you drop a ball while riding on a train moving at constant speed along a straight track. If you measure the ball's acceleration, will your result be greater than, less than, or equal to, the usual acceleration due to gravity?

9. Although everything is normal inside a smoothly moving jet airplane, things are not normal when the ride is bumpy; you spill your coffee, you experience moments of partial weightlessness, and so on. Does this conflict with Einstein's principle of relativity? Explain.

10. Think of several ways that you could determine from inside an airplane whether the plane was flying smoothly or parked on the runway. Do each of these ways involve some direct or indirect contact with the world outside the airplane?

11. How can you tell from inside a jet airplane that the plane is speeding up? Slowing down? Turning a corner? Does your ability to determine these things conflict with Einstein's principle of relativity?

12. How fast are you moving right now? What meaning does this question have?

THE PRINCIPLE OF THE CONSTANCY OF LIGHTSPEED

13. Does *every* observer measure the same speed for a light-beam? Explain.

14. A star headed toward Earth at 20% of lightspeed suddenly explodes as a bright supernova. With what speed does the light from the explosion leave the star? With what speed (as measured on Earth) does it approach Earth?

15. Is it physically possible (that is, would it be consistent with the principles of physics) for a person to move past Earth at 99% of lightspeed? If an Earth-based light beam were turned on, how fast would this person see it moving?

16. Is it physically possible for a person to move past Earth at exactly lightspeed? Explain.

17. Velma's spaceship approaches Earth at 0.75c. She turns on a laser and beams it toward Earth. How fast does she see the beam move away from her? How fast does an Earth-based observer see the beam approach Earth?

THE RELATIVITY OF TIME

18. Because of their relative motion, Mort says that Velma's clock runs slow, and Velma says that Mort's clock runs slow. Whose clock is right?

19. Your pulse could be considered to be a kind of clock. Think of several other natural (but not necessarily biological) clocks. Would an observer moving rapidly past these clocks observe all of them to run fast, slow, or on time?

20. Velma passes you at a high speed. According to you, she ages slowly. How does she age according to her own observations? How do you age according to her?

21. If you were moving away from Earth at nearly lightspeed (but with no acceleration), would you notice any change in your pulse? Any change in your breathing rate? Anything unusual about your clocks?

22. Suppose you have a twin brother. Is there anything you two could do that would make you older than him? Is there anything you two could do that would make you younger than him? Explain.

23. The center of our galaxy is about 30,000 light-years away. How long does light take to get here from there? Could a person possibly travel there in less than 30,000 years as measured on Earth? Could a person possibly travel there in less than 30,000 years of his or her own time? Could a person get there within his or her own lifetime? Explain.

24. A woman conceives a child while on a fast-moving space colony moving toward a distant planetary system. How long should it take before the baby is born, as measured by the woman? Would an Earth observer measure the same amount of time?

25. The average lifetime for an isolated neutron is 17 minutes, after which it decays by spontaneously splitting into three other particles: proton, electron, and antineutrino. Will the lifetime of a rapidly moving neutron still be 17 minutes? Explain.

26. A certain fast-moving particle is observed to have a lifetime of 2 seconds. If the same particle was at rest in the laboratory, would its lifetime still be 2 seconds, or would it be more, or less, than 2 seconds?

27. Does the special theory of relativity allow you to go on a trip and return older than your father?

28. Does the special theory of relativity allow your father to go on a trip and return younger than you?

29. Does the special theory of relativity allow you to go on a trip and return younger than you were when you left?

30. Does the special theory of relativity allow you to "time travel" to visit with Socrates in ancient Greece?

PROBLEMS

GALILEAN RELATIVITY

1. Velma bicycles northward at 4 m/s, and Mort jogs southward at 4 m/s. Are Velma and Mort in relative motion? What is Mort's velocity, relative to Velma?

2. Velma bicycles northward at 4 m/s, and Mort jogs northward at 4 m/s. Are Velma and Mort in relative motion? What is Mort's velocity, relative to Velma?

3. Velma bicycles northward at 4 m/s. Mort, standing by the side of the road, throws a ball northward at 10 m/s. What is the ball's speed and direction of motion, relative to Velma? What if Mort had instead thrown the ball southward at 10 m/s?

4. Velma bicycles northward at 4 m/s. She throws a ball northward at 10 m/s relative to herself. What is the ball's speed and direction of motion relative to Mort, who is standing beside the road? What if Velma had instead thrown the ball southward at 10 m/s? What if she had thrown the ball southward at 4 m/s?

5. A desperado riding on top of a train car fires a gun toward the front of the train. The gun's muzzle velocity (speed of the bullet relative to the gun) is 500 m/s, and the train's speed is 40 m/s. What is the bullet's speed and direction of motion as observed by the sheriff standing beside the tracks? What does a passenger on the train say about the bullet's speed? What if the desperado had instead pointed his gun toward the rear of the train?

6. In the preceding exercise, suppose that the sheriff fires his own gun at the desperado. The muzzle velocity of the sheriff's gun is 600 m/s. What is the bullet's speed relative to the desperado, if the sheriff is standing on the ground behind the receding train? What if the sheriff is instead standing in front of the train as it approaches him?

THE CONSTANCY OF LIGHTSPEED AND EINSTEIN'S LOGIC

7. A desperado riding on top of a freight-train car fires a *laser* gun pointed forward (compare with Exercise 5). What is this gun's "muzzle velocity"? Suppose the train is moving at 40 m/s (0.04 km/s). How fast does the tip of the laser beam move relative to the sheriff, who is standing on the ground beside the train? What answer would the Galilean theory of relativity have given to this question?

8. Velma's rocket ship is moving away from Mort at $0.75c$. She fires a laser beam in the *backward* direction, toward Mort. According to Galilean relativity, how fast does the laser beam move relative to Mort, assuming that Velma observes the beam to move away from her at speed c? How fast according to Einstein's relativity?

9. Earth orbits the sun at a speed of about 30 km/s. If we assume that ether fills all space, and that the sun is at rest in the ether, then it follows that Earth moves at about 30 km/s through the ether. Assuming also that light is a wave in the ether, what speed would the observer in Figure 10.8 observe for lightbeam A? For lightbeam B?

10. Michelson and Morley measured the *difference* between the speeds of two lightbeams moving in different directions. Under the assumptions stated in the preceding exercise, what should be the difference between the speeds of the two lightbeams in Figure 10.8? Express your answer as a fraction of c.

THE RELATIVITY OF TIME

11. Find the speed of a typical commercial jet aircraft (800 km/hr) as a fraction of lightspeed. Do the same for a typical rifle bullet (1 km/s, or 3600 km/hr) and for a low-orbit satellite (8 km/s). If you were traveling at any of these speeds, would you be able to notice the effects of time dilation (other than by the use of sophisticated measurement techniques)?

12. Velma passes Earth moving at 50% of lightspeed. She watches a TV program that runs one hour, as measured by her. How long does her TV program run as measured by an Earth-based observer?

13. Your fantastic rocketship moves at 30,000 km/s. If you took off, moved at this speed for 24 hours as measured by you, and returned to Earth, by how much time would your clock differ from Earth-based clocks? Would you have aged more than, or less than, people on Earth? By how much?

14. Your extraordinarily fantastic rocketship moves at 99% of lightspeed. If you took off, moved at this speed for 24 hours, as measured by you, and returned to Earth, by how much would your clock differ from Earth-based clocks? Would you have aged more than, or less than, people on Earth? How much?

CHAPTER 11

The Universe According to Einstein

Einstein's Pegasus

There's Einstein riding on a ray of light,
Holding a mirror up, at arm's length in his hand,
In which he cannot see his face in flight
Because his jesting image, I now understand,
Won't ever reach the mirror since its speed,
Too, is the speed of light. He rides, this fleeting day,
As if on Pegasus, immortal steed
Of bridled meditation, past the Milky Way,
Out to my mind's Andromeda, where I,
Also transported, staring at a windless pool,
Watch his repaired reflection whizzing by.
Though he can't see himself, this self-effacing fool
Who holds all motion steady in his head,
I won't forget his facing what he cannot see
In thought that binds the living and the dead,
And ride with him, outfacing fixed eternity.
ROBERT PACK, MIDDLEBURY COLLEGE, 1991

Humans stand about 1 or 2 meters tall and can run at a few meters per second. Since our brains have evolved to deal with these normal sizes and speeds, it is not surprising that phenomena at speeds approaching lightspeed, at intergalactic distances, and at atomic dimensions are hard to imagine. This chapter continues our study of Einstein's special theory of relativity, which is essentially the theory of high-speed phenomena. Also in this chapter we look at phenomena involving large distances or intense gravitational fields, ranges that are dealt with by Einstein's general theory of relativity.

Chapter 10 presented the foundations of the special theory of relativity: its two basic laws and its most fundamental prediction, time dilation. The first four sections of this chapter present four more extraordinary predictions: the relativity of space (Section 11.1), the non-Galilean relativity of velocity (Section 11.2), the relativity of mass (Section 11.3), and the equivalence of energy and mass (Section 11.4). The most fundamental of these, and the aspect of special relativity that Einstein himself thought was most important, is the equivalence of energy and mass. We discuss its philosophical significance in Section 11.5.

The special theory of relativity describes the observations of nonaccelerated observers. What about accelerated observers? Einstein found a new connection between acceleration and gravity. His general theory of relativity, which describes the observations of accelerated observers, is mostly a new theory of gravity. Sections 11.6 and 11.7 present this theory's key ideas and discuss its predictions about the shape and fate of the universe.

11.1 THE RELATIVITY OF SPACE: SPACE IS WHAT HUMANS MEASURE WITH RULERS

What is space? Just as time means "what is measured by clocks" (Section 10.6), space means "what is measured by rulers."

What operations should Mort (remember Mort? We met him in Chapter 10, along with Velma) perform to measure the length of, say, a table? For a table at rest relative to Mort, the prescription is to place a meter stick along the table (or several meter sticks if it is a long table) and compare the ends of the table with the marks on the meter stick.

But what if the table is moving past Mort? As we saw in Chapter 10, we must be careful because it is easy for unstated assumptions to slip in unnoticed. Mort should continue to use meter sticks that are fixed in his own reference frame, because he wants to know the length of the moving table as measured in his own reference frame. If the length being measured lies along the direction of motion, it is important for Mort to measure the positions of the two ends of the table simultaneously. Otherwise the table will move to a new location during the lag between measurements and Mort will not measure the table's true length.

In order to ensure that the front end and back end measurements are simultaneous, Mort must use two clocks—one at each end. This means that the question of measuring the length of a moving object is mixed up with the question of measuring time; *time and space are tangled up with each other*! Since time is relative, it then comes as no surprise to learn that space is relative too. We won't go through the argument that proves this result; it is similar to the argument in Section 10.6 by which we proved that moving clocks go slow.

More specifically, Einstein's theory predicts that the table's length along its direction of motion, as measured by Mort, will be shorter than its length as measured by Velma, who is traveling along with the table. This effect is called **length contraction**. There is no length contraction along directions perpendicular to the table's direction of motion. Figure 11.1 shows how this affects Mort's measurements of a window in Velma's spaceship that is 1 meter square as measured by Velma.

As with time dilation, length contraction works both ways: Velma finds that objects in Mort's reference frame are contracted along the direction of motion, just as Mort finds that Velma's objects are contracted. The general rule is that moving objects are contracted along their direction of motion.

(a)

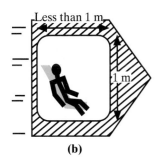

(b)

FIGURE 11.1
The window in Velma's spaceship as measured by (a) Velma and (b) Mort.

FIGURE 11.2
The relativity of space. The predicted length of a meter stick for various speeds of the meter stick relative to the observer.

A quantitative analysis leads to a formula, graphed in Figure 11.2.* It shows the predicted length of a 1-meter-long object such as a meter stick, held parallel to its motion, for various speeds of the object. Like time dilation, length contraction is barely detectable for speeds below about 0.1c but becomes large at higher speeds.

Recall that time dilation is not simply something that happens to clocks. Rather, it is time itself that is dilated. Similarly, length contraction is not just something that happens to meter sticks. Since space is defined by meter sticks, it is space itself that is contracted. Velma's space is different from Mort's space, just as Velma's time flow is different from Mort's time flow. Instead of a single, universal space, we must speak of "Velma's space" and "Mort's space."

Einstein's basic insight into space and time is that space and time are not abstractions that exist independently of observers and clocks and meter sticks. Space and time are *physical*; they exist only because physical objects such as observers, ice-cream cones, tables, clocks, and meter sticks exist.

■ **DIALOGUE 1**   Velma's spaceship is 100 meters long and 10 meters high as measured by Velma. (a) She passes Mort at 60% of lightspeed. How long and how high is her spaceship as measured by Mort? (b) What if she passes him at 90% of lightspeed?

■ **DIALOGUE 2**   (a) Velma observes Mort's automobile as she passes Mort at 60% of lightspeed. Mort's automobile is 4 meters long as measured by Mort. How long is it as measured by Velma? (b) What if she passes him at 90% of lightspeed?

■ **DIALOGUE 3**   Velma's spaceship is 100 meters long and 10 meters high as measured by Velma. (a) Roughly how fast must she be moving past Mort in order for Mort to measure the spaceship's length to be 50 meters? (b) Roughly

---

*The formula is $L = L_o\sqrt{(1 - s^2/c^2)}$, where L_o is the object's *rest length* (the length as measured by an observer for whom the object is at rest), and L is the length of the object when it is moving at speed s.

1. (a) 80 meters long and 10 meters high. (b) About 43 meters long and 10 meters high.

2. (a) 0.8 × 4 meters = 3.2 meters. (b) 0.43 × 4 meters =1.8 meters.

3. (a) About 0.87c. (b) About 0.99c.

how fast must she be moving past Mort in order for Mort to measure her space-ship to have a square shape?

11.2 THE RELATIVITY OF VELOCITY ACCORDING TO EINSTEIN

We have seen (Section 10.4) that when light beams are involved, velocities don't add up in the normal Galilean way. How do they add up in situations that don't involve light beams?

Consider Figure 11.3. Velma's spaceship passes Mort, and Velma fires a base-ball launcher capable of launching baseballs at a considerable fraction of light-speed. What is the baseball's speed relative to Mort? According to Galilean rela-tivity, the answer is found by simply adding the spaceship's speed relative to Mort and the baseball's speed relative to Velma. For example, if Velma passes Mort at $0.75c$ and if the launcher's muzzle velocity is also $0.75c$, Galilean relativ-ity predicts that the ball will pass Mort at $0.75c + 0.75c = 1.50c$.

Einstein's relativity gives a different prediction because speed measurements involve distance and time, which are subject to length contraction and time dila-tion. A quantitative analysis leads to a formula.* Table 11.1 gives some of the re-sults predicted by this formula. For example, if Velma passes Mort at $0.75c$ and if the baseball's launching speed is also $0.75c$, the table tells us that the ball's speed relative to Mort will be $0.960c$ rather than the $1.5c$ that Galilean relativity predicts. As we expected, the ball passes Mort at less than speed c.

Even if Velma is moving at nearly lightspeed and launches the ball at nearly lightspeed, Mort will observe the baseball pass him at less than lightspeed. For example, if Velma is moving at $0.99c$ and launches the ball at $0.99c$, Table 11.1 tells us that it will pass Mort at $0.99995c$. This is similar to the behavior of a light beam "launched" by Velma, shown in the bottom line of the table. It is not only

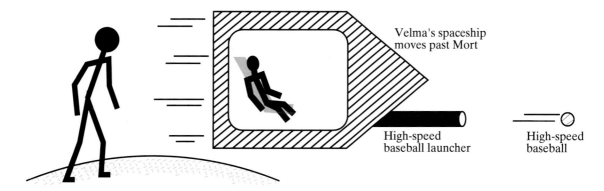

FIGURE 11.3
If Velma moves past Mort and she launches a baseball, how fast will it move relative to Mort?

---

*The formula is $S' = (s + S)/(1 + sS/c^2)$, where s is Velma's speed relative to Mort, S is the baseball's speed relative to Velma, and S' is the baseball's speed relative to Mort.

TABLE 11.1

The relativity of velocities: Some Quantitative Predictions.   The table gives the speed of the baseball relative to Mort (Figure 11.3) for various speeds of Velma relative to Mort and various speeds of the baseball relative to Velma.

		Velma's speed relative to Mort					
		0.10c	0.25c	0.50c	0.75c	0.90c	0.99c
Baseball's speed relative to Velma		**Baseball's speed relative to Mort**					
	0.10c	0.198c	0.341c	0.571c	0.791c	0.917c	0.992c
	0.25c	0.341c	0.470c	0.667c	0.842c	0.939c	0.994c
	0.50c	0.571c	0.667c	0.800c	0.909c	0.966c	0.997c
	0.75c	0.791c	0.842c	0.909c	0.960c	0.985c	0.9986c
	0.90c	0.917c	0.939c	0.966c	0.985c	0.994c	0.9995c
	0.99c	0.992c	0.994c	0.997c	0.9986c	0.9995c	0.99995c
(light beam):	c	c	c	c	c	c	c

light beams that have the strange behavior stated in the principle of the constancy of lightspeed; material objects such as baseballs also approximate this behavior at high speeds.

For slower speeds, Einstein's predictions are close to the normal Galilean predictions. For example, if Velma passes Mort at $0.1c$ and launches her baseball at $0.1c$, Table 11.1 predicts that the ball will pass Mort at $0.198c$, nearly as large as the $0.2c$ predicted by Galilean relativity. At normal speeds, Einstein's predictions are experimentally indistinguishable from the Galilean predictions. For example, if Velma's train passes Mort at 70 m/s and Velma throws a baseball forward at 20 m/s, Einstein's formula for relative velocities predicts that the ball will pass Mort at 89.999,999,999,999,982 m/s. This is so close to the Galilean prediction of 90 m/s that no known experiment could tell the difference.

■ **DIALOGUE 4**   Suppose that Velma passes Mort at $0.75c$ and that she launches a baseball at $0.5c$. According to Einstein's theory, how fast will the ball move relative to Mort? What does Galilean relativity predict? Which prediction agrees with nature?

11.3 THE RELATIVITY OF MASS: INERTIA IS RELATIVE

Einstein's new idea, the constancy of lightspeed, affects nearly everything in physics: time, space, relative velocities, and more. Relativity also alters Newton's law of motion (Chapter 4). This law states that an object's acceleration is equal to the net force exerted on the object divided by the object's mass, or in symbols

$$a = F/m$$

This implies that if you exert an unchanging force on an object, the object will maintain an unchanging acceleration. Eventually the object will be going at

---

4. $0.909c$. $1.25c$. Einstein's prediction agrees with nature.

lightspeed and still accelerating. An observer riding on such an object could keep up with a light beam.

So Newton's law of motion, the centerpiece of Newtonian physics, is not consistent with the theory of relativity! Apparently, relativity alters Newton's law in such a way as to prevent objects from being accelerated up to lightspeed. To describe this alteration of Newton's law of motion, let's imagine that Mort and Velma (who is moving past Mort) have identical 1-kilogram objects, 1-kilogram melons perhaps. If Mort pushes on his melon with a force of, say, 1 newton, he will find that his melon accelerates at 1 m/s^2, just as Newton's law of motion predicts. If he now pushes on Velma's melon (which is moving past him) with an identical 1-newton force, Newton's law of motion predicts that Velma's melon will accelerate at 1 m/s^2, but *relativity theory predicts that Velma's melon will accelerate at less than* 1 m/s^2. As was the case for other relativistic effects, this effect is negligibly small at normal speeds. But if Velma's melon is moving at a significant fraction of lightspeed, Mort will find that its acceleration is noticeably smaller than 1 m/s^2. Relativity theory correctly predicts this effect.

From Mort's point of view, a 1-newton force applied to each melon produces a smaller acceleration in Velma's melon than in his own melon. From Mort's point of view, Velma's melon has more **inertia** than does his own melon (recall that a body's inertia is its resistance to acceleration). But this is the same as saying that Velma's melon has more mass, because the fundamental meaning of **mass** is *amount of inertia* (see Chapter 4). In other words, Mort measures Velma's melon as having a larger mass than does his own melon, even though they are identical melons. As usual, the effect works the other way around: Relative to Velma, her own melon has a mass of 1 kg, but Mort's melon has a mass of more than 1 kg.

Mass is relative: An object's mass increases with its speed.

A quantitative analysis leads to a formula that predicts an object's mass for various speeds.* Figure 11.4 is a graph based on this formula, along with the graphs for time dilation (Figure 10.11) and length contraction (Figure 11.2). The formulas for mass increase and time dilation have identical forms, so these two graphs have identical shapes.

When we first met the idea of mass (Chapter 4), it was introduced in its most intuitive meaning as quantity of matter. We then found it necessary to redefine the word *mass* as *amount of inertia*, in order to be able to compare different masses. This was not a major shift in meaning at the time, because in Newtonian physics an object's inertia is precisely the same as its quantity of matter, so the two definitions of mass yield the same quantitative value for the mass of any object. But relativity implies that these two ways of looking at mass are different, because an object's inertia increases with its speed, but its quantity of matter doesn't change, because it still contains the same atoms and hence the same amount of matter.

Our words need to distinguish between these two different concepts of mass. An object's mass as measured by an observer for whom the object is at rest is called its **rest mass**. For example, Velma's and Mort's melons both have rest masses of 1 kg, regardless of who observes them. An object's rest mass measures its quantity of matter; it is a property of the object itself, regardless of who ob-

---

*The formula is $m = m_0 / \sqrt{(1 - s^2/c^2)}$, where m_0 is the object's *rest mass* (the mass as measured by an observer for whom the object is at rest), and m is the mass of the object when it is moving at speed s.

FIGURE 11.4

Relativistic mass increase, length contraction, and time dilation. The graph shows the duration of one clock tick (representing 1 second in the clock's reference frame) on a moving clock, the length of a moving meter stick, and the mass of a moving standard kilogram, for various speeds of the clock, meter stick, and kilogram relative to the observer.

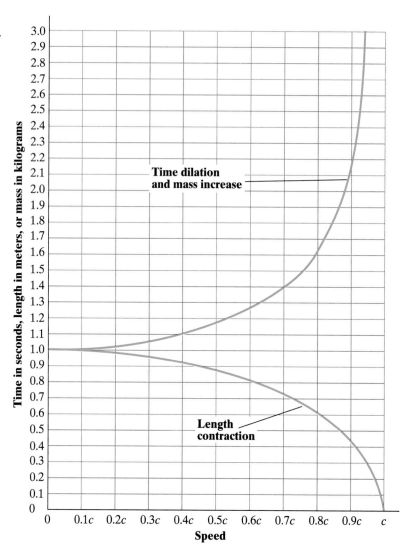

serves it. The amount of inertia possessed by an object is called its **inertial mass**. Inertial mass is relative, but rest mass is not relative. The inertial mass and rest mass of a slow-moving object are essentially the same, but the inertial mass of a high-speed object is significantly greater than its rest mass. Because the defining meaning of mass is *amount of inertia*, the unadorned word *mass* will henceforth always mean inertial mass.

H O W   D O   W E   K N O W ?

Relativistic mass increase is an everyday fact of life in high-energy physics labs. A subatomic particle can be accelerated to speeds in excess of 0.99c, speeds so close to lightspeed that relativity predicts its mass to be thousands of times greater than its rest mass. One way to check this prediction is by applying electric or magnetic forces to a fast-moving particle, in order to bend its path, and measuring the curvature of the resulting path. If fast-moving particles really do have larger masses, their paths should bend less than they otherwise would, because their larger inertia tends to keep them moving straight ahead. Measurements show that the curvatures of the paths of fast-moving particles are less bent than they would be in the absence of a relativistic mass increase and that the amount of bending agrees with the predictions of the theory of relativity.

Time, space, and mass are relative. Is everything relative? The answer is that many things are not relative. It is a fundamental principle of the theory of relativity that the speed of any light beam is not relative. And it is a fundamental principle that every law of physics is the same for every observer.

Relativistic mass increase explains why you cannot accelerate objects up to lightspeed. At high speeds, an object's mass becomes very large, increasing without limit as the speed approaches c (Figure 11.4). Eventually, the force needed for further acceleration becomes so large that the universe cannot provide it.

But there is something that moves as fast as lightspeed: light itself. In fact, light never moves slower than 300,000 km/s.* When you turn on a lightbulb, the light that is created does not accelerate from zero up to lightspeed; instead, it is moving at precisely lightspeed the instant it is created. Light is quite different from any material object. When you put a material object down in front of you, it has rest mass. Light beams must have no rest mass, because if they did, then relativistic mass increase would make their mass infinite while moving at lightspeed.

Anything, such as light or other forms of electromagnetic radiation, that moves at speed c and has no rest mass is classified as *radiation*. The distinction between **matter**, which has rest mass and always moves slower than lightspeed, and **radiation**, which has no rest mass and always moves at lightspeed, is a useful one.

■ **DIALOGUE 5**   Mort finds that Velma ages slowly, that her body is shortened along its direction of motion, and that she is more massive than her normal mass. What does Velma say about herself? What does Velma say about Mort?

■ **DIALOGUE 6**   Use Figure 11.4 to estimate how fast Velma must move, relative to Mort, for Mort to observe that her body's mass is 50% larger than normal. How fast must she move for Mort to observe that her spaceship is 50% shorter than normal?

■ **DIALOGUE 7**   Which of these is matter, and which is radiation: (a) red light, (b) the invisible waves drawn in Figure 9.5, (c) the invisible carbon-dioxide gas emitted by automobiles, (d) a radio wave, (e) chlorofluorocarbons drifting into the upper atmosphere from air conditioners, (f) X rays, (g) the electron beam that creates the picture on a TV tube (Section 8.6), (h) the so-called solar wind,

---

*However, light travels through a material substance such as water or glass at an average speed that is sometimes far less than lightspeed. When moving through matter, light momentarily vanishes when it is absorbed by an atom and is recreated when it is emitted by the atom. Whenever the light actually exists as light, it moves at 300,000 km/s.

5. To Velma, her aging process, body size, and mass all are normal. But she says that Mort ages slowly, that his body is shortened, and that he is more massive.

6. About 0.75c. About 0.87c.

7. (a) Radiation, (b) radiation, (c) matter, (d) radiation, (e) matter, (f) radiation, (g) matter, (h) matter, (i) radiation, and (j) matter.

which is a stream of protons emitted outward from the sun, (i) a gamma ray, (j) an electric current flowing in a copper wire (Section 8.7)?

11.4 E = MC²: ENERGY HAS MASS, AND MASS HAS ENERGY

The connection that Einstein found between mass and energy starts from the relativity of mass: If you accelerate a rock from rest, the work you do on the rock will increase the rock's kinetic energy (Chapter 6), and at the same time the rock's mass will increase because of relativistic mass increase. So energy increase and mass increase go hand in hand, or at least they do in this example.

Working from the theory of relativity and the law of conservation of energy, Einstein found that mass must be connected to energy in this fashion, regardless of the form of the energy that is given to an object. Instead of increasing the object's kinetic energy, you could give it thermal energy by warming it up, gravitational energy by lifting it, electromagnetic energy by charging it, or give it any other form of energy, and its mass would still increase.

This is surprising. If you lift a rock from the floor to the table, you don't expect its mass to increase. It is still the same rock, after all. Furthermore, since the rock on the table is at rest, its *rest* mass increases. This is a new result: An object's mass increases not only because of its motion but also because of any form of energy increase, even those such as gravitational energy that don't involve motion.

For another simple example, find a rubber band and stretch it. In stretching it, you give it elastic energy. The theory of relativity states that you also increase its mass!

Einstein's analysis yields a simple formula that quantitively relates the increase in mass to the increase in energy. The formula states that the amount of mass given to an object equals the amount of energy given to it divided by the square of lightspeed:

$$\text{change in mass} = \frac{\text{change in energy}}{\text{square of lightspeed}}$$

In symbols:

$$\Delta m = \frac{\Delta E}{c^2}$$

where Δm and ΔE mean *the change in* (or additional amount of) *mass* and *the change in energy*. If the standard metric units, kilograms and joules (J), are used for Δm and ΔE, then meters/second must be used for c. So $c = 3 \times 10^8$ m/s, and $c^2 = 9 \times 10^{16}$ m²/s².

To develop a feel for this result, let's look at examples. Suppose you accelerate a kilogram mass up to 100 m/s (360 km/hr, a fast train's speed). The kilogram's kinetic energy (Chapter 6) is then

$$\Delta E = (\tfrac{1}{2})\, ms^2 = (\tfrac{1}{2}) \times (1 \text{ kg}) \times (100 \text{ m/s})^2 = 5000 \text{ J}$$

Einstein saw the equation E = mc² as the most significant single result of the special theory of relativity.

RALPH BAIERLEIN, PHYSICIST AND AUTHOR OF *NEWTON TO EINSTEIN: THE TRAIL OF LIGHT*

Einstein's formula tells us that the kilogram's relativistic mass increase is

$$\Delta m = \frac{\Delta E}{c^2} = \frac{5000}{9 \times 10^{16}} = 5.6 \times 10^{-14} \text{ kg} = 0.000,000,000,000,056 \text{ kg}$$

Not much of an increase. The kilogram's mass is now 1.000,000,000,000,056 kg. The increase is small because c^2 is so large. This is why relativistic mass increase wasn't noticed before Einstein: At ordinary speeds, it is too small to notice.

Suppose you lift a kilogram. One kilogram on Earth weighs 10 newtons. So if you lift it by, say, 400 meters (about the height of New York's World Trade Center), the kilogram's gravitational energy will increase by

$$\Delta E = \text{weight} \times \text{height} = 10 \text{ N} \times 400 \text{ m} = 4000 \text{ J}$$

which is a little smaller than the 5000 J found in the previous example, and the mass increase is again too small to measure.

Heating an object typically involves larger amounts of energy. For example, in heating a pot of soup, you might give it 100,000 J of thermal energy. This is 20 times larger than the 5000 J in the first example, so the soup's mass gain is 20 times larger than the mass gain found in the first example, or about a trillionth of a kilogram. Still undetectable.

As another example, suppose that you have two bar magnets and that the north pole of one is joined to the south pole of the other so that they cling together (Figure 11.5(a)). Since you must do work on the two combined magnets to pull them apart, the separated magnets of Figure 11.5(b) must have more energy than do the joined magnets. But more energy means more mass. So the total mass of the two combined magnets is increased simply by pulling them apart! As the magnetic field lines shown in Figures 11.5(a) and (b) indicate, the work done in pulling the magnets apart creates a magnetic field that extends from one magnet to the other. The excess energy in the separated magnets resides in this invisible and nonmaterial magnetic field. We have encountered such "field energy" before, in the energy of electromagnetic radiation. But now we see that the energy of nonmaterial fields such as magnetic fields means that these fields also have *mass*.

The work involved quantitively in separating the two magnets is only a few joules, so the mass difference is again negligibly small compared with the original mass of the magnets. Nevertheless, it seems extraordinary that a system's mass should change simply by separating it into two parts.

This effect is more dramatic when we turn to examples involving nature's more powerful forces. **Nuclear reactions**, for example, involve nature's strongest forces, the forces acting within the atomic nucleus (Chapters 15 and 16). For now, all we need to know about nuclear reactions is that they are similar to chemical reactions but that they involve changes in the structure of the nuclei of one or more atoms instead of changes in the way that different atoms combine to form molecules.

For example, in nuclear power reactors and nuclear weapons, the element uranium undergoes a nuclear reaction known as **nuclear fission**, in which the nucleus of each uranium atom is altered.* The fission reaction is a little like com-

(a) Less energy

(b) More energy

FIGURE 11.5
The separated magnets of (b) have more energy, and hence more mass, than do the two joined magnets of (a). The excess energy and mass in (b) reside in the invisible and nonmaterial magnetic field, indicated by dashed lines. Compare this with Figures 9.1 and 9.2.

_____

*Each uranium nucleus is split to form two nuclei of various lighter-weight elements.

bustion, but the forces involved in fission are so strong that the thermal energy created is far larger than it is in any chemical reaction. So the mass loss, after removing the thermal energy, is far larger. If a kilogram of uranium is fissioned, the mass loss is about 0.01 kg (10 grams), which is a 1% mass decrease and easily detected. Measurements show that the predicted amount of mass actually is lost in fission experiments.

Nineteenth-century scientists believed matter to be indestructible, in other words, that rest mass is conserved in every physical process. This is certainly plausible. Since the days of the early Greek materialists (Section 2.1), scientists and others have felt that matter is indestructible—that although its form might change, its total amount cannot change. Nineteenth-century chemists performing high-precision mass measurements concluded that rest mass is conserved even in highly energetic chemical reactions.

But Einstein's relativity contradicts the conservation of matter. Matter, that is, rest mass, is not conserved in chemical reactions, in the stretching of a rubber band, and so forth. But these changes in rest mass are so small that they are experimentally undetectable. In high-energy processes such as nuclear fission, however, the changes are easily detected, and the results show that matter is not conserved.

MAKING ESTIMATES

The energy available in a few grams

We have seen that it takes a lot of energy to make even a little mass. It follows that even a small amount of mass can make a lot of energy. Suppose, for example, that 1 kilogram of uranium is entirely fissioned* and that all the energy "released" (converted to thermal and other forms of energy) during this process is used to do work. If this work were used to do some large task such as lifting the entire U.S. population, about how high would it be lifted?

The rest mass loss during the fissioning of 1 kg of uranium is 0.01 kg, or 10 grams—about 1/3 ounce. The amount of energy converted during this process is found from the formula $\Delta m = \Delta E/c^2$, only turned around to read $\Delta E = (\Delta m)c^2$:

$$\Delta E = (\Delta m)c^2 = (0.01 \text{ kg}) \times (9 \times 10^{16}) = 9 \times 10^{14} \text{ J}$$

How high could you lift the U.S. population with this much energy? The population is about 250×10^6, and a typical person's mass is perhaps 600 newtons (135 pounds). So

$$\text{weight of U.S. population} \approx 250 \times 10^6 \times 600 = 15 \times 10^{10} \text{ N}$$

Recall from Chapter 6 that gravitational energy equals weight times height. If the 9×10^{14} J of energy goes into lifting the population, then

$$9 \times 10^{14} \text{ J} = (15 \times 10^{10} \text{ N}) \times (\text{height})$$

---

*We assume, in this example, that our 1 kg of uranium is made entirely of the fissionable isotope of uranium, ^{235}U (see Chapter 16).

This means that the height to which you could lift the U.S. population is

$$\text{height} = \frac{9 \times 10^{14} \, \text{J}}{15 \times 10^{10} \, \text{N}} = 6000 \, \text{m}$$

This is 6 kilometers. That's a lot of work out of just 10 grams.

Einstein's result, $\Delta m = \Delta E/c^2$, says that any energy given to a body increases that body's mass. In other words, *energy has mass.* The preceding estimate shows, for example, that 9×10^{14} joules of energy has a mass of 10 grams: When you add this much energy to any body, you add 10 grams to it.

Now we take this reasoning one step further: Einstein believed that this result extended to any body's *entire* mass, so that a body's mass was entirely created by various forms of energy, such as kinetic energy, gravitational energy, and nuclear energy (energy due to the strong nuclear force). According to Einstein, not only did all energy have mass, but also all mass had energy.

This means that $\Delta m = \Delta E/c^2$ extends to the total mass and energy:

$$\text{total mass of any object} = \frac{\text{object's total energy}}{c^2}$$

or, in symbols,

$$m = \frac{E}{c^2}$$

This implies Einstein's famous formula,

$$\text{total energy of any object} = (\text{object's total mass}) \times (c^2)$$

$$E = mc^2$$

Since energy means the ability to do work, the practical meaning of $E = mc^2$ is that any object of mass m should be able to do mc^2 units of work.

H O W   D O   W E   K N O W ?

When I think of matter, I like to think mostly of fields. We are fields rather than particles.

FREEMAN DYSON, PHYSICIST

If matter turns out in the end to be altogether ephemeral, what difference can that make in the pain you feel when you kick a rock?

JOHN A. WHEELER

If Einstein's idea is correct, there should be some physical process by which mc^2 units of work can be obtained from any object of mass m. Such processes, known as **matter–antimatter annihilation**, have been discovered.

In addition to the protons, neutrons, and electrons that form ordinary matter, physicists have discovered three other material particles, known as *antiprotons, antineutrons,* and *antielectrons.* If one of these *antiparticles* is brought close to its corresponding particle, the two particles will vanish entirely, and high-energy radiation will be created. It's an extreme example of the nonconservation of matter: Matter entirely vanishes, to be replaced by radiation.

Since any proton, neutron, or electron can be annihilated in this manner, this experiment shows that in principle, any material object can be turned into radiation—although it would be difficult in practice to collect enough antiparticles to annihilate a macroscopic object. Since this radiation could then be used to do work, we have here a demonstration of Einstein's idea. Furthermore, when the radiation's energy is measured, it is found to equal the total mass of the pair of particles times c^2.

It is easy to misinterpret the mass–energy idea. For example, it is sometimes said, incorrectly, that "mass can be converted to energy." One problem with this statement is that $E = mc^2$ implies that mass (meaning *inertial* mass) is always conserved because energy is always conserved, so mass is never converted to anything else. In an annihilation event, for example, the mass of the particle–antiparticle pair is precisely equal to the mass of the created radiation because the energies are equal. Rather, it is *rest* mass, or matter, that is destroyed here. So it is correct to say that "matter, or rest mass, can be converted to radiation." Concerning mass (inertial mass), it is correct to say that "all mass has energy and all energy has mass" or even that "all mass is energy and all energy is mass" or, as Einstein put it, "mass and energy are equivalent." One must be careful with the word *mass.*

We will summarize Einstein's idea as follows:

> **THE PRINCIPLE OF MASS–ENERGY EQUIVALENCE**
> Energy has mass; that is, energy has inertia. And mass has energy; in other words, mass is capable of doing work. The quantitative relation between the amount of energy and the amount of mass is $E = mc^2$.

The visible world is neither matter nor spirit but the invisible organization of energy.
HEINZ PAGELS, PHYSICIST

■ **DIALOGUE 8**   In each of the following processes, does the system's mass change? Does its rest mass change? (a) An automobile speeds up from rest to 50 km/hr. (b) A rubber ball is squeezed. (c) Two positively charged objects at rest, such as two charged combs, are moved closer to each other and placed at rest. (d) An electron and an antielectron, at rest, spontaneously annihilate each other.

In the 21st century, the changing world view will be in a way to abandon materialism and to go to what's more abstract.
HERWIG SCHOPPER, DIRECTOR OF THE EUROPEAN CENTER FOR NUCLEAR RESEARCH (CERN)

11.5 RELATIVITY AND THE NEWTONIAN WORLDVIEW

Mass–energy equivalence represents a sharp break with the Newtonian worldview. Newtonian physics follows the Greek materialists in believing that the predictable interactions between indestructible atoms moving in empty space determine everything that happens in the physical universe. Toward the end of the nineteenth century, electromagnetic phenomena intruded on this worldview by implying the physical existence of nonmaterial, nonatomic, electromagnetic

---

8. (a) The mass increases, but the rest mass doesn't change. (b) The mass increases, and the rest mass increases (because the ball remains at rest). (c) The mass increases because work must be done to bring the two objects closer together against the repulsive electrical force. The rest mass increases also, because the system is at rest. (d) The system's mass is unchanged because no work is done by external agents, and so energy is conserved. But the entire rest mass vanishes, since radiation has no rest mass.

fields. Apparently, the universe was made of atoms and fields, interacting and moving in predictable ways.

Let's pursue the mass–energy relationship at the atomic level. Since all energy has mass, some of an atom's mass must be due simply to the kinetic energy of its moving electrons, protons and neutrons. In addition, some of an atom's mass must be due to the energy of its various electromagnetic and nuclear force fields, just as the magnetic field in Figure 11.5 possesses mass as a consequence of its energy. Thus, at least part of an atom's mass is due to fields and motion.

This suggests an interesting question: Is that all there is? Are atoms made *only* of fields and motion? Is there ultimately no such thing as "pure rest mass," apart from internal motions and nonmaterial force fields? If so, atoms are not only mostly empty space, they are *entirely* empty space and are made only of fields, similar to the magnetic fields in Figure 11.5.

Although the answer to this question is not yet known experimentally, Einstein appears to have believed that all mass arises from fields and the motion of those fields. The fundamental theories of contemporary physics, known as *relativistic quantum field theories* (Chapter 18), also lean in this direction. For example, Steven Weinberg, a leading high-energy theorist, stated:

> [According to the physical theories developed during the 1920s] there was supposed to be one field for each type of elementary particle. The inhabitants of the universe were conceived to be a set of fields—an electron field, a proton field, an electromagnetic field—and particles were reduced to mere epiphenomena. In its essentials, this point of view has survived to the present day, and forms the central dogma of quantum field theory: *the essential reality is a set of fields* [Weinberg's emphasis] subject to the rules of special relativity and quantum mechanics; all else is derived as a consequence of the quantum dynamics of these fields.

In this **field view of reality**, there is no "there" there (to quote the poet Gertrude Stein). There are no "things" at all. Electrons and other material particles are only force fields in space, just like the magnetic fields in Figure 11.5, and the particle is simply the place where the field is strongest. In this view, all mass is due only to the energy of force fields. Since fields are "possible forces" (Section 9.1), and forces are interactions, this view implies that every "thing," everything, is interactions and motion. It is the interactions and motion themselves that are fundamental rather than the material particles that we had always supposed were doing the interacting and the moving.

The theory of relativity tilts the philosophical balance farther from the Newtonian view that indestructible atoms moving in empty space are the cause of everything that happens. Indeed, the field view of reality stands materialism on its head by arguing that atoms are not the fundamental source of motion and interactions but, rather, that interactions and motion are the fundamental source of all matter, including atoms. As we will see in Chapters 13 and 14, quantum theory sweeps away most of what still remains of the Newtonian worldview.

11.6 EINSTEIN'S GRAVITY

The special theory of relativity begins with the idea that the laws of physics are the same for all unaccelerated observers. What about accelerated observers? This question is the starting point for the general theory of relativity.

Science has found no "things," only events. The universe has no nouns, only verbs.
R. BUCKMINSTER FULLER, ARCHITECT AND FUTURIST

There are no things, only processes.
DAVID BOHM, PHYSICIST

We still do not know what causes the electron to weigh.
ABRAHAM PAIS, PHYSICIST, IN *SUBTLE IS THE LORD*, A BIOGRAPHY OF EINSTEIN

There is no "there" there.
GERTRUDE STEIN, POET

Perhaps particles are essentially "boxes" that hold energy. Indeed, modern views of the origin of the mass of elementary particles have very much this character. Perhaps there is no meaning to the concept of mass separate from energy!
ROBERT ADAIR, PHYSICIST

We are such stuff As dreams are made on
SHAKESPEARE, *THE TEMPEST*

Perhaps Shakespeare understood this universe of ours better than we do ourselves!
JOHN A. WHEELER, PHYSICIST, *THE FRONTIERS OF TIME*, REFERRING TO THE PRECEDING LINES FROM *THE TEMPEST*

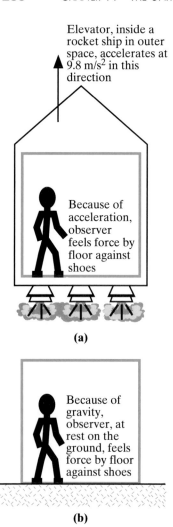

Elevator, inside a rocket ship in outer space, accelerates at 9.8 m/s² in this direction

Because of acceleration, observer feels force by floor against shoes

(a)

Because of gravity, observer, at rest on the ground, feels force by floor against shoes

(b)

FIGURE 11.6
The effects of acceleration are just like the effects of gravity. The observer inside the elevator cannot tell the difference.

Imagine that you are in an elevator accelerating upward from a building's ground floor. As you might have noticed in your own experience, the acceleration makes you feel squashed down, heavy, as though there were more gravitational pull on you than usual. This connection between acceleration and "apparent gravity" runs fairly deep. For example, imagine that you are inside an accelerating elevator in outer space, far from all planets and stars so that there are no gravitational forces (Figure 11.6). If the elevator's acceleration is 9.8 m/s² or 1g (pronounced "one gee"), you will feel the same as you do when you are stationary on Earth. The reason is that, according to Newton's law of motion, the elevator's floor must push upward on the bottoms of your shoes in order to provide the 1g acceleration. Because this push is quantitatively equal to your weight on Earth, you feel the same when you are accelerating through space at 1g as you do at rest on Earth.

If you are in an elevator accelerating at 1g through space, how can you tell, without communicating with the world outside the elevator, that you are actually in space and not at rest on Earth's surface? You might try dropping a stone, to see how it falls (Figure 11.7). But your elevator is accelerating at 9.8 m/s², so the floor accelerates upward to meet the stone. From your point of view in the elevator, the ball "falls" down to the floor with an acceleration of 9.8 m/s². Furthermore, from your point of view inside the elevator, Galileo's law of falling (Chapter 3) is valid: Both a large-mass stone and a small-mass stone, released together, will contact the floor at the same time.

You could even throw a stone horizontally to see whether it curves down to the elevator floor (Figure 11.8) the way it does when you throw a stone horizontally on Earth. Because of the elevator's upward acceleration, the stone gets closer to the floor as it moves across the elevator. As you view it from inside the elevator, the stone "falls" to the floor exactly as though it were thrown horizontally on Earth.

It seems that it is not easy to find an experiment that can tell you whether you are at rest on Earth's surface or moving through space with a 1g acceleration.

Einstein made this reasoning into a fundamental new principle. The new principle is similar to the principle of relativity, which says you cannot determine your own speed from inside your own laboratory. The new principle states that the laws of physics are unaffected by acceleration, so that there is no experiment that you can perform inside an elevator that can tell you whether you are accelerating or, on the other hand, sitting still and feeling gravitational effects.

We summarize this as follows:

THE PRINCIPLE OF GENERAL RELATIVITY*
Every accelerated observer experiences the same laws of nature. In other words, no experiment performed inside a sealed room can tell you whether you are accelerating in the absence of gravity or are at rest in the presence of gravity.

Light beams play a central role in the general theory of relativity, just as they do in the special theory. If you are accelerating through outer space and you turn on a flashlight horizontally, what will happen to the light beam? Just like

*Often called the principle of equivalence.

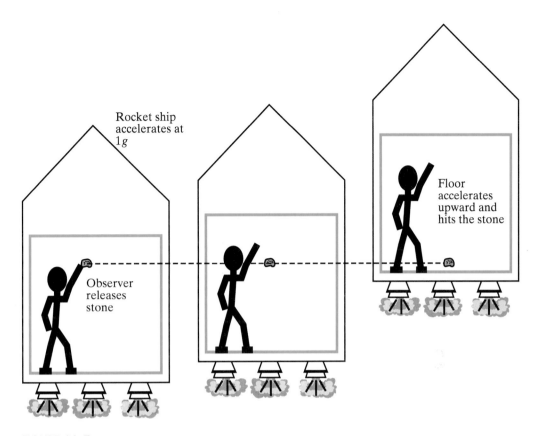

FIGURE 11.7
If you release a stone inside an accelerating elevator in outer space, it will appear to you that the stone falls "down" to the floor, just as though you were on Earth and feeling the effects of gravity.

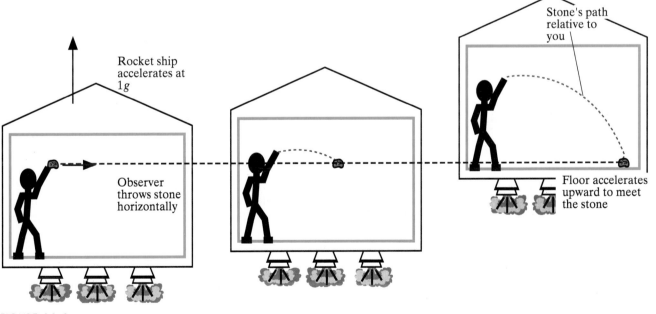

FIGURE 11.8
If you throw a stone inside an accelerating elevator in outer space, it will appear to you that the stone falls to the floor as though you were on Earth and feeling the effects of gravity.

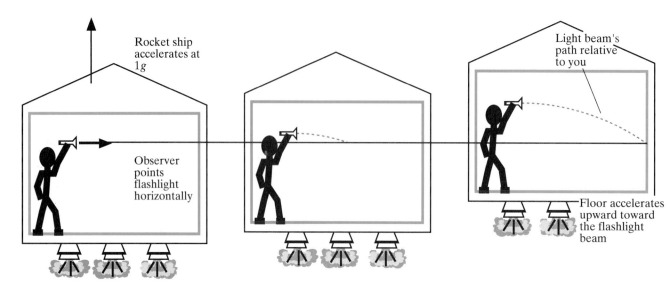

FIGURE 11.9

If you turn on a flashlight inside an accelerating elevator in outer space, the light beam will bend relative to you.

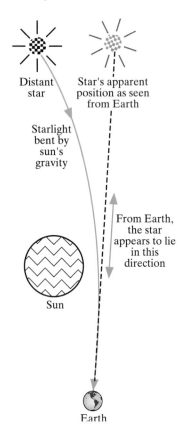

FIGURE 11.10

Because the sun bends light beams, we can see stars that are behind the sun.

the horizontally thrown stone (Figure 11.8), the light beam bends downward relative to you, because of your acceleration (Figure 11.9). The light beam curves far less than the stone does, but only because of the light beam's far greater speed.

The principle of general relativity implies that this experiment must come out the same way if performed in a stationary room in the presence of gravity. So *gravity must bend light beams.* This is a new, non-Newtonian, prediction of Einstein's theory.

H O W   D O   W E   K N O W ?

Earth's gravity is too weak to bend light beams very much. But the sun is massive enough to measurably bend light from the distant stars as the light beam passes close to the sun. The first measurement of this effect was made during a total eclipse of the sun in 1919, when astronomers could photograph the stars that appear near the edge of the sun (Figure 11.10). Measurements of these stars' positions showed that the starlight bends while passing the sun and that the amount of curvature agrees with the quantitative predictions of the general theory of relativity. Since 1919 there have been several more experiments of this type, often using radio telescopes to observe radio waves from stars observed near the edge of the sun. The advantage of using radio waves instead of light beams is that because the sun is a fairly "dim" radio star, stars near the edge of the sun can be detected by radio waves even when the sun is visible, so astronomers needn't wait for an eclipse. The results agree with Einstein's predictions.

Then I would have felt sorry for the dear Lord, for the theory is correct.

EINSTEIN'S REPLY WHEN ASKED HOW HE WOULD HAVE FELT IF THE 1919 SOLAR ECLIPSE OBSERVATIONS HAD DISAGREED WITH HIS GENERAL THEORY OF RELATIVITY

FIGURE 11.11
Stephen Hawking has made remarkable contributions to astrophysics and cosmology.

It is impossible to imagine a four-dimensional space. I personally find it hard enough to visualize three-dimensional space.

STEPHEN HAWKING, HOLDER OF ISAAC NEWTON'S CHAIR AS LUCASIAN PROFESSOR OF MATHEMATICS AT CAMBRIDGE UNIVERSITY

FIGURE 11.12
If it were extended forever in both length and width, a flat tabletop would be a flat two-dimensional space.

Einstein's special theory and general theory are laid out in similar ways. The two starting points, the principle of relativity and the principle of general relativity, state, respectively, that the laws of physics are unaffected by the reference frame's velocity and acceleration. In the special theory the next step is the principle of the constancy of lightspeed. Similarly, in the general theory the next step is the prediction that gravity must bend light.

Now recall that the constancy of lightspeed led Einstein to the surprising new conclusion that time is relative. Similarly, we will see that the gravitational bending of light implies a new and surprising property of space, related to the concept of *straightness*. Just as time can be defined by light clocks, straightness can be defined as the path followed by a light beam. In fact, surveyors often use laser light beams to determine straightness, and you use light beams to determine straightness when you aim a gun by sighting along its barrel.

But what can it mean to say that gravity bends light beams, when light beams themselves are the definition of straightness? Einstein saw the answer. Just as the slowing down of moving light clocks implies that time itself slows down, the bending of light beams means that *space itself is bent by gravity*. Just as Einstein saw that time is physical and is defined by physical objects such as light clocks, he saw that straightness is physical and is defined by physical objects such as light beams.

It turns out that gravity affects not only space but also time, in a way that is best described as the **bending of space–time**. The idea of space–time arises because, as we saw in Section 11.1, space and time are intertwined. It turns out that time is intertwined with space in a way that is mathematically analogous to the way that each of the three spatial dimensions (length, width, height) are intertwined with one another. Because of this, time is sometimes thought of as a "fourth dimension," even though neither Einstein nor anybody else can visualize four physical dimensions. As Stephen Hawking (Figure 11.11), one of today's most profound relativity and quantum theorists, remarked, "It is hard enough to visualize three-dimensional space," let alone four. Einstein's general theory mathematically describes this imagined **four-dimensional space–time**.

What does a curved three-dimensional space mean? Just as it is impossible to visualize four dimensions, it also is impossible to visualize a three-dimensional space that is curved. The reason is that space has only three dimensions (length, width, and height), and we exist within these three dimensions. There is no other dimension, no higher dimensionality, from which we can visualize the curvature of our three-dimensional space. Furthermore, the phenomenon of curved space is entirely outside our intuitions, because the bending of space is too small to be noticeable over the relatively short distances on Earth with which we are familiar.

The best we can do is visualize an analogy to this important idea: Instead of visualizing a curved three-dimensional space, we will now visualize a curved *two*-dimensional space, and this will give us insight into a curved three-dimensional space.

First, what do we mean by a *space*? We call our three-dimensional space a "space" because we can continue through it forever and never come to a boundary or edge beyond which there is no more space. This is the essential idea of a **space**: an extended region that has no edges or boundaries. A two-dimensional space, then, is a two-dimensional region that has no edges. For example, a flat tabletoplike area that is extended forever in both directions is a two-dimensional space (Figure 11.12).

FIGURE 11.13
If you put a warp in a flat two-dimensional space, it becomes a curved two-dimensional space.

FIGURE 11.14
The surface of a sphere is a curved two-dimensional space of finite total extent.

FIGURE 11.15
In a two-dimensional spherical space, two lines that start out parallel and extend as "straight" (or straightest) lines will eventually meet.

Now imagine that we begin with this infinite flat tabletop, this *flat* two-dimensional space, and we put a warp in it, a depression perhaps (Figure 11.13). This makes it into a *curved* two-dimensional space.

A flat two-dimensional space must have infinite extent, infinite total area, in order not to have edges. A *finite* (meaning *not infinite*) tabletop is not a space because it has edges. But if you imagine squeezing a flat tabletop and bending it around on itself to form the surface of a sphere, the edges will vanish (Figure 11.14). This spherical surface is a curved two-dimensional space of finite extent. Note that this *spherical two-dimensional space* consists of only the sphere's surface; the sphere's interior and exterior are not part of the space.

Suppose you are a two-dimensional creature inhabiting a two-dimensional spherical space, something like a flat ant walking around on the surface of a beach ball. How can you tell that your space is curved? You couldn't stand outside or inside the sphere's surface, in the third dimension, and see that your space is a spherical surface, because you are confined within your two dimensions. One way you can learn that your space is curved is by performing geometry experiments. For instance, two lines, beginning as parallel and extending as straight (or straightest) lines, should eventually meet (Figure 11.15).

Similarly, we cannot visualize the curvature of three-dimensional space, but we can visualize experiments to determine whether or not our space is curved. The 1919 experiment that measured the curvature of light near the sun was just such an experiment. It found that even the straightest path, the path of a light beam, curves near the sun. We conclude that three-dimensional space is curved. Like two-dimensional creatures living in a curved two-dimensional space, we three-dimensional creatures live in a curved three-dimensional space, but we cannot visualize the curvature because we cannot stand outside our three dimensions. Nevertheless, the curvature is real because we can observe (but not visualize) and measure it.

HOW DO WE KNOW?

Despite the bending of light beams, one might still imagine that this bending doesn't really show space to be curved but instead shows only that light beams curve in a "flat" space. This possibility is ruled out by an experiment in 1972 in which a spacecraft orbiting Mars beamed back radar signals sent from Earth (Figure 11.16). When the line of sight from Earth to Mars passed near the sun, the radar beam's travel time was measured. This travel-time measurement can tell whether the curved light beam is traveling through a flat space or through a curved space: It is easy to use the curved path to predict the travel time in a flat space by making a scaled-down drawing of the curved path of the radar beam on a flat sheet of paper and seeing how much longer it is than a straight line. In the experiment, the answer was about 10 m, so in a flat space the signal should have been delayed by about 30 billionths of a second, the time that it takes light to travel 10 m.

But you cannot use a flat sheet of paper to measure distances in a curved space, for the same reason that you cannot accurately determine the distance from Los Angeles to London by making measurements on a flat map: The "scale" keeps changing because of the curvature. Einstein's formulas for curved space–time predict a delay of 200 millionths of a second, about 7000 times longer

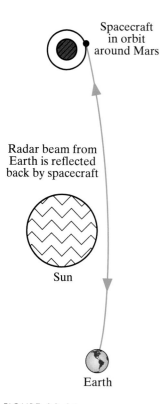

Spacecraft
in orbit
around Mars

Radar beam from
Earth is reflected
back by spacecraft

Sun

Earth

FIGURE 11.16
An experiment to measure the total travel time for a radar beam to get to Mars and back.

than the predicted delay through a flat space–time. The experiment confirmed the longer delay.

There still are scientific doubts about the validity of Einstein's general theory of relativity, although there no serious doubts about the special theory. One reason for the doubts is that unlike the special theory, it is difficult to find practical experiments that can test the general theory. Despite these doubts, all the other possible theories that have been proposed as competitors to the general theory are curved-space theories. The consensus is that space really is curved.

■ **DIALOGUE 9**   The surface of a sphere is a two-dimensional space. (a) Is the interior of a sphere a three-dimensional space? (b) Is it a two-dimensional space? (c) How many dimensions does the interior of a sphere have? (d) Is the interior finite or infinite in extent?

■ **DIALOGUE 10**   (a) How many dimensions does a circle (meaning the perimeter, or edge, of a circular area) have? (b) How many dimensions does the interior of a circle have? (c) Is a circle a space? (d) Is the interior of a circle a space?

11.7 THE SHAPE AND FATE OF THE UNIVERSE

The general theory of relativity is primarily a new theory of gravity. Conceptually, it is radically different from Newton's theory of gravity, although its predictions for normal situations such as the fall of a stone to Earth or the orbits of the planets are nearly identical to Newton's predictions. As we have seen, general relativity predicts that gravity bends space–time; in other words, masses bend space–time (Figure 11.17). According to the theory, gravity *is* this bending of space–time. In Einstein's theory, gravitational effects such as Earth's circular motion around the sun are not caused by forces at all but are instead entirely due to the curvature of space–time.

When Einstein's theory is applied to space–time as a whole, it describes the possible ways that our curved three-dimensional universe could evolve throughout past and future time. Einstein's theory and several significant pieces of experimental information about the origin and evolution of the universe (Section 16.3) lead to the following picture of the evolution of the universe: The universe began in a single event some 12 billion years ago, a violent explosion called the **big bang*** that created the different forms of energy and matter and that hurled matter and energy outward. Although the universe is expanding, it is not expanding into anything because there is no "outside" of the universe. And just as there is no space outside the universe, there was no time before the big bang. Questions such as What is outside our universe? and What happened before the

---

9. (a) No, because it has a boundary. (b) No, because it's not even two dimensional. (c) Three. (d) Finite.

10. (a) One. (b) Two. (c) Yes. (d) No.

*See Section 16.3 for four separate lines of evidence supporting the big bang theory. See Section 18.6 for a discussion of the early moments of the big bang as described by general relativity, quantum theory, and high-energy physics.

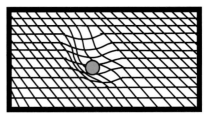

FIGURE 11.17
Masses such as the sun cause bends in space-time.

big bang? may be meaningless. The big bang was not an event, like the explosion of a firecracker, that happened in time and space. Rather, the big bang *created* time and space.

The general theory of relativity predicts that the universe's three-dimensional space must either expand or contract. It is remarkable that space and time themselves are in a state of continual change. Everything, it seems, changes: Time and space expand, the stars are born and die, life on Earth evolves, you and I are born and will die, the atoms perpetually change (Chapters 14 and 18), and even so-called empty space or vacuum is full of activity (Chapter 18).

The earliest evidence for the **expansion of the universe** and for the big bang came from astronomical observations of other galaxies outside our own Milky Way galaxy. Distant galaxies are moving away from us, and the more distant galaxies move away faster. But the galaxies are not just moving away from our particular galaxy; they all are moving away from one another. Regardless of which galaxy you live in, you will observe the other galaxies moving away from you. It's like a loaf of raisin bread expanding as it bakes: If you were standing on any one of the raisins observing the other raisins, you would observe all of them moving away from you, and more distant raisins would move away from you faster.

According to general relativity, there are three possible overall shapes or *geometries* for the large-scale structure of the three-dimensional universe. Figure 11.18 shows the two-dimensional analogs of these three-dimensional geometries. A **closed universe** bends back on itself to form a three-dimensional spherical space in which the angles of a triangle add up to more than the normal 180°. As you can see from the two-dimensional analog, such a universe has a finite (not infinite) total volume. An **open universe** is analogous to a saddle-shaped surface; in such a universe, the angles of a triangle add up to less than 180°. It's total volume is infinite. A **flat universe** has no overall large-scale curvature (although in all three geometries there will be smaller-scale curvatures caused by stars, black holes, and other objects) and has the normal Euclidean geometry with which we are familiar—for instance, the angles of a triangle add up to 180°. Its total volume is infinite.

Two factors determine the large-scale geometry: the universe's expansion rate and its average mass density (the average amount of mass per unit of volume throughout the universe). Slower expansion rates or greater mass densities cause the universe to bend back on itself and form a closed universe, while faster expansion rates or smaller mass densities cause the universe to spread out to form an open universe. The flat geometry lies between these two extremes and can occur only if the expansion rate and mass density are adjusted to each other in just the right way.

The factors determining the universe's geometry—expansion rate and density—also determine its long-term fate. If the expansion rate is slow enough, or the mass density high enough, for the universe to be closed, then the expansion

A straight road really curves, because it is on Earth and Earth is curved.

REMARK MADE TO ME BY A GRADE-SCHOOL GIRL

Matter tells space how to curve, and space tells matter how to move.

JOHN WHEELER

(a) Closed geometry

(b) Open geometry

(c) Flat geometry

FIGURE 11.18
Two-dimensional analogs of the possible large-scale geometries of the three-dimensional universe, as predicted by general relativity. A closed universe bends back on itself to form a three-dimensional spherical space in which the angles of a triangle add up to more than the normal 180° and has a finite total volume. An open universe is analogous to a saddle-shaped surface; in such a universe, the angles of a triangle add up to less than 180°, and the total volume is infinite. A flat universe has no overall large-scale curvature; it has the normal Euclidean geometry with which we are familiar (e.g., the angles of a triangle add up to 180°) and an infinite total volume.

will eventually cease and become a contraction. It's just like throwing a rock upward: Unless the rock is thrown very fast (11 km/s), Earth's gravitational pull will pull the rock back down.

Similarly, if the expansion rate is fast enough, or the mass density low enough, for the universe to be open, then the expansion will slow somewhat with time and eventually settle into a permanent expansion that goes on at an unchanging rate forever. This is like a rock that is thrown upward at faster than 11 km/s.

Finally, if the expansion rate is adjusted to the mass density in just the manner needed to create a flat universe, then the universe will continue expanding forever but the expansion rate will become slower and slower, tending toward zero at long times. This is analogous to a rock that is thrown upward at that precise speed—11 km/s—needed for it to just barely escape from Earth.

Figure 11.19 graphs the history of the three large-scale geometries. The 3 graphs show the "size" of the universe at various past and future times. But since the open and flat geometries are of infinite size, we graph the average separation between galaxies instead of the overall size. Since the universe is known to be expanding at the present time, all three graphs show an increasing separation at the time marked "now." The closed geometry eventually stops expanding at some future time and then collapses back to zero size, while the open and flat geometries continue expanding forever. The difference between the open and flat geometries is that, as time goes on, the former approaches a situation of continued expansion at an unchanging rate (the graph becomes a sloped but straight line), while the latter approaches a situation of zero expansion (the graph becomes level).

To make this more concrete, let's follow the history of the closed geometry in more detail. Here, either slow expansion or large mass density, or both, causes the universe to bend in on itself to form the three-dimensional analog of a two-dimensional spherical surface. The universe has a finite total volume but no edge. The expanding three-dimensional universe is like the two-dimensional surface of a balloon that is being inflated (Figure 11.20), with the galaxies represented by small dots on the balloon's surface.* As the balloon expands, the distance between any pair of dots increases. In agreement with the philosophy of the Copernican revolution (Chapter 1), the universe is the same all over: no galaxy is at the center, and no galaxy is at the edge, because the universe has no center and no edge, just as the surface of a spherical balloon has no center and no edge. The expansion continues until some maximum total volume is reached, whereupon contraction takes over and squeezes the universe back down toward zero volume.

H O W   D O   W E   K N O W ?

Dark matter and cosmic inflation

What do observations and theories indicate about the shape and ultimate fate of our actual universe? The answer remains open because the two crucial factors that determine the answer are so difficult to measure. The present expansion

---

*This analogy is not perfect because dots on a balloon expand as the balloon expands, while gravitational attraction prevents the galaxies from expanding, even as the space between the galaxies increases.

FIGURE 11.19
The predicted evolutions of the three possible large-scale geometries of the universe. In agreement with observations, the graph shows all three geometries expanding at the present time. In a closed universe, the expansion turns into a contraction at some future time. In both an open universe and a flat universe, the expansion continues forever. The difference between the two is that, as time goes on, the former approaches a situation of expansion at an unchanging rate (the graph becomes a sloped but straight line), while the latter approaches a situation of zero expansion (the graph becomes level).

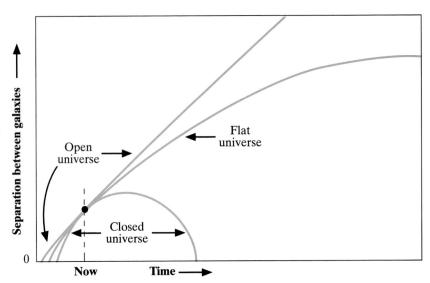

speed of the universe is known to within only about 20%, and the mass density is far more uncertain than this.

Although astronomers and physicists broadly support the notion of a big-bang creation of our universe, there is no consensus yet on the details of how and why the big bang occurred. Probably today's most widely respected theory regarding these details is **cosmic inflation**, devised in 1979 by Alan Guth, a physicist with a background in high-energy subatomic physics. Guth combined quantum theory (Chapters 13 and 14), general relativity, and recent results in high-energy physics (Chapter 18), to explain how the universe could have been created out of "nothingness" by high-energy quantum processes occurring "outside" of our normal space and time. See Chapter 18 for more about cosmic inflation. For our present purposes, we need only note that the most plausible version of Guth's theory predicts that the large-scale geometry of the universe should be *flat*.

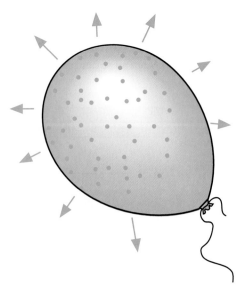

FIGURE 11.20
A two-dimensional representation of the expansion of the three-dimensional universe. As space, represented by the balloon's surface, expands, the galaxies, represented by the dots, move farther apart. Although this two-dimensional analogy shows the universe expanding into the empty space outside the balloon, there is no space outside the real three-dimensional universe. Our universe is all there is.

What do observations tell us? For decades, astronomers have been able to measure the speed with which distant galaxies are receding by measuring the *red shift* of their light. Light waves from a receding object are shifted in frequency, toward lower frequencies, for the same reason that sound waves from a receding ambulance are shifted toward lower frequencies: The waves get "stretched out" by the receding source. You can observe this *Doppler effect* in the sound from a passing ambulance. The Doppler effect for light tells astronomers, who measure the amount of shifting toward redder frequencies, how fast the source is receding. Red shifts in the light from distant galaxies, discovered during the 1920s, were in fact the earliest evidence of the big bang origin of the universe.

To know the overall rate at which the universe is expanding, we must know not only the recession speed of distant galaxies but also the distance to these galaxies. This distance is not easy to determine. One cannot just stretch a tape measure out to a distant galaxy! Here, the situation has improved recently, because powerful modern telescopes have been able to detect a particular type of supernova explosion (an explosion of a star—see Section 5.4) in far distant galaxies. Because it is thought that the inherent brightness (as it would be observed near the explosion) of every supernovae of this particular type is about the same, the brightness of these explosions as observed on Earth gives a good indication of their distance from Earth: Dimmer explosions are farther away.

Measurement of the universe's average mass density is even more difficult. It is not difficult to approximately measure the mass density of the *visible* matter in the universe by surveying distant galaxies with telescopes. But there is more to the universe than meets the eye: Two types of observations show directly that much of the universe's mass is invisible. In one type of observation, astronomers measure the speeds at which gas clouds orbit the centers of distant galaxies. The speeds turn out to be too high for the gravitational pull of the visible stars in the galaxies to hold the clouds in their orbits, implying that a large but invisible source of gravitational pull must be at work. The second type of observation is called *gravitational lensing*. Light reaching Earth from distant galaxies is sometimes bent by the gravitational pull of galaxies that lie in the path of the light. By analyzing this bending, astronomers can determine the mass of the intervening galaxies and show that they must contain a lot of invisible matter.

So for at least the past two decades, astronomers have been searching for so-called **dark matter**: matter that does not glow. The search has met with some success in the past few years, mainly through the discovery of "Massive Compact Halo Objects," or **MACHOs**. These are small nonglowing objects, such as white dwarf stars (Section 5.3) that have cooled to the point that they no longer glow, that lurk in the outer or "halo" region of our own Milky Way galaxy. Such objects are, as you can imagine, extremely hard to detect. They are detected when they pass directly in front of a distant glowing star, and briefly (for a few days, as the dark star passes by) cause gravitational lensing of the light from the distant star. Although this research is still in its early stages, it appears possible that 50%, and perhaps even 100%, of the unseen matter in our own galaxy can be accounted for by such MACHOs.

But even if MACHOs turn out to account for all of the dark matter within galaxies, this does not resolve the dark matter question. Widely accepted principles of nuclear physics and of the early moments of the big bang indicate that the amount of "ordinary" matter—matter made of protons, neutrons, and electrons—can constitute, at most, only some 10% of the amount needed to cause the universe to be either flat or closed. That is, if the universe consists of only ordinary matter, then we live in an open universe. But we have already noted that the theory of cosmic inflation predicts that the universe is flat. If cosmic inflation is correct, then some 90% of the matter in the universe must be in the form of matter that is not only dark (invisible), but that also is not ordinary matter. What, then, might be the form of this "missing matter"? Neutrinos (Section 6.5, also see Chapter 18) offer one possibility for such **exotic dark matter**: Neutrinos have been thought to have no mass, but some scientists believe they could have a very small mass, tens of thousands of times smaller than an electron's mass but still larger than zero. Several on-going experiments are looking for the answer to this question. Since the universe contains an enormous number of neutrinos, their total mass could explain the missing matter. Or it is quite possible that entirely new forms of matter will be discovered in the universe. This possibility has inspired a number of experimental searches for hard-to-detect particles of a yet-unknown variety entering Earth from space. A parallel situation existed during 1914 to 1955 (Section 6.5) when theory suggested that a then-unobserved particle was created during beta-decay, but no such particle could then be detected. Despite decades of unsuccessful searches, the ghostly new particle was not found. Finally, in 1955, physicists discovered the neutrino. It remains to be seen whether the search for exotic dark matter will yield a similar result. If it exists, it could be highly significant: 90% of the universe might be made of it.

Recently, astronomers have used red shifts and the supernova explosion distance-measuring technique, described above, to measure the speeds and distances of galaxies that are so far away that their light was emitted during earlier epochs of cosmic history, when the expansion rate was larger than it is now. Such measurements can determine the rate at which the expansion is slowing down—the deacceleration of the universe. These measurements are still inconclusive, but they are expected to eventually tell us directly whether the deacceleration is large enough to eventually bring the expansion to a halt, and so to tell us whether we live in an open, closed, or flat universe.

Cosmology is experiencing an exciting period today!

The general theory of relativity holds all sorts of fascinating possibilities, which are, to quote philosopher-scientist John Haldane, "not only queerer than we suppose, but queerer than we can suppose." Perhaps our universe is one among many universes, each having its own space and time, perhaps different space–time dimensionality—life might be fairly complicated in an 11-dimensional universe—and its own physical laws. Perhaps, over many different "times," an infinity of different universes passes into and out of existence, forming collectively a reality that occurs not in space and time at all but that is in some sense beyond space and time.

In one such universe, in one galaxy called Milky Way, on one planet called Earth, you who read these words and I who write them are privileged to hold such ideas in our minds.

SUMMARY OF IDEAS AND TERMS

Relativity of space or length contraction   Moving objects are contracted along their direction of motion, so an object's length is different for different observers.

Relativity of velocities   Both Galileo's and Einstein's theories predict that velocities are relative, but velocities "add" differently in Einstein's theory.

Relativity of mass   An object's inertia (in other words, its mass) increases with its speed, so its mass is different for different observers.

Rest mass   The mass of an object at rest. Rest mass represents quantity of matter.

Inertial mass or **mass**   The amount of inertia possessed by an object, measured in kilograms.

Matter   has rest mass, but **radiation** has no rest mass. Matter moves at less than lightspeed, but radiation always moves at lightspeed.

Principle of mass–energy equivalence, or E = mc²   All mass has energy, and all energy has mass. A system with m units of mass has mc^2 units of energy. A system with E units of energy has E/c^2 units of mass.

Nuclear reaction   A change in nuclear structure. **Nuclear fission** is one type of nuclear reaction.

Matter–antimatter annihilation   The transformation of matter into high-energy radiation that occurs when a subatomic particle such as an electron is brought close to its antiparticle (antielectron).

Field view of reality   The view that reality is a set of fields in motion and that material objects arise from the energy of these fields.

Principle of general relativity   No experiment performed inside a sealed room can tell you whether your room is accelerating in the absence of gravity or is at rest in the presence of gravity. In other words, every accelerated observer experiences the same laws of nature.

General theory of relativity   Describes the observations of accelerated observers, while the **special theory of relativity** applies to nonaccelerated observers.

Four-dimensional space–time   The intertwined combination of space and time described by Einstein's relativity. Gravity affects space and time in a manner best described as the **bending of space–time**.

Gravity, light beams, and curved space   Gravity bends light beams, and so gravity must bend space. In the general theory of relativity, gravity is this curvature of space–time caused by masses.

Space   A space is a region without edges. A space may have any number of dimensions and may be curved or flat, infinite or finite.

Big bang   An event some 12 billion years ago that created time, space, energy, and matter and sent them hurtling outward. The universe has been expanding ever since.

Expansion of the universe   The general theory of relativity predicts that the universe's three-dimensional space must either expand or contract. Observations show that expansion is, in fact, occurring, as evidenced by the fact that distant galazies are all moving away from each other.

Shape, or geometry, of the universe   According to general relativity, the large-scale structure of the three-dimensional universe must have one of three possible shapes: A **closed universe** bends back on itself to form a three-dimensional space that is analogous to the two-dimensional surface of a sphere; it has a finite total volume, and will eventually stop expanding and collapse back down to zero size. An **open universe** is analogous to a two-dimensional saddle-shaped surface; it has an infinite total volume and will continue expanding forever. A **flat universe** has no overall large-scale curvature, and is analogous to a flat two-dimensional surface; it has infinite total volume and will continue expanding forever but with an expansion speed that approaches zero at long times.

Cosmic inflation   A widely respected theory of the details of how and why the big bang occurred. The most plausible version of this theory predicts that the universe is flat.

Dark matter   Astronomical material that is not visible. Observation of gravitational effects in our own and other galaxies shows that much of the universe's mass is invisible. Dark matter has been discovered in the outer halo of our galaxy in the form of non-glowing star-like or planet-like objects called **MACHOs** (massive compact halo objects). Other forms of dark matter might not be made of the ordinary matter (protons, neutrons, and electrons) that we are familiar with on Earth. Candidates for such **exotic dark matter** include neutrinos with mass, and entirely new forms of matter. Widely accepted principles of the big bang and nuclear physics imply that the ordinary matter in the universe can constitute only about 10% of the mass that would be needed to cause the universe to be either flat or closed. Thus, if cosmic inflation's prediction of a flat universe is correct, some 90% of the universe must be made of exotic dark matter.

REVIEW QUESTIONS

1. Name the two basic laws of the special theory of relativity, and list four phenomena predicted by this theory.
2. What distinguishes the special theory from the general theory of relativity?

THE RELATIVITY OF SPACE

3. What do we mean by "space" or "distance"?
4. What does "space is relative" mean?
5. Velma passes Mort at a high speed. Each of them holds a meter stick parallel to the direction of motion. What does each observer say about Velma's meter stick? What does each say about Mort's meter stick?
6. In the preceding question, suppose that Mort and Velma hold their meter sticks perpendicular to the motion. Now what does each observer say?

THE RELATIVITY OF VELOCITY

7. Velma passes Mort at a speed of $0.75c$, and she launches a high-speed projectile in the forward direction at $0.75c$ relative to herself. According to Galilean relativity, what is the projectile's speed relative to Mort?

8. In the preceding question, how does Einstein's prediction compare with Galileo's: faster than Galileo's prediction, equal, or slower? How does Einstein's prediction compare with lightspeed: faster, equal, or slower?

THE RELATIVITY OF MASS

9. According to Galilean relativity, which of the following items are relative (different for different observers): time, space (distance), velocity, rest mass, inertial mass, lightspeed?
10. In the preceding question, which items are relative according to Einstein's theory?
11. Velma passes Mort at a high speed. Both observers carry a standard kilogram. What does Mort say about the mass of each of the standard kilograms? What does Velma say?
12. Mort exerts a 1-newton force on his standard kilogram. What acceleration does this give to the kilogram? What will he find if he exerts the same force on Velma's standard kilogram if Velma is passing him at a high speed?

13. What is the distinction, if any, between rest mass, inertial mass, and mass? Which ones increase with speed?
14. What is the distinction between matter and radiation?
15. Why can't material objects be sped up to lightspeed? Does anything move at lightspeed?
16. According to the theory of relativity, is there anything that is not relative? If so, what?

$E = MC^2$ AND THE NEWTONIAN WORLDVIEW

17. What does $E = mc^2$ mean? Does it mean that mass can be converted to energy? Explain.
18. Is matter always conserved? Is mass always conserved? Is rest mass always conserved? Is energy always conserved?
19. Name the process in which about 1% of the rest mass vanishes.
20. According to Einstein's relativity, is rest mass precisely conserved in chemical reactions? Can this effect be measured?
21. Describe an experiment in which a system's entire rest mass vanishes. Is matter conserved here? Mass? Energy?
22. Special relativity suggests that the physical world is made of fields in motion. List some similarities and differences between this view and the Newtonian view.

EINSTEIN'S GRAVITY AND THE SHAPE OF THE UNIVERSE

23. List two experiments you could do in a spaceship accelerating at $1g$ through outer space, which might make you think you are at rest on Earth.
24. What is acceleration equivalent to?
25. As observed in an accelerating reference frame, does a light beam bend? What does this tell us about the effect of gravity on light beams?
26. Give an example of a flat two-dimensional space, a curved two-dimensional space, and a two-dimensional space of finite extent.
27. How can we tell from inside our actual three-dimensional space whether space is curved?
28. According to Newton, gravity is a force exerted by material objects on other material objects. What is gravity according to Einstein?
29. What are the three possible long-term fates of the universe, and what physical properties of the universe determine which way it will go?

HOME PROJECT

Try geometry on a spherical surface. Use a sturdy balloon or a smooth ball and a marker pen. Draw a short "straight" (straightest possible) line. Draw a second short line parallel and next to the first. Extend both line segments as straight lines. Try extending them all the way around the sphere. Draw a large triangle, with "straight" sides. What can you say about the sum of the triangle's angles (*Hint:* What is the sum of the angles of a triangle in a flat space)? Can you draw a triangle whose angles are each 90 degrees? What if the triangle is far smaller than the size of the sphere—What is the sum of the angles?

FOR DISCUSSION

In considering the birth, shape, and fate of our universe, is science trying to answer a basically religious question? Can science answer such questions? Should it try?

EXERCISES

THE RELATIVITY OF SPACE

1. How fast must Velma move past Mort if Mort is to observe her spaceship's length to be reduced by 50%? If Velma is flying over the United States (about 5000 km wide) at this speed, how wide will she observe the United States to be?
2. Mort's swimming pool is 20 m long and 10 m wide. If Velma flies lengthwise over the pool at 60% of lightspeed, how long and how wide will she observe it to be?
3. How fast would Velma have to fly across Chicago in order to observe the city's width to be reduced 10% below the width measured on the ground?
4. Velma, who is carrying a clock and a meter stick, passes Mort. Is it possible that Mort could observe length contraction of Velma's meter stick but yet observe no time dilation of her clock? If so, how?
5. Velma, who is carrying a clock and a meter stick, passes Mort. Is it possible that Mort could observe time dilation of Velma's clock but yet observe no length contraction of her meter stick? If so, how?
6. Velma drives a really fast rocket train northward past Mort, who is standing beside the tracks. Two posts are driven into the ground along the tracks. How does

Mort's measurement of the distance between the posts compare with Velma's: longer, shorter, or the same?

7. Velma drives a really fast rocket train northward past Mort, who is standing beside the tracks. Two posts are driven into the ground on directly opposite sides of the tracks. How does Mort's measurement of the distance between the posts compare with Velma's measurement: longer, shorter, or the same?

8. In the preceding exercise, how does Mort's measurement of the length of the train compare with Velma's: longer, shorter, or the same?

9. In Exercise 7, how does Mort's measurement of the width of the train compare with Velma's: longer, shorter, or the same?

THE RELATIVITY OF VELOCITY

10. Velma, passing Mort at 0.1c, launches an apple forward at a launch speed of 0.1c. How fast does the apple move relative to Mort?

11. Velma, passing Mort at 0.25c, launches a rocket forward at 0.5c. According to Galileo's relativity, how fast does the rocket move relative to Mort? How fast according to Einstein? Which answer is correct?

THE RELATIVITY OF MASS

12. If Velma passes Mort at a high speed, Mort will find her mass to be larger than normal. Will he also find her to be larger in size?

13. Velma's spaceship has a rest mass of 10,000 kg, and she measures its length to be 100 m. She moves past Mort at 0.8c. According to Mort's measurements, what are the mass and the length of her spaceship?

14. How fast must Velma move past Mort if Mort is to observe her spaceship's mass to be increased by 50%? How fast must she move if Mort is to observe her spaceship's length to be reduced by 50%?

15. A meter stick with a rest mass of 1 kg moves past you. Your measurements show it to have a mass of 2 kg and a length of 1 m. What is the orientation of the stick, and how fast is it moving?

16. A meter stick with a rest mass of 1 kg moves lengthwise past you. Your measurements show it to have a mass of 2 kg. How long is this meter stick relative to you?

17. You are in a spaceship moving past Earth at nearly lightspeed. You measure your own mass, pulse rate, and size. How have they changed?

18. You are in a spaceship moving past Earth at nearly lightspeed and you observe Mort, who is on Earth. You measure his mass, pulse rate, and size. How have they changed?

$E = MC^2$

19. When you throw a stone, does its mass increase, decrease, or neither? Can this effect be detected?

20. A red-hot chunk of coal is placed in a large air-filled container where it completely burns up. The container is a perfect thermal insulator—in other words, thermal energy is unable to pass through the container's walls. According to $E = mc^2$, does the total mass of the container and its contents change during the burning process? If so, does the mass increase, or decrease?

21. Referring to the previous problem: Suppose that the container is not a thermal insulator—in other words, thermal energy is able to pass through the walls. In this case, does the total mass of the container and its contents change during the burning process? If so, does the mass increase, or decrease?

22. An electron and an antielectron annihilate each other. In this process, is energy conserved? Is mass conserved? Is rest-mass conserved?

23. Two mouse traps are identical except that one of them is set to spring shut when the trigger is released, and the other is not set. They are placed in identical vats of acid. After they are completely dissolved, what, if anything, are the differences between the two vats? Will the masses differ?

24. In a physics laboratory, an electron is accelerated to nearly lightspeed. If you were riding on the electron, would you notice that the electron's mass had increased? If you were standing in the laboratory, what would you notice concerning the electron's mass and energy?

EINSTEIN'S GRAVITY

25. If you were in a rocket ship in space (far from all planets and stars) accelerating at 2g, how heavy would you feel? What if your acceleration were, instead, 0.5g? What if you were not accelerating at all?

26. If you were in a rocket ship in space accelerating at 2g and you dropped a ball, how would it move as observed by you? What if your acceleration were instead 0.5g? What if you were not accelerating at all?

27. Longitudinal (north–south) lines on Earth are "straightest" lines that eventually meet. What about latitudinal (east–west) lines? Are they circles? Are they "straightest" lines? Do they eventually meet?

28. Is a circle (meaning the perimeter of a flat circular area) a space? How many dimensions does it have? Is it a curved space, or is it flat? Give an example of a flat one-dimensional space.

THE SHAPE OF THE UNIVERSE

29. Since the universe is about 12 billion years old, what would we expect to see if we could see a distance of 12 billion light-years (recall that a light-year is the distance that light travels in a year)?

30. If you could miraculously travel much faster than light and traveled in a straight line, would you get to the edge of the universe? Where would you get?

PROBLEMS

$E = MC^2$

1. You impart 90 J of kinetic energy to a 1-kg stone when you throw it. By how much do you increase its mass?

2. A large nuclear power plant generates electric energy at the rate of 1000 MW. How many joules of electricity does the plant generate in one day? What is the mass of this much energy?

3. If you had two shoes, an ordinary shoe and an "anti-shoe" made of antiparticles, and you annihilated them together, by how far could you lift the U.S. population? A person weighs about 600 newtons (about 150 pounds). Assume that each shoe's rest-mass is 0.5 kg, and that all the energy goes into lifting.

4. Making estimates. Show that, if all the energy released (transformed) in fissioning 1 kg of uranium were used to heat water, about 2 billion kg of water could be heated from freezing up to boiling. [Assume that the uranium's rest-mass is reduced by about 1%. Roughly 4 J of thermal energy are needed to raise the temperature of 1 gram of water by 1°C.] How many tonnes of water is this (a tonne is 1000 kg). How many large highway trucks, each loaded to about 30 tonnes, would be needed to carry this much water?

5. Solar radiation reaches Earth at the rate of 1400 watts for every square meter directly facing the sun. Using the formula πR^2 for the area of a circle of radius R, find the amount of solar energy entering Earth's atmosphere every second. Earth's radius is 6400 km. Use your answer to find how many kilograms of sunlight hit Earth every second.

Are We Alone?

CHAPTER 12

the search for extraterrestrial intelligence

For this world was created by Nature after atoms had collided spontaneously and at random in a thousand ways, driven together blindly, uselessly, without any results, when at last suddenly the particular ones combined which could become the perpetual starting points of things we know—earth, sea, sky, and the various kinds of living things. Therefore, we must acknowledge that such combinations of other atoms happen elsewhere in the universe to make worlds such as this one. . . . So we must realize that there are other worlds in other parts of the universe, with races of different men and different animals.

LUCRETIUS, ROMAN POET, FIRST CENTURY B.C.

Sometimes I think we're alone. Sometimes I think we're not. In either case, the thought is staggering.

BUCKMINSTER FULLER, ARCHITECT AND FUTURIST

One of science's most intriguing questions is whether the universe harbors other instances of what we are pleased to call "intelligent life." Either answer to this question has immense consequences. If intelligent aliens exist, our culture is likely to eventually be influenced by them. Some anthropologists believe that the discovery of alien intelligence would irreversibly change humankind's self-image. The possibility of such a discovery has inspired many scientists to devote much of their professional lives to the search for extraterrestrial intelligence, or **SETI**.

On the other hand, the conclusion that intelligence evolved only on Earth would challenge science's general Copernican view that nature is basically the same everywhere, even though the details differ (see Section 1.7).

A survey of the universe certainly indicates nothing strikingly unique about our own location. Even within our own planetary system, there are five other bodies (Mars, Jupiter, a moon of Jupiter named Europa, Saturn, and a moon of Saturn named Titan) that may have life-supporting potential or may have had it in the past. Furthermore, in our Milky Way galaxy alone, there are hundreds of billions of stars, many of them similar to our sun. And there are billions of other galaxies. It seems plausible that among so many stars in so many galaxies, there are some with planets similar enough to Earth that life and intelligence evolved there.

How can we bring scientific reasoning to bear on a question like this? One feature that makes this topic both difficult and fascinating is its interdisciplinary character. It involves lots of physics, as well as astronomy, geology, chemistry, biology, and even anthropology and sociology, and it teaches us a good deal about the methods of science as well as the dangers of pseudoscience. This is a good place to discuss SETI because we now have the physics tools—Newton's laws of motion and gravity, an understanding of electromagnetic radiation, and special relativity—needed for it.

Any discussion of extraterrestrial life must be based on incomplete evidence. There are many science-related issues for which the evidence is incomplete. Examples include global warming (Section 9.7), health risks from low-level nuclear radiation (Sections 15.6 and 15.7), and future energy technologies (Chapter 17). We must draw conclusions about such pressing issues despite the incomplete information. This chapter provides an example of reasoning from incomplete evidence. We will find that the evidence allows us to draw conclusions about some questions but not about others. In dealing with speculative topics, we must be careful to stay grounded in the observed evidence and to take note of the uncertainties in our conclusions.

This chapter follows a general method of thinking that has become standard in discussions about the likelihood of extraterrestrial life* and that is a good model for thinking about other speculative topics. We break the question into subquestions, each interesting in its own right: What is the likelihood that other Earth-like planets exist (Section 12.1)? Given an Earth-like planet, what is the likelihood that life would actually have started there (Section 12.2)? Given life, what is the likelihood that intelligence would have evolved (Section 12.3)? Given intelligence, what is the likelihood that a life-form would develop technology (Section 12.4)? And given technology, what is the likelihood that a life-form would make contact with us, or we with them (Section 12.5)?

Section 12.6 looks at a question that might be disturbingly relevant to *Homo sapiens*: If technology-based civilizations do develop on other planets, do they endure for long? Finally, Section 12.7 asks whether we have actually ever been visited by alien life and puts this question into the context of a topic that goes to the heart of human culture in our scientific age: pseudoscience.

■ **DIALOGUE 1** Based on the history presented in Chapter 1, at what time in history would you guess that many people began actively speculating about the possibility of extraterrestrial beings?

12.1 ARE THERE OTHER "GOOD" PLACES FOR LIFE?

We begin from an obvious piece of evidence: If *intelligence* means the ability to solve new problems and use abstract reasoning, then intelligent life did develop on Earth. This, along with science's general Copernican view that the principles of physics are the same throughout the universe, suggests that Earth-like conditions elsewhere should lead to intelligent life elsewhere.

We adopt the conservative attitude of inquiring only about the possibility of intelligence that developed under Earth-like conditions. We will not speculate on such possibilities as, for example, life developing in interstellar space. Such possibilities seem implausible, and in any case it is hard to see how to even begin thinking about such possibilities.

Science's successes are due partly to the method known as *analysis*: breaking a question into parts that are examined separately. Extraterrestrial intelligence is an especially good topic for analysis because it breaks up naturally into several distinct subquestions, as outlined in the introduction. Each subquestion corresponds to one or two particular scientific disciplines. Our first subquestion is an astrophysical one: How many Earth-like or "good" places are there, places that are sufficiently similar to Earth that life could have started there? Here and in the rest of the chapter we restrict our analysis to our Milky Way galaxy alone,

---

*Although few physics texts discuss this topic, most astronomy texts do.

1. People began such speculation around 1600, as a result of the publication of Copernicus's theory in 1543.

because our galaxy is thought to be typical of galaxies throughout the universe. Our conclusions can then be extended to the universe by multiplying by the number of galaxies—at least 1 billion (the number within reach of present telescopes) and perhaps 100 billion.

Astronomers believe that there are roughly 4×10^{11} (400 billion or 400,000 million!) stars in our galaxy alone. How many good planets or moons might orbit these stars? The theory and observations of star formation (Section 5.3) indicate that other stars developed the way that our sun developed, by the gravitational collapse of interstellar gas and dust. Planets should form from these clouds, just as planets formed around the sun. But many developing stars break up into double-star or triple-star systems rather than collapsing to become a single star. Observations of the stars in the sun's neighborhood show that about 50% are not the single points of light that they appear to be to the unaided eye but are systems of two or more stars orbiting around each other. Because binary stars are thought to disrupt the planet formation process, multiple stars are less likely to have planets. To simplify our argument, we will neglect the possibility of life around multiple stars.

Many single stars have either too much or too little mass to have planets with conditions conducive to life. Stars much more massive than our sun "burn"* so brightly that they use up their fuel in only a few hundred million years. Life on Earth needed longer than that even to get started, and once it did get started, several billion years elapsed before it evolved beyond primitive single-cell forms. So we assume that very massive stars do not have good planets.

Stars much less massive than the sun burn so faintly that a planet would have to orbit close to the star in order to be warm enough for life. But any such close planet would experience gravitational forces that would lock it into a situation in which the same side of the planet would always face the star (in the way that the moon is locked into Earth). This would force any atmosphere that the planet might have around to the planet's dark side, where it would freeze out. Because Earth's atmosphere is essential to life here, we rule out low-mass stars as candidates.

Astronomers estimate that about 10% of the single stars in our galaxy have masses close enough to the sun's mass to be candidates for having good planets. Of these sunlike stars, how many actually have planets, and how many of these planets are actually Earth-like?

Ten years ago, it was possible to argue that the solar system is unique. Today the evidence strongly suggests that planetary systems are abundant in the Galaxy. From statistical arguments alone, the likelihood is that many of them will have conditions favorable for life.

ANNEILA SARGEN AND STEVEN BECKWITH, ASTRONOMERS, IN AN APRIL 1993 ARTICLE IN *PHYSICS TODAY*

H O W   D O   W E   K N O W ?

Planets around other stars

It is difficult to see planets around other stars because their faint light is overwhelmed by the light from their central sun. But both the theory of star formation and the available evidence indicate that many single stars have planetary systems.

When planets orbit a star, the star also orbits, or wobbles, slightly because of the gravitational pull of the planets. Astronomers can directly detect this wob-

---

*Stars don't really burn; they "fuse" (Chapter 16).

bling by carefully tracking the position of the star. Using this method, astronomers have discovered two or three Earth-sized planets orbiting a star very near to our sun. A second, easier, method is observation of the frequencies of the light emitted by the star. As the star wobbles toward and then away from Earth, these frequencies "wobble" to higher and then lower frequencies due to the Doppler effect (Section 11.7), akin to the changes in the frequency of a car's horn as it drives toward or away from us. With this method, astronomers have discovered Jupiter-sized planets orbiting four nearby sunlike stars. Look for more evidence of this sort in the near future, as more accurate telescopes and new observational techniques extend these methods to fainter stars and smaller planets orbiting those stars.

Most single sunlike stars are observed to rotate slowly rather than rapidly. A single star forming by the collapse of a gas cloud would be expected to spin faster and faster as the cloud collapses, just as a spinning ice-skater spins faster as she pulls in her arms. The only plausible mechanism for preventing a single star from rotating rapidly is the formation of a rapidly rotating flattened disk that is left behind as the central star collapses (Section 5.3), and such a rotating disk should eventually coalesce into planets.

Further evidence comes from thick gas clouds such as the one shown in Figure 5.12, where astronomers have found thousands of newly minted stars. Nearly all of these new stars are wrapped in flattened disks of dust grains and gas, just the sort of formation that the theory predicts will eventually coalesce into planets. Furthermore, searches have found that roughly half of all young sunlike stars show evidence for such disks. This provides perhaps the strongest evidence that planetary systems are common.

Since there is room for reasonable disagreement about the number of stars that have planetary systems, the most honest way to proceed is to establish a range of uncertainty. We will make two estimates, a high estimate that is plausible but optimistic about the chances for extraterrestrial intelligence, and a low estimate that is at the pessimistic end of the plausible range. A reasonable optimistic estimate is that nearly all of the sunlike stars have planets, and a reasonable but pessimistic estimate is that only about 10% of such stars have planets. A much smaller estimate, such as 1%, seems unreasonable, given the evidence.

Given a sunlike star with a planetary system, how many planets (or moons) are likely to have conditions conducive to the emergence of life? In addition to Earth, such conditions might also exist in our own solar system on Mars, in the gaseous atmospheres of Jupiter and Saturn, on a moon of Jupiter named Europa, and on Titan, a moon of Saturn that has an atmosphere resembling the early atmosphere of Earth.

Mars probably does not contain life today. The United States launched two spacecraft that landed there in 1975 and, among other things, tried to detect signs of life. It was thought that primitive microscopic single-celled life-forms, similar to those that existed for billions of years on Earth before the emergence of multicelled plants and animals, might exist in the Martian soil. But a search for organic molecules in the Martian soil was negative.

The Mars meteorite

However, life might have existed on Mars in the past. Although it is a dry and frozen desert today, all over the planet there are ancient river valleys and signs of ancient lakes and perhaps oceans, indicating a warmer and wetter planet in the past. From the cratering of the terrain, we can estimate when Mars was warmer and wetter. The answer is about 4 billion years ago.

Recently, interesting evidence was found for ancient life on Mars. Many meteorites—rocks from space—have been found in the Antarctic, sitting on top of the ice sheet where they are easy to see. Most of these meteorites fell to Earth thousands of years ago. Judging by the Martian atmospheric signature hidden in their minerals, 12 of them come from Mars. Scientists have carefully examined one of them, a brownish potato-sized meteorite. Radioactive dating (Chapter 15) showed that the rock first crystallized on Mars about 4.5 billion years ago. Subsequently, an impact fractured the rock, and carbon-rich fluids deposited solid material in the fractures. In a relatively recent event 16 million years ago, another impact launched the rock fragment on a celestial journey to Antarctica, where it fell to Earth 13,000 years ago. The scientists examining this rock discovered a certain type of organic molecule that is often (but not exclusively) associated with life. These molecules were more frequent deeper within the meteorite, indicating that they did not come from Earthly contamination. Most provocatively, they found what some scientists are calling *nanofossils* (Figure 12.1), segmented structures and tiny spheres attached one to another, reminiscent of, but 100 times smaller than, bacterial colonies on Earth.

How sure are we that these are the remains of ancient Martian life? Not sure enough, in the opinion of many scientists. Because confirmation of ancient life on Mars would be extraordinary, scientists demand that the evidence for it be extraordinarily strong. Many scientists feel that, although the claim might eventually be confirmed by further investigation, the evidence is not yet sufficiently convincing. Such skepticism is characteristic of scientific methodology and has proved quite healthy for science.

Many nations plan to send robot orbiters, landers, roving vehicles, and subsurface penetrators to Mars in the near future. And, perhaps in 2005, a robotic mission may return surface and subsurface samples from Mars to Earth. Be on the lookout for further news!

Optimists argue that there are probably one or more Earth-like planets or moons orbiting each sunlike star, because there is at least one orbiting our sun and there are five others that are candidates for past or present life. Pessimists argue that the other four places in our solar system aren't really conducive to life and that perhaps as few as 10% of the planetary systems have even a single Earth-like place.

To summarize, an optimistic estimate of the number of Earth-like planets in our galaxy is

FIGURE 12.1
(a) Electron microscope images of tube-like structures found in a meteorite that is believed to be of Martian origin. While the exact nature of these structures is not known, one interpretation is that they may be microscopic fossils of primitive, bacteria-like organisms that may have lived on Mars more than 3.6 billion years ago. The largest possible fossils, such as the structure shown in (b), are about one-tenth of a micrometer (one micrometer = one-millionth of a meter), or less than 1/100th the diameter of a human hair in size.

$$
\begin{aligned}
&\text{(number of stars in galaxy)}\\
&\times \text{(fraction that are single stars)}\\
&\times \text{(fraction of single stars that are sunlike)}\\
&\times \text{(fraction of sunlike stars that have planetary systems)}\\
&\times \text{(number of Earth-like planets per planetary system)}\\
&= (4 \times 10^{11}) \times 0.5 \times 0.1 \times 1 \times 1 = 2 \times 10^{10}
\end{aligned}
$$

A pessimistic estimate is

$$(4 \times 10^{11}) \times 0.5 \times 0.1 \times 0.1 \times 0.1 = 2 \times 10^{8}$$

We estimate that there are some 200 million to 20 billion Earth-like planets in our galaxy. This result reflects the consensus among scientists that there are many places where life might have arisen.

In our time this search [for extraterrestrial life] will eventually change our laws, our religions, our philosophies, our arts, our recreations, as well as our sciences. Space, the mirror, waits for life to come look for itself there.

RAY BRADBURY, SCIENCE
FICTION WRITER

■ **DIALOGUE 2** Suppose that some other galaxy contains 8×10^{12} stars, that only 25% of these are single stars, that 10% of the single stars are similar to the sun, that 50% of these sunlike stars have planetary systems, and that 1% of these planetary systems contain one Earth-like planet. How many Earth-like planets does this galaxy contain?

2. $(8 \times 10^{12}) \times 0.25 \times 0.1 \times 0.5 \times 0.01 = 10^{9}$ (1 billion).

12.2 DOES LIFE DEVELOP ON "GOOD" PLANETS? HOW DID LIFE DEVELOP ON EARTH?

Scientists have studied the origin of life on Earth for more than a century and now know enough to form plausible hypotheses. We can explore the likelihood of life elsewhere by studying its origin on Earth.

Earth developed from the material left behind as gas and dust collapsed gravitationally to form the sun (Section 5.3). When Earth emerged from this process 4.6 billion years ago,* it was a solid, rocky planet having essentially no gaseous atmosphere.

Some of the materials that formed Earth were radioactive (Chapter 15), and their radioactivity heated Earth's interior, causing gases that had been trapped within Earth to rise to the surface during volcanic and other activity. These gases, mostly steam, CO_2, and N_2, formed Earth's earliest atmosphere. Once in the atmosphere, most of the steam cooled, condensed, and formed the oceans, with some of it remaining in the atmosphere as water vapor. Chemical reactions involving such surface materials as iron caused some H_2O and CO_2 to decompose, releasing carbon and hydrogen. Hydrogen gas is so light that most of it eventually escaped Earth's gravitational grip, but some of it participated in further reactions to create gaseous methane (CH_4) and ammonia (NH_3).

So before the emergence of life, Earth was partly covered by oceans of liquid water and had an atmosphere composed of water, carbon dioxide, nitrogen, methane, ammonia, and hydrogen. This primitive Earth was subjected to a variety of energy sources, such as intense ultraviolet radiation from the sun (at that time, there was no ozone to shield the surface from ultraviolet), lightning, heating by volcanic activity, radioactivity, and radiations from outer space. Each of these energy sources can make and break chemical bonds and so promote new chemical reactions.

In the beginning, this world was nothing at all,
heaven was not, nor earth, nor space.
Because it was not, it bethought itself:
I will be. It emitted heat.

ANCIENT EGYPTIAN TEXT

H O W   D O   W E   K N O W ?

Scientists have performed a variety of simulation experiments to determine what happens when a primitive Earth-like environment is subjected to these energy inputs. One of the best-known experiments was performed in 1953 by Stanley Miller and Harold Urey. Their experiment was simple: They reproduced a plausible early Earth atmosphere in a jar, placed it in contact with liquid water, and subjected it to electric-spark discharges to simulate lightning. The results were dramatic. Within a few days the liquid water turned brown, and chemical analysis revealed a rich array of amino acids and nucleic acids—the building blocks of life. Biological proteins are formed from amino acids linked together into long chains, and DNA molecules are formed from nucleic acid bases strung together into a long *double helix*.

Since the Miller–Urey experiment, numerous workers have performed similar experiments using a variety of gas mixtures and energy sources. As long as

---

*Section 15.5 presents evidence for this date.

these gases contain water, hydrogen, nitrogen, and a gaseous form of carbon, the outcome is consistent: Many biological building blocks are formed. It thus seems likely that energy interacted with the early Earth atmosphere to form the chemical building blocks of life.

Amino acids also appear to have formed spontaneously beyond our solar system. They have been found in meteorites and in moon rocks. Radio telescopes searching a star-forming region near the center of our galaxy have detected the radio frequencies characteristic of the 10-atom amino acid known as glycine. The possibility that such biological building blocks are scattered throughout our galaxy enhances the chances that life exists elsewhere.

For those who are studying aspects of the origin of life, the question no longer seems to be whether life could have originated by chemical processes involving nonbiological components but, rather, what pathway might have been followed.
NATIONAL ACADEMY OF SCIENCES

In the caldron boil and bake
Eye of newt and toe of frog,
Wool of bat and tongue of dog,
Adder's fork and blindworm's sting,
Lizard's leg and howlet's wing
SHAKESPEARE, *MACBETH*

Once amino acids and other organic molecules formed, there would have been many opportunities for them to aggregate to form a dense organic "soup." For example, if the water in a tidewater pool on the fringe of an ocean contained organic molecules and if the water then partially or completely evaporated, the organic molecules would be highly concentrated in the remaining water or on the dried-out land (Figure 12.2).

The remaining steps are less certain. Concentrated organic material may have dried out entirely and undergone dry reactions, or it may have remained suspended in liquid water and undergone wet reactions. Laboratory experiments of both types have been performed. In the dry experiments, amino acids react by linking up to form long protein molecules! When placed back into water, these proteins assume the shape of round objects similar to simple biological cells (Figure 12.3). Similar results are obtained in the wet experiments. These objects, called **protocells**, absorb material from the surrounding liquid, grow by attaching to one another, and divide to produce additional such objects.

Biochemical theories indicate that if protocells are placed in water containing proteins and other nutrients, they can grow by absorbing the nutrients and can multiply by simple division. Today, single-celled organisms, such as the amoeba, perform essential life functions by just such mechanisms (Figure 12.4). Protocells could have undergone a process of "chemical evolution" in which those that were more efficient at these processes prospered at the expense of the less

FIGURE 12.2
On the primitive Earth, tidewater pools such as this might have contained a rich soup of organic molecules that became more highly concentrated as the water evaporated, allowing complex reactions whose products could then have "fertilized" the oceans with protocells or living organisms.

FIGURE 12.3
Electron microscope image of "protein microspheres." These are formed by heating a dry mixture of amino acids that link up to form long protein molecules which, when placed back into water, assume the spherical shape shown. These protocells are similar to biological cells; they can even grow at the expense of the protein dissolved in their surroundings and can bud.

Looking back over the geological record it would seem that Nature made nearly every possible mistake before she reached her greatest achievement, Man; or perhaps some would say her worst mistake of all. At last she tried a being of no great size, almost defenseless; one gift she bestowed to save him from threatened extinction—a certain stirring, a restlessness, in the organ called the brain.

ARTHUR STANLEY EDDINGTON, ASTRONOMER AND PHYSICIST

efficient protocells. Chemical evolution favors the development of ever more efficient and more elaborately organized protocells that could eventually have evolved into primitive single-celled life-forms.

Life arose surprisingly early in Earth's 4.6-billion-year history. The earliest fossils are dated at over 3 billion years (Figure 12.5), and there is chemical evidence for life at least as early as 3.85 billion years ago. For about the next 2 billion years, the only life-forms were simple one-celled organisms with no central cell nucleus. It was only after the biological evolution of more complex cells containing cell nuclei that multicelled organisms, and higher forms of life, became possible.

Biologists believe that life on Earth developed through a sequence of steps similar to those just outlined. We summarize this hypothesis as follows:

THE HYPOTHESIS OF A CHEMICAL ORIGIN OF LIFE
Life originated on Earth by physical and chemical processes, beginning with the action of various energy sources in the prelife atmosphere to create the organic chemicals that are the building blocks of life.

There is nothing that is necessarily unique to Earth in this scenario. What is needed is a good planet with water. Most biochemists agree that given conditions similar to those of the primitive Earth and given something like a billion years, life would develop elsewhere.

An optimistic estimate is that given a good planet and time, life is nearly certain to appear. A pessimistic view might assume that because of periodic catastrophes (such as collisions with another astronomical object) or because of a lack of water or other necessary substances, life might develop on only some fraction, such as 10%, of the good planets. Because of the ease with which the precursors of life can be created in the laboratory, any much smaller estimate seems unreasonable.

Putting this together with our previous estimates, our optimistic estimate is that life developed at 2×10^{10} (20 billion) places in our galaxy alone, and our pessimistic estimate is a "mere" 2×10^7 (20 million) places. It appears that life is not a stranger in the universe.

FIGURE 12.4
An amoeba, a living single-celled organism, flows to surround a nearby food particle.

FIGURE 12.5
The wavy rock depositions result from the actions of blue-green algae and bacteria. Such biologically induced rock formations are a clue to early biological activities. This sample is from the earliest known rock formation of this type, in Rhodesia, and is dated at between 3.1 billion and 2.7 billion years old. A 5-millimeter-long ruler is shown for scale.

■ **DIALOGUE 3**   Earth's prelife atmosphere contained very little oxygen (O_2), so where did Earth's oxygen come from?

12.3 IS INTELLIGENCE A CHARACTERISTIC FEATURE OF LIFE?

Are we the universe's only thinking creatures? If we are alone in being able to understand how nature operates, then it is only here on Earth that nature has evolved the ability to understand itself. This appears to violate the Copernican viewpoint.

Although there is little data to help us evaluate the likelihood of intelligence emerging from life, it appears that the concept of "biological convergence" is central to this question. On Earth, biological features sometimes emerge independently in two or more widely different organisms. For example, wings have appeared in insects, in the extinct reptile called the *pterodactyl*, in birds, and in bats. The reason wings evolved so many times is that flying has survival value because flying animals can escape their enemies and can find food more easily. Such features that tend to evolve again and again are called **biologically convergent properties**.

There are many examples of biological convergence. For instance, North American hummingbirds are similar to Hawaii's hummingbird moth, although one is a bird and the other is an insect. Fishlike bodies developed independently in fish, in the water mammals (dolphins, porpoises, whales), and in an extinct reptile called the *ichthyosaur*. Catlike and doglike shapes developed many times in unrelated creatures.

---

3. Green plants created O_2 from CO_2 and H_2O during photosynthesis (Section 2.8).

We are the children of the eighth day.

THORNTON WILDER

Is intelligence biologically convergent? If so, it seems likely that intelligence would evolve on other Earth-like planets having life. Let us examine the arguments on both sides.

Those who argue that intelligence is not biologically convergent point out that intelligence evolved only once on Earth and that it emerged only a few million years ago, a brief moment ago in Earth's 4600-million-year history. If intelligence is biologically convergent, then why didn't it emerge much sooner? A second argument is that the emergence of *Homo sapiens* depended on a series of unlikely events. One such event occurred 65 million years ago when a large astronomical body slammed into Earth, helping make the dinosaurs extinct and providing an opportunity for the mammals to evolve in many new directions.

Those who regard intelligence as a convergent property reply that it is not surprising that 4.6 billion years were needed for intelligence to develop, because after all it took some 3 billion years just to get to multicelled organisms. Concerning the second argument, if the dinosaurs had not become extinct and if intelligence actually is a convergent property, there should eventually have appeared an intelligent reptile descended from the dinosaurs. Furthermore, intelligence does seem to have survival value. How could a slow-moving, weak, soft-skinned, ground-dwelling creature that requires nearly two decades to rear its offspring have possibly survived and prospered for the past few million years if not by its wits? Since intelligence confers such a biological advantage, it should be a convergent property for the same reason that wings are a convergent property.

It appears that near intelligence has evolved independently in another animal: the dolphin. Dolphins can respond accurately to complex four- or five-element "sentences." They apparently communicate with one another via sound waves. They seem to be as intelligent as apes and more like human beings in their curiosity and eagerness to communicate. The apes are also nearly intelligent, but they are not an example of the independent evolution of intelligence, because apes and humans share recent common ancestors. Dolphins, on the other hand, are as far removed from humans as are dogs, bats, rabbits, and horses. How could dolphins have developed near intelligence if intelligence is not a convergent property?

So optimists contend that intelligence is a convergent property and should eventually appear on nearly all Earth-like planets that support life. Pessimists, reasoning that intelligence can arise only after a particular sequence of highly improbable events, estimate that life has a very low probability of evolving into intelligent life, perhaps only one chance in a million (10^{-6}). Combining these figures with our previous estimates, our optimistic estimate is that intelligent life has arisen (or will arise) at some 2×10^{10} (20 billion) places in our galaxy, and our pessimistic estimate is that it has arisen or will arise in only 20 places. The billion-fold spread between these estimates reflects the lack of consensus regarding intelligence.

Like a fisherman, we have cast our net into the cosmic ocean. I cannot help feeling the net will not be empty once we have finished.

SAM GULKIS, NASA SCIENTIST, SPEAKING AT THE OPENING OF THE NEW SETI PROJECT

■ **DIALOGUE 4**   Suppose that the optimistic estimate is correct. Then how many stars would we need to survey, on the average, before finding one having a planet on which intelligence has arisen or will arise?

---

4. Since there are some 400 billion stars in our galaxy, we would need to survey about (400 billion)/(20 billion), or about 20 stars.

12.4 DOES INTELLIGENT LIFE DEVELOP TECHNOLOGY?

According to our optimistic estimate, one-twentieth of the stars have a planet where intelligent life has arisen or will arise (see Dialogue 4). We could experimentally check this estimate by surveying a few tens of other stars for signs of intelligent life. But what signs would we look for? We know only one way of detecting intelligent life on planets at other stars: detecting electromagnetic radiation sent out from those planets by radio transmitters or other devices.

So from an observational point of view, the relevant question is not how many places harbor intelligent life but, rather, how many places harbor life that is capable of sending out radio signals or some other form of communication. We will refer to any such form of life, having sufficient technology to be capable of sending out radio signals, as **technological**.

It is difficult to form any credible estimate of the likelihood of technological life. Humans first stood upright about 4 million years ago, began using tools 2.5 million years ago, developed agriculture 10,000 years ago, formed the first cities 5000 years ago, and invented radio only during the twentieth century. Would other forms of intelligent life go through a similar development? The example of the dolphins shows that life can become intelligent without becoming technological—dolphins cannot develop technology because they don't have hands. There is wide disagreement on this question because there is no evidence as to what form of life might develop elsewhere. There is certainly no reason to believe that multicelled life-forms on other planets would resemble plants or animals on Earth.

Although we cannot draw any conclusion about the likelihood of technological life elsewhere, our discussion has led to the interesting hypothesis that life exists all over the universe, and this has provided a framework of questions for further thinking about this issue. This is the best we can hope for when treading such uncertain ground: Some of the questions are tentatively answered, and a framework is established for further study. Table 12.1 summarizes our conclusions.

TABLE 12.1

Extraterrestrial life in our galaxy, pessimistic and optimistic estimates

	Plausible upper estimate	Plausible lower estimate
Number of stars in our galaxy	4×10^{11}	4×10^{11}
Fraction of those that are single stars	0.5	0.5
Fraction of those that are sunlike	0.1	0.1
Fraction of those that have planetary systems	1.0	0.1
Number of Earth-like planets per planetary system	1.0	0.1
Number of Earth-like planets in our galaxy	2×10^{10}	2×10^{8}
Fraction of those on which life has arisen or will arise	1.0	0.1
Number of planets where life has arisen or will arise	2×10^{10}	2×10^{7}
Fraction of those where intelligence develops	1.0	small, perhaps 10^{-6}
Number of planets where intelligence arises	2×10^{10}	few, perhaps 20
Fraction of those that become technological	????	????

12.5 INTERSTELLAR COMMUNICATION AND TRAVEL: MIGHT WE MAKE CONTACT?

We have unintentionally sent out electromagnetic signals to space for most of this century, as radio and television communication. This evidence of our own technological civilization has now traveled some 100 light-years (the distance that light travels in 100 years) from Earth and has already arrived at hundreds of other stars. If technological life exists around these stars, they might—for better or worse—be tuning in our old TV shows right now!

We have also sent out intentional signals. The first such radio message was sent from the large radio telescope at Arecibo, Puerto Rico, in 1974 and was beamed toward a cluster of stars 27,000 light-years away. Since the message will be 27,000 years in transit, nobody is holding his or her breath for an immediate reply.

Other communications have been sent out on space vehicles. The first was a plaque carried by *Pioneer 10* (Figure 12.6), launched toward Jupiter in 1972. The spacecraft was accelerated by the gravitational effects of Jupiter and its moons so that it became, in 1983, the first human artifact to leave the solar system. Since then, *Pioneer 11* and two *Voyager* spacecraft have exited the solar system. Because they are not moving toward any particular star, there is little expectation that their messages will ever be received. In any case, it would be hundreds of thousands of years before they could reach any distant star.

A new project got under way in 1992 to analyze radio-frequency radiation from many stars. The signals are detected by several radio telescopes that are used mostly for other non-SETI purposes and fed into an advanced radio receiver that simultaneously analyzes millions of frequency channels. Although this project is far more sensitive and comprehensive than any previous SETI project, it is expected to succeed only if there is an alien civilization in our part of the galaxy and then only if that civilization is purposely beaming a powerful signal at us.

> [The SETI project is] a legitimate part of astronomy. It is hard to imagine a discovery that would have greater impact on human perceptions than the detection of extraterrestrial intelligence.
>
> NATIONAL ACADEMY OF SCIENCES

> Like Columbus, the only thing we can be certain of is that there is something beyond our shores.
>
> EDWARD STONE, PHYSICIST

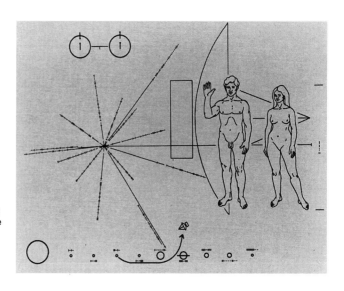

FIGURE 12.6
The plaque aboard the *Pioneer* spacecraft. The message is written in the only language we share with possible recipients: science. It shows humans in front of the spacecraft for scale; the path of the spacecraft from Earth past Jupiter; a process occurring in a hydrogen atom, to provide units of distance and time (upper left); and a map showing directions and frequencies (in binary numbers—the language of computers) of pulsating neutron stars as measured from the solar system. Any extraterrestrials discovering the plaque could deduce where and when we were.

It [SETI] is an essential part of understanding who we are. You have to be made out of wood not to be at least a little interested in the answer.

CARL SAGAN, ASTRONOMER

Especially allowing for a modicum of scientific and technological progress within the next few centuries, I believe that interstellar spaceflight at relativistic velocities to the farthest reaches of our Galaxy is a feasible objective for humanity.

CARL SAGAN

Could extraterrestrial civilizations visit Earth, or could we someday visit them? It would be a long trip. It is 4 light-years to even the nearest other star, Alpha Centauri. This is 100 million times farther then the distance to the moon, the farthest that humans have yet traveled. Travel to Alpha Centauri compares with travel to the moon in the same ratio that a trip to the moon compares with a trip across your living room.

Despite the huge distances, interstellar travel may be feasible. It is interesting to see what might be done using only current technologies. One approach to interstellar travel involves spaceships that move much slower than light and that would take hundreds or thousands of years to reach their destination. This approach could be used for robotic probes that would radio back information about a particular star's planets, including signs of life. A colony of humans could make a long-duration trip in several generations aboard a large "world-ship" designed as a complete living environment (Figure 12.7). Refrigeration techniques might allow humans to be in suspended animation during the trip.

It is not difficult to conceive of technologies that could accomplish such slow trips. The *Pioneer* spaceships, launched from Earth with conventional chemical rockets, used Jupiter and its moons as a slingshot to give them enough kinetic

(b)

(a)

FIGURE 12.7
(a) Artist's concept of a space habitat for 10,000 people. To provide artificial gravity, the colony would spin, like a bicycle wheel, around an axis through its center. The wheel might be two kilometers in diameter. (b) Interior view of the habitat. It could contain air, clouds, lakes, mountains, and towns. Gravity would feel normal for inhabitants on the inside of the outer surface.

energy to exit the solar system. This produces rather slow speeds, however. It will take 80,000 years for *Pioneer* to reach a distance as far as the nearest star. Controlled nuclear fission, which powers today's nuclear power plants (Chapters 16 and 17), could accelerate a rocket for several years until it attained a speed high enough to exit the solar system and to reach the nearest star in about 10,000 years. Nuclear *fusion* bombs or hydrogen bombs (Chapter 16) could provide still more energy. According to physicist Freeman Dyson, a spaceship supporting several hundred crew members, with a fuel supply of 300,000 fusion bombs(!), could explode 1 bomb every 3 seconds to maintain an acceleration of 1g (9.8 m/s^2) for 10 days. This would accelerate the spaceship to one-thirtieth of light-speed, enabling it to travel to Alpha Centauri in 130 years.

Much faster trips, well within a human lifetime, are possible using more futuristic technologies. Although these technologies are very demanding, they are based on technology currently being developed for other purposes and do not appear to violate any scientific principles. One possibility would use antimatter (Section 11.4), a material that is not normally found on Earth but that has been created and stored for days at a time in high-energy physics facilities. The coming decades will probably see the production and storage of significant quantities of antimatter. With enough time and enough energy (from the sun, perhaps), it should be possible in the future to create and store as much as a few kilograms of antimatter. When combined with a few tonnes of liquid hydrogen, this would be enough to fuel a spaceship that could reach 10 to 50% of lightspeed and reach Alpha Centauri in 8 to 40 years.

Another concept for accelerating a spaceship to relativistic speeds is the *interstellar ramjet*. Its fuel is the dilute hydrogen gas that fills our galaxy. The spaceship consists of a payload, a nuclear fusion reactor (see Chapters 16 and 17), and a scoop some 100 km in diameter to collect hydrogen atoms in space. This hydrogen fuels the reactor, and the reactor's output (probably high-speed helium atoms) provides the vehicle's thrust. The beauty of this scheme is that the spaceship needn't carry its fuel, and so it can accelerate for as long as desired. An unchanging 1g acceleration creates apparent Earth-like gravity for the inhabitants and takes the spaceship to Alpha Centauri (4 light-years away) in 7 years as measured on Earth. But relativistic time dilation (Chapter 10) would cause the travelers to experience a time duration of only 4 years.

If the 1g acceleration could be maintained, much longer trips would be feasible, because the spaceship would now be moving at nearly lightspeed for nearly the entire trip and time dilation would become significant. For example, it would take only 21 ship-time years to travel to the center of our galaxy, a distance of about 30,000 light-years, whereas the elapsed time on Earth during this trip would be a little more than 30,000 years. Spaceship inhabitants would truly become time-travelers, who could not go home again to the era of their birth.

Spaceships could be pushed with particle or radiation beams sent out from our solar system. For instance, a spaceship carrying light-reflecting sails could be pushed by a powerful light beam sent out from a laser array in our solar system. The lasers could be in orbit around the planet Mercury and use the abundant sunlight there to produce a single combined beam that could travel 40 light-years before spreading significantly. Within this distance, the spaceship could accelerate to nearly lightspeed in a few years and could reach any of the 20 nearest stars within about 17 years ship time and 20 years Earth time.

So travel to the stars appears possible. On the other hand, interstellar travel might run into unforeseen difficulties. For example, collisions of spaceships with pebbles, dust, or gas might be catastrophic.

■ **DIALOGUE 5**   One of the closest stars that is a candidate for having life around it is Tau Ceti, 11 light-years away. At one-thirtieth of lightspeed, how long would a one-way trip to Tau Ceti take? How long would this trip take (as measured on Earth) if the spaceship moved at one-half of lightspeed? Would the crew members experience this same trip time?

■ **DIALOGUE 6**   Have our old radio or TV shows passed over Tau Ceti (Dialogue 5) yet? If we received a message from Tau Ceti today, when would it have been sent? How long might it take them to get here, assuming that the technologies discussed in this Section were feasible?

12.6 WHERE IS EVERYBODY? DO TECHNOLOGICAL CIVILIZATIONS ENDURE?

If interstellar travel is possible, then any civilization having the drive to develop technology in the first place might be expected eventually to travel to other stars or at least send robotic probes. Many SETI specialists believe that such a civilization would, in only a few million years, explore the entire galaxy and colonize large parts of it. If even one such "spacefaring" civilization arose during the past history of our galaxy, it should have reached Earth by now. But as is pointed out in Section 12.7, there is no credible evidence that we have been visited; indeed, the overwhelming scientific consensus is that we have never been visited.

Where is everybody? This question was first asked by the great physicist Enrico Fermi in 1950. Fermi's reasoning was similar to ours: It is plausible that technological civilizations exist, and if they do exist, they should be here by now. But they are not here. So where is everybody? Before reading on, you might want to stop and speculate about that.

One suggestion is that the pessimistic estimates are correct and that technology has arisen only here. This is certainly possible. But it does seem wondrously strange that we should be the proud possessor of the only technology in our galaxy or in the universe. It certainly seems to violate the Copernican attitude underlying modern science.

Perhaps civilizations tire of technology, in the way that a child outgrows its toys, and enter a less adventurous phase that is not concerned with interstellar communication.

Perhaps aliens have long known about developments on Earth and are refraining from communicating because they do not want to disturb this experiment in intelligence.

Enrico Fermi was a frequent visitor to Los Alamos. . . . A lunchtime conversation took place in 1950 when someone brought up the question of flying saucers. . . . They all agreed that flying saucers were not real alien spacecraft. . . . Fermi said "Don't you wonder where everybody is?" . . . He followed up with a series of calculations. . . . He concluded that we ought to have been visited long ago and many times over. . . . He went on to conclude that the reason we hadn't been visited might be that interstellar travel is impossible, or if possible, it is always judged not to be worth the effort, or technological civilization doesn't last long enough for it to happen.

"FERMI'S QUESTION," IN *EXTRATERRESTRIAL CIVILIZATION*, AMERICAN ASSOCIATION OF PHYSICS TEACHERS

5. $30 \times 11 = 330$ years; $2 \times 11 = 22$ years. The crew members would experience a shorter trip time, owing to relativistic time dilation.

6. Yes. 11 years ago. If they could travel at close to lightspeed, it would take them just a little over 11 years (Earth time) to get here.

[Civilization] is a highly complicated invention which has probably been made only once. If it perished it might never be made again. . . . But it is a poor thing. And if it is to be improved there is no hope save in science.

J. B. S. HALDANE, BIOLOGIST AND PHILOSOPHER

Can a society in which thought and technique are scientific persist for a long period, as, for example, ancient Egypt persisted, or does it necessarily contain within itself forces which must bring either decay or explosion?

BERTRAND RUSSELL, PHILOSOPHER, IN 1949

If this [humankind's extinction] happens I venture to hope that we shall not have destroyed the rat, an animal of considerable enterprise which stands as good a chance as any . . . of evolving towards intelligence.

J. B. S. HALDANE

There are worlds on which life has never arisen. There are worlds that have been charred and ruined by cosmic catastrophes. We are fortunate. We are alive. We are powerful; the welfare of our civilization and our species is in our hands.

CARL SAGAN

Perhaps interstellar travel is far less feasible than it seems. If so and if many technological civilizations exist, the radio searches now under way should be successful before long.

Many astronomers and SETI specialists feel that the most reasonable explanation is the **short-lifetime hypothesis**: *Perhaps technological civilizations do not survive for long because overindulgence in their own technology destroys their capacity to continue functioning as a technological civilization.* This explanation is more plausible than many of the others because it is suggested by our own experience on Earth and because it is consistent with the Copernican attitude that we are not unique. Unfortunately, it leads to the pessimistic conclusion that we, too, could have a short lifetime.

To make the short-lifetime hypothesis more concrete, let's return to our estimate of the number of places in which life arose: between 20 million and 20 billion. Let's also assume, somewhat optimistically, that technological civilizations developed at between a thousand and a million of these places. Now we take note of an additional point: Our galaxy is about 11 billion years old.

H O W   D O   W E   K N O W ?

The age of our galaxy is determined from the age of its oldest stars. To determine the age of a star, we must first determine its mass. For binary-star systems (two stars orbiting each other), this can be done by observing the gravitational effect that each star has on the motion of the other. Since 50% of all stars are binary, this method can determine the mass of about half the stars. Once the mass is known, the established theory of stellar evolution (Section 5.3) can predict the star's total lifetime. As explained in Section 12.1, lower-mass stars have longer lifetimes. Astronomers have observed stars in our galaxy whose mass is so low that they should have lifetimes as long as 11 billion years. As such a star begins to exhaust its fuel at the end of its life, it expands and cools and glows redder than its normal white or blue. Observation of the color of these low-mass stars shows that some of them, in our galaxy, are near the end of their life. These stars must be 11 billion years old. Since no older stars have been observed, we conclude that this is the age of our galaxy.

So stars formed, burned, and died in our galaxy for about 6 billion years before our sun was created. If there have been thousands of technological civilizations, most of them should have become technological long ago. It is probable that ours, a mere century old, would be the galaxy's most recent newly born technological civilization. Calculation* shows that if a thousand to a million such civilizations have cropped up at various times during the entire 11-billion-year history of our galaxy, on the average one such civilization would start up every 15,000 to 15 million years. This means that the most recent technological civilization preceding ours would have started some 15,000 to 15 million years ago!

These numbers, plus our own experiences with technology on Earth, give concrete meaning to the short-lifetime hypothesis. Unless the typical technologi-

---

*$(11 \times 10^9 \text{ years})/(10^3 \text{ to } 10^6 \text{ tech civs}) = 11,000$ to 11 million years.

We are changing our planet faster than our ability to understand or undo many of the critical changes. . . . The unprecedented rates and nature of these changes make it difficult to predict their consequences. . . . As scientists, we have to become more active in addressing environmental issues . . . [such as] the exponential growth of the human population, our unsustainable use of resources and generation of wastes, increasing poverty, inequity, and social injustice. . . . We must bring the full power of science to bear in solving these problems. . . . All too often, our nation's policies do not take advantage of the current scientific understanding.

To have arrived on this Earth as the product of a biological accident, only to depart through human arrogance, would be the ultimate irony.

Our obligation to survive is owed not just to ourselves but also to that Cosmos, ancient and vast, from which we spring.

cal civilization survives through some 15,000 to 15 million years of technology, we would expect no other such civilizations still surviving in our galaxy, even assuming some rather optimistic numbers about the number of technological civilizations that have arisen. Do technological civilizations generally last this long? Will ours? It's a good question.*

Consider the only evidence we have, our own experience with technology. Our most vigorous technological efforts have gone into weaponry. For decades humans have threatened themselves with intercontinental nuclear weapons that could have bombed the world back into a stone age. We have emerged, at least for now, from this threat, but smaller nuclear threats are now even more plausible than they were. Using conventional high-tech and low-tech weapons, humans continue the organized killing and brutalization of one another all over the world, both within and between nations.

But today the threat of war pales beside more insidious technology-related environmental problems. Driving all the other problems is the exponential growth of our species (Section 7.8). We are now 6 billion strong, and it is by no means clear that Earth can support us. Yet our numbers will probably exceed 10 billion during the twenty-first century.

Other known problems include poverty, illiteracy, deforestation, animal extinctions, large-scale plant extinctions for the first time in Earth's history, global warming (Section 9.7), ozone depletion (Section 9.6), abuse of legal and illegal drugs, resource depletion, urban decay, air pollution, water pollution, solid-waste disposal, famine, desertification, subtle new epidemics, the gap between rich and poor nations, groundwater depletion, other water shortages, loss of soil fertility, overfishing, toxic algal blooms, weakened immune systems due to pollution, rising sea levels, new drug-resistant bacterial strains from legal drug overuse, and the problem-multiplier effect of an expected 10-fold increase in global economic activity to meet the demands and aspirations of our exploding population. The unknown problems are anybody's guess.

Any one of these problems could prevent our development as a mature technological civilization. The combined long-term effect of all of them is difficult to predict. These problems are subtle and pervasive. They are caused by the small and apparently inconsequential daily actions of each one of us. Few of us pause to consider the global effects of, say, our own driving or air conditioning.

One root of the problem is continued reliance on age-old instincts in an age of powerful technologies. For example, because we have enthusiastically accepted technology's ability to control disease while doing little to control our instinctive desire to have children, we have the population explosion. This is but one example of our willingness to accept the easy fruits of technology without accepting the hard choices and new thinking that must go with it if we are to survive. Intelligence-based technology demands that we apply our intelligence, rather than our instincts, to the uses of technology. It is difficult for our entire species to make such a cultural change during the short, explosive birth of the technological age. It is the fundamental contradiction of the technological age. A similar conflict between traditional biological behavior and the sudden new

*Astrophysicist Richard J. Gott combined the general Copernican idea that "our time and our place are not special in the universe" with the fact that humans have been on Earth for only a brief portion of Earth's history to show that the mathematical probability of an intelligent civilization achieving a long lifetime is low. See *Nature*, May 27, 1993, pp. 315–319.

requirements of technology could have caused the demise of most technological civilizations.

If the short-lifetime hypothesis is correct, then humankind is faced with a challenging project, one of cosmological proportions: the challenge of becoming one of the breakthrough civilizations, a wise civilization that knows better than to use technology self-indulgently.

Human beings have been described as natural systems having the curious property of self-awareness. We are perhaps among nature's most advanced attempts to become aware of itself. Will we meet the challenge of extending that awareness for millions of years following the explosive recent birth of technology on our planet? This question is being decided as you read these words.

12.7 Have We Been Visited? UFOs and Pseudoscience

Pseudoscience, the dogmatic belief in an appealing idea that purports to be scientific but that is not supported by scientific methods, comes in many guises: parapsychology, spoon bending, the Bermuda Triangle, dowsing, levitation, occult chemistry, psychokinesis, pyramid power, crystal power, creationism, fortune-telling, ancient astronauts, flying saucers, dianetics, astrology, Velikovsky's colliding worlds, poltergeists, orgone boxes, and so forth. Although some of these supposed phenomena might be possible (i.e., not entirely ruled out by the evidence) there is no real scientific evidence for any of them, so they are, at the least, outside of science.

It is a significant issue. For example, two-thirds of American adults read astrology reports periodically; 26 million read them often; 66 million believe that astrology is scientific; and 12 million say they sometimes change their plans after reading their horoscope. One and a half centuries after Darwin's *Origin of the Species*, 46% believe that human beings did not develop from earlier species of animals. And 43% believe it likely that some of the reported unidentified flying objects are really space vehicles from other civilizations, while tabloid newspapers bearing such headlines as "I bore an alien's child" continue to prosper. And 42% believe they have been in contact with someone who has died.

More fundamentally, because rational thought is a threat to pseudoscientific beliefs, such beliefs create an anti-intellectual climate that inhibits the free expression of rational thought. Habitually dwelling on pseudoscientific ideas weakens one's own ability to think rationally and thus strikes at the heart of that which makes us human.

Let's look at three typical pseudoscientific beliefs: extraterrestrial visitations, astrology, and creationism.

UFOs are unidentified objects in the sky. Two **UFO beliefs** have gained a following in the popular media: The first is that some UFOs are visitations by aliens today; the second is that aliens visited Earth within the past few thousand years.

The problem with these ideas is not that UFO beliefs themselves are inherently antiscientific. In fact, our preceding discussion shows that visits to Earth by extraterrestrial beings are plausible. The problem, instead, is in the way in which these beliefs are supported. Let's examine the evidence.

There have been thousands of recent reports of sightings of strange lights, strange aircraft, and people being captured by aliens. Upon investigation, these reports fall into three categories. Most have normal explanations: automobile

And yet we cannot in a direct way fight the antiscientific trends. They are symptoms of a malaise lying at deeper levels, and social maladjustments are not so simply treated. A sustained and systematic effort in propagating science as part of modern culture is critically important for the future, for the generations yet to come, rather than for those who are here now.

PETER KAPITZA, PROFESSOR AT THE MOSCOW INSTITUTE FOR PHYSICS AND TECHNOLOGY

SETI is always burdened with the threat of being declared flaky, or fringe, or pseudo science. So you have to be very careful that very qualified, right-thinking people are involved.

FRANK DRAKE, ASTRONOMER AND FOUNDER OF THE FIELD OF EXTRATERRESTRIAL SEARCHES

The difficulty in any investigation of phenomena such as psychic phenomena or UFOs is that it is impossible for science to prove a universal negative. There will be cases which remain unexplained because of lack of data, lack of repeatability, false reporting, wishful thinking, deluded observers, rumors, lies, and fraud. A residue of unexplained cases is not a justification for continuing an investigation after overwhelming evidence has disposed of hypotheses of supernormality, such as beings from outer space or communications from the dead. Unexplained cases can never constitute evidence for any hypothesis.

JOHN A. WHEELER, PHYSICIST

To the best of my knowledge there are no instances out of the hundreds of thousands of UFO reports filed since 1947—not a single one—in which many people independently and reliably report a close encounter with what is clearly an alien spacecraft.

CARL SAGAN

headlights reflected off high-altitude clouds, a flight of luminescent insects, unconventional atmospheric effects, unconventional aircraft, aircraft using searchlights for meteorological observations, aerial refueling operations, the setting planet Venus distorted by the atmosphere. Although these are honest reports, self-deception or "seeing what you want to believe" is common. For example, in 1968, the U.S. Air Force collected 30 UFO reports when a satellite reentered the atmosphere and broke into burning pieces in the night sky. Of these, 57% reported that the objects were flying in formation, implying intelligent control, and 17% claimed that the glowing objects were attached to a black "cigar-shaped" or "rocket-shaped" object, sometimes with glowing windows.

Other reports are hoaxes, often for profit. For example, a 1968 University of Colorado study, headed by physicist Edward Condon, established that many of the classic UFO photos are either fakes or photos of known natural phenomena. Nevertheless, these photos continue to reappear in new UFO publications. Great Britain's widely publicized "crop circle" phenomenon reported around 1990 was caused by pranksters.

Finally, a few UFO reports cannot be explained. In any investigation of unusual phenomena, there will always be cases that remain unexplained because of lack of data, false reporting, self-deception, and so forth. The unexplained UFO reports offer no positive evidence, such as unambiguous photographs or unambiguous sightings by many observers or an artifact (a tool or piece of material) left behind by aliens. Such a residue of unexplained cases, with no positive evidence, is not surprising and offers no support for UFO beliefs.

In fact, the evidence points the other way. The only real evidence we have is negative: Extraterrestrials have not come right out and revealed themselves to us. So if they exist, they prefer to conceal themselves. We have seen that any extraterrestrial civilization that visits Earth is overwhelmingly likely to be millions of years in advance of us. Any such civilization would surely be able to conceal themselves from us if they wanted to. They would not make simple mistakes like flying around in visually observable vehicles. So reports of UFO sightings are inherently implausible.

Furthermore, it is surprising that such beings would want to conceal themselves. It seems more reasonable that they would want to contact and investigate us. At least, this is what human explorers have done when they discovered new cultures.

A common fallacy of many UFO reports is that far from being overly fantastic, they are not nearly fantastic enough to be believable. The reported technologies are always just a little in advance of, or even behind, the current technology on Earth. The aliens are reported to have curiously humanlike features. But there is little reason to expect that alien technologies, or alien bodily features, would resemble ours.

These are some of the reasons that scientists who have thought about this matter overwhelmingly reject the hypothesis that we are being visited.

The second UFO belief, that we have been visited in the past, has even less supporting evidence. Ancient legends of superior beings mean little: Most cultures have had such legends, based on either real humans or stories promoted by the priesthood. Only an ancient legend containing "futuristic" information, such as instructions for an electronic circuit, would be convincing. Also convincing would be an ancient artifact that could not have been made by the ancient civilization, like an electronic microchip or an advanced metallic alloy.

There is a set of enormous geometrical figures in the high plains of Peru, not easily discernible from the ground but discernible from the air. Several books have reaped large profits by portraying these and other similar figures as a mystery that can be explained only by assuming that the figures were constructed by supertechnologists. But scientists visiting these markings found that they were made by simply pushing aside an inch-thick covering of dark stones. The simplest explanation is that they were intended for gods in the sky. Many cultures have believed in gods in the sky. The markings may be a graphical prayer; there is no reason to assume that they represent messages to ancient astronauts.

The idea that ancient astronauts taught early civilizations much of what they knew not only borders on deliberate fraud for profit; it also insults the accomplishments of past civilizations. Humans living many thousands of years ago were quite capable of building the sophisticated structures that archaeologists have discovered. To attribute these accomplishments to mythical visitations from space is dishonorable to both the living and the dead.

Popular UFO mythology illustrates several common features of pseudoscience: mistaken observations attributed to exotic causes, deliberate fraud, the belief in unexplained cases as proof of an exotic hypothesis, and self-deception caused by a desire to believe. A belief in extraterrestrial visitations can offer people excitement and psychological comfort. Believers are easily convinced by dubious sightings and outlandish claims. It is precisely for such reasons that scientists always ask, How do we know? What is the evidence?

Astrology, the belief that events on Earth are influenced by the positions and motions of the planets, began in ancient Babylonia. It seemed reasonable in an era that believed the planets existed for human purposes. Its central belief is that the configuration of the sun, moon, and planets at the moment of a person's birth affects his or her personality or fortune. A simplified form of astrology, based only on the position of the sun, is the mainstay of newspaper astrology columns.

Today, astrology is scientifically implausible, to say the least. The only known physical influences exerted on Earth by the planets are gravitational effects and electromagnetic radiation. It is hard to imagine how these effects at birth could influence our lives. For example, the gravitational effects* exerted by the doctor and nurse and furniture in the delivery room far outweigh the effects of the planets; the walls of the delivery room shield us from many radiations; and the variations in the sun's radiation output (variations that are unrelated to a person's astrological sign) are far larger than the total radiation received from the moon and all the planets added together.

Despite its scientific implausibility, can we find any observational evidence that astrological predictions are correct? A number of researchers have studied this question, some of them using astrological predictions based on horoscopes (charts showing the orientation of the planets at the moment of a person's birth) calculated for thousands of people, and found no evidence that astrology has any predictive power. There is no correlation, even in an average statistical sense, between astrological predictions and people's personalities or lives. If there is any validity to astrology, such correlations should be easy to find.

Even though astrology is incredible theoretically and disproved observationally, more than half of America's teenagers say they "believe in astrology"; newspapers continue their daily astrological predictions; and in the United

---

*The effects referred to here are "tidal effects," by which the moon and sun cause tides in bodies of water on Earth. Tidal effects cause similar, but smaller, distortions in all objects on Earth.

It is our conclusion that creationism . . . is not science. It subordinates evidence to statements based on authority and revelation. . . . Its central hypothesis is not subject to change in light of new data or demonstration of error. Moreover, when the evidence for creationism has been subjected to the tests of the scientific method, it has been found invalid. No body of beliefs that has its origin in doctrinal material rather than scientific observation should be admissible as science in any science course.

NATIONAL ACADEMY OF SCIENCES, COMMITTEE ON SCIENCE AND CREATIONISM

The gospel of evolution is the enemy of the Gospel of Christ. The Gospel of Christ leads to salvation, righteousness, peace, and meaning of life. Evolution's gospel yields materialism, collectivism, anarchism, atheism, and despair in death. Evolutionary thinking dominates our schools today—our news media, our entertainment, our politics, our entire lives. But evolution is false and absurd scientifically! How long will Christian people and churches remain ignorant and apathetic concerning it?

HENRY M. MORRIS, PRESIDENT OF THE INSTITUTE FOR CREATION RESEARCH, SAN DIEGO, IN *THE TROUBLED WATERS OF EVOLUTION*

My deeply held belief is that if a god of anything like the traditional sort exists, our curiosity and intelligence is provided by such a God. We would be unappreciative of the gift if we suppressed our passion to explore the universe and ourselves. On the other hand, if such a traditional god does not exist, our curiosity and our intelligence are the essential tools for managing our survival. In either case, the enterprise of knowledge is consistent with both science and religion, and is essential for the welfare of the human species.

CARL SAGAN

States there are 20,000 professional astrologers, compared with only 2000 astronomers. Will humankind outgrow its most harmful instincts and develop into a mature culture that is able to control its own technology? Figures such as these give little cause for optimism.

Creationism is the belief that the Bible's Old Testament can be read literally, as scientific and historical truth, and that Earth and the main types of biological organisms, including humans, all were created separately and at roughly the same time, just a few thousand years ago. In the United States, creationism is perhaps the most important of the pseudoscience issues because it is believed by so many people, it has religious implications that are important to many people, and it affects the way that science is taught in the public schools. Like astrology, creationism was credible until a few centuries ago, and many scientists believed it. But today it conflicts with the observations and principles of astronomy, physics, chemistry, geology, biology, paleontology, and archaeology. There is a broad scientific consensus, supported by evidence from many sciences, that Earth is billions of years old, that humankind is millions of years old, and that humans are related through biological evolution to the other animals. Scientific ideas are never certain, however, and honest doubts about established theories should not be arbitrarily dismissed. But so far the creationist arguments have found essentially no scientific support.

In physics, the strongest evidence against creationism comes from radioactive dating and other methods of determining the ages of things. We will discuss some of this evidence in Chapter 15.

As with other pseudoscientific beliefs, the main danger lies not in creationist beliefs themselves but in the way that they are defended. For example, a leading creationist, Henry M. Morris, stated:

> It should be emphasized that this order is followed not because the scientific data are considered more relevent than Biblical doctrine. To the contrary, it is precisely because Biblical revelation is absolutely authoritative and perspicuous that the scientific facts, rightly interpreted, will give the same testimony as that of Scripture. There is not the slightest possibility that the facts of science can contradict the Bible and, therefore, there is no need to fear that a truly scientific comparison of any aspect of the two models of origins [evolution and creationism] can ever yield a verdict in favor of evolution.*

The assumption here is that the answers are already known from the Bible and that science merely studies nature to verify those answers. This attitude precludes any true dialogue with nature. To scientists, it is a sin of pride to presume to tell nature the answer.

■ **DIALOGUE 7** One common creationist argument is that all the fossils and other objects that appear old were created 5000 years ago in order to make Earth appear old without really being old. Is it possible to disprove this argument? Can this statement be investigated scientifically, or does it fall outside science? Consider the proposition that Earth was created just 1 hour ago and that all the fossils and other objects, including your memory, were created at that time, just to make everything seem old. Is it possible to disprove this proposition?

---

*Henry M. Morris, *Scientific Creationism* (San Diego: Creation-Life Publishers, 1974), pp. 15–16.

7. No, it is not possible to disprove this statement. Since observations cannot disprove it, it is not a scientific statement at all; it falls outside science.

"HOWEVER, IT'S EXCELLENT PSEUDOSCIENCE."

SUMMARY OF IDEAS AND TERMS

SETI   The search for extraterrestrial intelligence. Usually conducted with **radio telescopes**, receivers that detect radio waves from space.

The Copernican viewpoint   The view that Earth is not a unique place in the universe, that the same principles of nature apply everywhere. Since intelligent life occurred here, this view argues that it also should have occurred elsewhere.

Analysis   Breaking a question into parts.

Protocells   Nonliving precursors of single-celled organisms.

The hypothesis of a chemical origin of life   Life originated on Earth by means of chemical processes, beginning with the action of various energy sources on the prelife atmosphere to create the organic chemicals that are the building blocks of life.

Biologically convergent property   A biological feature that, because of its survival value, tends to occur again and again during biological evolution.

Technological life   Any form of life having a sufficiently well developed technology to be capable of sending out radio signals.

Short-lifetime hypothesis   Technological civilizations do not survive for long because overindulgence destroys their capacity to continue being technological. This is a possible answer to the question, Why haven't aliens visited Earth by now?

Pseudoscience The dogmatic belief in an appealing idea that purports to be scientific but that is not supported by scientific methods.

UFO Unidentified object in the sky. Two popular **UFO beliefs** are that (1) UFOs are visitations by aliens today and (2) aliens visited Earth within the past few thousand years. There is no evidence to support either belief. Scientists overwhelmingly reject them as pseudoscientific and false.

Astrology The belief that events on Earth are influenced by the positions and motions of the planets. Scientists overwhelmingly reject astrology as pseudoscientific and false.

Creationism The belief that the Bible's Old Testament can be read literally as scientific and historical truth and that Earth and the biological organisms, including humans, were created separately just a few thousand years ago. Scientists overwhelmingly reject creationism as pseudoscientific and false.

REVIEW QUESTIONS

ARE THERE OTHER GOOD PLACES?

1. This chapter analyzes the question of alien technological life by breaking it into four questions. What are these questions?
2. Is a multiple-star system likely to support life? Explain.
3. Is it likely that there are planets around other stars? What is the evidence?
4. Are there likely to be Earth-like planets orbiting very massive stars? Explain. What about very low-mass stars?
5. Which one or two of the following are the most plausible estimates of the number of Earth-like planets in our galaxy: billions, millions, thousands, less than one thousand, one? Or is there too little evidence to decide?

DOES LIFE DEVELOP ON GOOD PLANETS?

6. Describe an experiment, and its results, to determine what happens when a typical primitive Earth environment is subjected to typical early Earth energy inputs.
7. What are protocells?
8. Describe the most widely accepted (by scientists) hypothesis concerning the origin of life on Earth.
9. Which one or two of the following are the most reasonable estimates of the number of planets in our galaxy where life has emerged: billions, millions, thousands, less than one thousand, one? Or is there is too little evidence to decide?
10. How did Earth's molecular oxygen (O_2) get here?

DOES LIFE DEVELOP INTELLIGENCE OR TECHNOLOGY?

11. What is a biologically convergent property? Give an example.
12. Give one argument for and one against the proposition that intelligence is a biologically convergent property.
13. Which one or two of the following are the most credible estimates of the number of planets in our galaxy where intelligence has emerged: billions, millions, thousands, less than one thousand, one, too little evidence to decide?
14. How did we define *technological life* in this chapter?
15. Which one or two of the following are the most plausible estimates of the number of planets in our galaxy where technology has emerged: billions, millions, thousands, less than one thousand, one, too little evidence to decide?

INTERSTELLAR COMMUNICATION AND TRAVEL

16. What is SETI?
17. Describe one slow way and one fast way for humans to travel to the stars.
18. Give an argument supporting the notion that aliens should have visited here by now. Give at least two possible explanations of why they have not visited here.
19. What is the short-lifetime hypothesis?

UFOS AND PSEUDOSCIENCE

20. What is pseudoscience? List several examples.
21. What are the two popular UFO beliefs? What is the scientific consensus about them, and why?
22. Is the notion that aliens have visited Earth inherently implausible?
23. What is astrology? What is creationism?
24. Why do scientists consider UFO beliefs to be pseudoscientific? Answer the same question for astrology and for creationism.

HOME PROJECT

Former President Ronald Reagan stated in a 1980 newspaper interview that he consulted astrologer Carroll Righter's horoscope column every day, and he stated, "I believe you'll find that 80% of the people in New York's Hall of Fame are Aquarians" (Reagan, born on February 6, 1911, is an Aquarian). Perform an experimental check of sun-sign predictions of this sort by finding the birth dates of 50 persons who are outstanding in a single way. For instance, you might choose 50 musicians, athletes, or hall-of-famers from New York. How many of them were born under each of the 12 astrological signs? Do they fall preferentially into specific signs (for example, are 80% of them Aquarians?), or do your results appear to reflect a random distribution?

FOR DISCUSSION

1. What would be the short- and long-term consequences of (a) aliens arriving at the UN building? (b) Radio signals from a star 10 light-years away, asking for two-way communication? (c) Communication from a civilization claiming to be one of the universe's few long-term technological civilizations, stating that there have been millions of technological civilizations but that only a few survived their own technology for more than a few centuries?
2. In view of the devastation wrought on many of Earth's cultures by contact with more advanced cultures, is it safe for us to (a) deliberately send out signals to possible aliens, (b) allow TV and radio signals to escape from our planet, and (c) search for extraterrestrial life with radio telescopes? How would the results of our listening program be affected if alien civilizations reached the conclusion that either (a) or (b) is not safe?
3. The sun will eventually become a red giant star (Section 5.3). When will this happen? How does this time compare with the roughly 10 million years that it took for apelike creatures to evolve into modern humans? Would

you expect the human species to be recognizable by the time the sun turns into a red giant?
4. Suppose that an earthquake occurs and that the following day five people report dreaming about an earthquake just the night before. Is this evidence that dreams can foretell catastrophic events? About how many dreams does the U.S. population have every night? Is it likely that five of them would be about earthquakes?
5. It has been said that "humankind cannot stand too much reality." Historically, science has caused some mental anguish by confronting cherished beliefs, such as an Earth-centered universe or creationism, with scientific findings. Are scientific ideas ultimately difficult to accept?
6. Beginning in the 1980s, lawsuits have attempted to require that every science course that teaches evolution also teach creationism, on the grounds that students should learn all of the competing theories. What is your opinion, and why? If you disagree, is there any other place in the public school system where it might be appropriate to teach creationism?

EXERCISES

ARE THERE OTHER GOOD PLACES?
1. Would life be likely on a planet associated with a white dwarf, neutron star, or black hole (Section 5.4)? Why?
2. Would life be likely on a planet associated with a double (or binary) star? Explain.

DOES LIFE DEVELOP ON GOOD PLANETS?
3. Suppose that the soil on Mars contained organisms that undergo photosynthesis (Section 2.8). What type of organic molecule would you expect to find in the soil?
4. Name two chemical compounds that a planet's atmosphere or surface would need in order for the planet to support organisms that undergo photosynthesis.

DOES LIFE DEVELOP INTELLIGENCE OR TECHNOLOGY?
5. Redo this chapter's calculation of the number of places supporting life, supporting intelligent life, and supporting technological life, substituting numbers that you think are most reasonable.
6. On which of the following five questions is there a rough consensus: How many good planets are there for life in our galaxy? On how many of these did life emerge? On how many of these did intelligence emerge? On how many of these did technology emerge? On how many of these does technology still survive?

INTERSTELLAR COMMUNICATION AND TRAVEL
7. If we listen at many frequencies and pick up no artificial signals, does that prove that life has not evolved elsewhere? Explain.
8. If we listen at many frequencies and pick up no artificial signals, does that prove that technological life has not evolved elsewhere? Explain.
9. In traveling from Earth to the star Alpha Centauri, you first accelerate to half of lightspeed, then travel most of the distance at this speed, and then slow down at your destination. Which phases of the trip require rocket power? What is the direction of the rocket's thrust during any powered phases?
10. You make a fast trip from Earth to the star Alpha Centauri and back, by first accelerating to a high speed, then traveling there and back at that speed while executing a 180-degree turn near Alpha Centauri, and then slowing down as you approach Earth. Which phases of the trip require rocket power? What is the direction of the rocket's thrust during any powered phases?
11. Referring to Exercise 9, during which phases of the trip would you experience apparent gravitational forces (Chapter 5)? What would be the direction of the apparent gravitational force (i.e., which direction would feel like "downward")?

12. Referring to Exercise 10, during which phases of the trip would you experience apparent gravitational forces (Chapter 5)? What would be the direction of the apparent gravitational force (i.e., which direction would feel like "downward")?

PSEUDOSCIENCE

13. Jeane Dixon, who also claims to have forecast John F. Kennedy's assassination, once claimed that an incredible vision informed her that aliens from another planet in our solar system would visit Earth the following August and announce their arrival to the entire Earth. She claimed that this other planet lies directly on the other side of the sun, which is why we have never seen it. Give one good scientific argument against the existence of any such planet.

14. Continuing Exercise 13: One answer is that any such planet should have a gravitational effect on the other planets and that this effect has not been observed. Suppose that Jeane Dixon then replied, "But these aliens are so advanced that they have been able to completely mask the effects of their planet's gravity, as well as all other observable effects of their planet." What is your response to this explanation? Does this supposed planet fall within the realm of science?

15. Continuing Exercise 14: Jeane Dixon's forecast was published on the front page of the *National Enquirer* on September 14, 1976. Have you heard of any reports, the following August, that her forecast was correct? Do you suppose that *National Enquirer* then printed a front-page story reporting that her forecast was wrong? Can you recall any instance when such forecasts were later reported as false when they turned out to be false?

16. Some supporters of ESP (extrasensory perception, or mind reading, causing objects such as spoons to move by means of mental concentration and the like) claim that ESP really exists but that it cannot be checked scientifically because scientific experiments always cause the ESP effect to vanish. What is your response to this argument? As far as science is concerned, then, would ESP exist or not?

PROBLEMS

DOES LIFE EVOLVE ON GOOD PLANETS?

1. Suppose we make the following pessimistic (conservative) assumptions of the likelihood of life in our 400-billion-star galaxy: 0.1% of the stars are sunlike with planetary systems, 1% of these planetary systems have one Earth-like planet, and life emerged on 1% of these Earth-like planets. In this case, on how many planets in our galaxy would life have emerged? What overall fraction of the stars would have life on one if its planets? How many stars would need to be surveyed at random in order to have a good chance of hitting upon one that had a planet with life?

2. Suppose we make the following optimistic assumptions about the chances of technology in our 400-billion-star galaxy: 1 star in 20 has an Earth-like planet, life arises on half of these, intelligent life arises on one-tenth of these, and technology develops on one-tenth of these. Then at how many places would technology arise in our galaxy? What overall fraction of the stars would then give rise to technological life on one of its planets? How many stars would need to be surveyed at random in order to have a good chance of hitting upon one that had a planet where technology develops?

INTERSTELLAR COMMUNICATION AND TRAVEL

3. There are about 20 stars within 12 light-years from Earth. Of these 20, one of the stronger candidates for having a planet with life is Tau Ceti, 11 light-years away. How does this distance compare with the distance to our sun, 8 light-*minutes* away?

4. There are about 400 billion stars in our galaxy. On the assumption that there are 10,000 technological civilizations in our galaxy today, roughly how many stars would astronomers have to scan to have a reasonable chance of finding one with technological life?

5. Velma makes a round-trip to the star Tau Ceti, 11 light-years away. If her average speed is $0.5c$, how many Earth years does the round-trip take? How does time dilation (Chapter 10) affect the duration of the trip as measured by Velma: Does she measure more years, fewer years, or the same number of years as is measured from Earth? Using Figure 10.11, calculate Velma's amount of biological aging during the trip.

6. Referring to Problem 5, suppose that Velma makes the round-trip to Tau Ceti at an average speed of $0.99c$. Approximately how long does the trip take, as measured from Earth? Using Table 10.1, calculate Velma's amount of biological aging during the trip.

7. The nearest star beyond the sun is Alpha Centauri, 4 light-years away. How fast would you have to travel in order to get there in 5 years of Earth time?

8. Referring to the previous question and also to Figure 10.11, how many years would you measure within your spaceship during a trip to Alpha Centauri that took 5 years of Earth time?

CHAPTER 13 Quantum Theory

O amazement of things—even the least particle!

WALT WHITMAN

This and the next chapter discuss science's most accurate and complete description of physical reality: quantum theory. Originating during 1900 to 1930 and still under active development today, quantum theory is the set of ideas that scientists use to study the microscopic world. Its central notion is that nature is discontinuous or "quantized" at the microscopic level, in contrast to the Newtonian assumption that nature is smooth or continuous at all levels. Quantization represents a radical break with Newtonian physics, leading to fundamental new developments not only in physics but also in the philosophical and cultural impact of physics.

We set the stage in Section 13.1 with a broad description of the general nature, aims, and cultural role of quantum theory. Section 13.2 uses a specific experiment, the photoelectric effect, to introduce quantization, at least as it applies to radiation. Quantization leads to some paradoxical questions about just what radiation really is, questions that are first raised in Section 13.3 and that will follow us all the way through these two chapters. Section 13.4 uses another specific experiment, the electron interference effect, to extend quantization to the study of matter. We find that matter at the microscopic level behaves a lot like radiation, that the concepts of quantum theory apply in much the same way to both matter and radiation. So questions arise about what matter really is, similar to the questions raised about radiation. We discuss these questions in Section 13.5. Section 13.6 introduces the theory's central concept, the psi field. As its name indicates, this is a field, similar in some ways to the electromagnetic or gravitational field. Section 13.7 summarizes what we have found out so far, as well as some perplexing questions about radiation, matter, and the psi field.

13.1 THE POST-NEWTONIAN REVOLUTION

In 1900 the quantum idea slipped nearly unnoticed into physics. Although nobody realized it at the time, it was the dawn of the post-Newtonian era.

It is remarkable that the other big post-Newtonian idea, Einstein's relativity, was announced only 5 years later, in 1905. As we saw in Chapter 10, relativity altered Newtonian physics in some fundamental ways: It implied different spaces and times for differently moving observers. Matter was not an indestructible substance but was instead made of energy, motion, and fields. Following three centuries of clockwork determinism and impersonal objectivity, the universe now looked far less like a clock, and reality was somewhat dependent on the observer.

Einstein's special theory of relativity was fairly complete when first announced in 1905, and its revolutionary nature was already clear. Today, the special theory is accepted by all physicists and is not being actively evaluated.

Quantum theory is the detailed description of the behavior of matter and radiation and, in particular, of their behavior at the microscopic level. It started quietly in 1900 and developed slowly, but its impact ultimately went far beyond special relativity's impact, and today it is far from being a closed book. Although the theory's main principles had appeared by 1930, and despite the theory's wide testing and application, it is still not clear what the theory really means.

Because it predicts such a wide variety of phenomena so accurately, quantum theory is probably the most successful scientific theory ever invented. Its practical impact extends to every device or idea that depends on the details of the microscopic world: electronic devices such as transistors, silicon chips, and integrated circuits, and so all the information and communication technologies such as television and computers; most of modern chemistry and biology; lasers; our understanding of different types of matter ranging from superconductors to neutron stars; and nuclear physics, nuclear power, and nuclear weapons. Central to the entire high-tech world is an elusive and highly "non-Newtonian" particle: the electron.

Anyone who has not been shocked by quantum physics has not understood it.

NIELS BOHR

Perhaps more significant but certainly less appreciated is quantum theory's philosophical impact. Quantum physics represents a more radical undoing of the Newtonian worldview than does relativity theory. We have emphasized throughout this book that a scientific worldview is by no means a trivial academic matter. Newtonianism is woven subtly into the fabric of Western civilization. The mechanical worldview has dominated Western culture for centuries, has been assimilated so deeply that it is accepted without even realizing that it is a particular worldview.

We will discover that contrary to the Newtonian worldview, quantum physics implies that randomness, or chance, is built into nature at the microscopic level: Nature doesn't know what she will do next! No longer can the universe be a predictable machine.

Also contrary to the Newtonian worldview, quantum physics implies that nature is deeply connected, that such parts of nature as electrons, protons, and light waves cannot be separated from their surroundings without fundamentally altering their character. No longer can the universe be viewed as a machine at all, even an unpredictable one, for the most basic feature of the machine metaphor has always been its separable parts.

Quantum physics holds that changes in nature occur discontinuously, rather than continuously as Newtonian physics predicts. Here is an example: Imagine swinging in a child's swing, and "dying down" (without pumping the swing) to smaller and smaller oscillations. The process of dying down is continuous, gradual, and this continuous process is exactly what Newtonian physics predicts. It would be surprising if, without pumping, you maintained an amplitude (width of oscillation) of say 4 m for several oscillations and then instantaneously "jumped" to an amplitude of only 2 m, where you remained for several more oscillations without pumping, after which your swing suddenly stopped. Such a discontinuous process is not predicted by Newtonian physics, and it is not observed in the macroscopic world around us.

Yet such discontinuous non-Newtonian processes are the norm at the microscopic level. For example, nature requires an atom to vibrate at only certain precise energies, just as our imaginary swing could oscillate only at amplitudes of 4 m, 2 m, or 0 m. When an atom loses energy, it must do so in sudden jumps from one of its "allowed" energies to a lower one. In doing so, it must release an instantaneous burst or quantity or *quantum* of energy. This is the central new feature of the theory, and it is the origin of the term *quantum theory*.

The quantum revolution challenges not only the Newtonian assumptions but even more general scientific assumptions. An attitude underlying science itself has been the notion that there is an independent reality "out there" in nature, a reality that would be essentially unchanged even if we did not observe it. Scientists have always assumed that they were learning about this independent reality. But quantum physics describes a microscopic world so delicate that it is altered significantly by the mere act of observation, even by observations that could not be said to physically disturb the measured object, and even by the mere possibility of observation. This challenges the notion of an independent and knowable microscopic reality, suggesting instead a participatory reality that includes macroscopic observers.

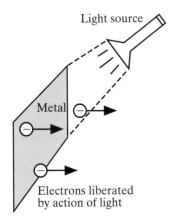

Light source

Metal

Electrons liberated
by action of light

FIGURE 13.1
The photoelectric effect.

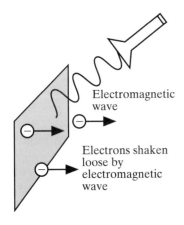

Electromagnetic
wave

Electrons shaken
loose by
electromagnetic
wave

FIGURE 13.2
The photoelectric effect as described
by the electromagnetic wave theory
of radiation.

13.2 QUANTIZATION: THE PHOTOELECTRIC EFFECT

Certain experimental observations made in the late nineteenth century resisted traditional Newtonian explanations. We have already discussed Michelson and Morley's experiment (Section 10.5), which was explained only by Einstein's non-Newtonian proposals concerning light beams, time, and space.

Other phenomena indicated that new ideas were needed to understand the microscopic world. The photoelectric effect was one such phenomenon.

H O W   D O   W E   K N O W ?

The photoelectric effect

Around 1890, scientists began noticing that when light and other electromagnetic radiations shine on a metal surface, they can eject electrons from the surface (Figure 13.1). This **photoelectric effect** is the basis of many practical devices, most notably the *photovoltaic cells* that can help resolve society's energy problems (Chapter 17). Here we are interested in the scientific, rather than the technological, impact of this effect.

It isn't difficult to explain the photoelectric effect in general qualitative terms, on the basis of the standard electromagnetic wave theory of radiation (Figure 13.2). According to this theory, light arrives at the metal surface in the form of a broad, continuous wave of electromagnetic field vibrations. The vibrating field exerts a vibrational force on all the charged particles (protons and electrons) in the metal surface. The field shakes these particles harder and harder until the most loosely bound electrons shake loose.

This explanation seems credible until one looks at the observed details. Here are two that are troubling:

1. There is no experimentally observable delay between the arrival of radiation at the surface and the emission of the first electrons. Even if the light is very dim, electrons are still ejected the instant that light strikes the surface.
2. Emission occurs only if the light's frequency is above a certain *threshold frequency.** Light whose frequency is below the threshold ejects no electrons, regardless of the light's intensity (brightness) and regardless of how long one waits. But light whose frequency is above the threshold ejects electrons immediately, even if the light is very dim.

Both features contradict the electromagnetic wave theory of radiation (Section 9.2). An electromagnetic wave arriving at the metal surface would be spread out smoothly over the portion of the surface illuminated by the light. Only a small portion of the wave's energy could be absorbed by any single atom in the surface. For sufficiently dim light, then, the energy absorbed by any single atom should be so small that a considerable time must elapse before an electron could acquire enough energy to shake loose. Calculations based on the electromagnetic wave theory show that in dim light, the delay should be as long as several seconds. Yet no delay is observed, even though a delay as short as a few billionths of a second should be detectable.

There is nothing in the electromagnetic wave theory to suggest the second troubling feature. Since higher-frequency light has higher energy, we might expect that it would eject electrons sooner. But we would expect—and wave-theory calculations predict—that even low-frequency light would eventually transmit enough energy to shake an electron loose. Why the sudden cutoff at a specific frequency?

As an analogy to the first feature, imagine a string of ripples on a pond rolling in to a beach filled with small pebbles. We might expect the ripples to disturb the pebbles a little. But imagine our surprise if as soon as the ripples arrived at the beach, a few pebbles were violently knocked off the beach, landing hundreds of yards away! It is as though all the energy of the ripples were concentrated on just a few pebbles.

As an analogy to the second feature, suppose that the "ripple–pebble" effect occurred only if the ripple frequency were above some threshold frequency. At lower ripple frequencies, no pebbles would be ejected, even though the ripples might be huge (large amplitude), whereas at ripple frequencies above this threshold, pebbles would be violently ejected even by tiny ripples.

This was the paradox that faced anyone wanting to explain the photoelectric effect using the wave theory of radiation.

FIGURE 13.3
Max Planck. His introduction of the formula $E = hf$ at a meeting of the German Physical Society on December 14, 1900, is usually taken as the birthdate of the quantum theory. In Planck's theory, this formula represented the smallest unit of exchange of thermal energy into radiant energy, that is, the smallest amount of microscopic kinetic energy that a microscopic particle could give up in order to produce light.

In 1900, German physicist Max Planck (Figure 13.3) had a new idea concerning vibrating objects. He had been thinking about the emission of radiation by

*This threshold frequency is different for different metals; that is, copper's threshold differs from silver's, and so forth.

vibrating charged particles. According to the traditional theory of radiation (Section 9.2), radiation was a wave emitted by charged objects when they vibrate. For example, the light from a flashlight was supposed to be an electromagnetic wave emitted by vibrating electrons and protons within the atoms of the flashlight bulb's filament.

Planck's study of radiation led him to the hypothesis that every vibrating object's energy was restricted in a way that had no explanation in Newtonian physics. He suggested that a vibrating object's energy must for some reason be chosen from a specific collection of allowed, or possible, values. To describe this restriction, we say that a vibrating object's energy is **quantized**, meaning *restricted to specific allowed quantities.*

It is an odd hypothesis. Why should a vibrating object be restricted to just a particular set of allowed energies? Once you inject a really new idea into a preexisting scientific theory, there is no telling what the ramifications might be. Quantization was not much noticed for a few years, but it eventually swept across all of physics.

1905 was a busy year for Einstein: He published his special theory of relativity and his paper on Brownian motion that established the reality of atoms (Section 2.1), and he gave the now accepted explanation of the photoelectric effect by applying Planck's quantization idea.

Einstein reasoned that if a vibrating charge's energy is quantized, it can change its energy only by instantaneously jumping from one allowed energy to another, because it is never allowed to have any intermediate amount of energy. The conservation of energy then implies that the emitted radiation itself must emerge from a vibrating charge as an instantaneous burst of radiation rather than as the long continuous wave that the electromagnetic wave theory predicts.

Einstein concluded that radiation always appears as tiny bundles, tiny particles, but not particles of matter like protons, neutrons, and electrons. These new particles are made of radiation; they are light particles, infrared particles, X-ray particles, and the like. These particles of radiation are called **photons** (photo particles). Photons are different from material particles: They always move at lightspeed; they have a rest mass of zero; and vibrating charged particles create photons.

Einstein pointed out that if a light beam's energy is concentrated in tiny photons rather than spread out across an entire light wave, there is a simple explanation of the photoelectric effect (Figure 13.4): Photons simply rain down on the metal surface and knock out electrons the way machine-gun bullets can hit a concrete wall and knock out pieces of concrete. According to this view, we would expect electrons to come flying out of the metal as soon as the first photons hit.

To explain the second observed feature of the photoelectric effect, the threshold frequency, Einstein turned to a formula that appeared in Planck's quantization hypothesis. Planck had devised a formula specifying the energy spacings, the gaps between the allowed energy values of a vibrating particle. Einstein reasoned that a photon must carry an amount of energy equal to one of these energy spacings, because a vibrating charge would have to suddenly jump downward in energy by at least one energy spacing whenever the charge emitted radiation. Planck's formula, as interpreted by Einstein, is

$$\text{energy of photon} = (6.6 \times 10^{-34}) \times (\text{frequency of radiation})$$
$$E = hf$$

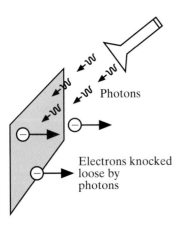

Photons

Electrons knocked loose by photons

FIGURE 13.4
The photoelectric effect as described by the particle theory of radiation.

where h equals 6.6×10^{-34} joule-seconds* and where the frequency f is measured in vibrations/second, or hertz. The very small number, 6.6×10^{-34} joule-seconds, obviously plays an important role here. Planck chose this particular number in order to get his theory to agree with the experimental data. This number came to be called **Planck's constant** and to be symbolized as h. Because h is so small, the energy of a photon is small too. The allowed quantity of energy carried by a photon is called a *quantum of energy*. Think of a quantum of energy as a burst or chunk of energy.

Since Einstein's idea is that radiation is made of photons rather than waves, you might wonder what f, the frequency of the radiation, can mean in the preceding formula. How can individual particles, or photons, have a frequency? It's an excellent question. We will discuss it in the next section.

Planck's formula tells us that a photon's energy is smaller or larger depending on whether the radiation's frequency is smaller or larger. So lower-frequency radiation should be accompanied by lower-energy photons. Suppose that in the photoelectric effect, the light's frequency is so low that individual photon energies are too small to jar loose even the most loosely bound electrons from the metal. Then no electrons at all will be knocked loose, no matter how long one waits and no matter how intense the low-frequency light is. So Einstein's photon idea explains the threshold frequency.

Once Einstein hit on using photons to explain the photoelectric effect, the explanation became surprisingly simple and obvious. Today, science students work through the details as an introductory exercise. But in 1905, scientists were uncomfortable with quantization. Just as in his invention of the theory of relativity, it was Einstein's originality, intellectual courage, physical insight, and search for clean and simple explanations that won the day. His quantitative predictions were beautifully confirmed by experiments more than a decade later, in 1916.

We summarize the ideas of Planck and Einstein as follows:

> ## THE PARTICLE THEORY OF RADIATION
> Electromagnetic radiation is created by vibrating charged particles whose energy is quantized, that is, restricted to certain allowed values. A particle of radiation called a *photon* is created when a charged particle jumps instantaneously from one allowed energy to another lower energy. Each photon has energy hf, where h is Planck's constant and f is the radiation's frequency.

This theory of radiation is analogous to the atomic theory of matter. It says that radiation is grainy in the way that matter is grainy. The grains of matter are called *atoms* and the grains of radiation are called *photons*.

Neither Planck nor Einstein had any idea why a vibrating particle's energy must be quantized. Today we still do not know. For unfathomable reasons, nature requires a vibrating particle to choose its energy from among only particular allowed energies, requiring radiation to be emitted in lumps.

And we still do not know today why Planck's constant has the particular value it has. The small number h plays a role in quantum theory that is analo-

---

*The units joule-seconds need to be attached so that when multiplied by a frequency, the result will be an energy measured in joules.

gous to the role played by the large number c in relativity theory. Both numbers show up throughout their respective theories. Although we don't know why either one has the value it has, we know that the universe would be quite different if either one had a very different value.

■ **DIALOGUE 1**   Light of two colors (two frequencies) shines on a metal surface whose photoelectric threshold frequency is 6.2×10^{14} Hz. The two frequencies are 5×10^{14} Hz (orange) and 7×10^{14} Hz (violet). Will this light eject electrons from this surface? If so, is it the orange light or the violet light or both that does the job?

■ **DIALOGUE 2**   Find the energies of the two types of photons in Dialogue 1. How much energy does a photon need to knock electrons out of this surface?

13.3 RADIATION: WAVES OR PARTICLES? THE DUALITY OF NATURE

Section 13.2 presented evidence supporting a new particle theory of radiation and contradicting the older wave theory. You might suppose, then, that scientists would reject the wave theory and replace it with the particle theory.

But it turns out that nature won't let us reject the wave theory. Recall the evidence (Section 8.3): A wave-interference pattern like that in Figure 8.17 is direct evidence for light waves. Because nobody has found a way to explain such patterns except by assuming that radiation is made of waves, we cannot reject the electromagnetic wave theory. Yet the photoelectric effect requires a particle theory.

So we have two quite different theories of electromagnetic radiation: a wave theory and a particle theory. The wave theory is required by wave-interference experiments but is contradicted by the photoelectric effect, and the particle theory is required by the photoelectric effect but is contradicted by interference experiments. Although both theories are required, they seem mutually inconsistent. It is hard to imagine how radiation could be both waves and particles at the same time.

This paradoxical, wave-particle nature of radiation is evident in Einstein's proposal that light is made of photons whose energy is Planck's constant multiplied by their frequency. How can a photon have a frequency? A frequency is something that a wave has. A photon is not a wave. It simply moves in a straight line at lightspeed, without vibrating or waving in any way. What Einstein meant in his formula $E = hf$ is that f is the frequency that the radiation would have if we were thinking in terms of the old wave theory of radiation. But this doesn't answer the question of how a particle, a photon, can have a frequency.

To study this **wave-particle duality of radiation** in detail, let's look again at the best evidence for light waves: the double-slit interference experiment

---

1. The violet light is above the threshold frequency, so its photons have sufficient energy to eject electrons.
2. $E = hf = (6.6 \times 10^{-34}) \times (5 \times 10^{14}) = 33 \times 10^{-20}$ J (orange), $E = 46 \times 10^{-20}$ J (violet). For the photoelectric effect, a photon must have energy $E = hf = (6.6 \times 10^{-34}) \times (6.2 \times 10^{14}) = 41 \times 10^{-20}$ J.

FIGURE 13.5
The double-slit experiment with light: the experimental setup and the results.

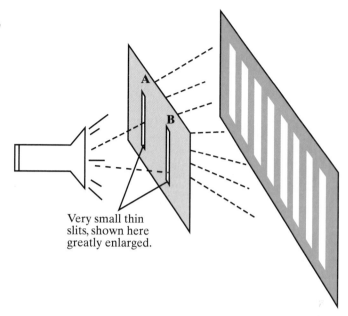

Very small thin slits, shown here greatly enlarged.

with light (Chapter 8). Figure 13.5 reproduces the experimental arrangement. Light from a single source goes through two thin slits and then forms a wave-interference pattern of bright lines and dark spaces on a screen. The bright lines are regions of constructive interference, where wave crests coming from one slit meet crests from the other slit and valleys meet valleys. The dark spaces are regions of destructive interference, where crests from one slit meet valleys from the other slit and cancel each other out.

There seems to be no trace of photons here. But if we perform the experiment using very dim light and record the results on photographic film placed on the receiving screen and exposed very briefly to the light from the slits, a surprising result emerges. Instead of the wave-interference pattern, tiny individual flashes appear randomly all over the screen. Figure 13.6(a) is a drawing that simulates the results of an experiment of this sort. The flashes cannot be explained by the wave theory of light; they are in fact evidence that light is made of photons that create each flash as each photon hits the screen. If we expose the film for a longer time, an interference pattern begins to emerge, formed by the impacts of a large number of individual photons (Figure 13.6(b)). The pattern becomes clearer the longer we wait (Figure 13.6(c)).

FIGURE 13.6
Close inspection shows that an interference pattern is formed by individual photons hitting all over a screen or film. In (a) a film briefly exposed shows 14 photon impacts. In (b) a longer exposure time allows about 150 photons to hit the film. In (c) a few hundred photons have hit.

(a)   (b)   (c)

(a)   **(b)**

(c)   **(d)**

(e)   **(f)**

FIGURE 13.7
Light is made of photons striking
randomly all over the lit surface. Here
a photo emerges from individual
photon impacts. Each photo in the
sequence has a longer exposure time.
The approximate number of photons
in each photo is (a) 3×10^3, (b) 10^4,
(c) 10^5, (d) 8×10^5, (e) 4×10^6,
and (f) 3×10^7.

HOW DO WE KNOW?

Ordinary photos are also made by photon impacts. Usually we think of photography in terms of light waves that come from the photographed object and that are focused onto the film by the camera's lens system. But dim light and a short exposure time allow us to see individual photon impacts. Figure 13.7 shows the photo emerging from individual impacts.

So the patterns made by light waves are due to large numbers of photons. In the double-slit experiment, each photon strikes the screen fairly randomly, the first hitting in one place, the second in quite another place, and so forth. But there is a pattern in this randomness. A photon never hits in the regions that will emerge as dark regions in the interference pattern but instead strikes preferentially in the regions that will emerge as bright lines. The interference pattern is best described as a statistical pattern formed by large numbers of individual impacts. We will discuss the meaning of this in greater detail later.

We do not normally notice that light is made of photons, because each photon carries such a small energy. A light beam striking a surface seems to arrive continuously and seems spread out evenly over the surface, because it contains so many photons that individual ones cannot be detected without special apparatus. Because the energy of one photon is hf, the smallness of each photon's energy is a consequence of the smallness of Planck's constant. If Planck's constant were larger, light would be obviously lumpy.

These experimental results highlight, without resolving, the wave–particle duality of light. What is it that directs an individual photon to strike the screen at a point that will contribute to the appropriate interference pattern while still allowing each photon to strike rather randomly all over the screen? Although we will say more about this question later, we will find that ultimately science does not know the answer.

■ **DIALOGUE 3** Name at least one common technological device whose operation is based on the wave aspect of light and at least one whose operation is based on the particle aspect of light.

■ **DIALOGUE 4 MAKING ESTIMATES** On a clear day, the sunlight striking each square meter of Earth's surface during each second carries 1000 joules of energy. Estimate the number of photons striking 1 square meter during 1 second. Estimate the number of photons striking your hand during 1 second when you hold your hand open to bright sunlight.

---

3. Any device that uses a lens to focus light is based on the wave aspect of light: cameras, eyeglasses, microscopes, telescopes, and the like. Photovoltaic cells are based on the particle aspect of radiation. More generally, devices that convert light to an electric current are based on the photoelectric effect and so on photons: light meters and electric eyes, for example.

4. According to Chapter 9, sunlight is made mainly of radiation in or near the visible region, having a frequency around 10^{15} Hz (see Figure 9.7). The energy of one photon of this radiation is $(6.6 \times 10^{-34}) \times 10^{15}$, or 6.6×10^{-19} joules, or about 10^{-18} joules. To get 1000 joules of energy, we would then need $1000/10^{-18}$ or 10^{21} photons. This is a billion trillion photons in just 1 second. About how large is your hand? I measure my hand to be roughly 9 cm $\times$ 18 cm, or about 200 cm^2. One square meter is 100 $\times$ 100 cm, or 10,000 cm^2, in area. So a hand is about 200/10,000 m^2, or 0.02 m^2, and the number of photons falling on a hand in 1 second is about 0.02×10^{21}, or 2×10^{19}. This is 20,000,000,000,000,000,000 (20 billion billion) photons every second.

13.4 THE ELECTRON WAVE INTERFERENCE EFFECT

As far as we know, everything is made of matter and radiation. The conventional view was that radiation was made of waves, but, as we have just seen, we have found more recently that radiation is in some sense also made of particles.

What about matter? Since the early Greeks, scientists have believed that matter is made of particles. Today we recognize many varieties of material particles: molecules, atoms, protons, neutrons, electrons, and so forth.

Louis de Broglie (pronounced "de Broy"; Figure 13.8), a Ph.D. student at the University of Paris in 1923, felt that there should be a kind of symmetry, or balance, between matter and radiation. He thought it unsymmetric that radiation should exhibit wave–particle duality while matter behaved always as particles. He suggested that if radiation had a dual wave–particle nature, then so should matter. Despite the lack of experimental evidence at that time to support it, de Broglie found this bold idea so beautiful that he included it in his Ph.D. dissertation, submitted in 1924. De Broglie's Ph.D. committee didn't know what to make of it and sent the dissertation to Einstein for his opinion. Einstein was impressed and commented later that "it is a first feeble ray of light on this worst of our physics enigmas." The committee approved de Broglie's dissertation.

Waves of matter? De Broglie's idea was even odder than Planck's and Einstein's particles of radiation. How could individual particles of matter—tiny particles such as electrons and atoms—also be spatially extended waves? Nevertheless, de Broglie pursued his notion. Based on the symmetry that he envisioned between radiation and matter and working from the formula $E = hf$ that connects the wave and particle aspects of radiation, he deduced a formula that predicted the wavelength of the wave that he believed was associated with any material particle:

$$\text{wavelength of material particle} = \frac{\text{Planck's constant}}{(\text{particle's mass}) \times (\text{particle's speed})}$$

$$\lambda = \frac{h}{ms}$$

This formula for waves of matter is analogous to the formula for particles of radiation. Both connect a particle property to a wave property. Planck's constant plays an important role in both formulas. The smallness of h implies that the wavelength λ of a material particle is very small, just as it implies that the energy E of a photon is very small. The smallness of λ means that the wave aspects of matter are difficult to detect, just as the smallness of E means that the particle aspects of radiation are difficult to detect. That is why we normally think of matter as made of particles but of radiation as made of waves.

If we apply de Broglie's formula to a typical macroscopic material object like a 1 kg baseball rolling across the floor at 1 m/s, we get a wavelength of

$$\lambda = \frac{6.6 \times 10^{-34}\,\text{J} \cdot \text{s}}{(1\,\text{kg}) \times (1\,\text{m/s})} = 6.6 \times 10^{-34}\,\text{m}$$

FIGURE 13.8
Louis de Broglie. Feeling that there should be a symmetry between matter and radiation, he predicted that matter should display the same dual wave-particle nature that radiation was known to display.

The baseball's wavelength is about a billionth of a trillionth of a trillionth of a meter. This is far smaller than an atom and far too small to be detected. It is no wonder that we have never noticed the wave aspects of ordinary objects.

The wavelengths of microscopic particles are much longer. Since mass shows up in the denominator of de Broglie's formula, the least massive material particles generally have the largest wavelengths. The least massive material particle known today is the electron.* In experiments involving streams of electrons, such as the experiment described in the next paragraph, electrons typically move at speeds of 10^7 or 10^8 m/s. At these speeds, de Broglie's formula predicts an electron's wavelength to be about 10^{-11} m. Although this is very small, being about one-tenth the size of a typical atom, it is large enough to be detected in careful experiments.

H O W   D O   W E   K N O W ?

De Broglie's hypothesis was confirmed in two independent experiments in 1927. In one of these, British physicist G. P. Thomson passed a beam of electrons through a thin metal foil. The pattern formed by the impacts of large numbers of electrons on a screen behind the foil was a wave-interference pattern that could have been caused only by some unknown kind of wave accompanying the electrons as they passed through the foil. Apparently the gaps between the atoms in the foil were acting like the slits in a wave-interference experiment, causing the waves that passed through different gaps to interfere on the far side of the foil. We call this effect **electron wave interference**.

Although the physical nature of the mysterious waves was not known, it was possible to determine their wavelength by measuring the interference pattern on the screen. The results agreed with de Broglie's formula.

It is interesting that G. P. Thomson, who showed that electrons are waves, was the son of J. J. Thomson (Section 8.6), who showed that they are particles.

We summarize de Broglie's idea as:

THE WAVE THEORY OF MATTER
Every material particle is accompanied by a wave whose wavelength is h/ms, where m is the particle's mass and s is its speed.

As if the wave-particle confusion concerning radiation were not bad enough, de Broglie's idea now presents us with a paradoxical **wave–particle duality for matter**. How can a material particle such as an electron have a wavelength? An electron is not spread out in space, and it doesn't normally vibrate or wave as it moves. Rather, an electron is a single tiny object that moves in a straight line at a constant speed unless acted on by external forces. Let's turn to experiment for guidance.

---

*It is possible that particles known as *neutrinos* are material particles with a rest mass far smaller than the electron's rest mass. Currently, however, neutrinos are believed to be nonmaterial, that is, to have no rest mass at all (Chapter 18).

[The double-slit experiment is] a phenomenon which is impossible, absolutely impossible, to explain in any classical [Newtonian] way, and which has in it the heart of quantum mechanics. In reality, it contains the only mystery. We cannot make the mystery go away by explaining how it works.

RICHARD FEYNMAN

Figure 13.9 shows the experimental arrangement for a two-slit experiment, but using matter instead of light. The arrangement is identical to the two-slit experiment using light, Figure 8.16, except that instead of a light source this experiment uses a source of material particles. We will assume that the experiment uses electrons, although any other material particles such as neutrons, protons, atoms, or molecules could be used and the results would, as far as we know, be the same (although the wavelength would be different).

Figure 13.9 pictures electrons emerging from the source as lots of tiny particles, since that is what we ordinarily suppose electrons to be. Electrons are sprayed all over a partition containing two thin parallel slits. We observe the pattern formed by the electrons as they strike a screen placed behind the partition. The experiment is analogous to shooting machine-gun bullets at a pair of slits in a partition and observing the pattern made by the bullets as they hit a screen behind the partition (Figure 13.10). We will find this analogy with bullets helpful in understanding the experiment with electrons. The difference between the two experiments is that electrons are much smaller than bullets.

Suppose we perform both experiments with only slit A open. Figure 13.11(a) shows the experiment with bullets, viewed from above, looking down at the thin dimension of the slits, while Figure 13.11(b) shows the experiment with electrons. We get just the result that we expect based on the "tiny particle" model of electrons: Just like bullets, each electron makes an individual impact (observable as a tiny flash of light on the screen), while the overall pattern of impacts is spread out over an area behind the slit. We expect such a spread of impacts because of interactions of the bullets or electrons with the sides of the slits, slightly differing initial directions of motion, and other reasons. The results of firing a large number of bullets or electrons are shown in the graphs drawn behind the screens, showing the average number of impacts versus the position on the screen.

Naturally, if we perform the two experiments with only the other slit, slit B, open, we will get the same result, only now the patterns will appear behind B instead of A (Figure 13.12).

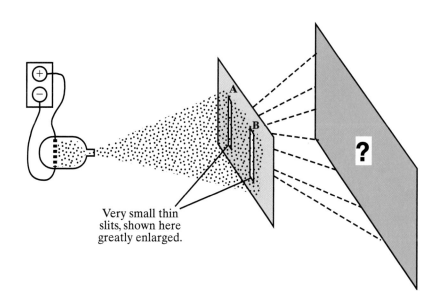

FIGURE 13.9
The double-slit experiment with particles of matter, such as electrons. The electron source is a thin tungsten metal wire that is heated electrically until electrons "boil" off it. A similar electron beam is central to TV picture tubes. In the experiment shown, what will we see on the screen?

Very small thin slits, shown here greatly enlarged.

FIGURE 13.10
The double-slit experiment with machine-gun bullets.

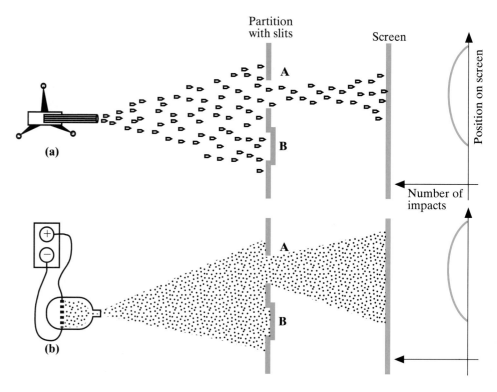

FIGURE 13.11
The double-slit experiment but with only slit A open, using (a) machine-gun bullets and (b) electrons. The average number of impacts at each point on the screen is graphed versus the position on the screen.

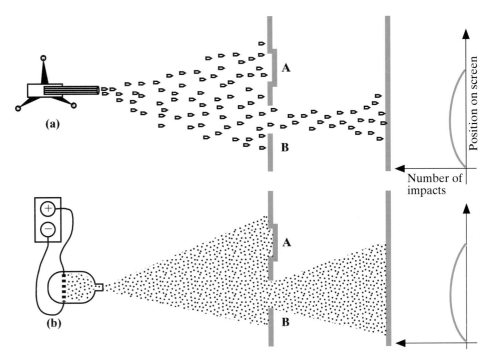

FIGURE 13.12
The double-slit experiment but with only slit B open, using (a) machine-gun bullets and (b) electrons.

What happens with both slits open? It's easy to predict the result of this experiment with bullets. Any bullet that passes through slit A will do precisely what it would do if B were closed, and any bullet that passes through B will do precisely what it would do if A were closed. Note that a complication would arise if bullets passing through one slit could influence the bullets passing through the other slit, for instance by bouncing off of a bullet passing through the other slit. To prevent this, we fire the bullets so infrequently—at say one bullet every other second—that only one bullet is in transit at a time. In this case, the pattern of impacts should be just the two single-slit patterns added together, as shown in Figure 13.13(a). This predicted result is exactly what we get when we actually do this experiment with bullets.

Now we try the two-slit experiment with electrons. As with the bullets, we fire individual electrons one at a time so that they cannot interact with one another. As in the single-slit experiment with electrons, we will observe individual electron impacts on the screen, showing again that electrons are tiny particles. Since it is a tiny particle, far smaller than one slit, it seems that each electron must pass through either slit A or slit B and then do just what an electron did in the single-slit experiments when it passed through one slit. After all, an electron coming through one slit cannot behave differently just because the other slit happens to also be open rather than closed. So we expect that the double-slit impact pattern will be simply the sum of the two previous single-slit patterns, as was the case with bullets.

But as you may have suspected from all this buildup, a most curious result emerges: The overall pattern is a wave-interference pattern (Figure 13.13(b)).

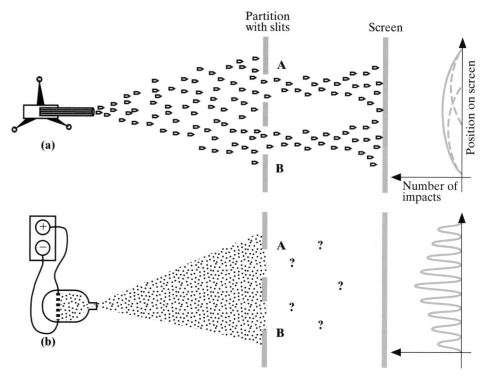

Partition with slits

Screen

Position on screen

(a)

A

B

Number of impacts

(b)

A

?

?

?

?

B

?

FIGURE 13.13
The double-slit experiment with both slits open, using (a) machine-gun bullets and (b) electrons. The impact pattern in (a) is just the sum of the two single-slit impact patterns (shown as the dashed graphs), as expected. But in (b), how can electrons create the wave-interference pattern shown in the graph? What is happening in the region between the slits and the screen?

The figure does not picture the electrons as they move through the space between the partition and the screen because, as we will see, there is a big question about how we should visualize the electrons in this region and even about whether we can visualize them.

H O W   D O   W E   K N O W ?

The double-slit experiment using electrons was performed by Claus Jonsson in 1961, although electron wave interference was demonstrated much earlier by other experiments, such as G. P. Thomson's 1927 experiment. Figure 13.14 is a photograph of the pattern formed on the screen behind the slits. Comparison with a typical interference pattern for light, Figure 8.17, indicates that we are indeed dealing here with wave interference. But Figure 8.17 is made by light, a form of radiation, and Figure 13.14 is made by something quite different: electrons, a form of matter.

An experiment by A. Tonomura and other Japanese physicists, reported in 1989, shows in detail the manner in which the alternating bright and dark lines are formed in a pattern such as that in Figure 13.14. Figure 13.15 shows the gradual buildup of a wave-interference pattern from the impacts of individual elec-

FIGURE 13.14
A wave-interference pattern made by electrons.

FIGURE 13.15
The buildup of an interference pattern in the electron wave-interference experiment by individual impacts of electrons. The five photos use exposures of 0.01 s (when only 10 electrons have hit), 0.1 s (100 electrons), 3 s (3000 electrons), 20 s (20,000 electrons), and 70 s (70,000 electrons).

trons in a double-slit experiment. In the experiment, about 1000 electrons per second passed through the two slits and then hit a fluorescent surface similar to the inside of a TV tube. A time-exposure photograph was made of the results. The top photograph shows this time exposure after about 0.01 seconds when only 10 electrons had hit. The other photos show the time exposure after longer times. The electrons make individual impacts on the screen, confirming their particle nature. The first few electrons appear to arrive randomly. But it becomes apparent by the third photo that electrons are arriving preferentially in regions that turn out to be regions of constructive interference and are avoiding regions of destructive interference. This result is precisely like the buildup, from individual photon impacts, of an interference pattern for radiation; compare Figure 13.15 with Figure 13.6.

■ **DIALOGUE 5**   Arrange these from shortest to longest wavelength, assuming that all six have the same speed: helium atom, automobile, DNA molecule, electron, neutron, baseball.

13.5 MATTER: PARTICLES OR WAVES? THE QUANTUM MYSTERY

Electron wave interference is strong evidence that matter is made of waves. These waves are exploited every day in such devices as the **electron microscope**. Using electromagnetic fields instead of the glass lenses used by visible-light microscopes, electron microscopes bend and focus the waves associated with electrons to form electron images of microscopic phenomena. Since electron wavelengths can be smaller than an individual atom, electron microscopes can form images of atoms (Figure 2.12), something that visible-light microscopes cannot do because visible wavelengths are thousands of times larger than an atom.

Even though matter acts as if it were made of waves, we know from the tiny flashes that material particles make on screens, as well as from such phenomena as Brownian motion (Chapter 2), that matter also acts as if it were made of particles. Like radiation, matter has a paradoxical wave–particle nature.

The similarity between wave–particle duality for radiation and for matter goes fairly deep. Radiation is quantized into discrete (separated) photons; a beam of radiation makes tiny flashes on a screen; individual photons strike a screen in a random pattern that is somehow directed by an underlying wave; and the wave pattern emerges as a consequence of many individual impacts. All of this is true for matter, too, except that matter is quantized into electrons, protons, and other material particles, instead of into photons. Although the remainder of this chapter concentrates mostly on matter, nearly everything that is said here is equally true for radiation.

Let's look closely at the most paradoxical feature of electron wave interference. Figure 13.16 reproduces the graphs of the patterns formed in the two-slit experiment: (a) is the pattern with only slit A open; (b) is with only slit B open;

5. Automobile, baseball, DNA molecule, helium atom, neutron, electron.

FIGURE 13.16
Four impact patterns in the double-slit
experiment. In (a) only slit A is open. In
(b) only slit B is open. We expect (c) with
both slits open, but we actually get (d) when
using electrons or other microscopic
particles.

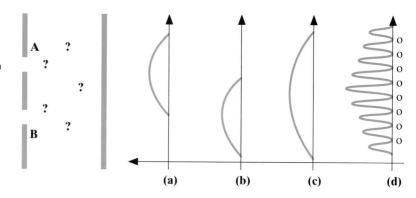

(c) is the pattern we would expect to get with both slits open; and (d) is the wave-interference pattern that we actually get. In graph (d), look at the points of destructive interference, the points marked o in the figure. These are points where no electrons hit the screen when both slits are open. Somehow, each electron "knows" that both slits are open and that it is supposed to avoid the places marked o. How could an individual electron "know" this? An electron is far smaller than either slit. It is hard to imagine how it could come through both slits or "sense" that the other slit is open.

When we close one slit, however, we destroy the interference pattern. Electrons simply pass through the open slit and hit all over the screen, forming patterns (a) or (b), depending on which slit is open. With only one slit open, electrons strike at places like the o's as frequently as at any other points. One expects that opening the other slit should *increase* the number of electron impacts at every point on the screen, yet the mere act of opening slit B somehow informs every electron—no matter which slit it goes through—that it is not to strike at the o's.

We can conclude that when both slits are open and a single electron goes through the apparatus, something goes through both slits. Something associated with each electron is spread out over both slits when both slits are open. Whatever this something is, it interferes in the way that a wave interferes after passing through the two slits. It is as though the electron were interfering with another electron—although there is no other electron. Somehow, an individual electron interferes with *itself* when both slits are open.

This paradoxical phenomenon was predicted by de Broglie in 1923 and was first observed in 1927, yet physicists today are as perplexed by it as ever. If an electron (or a proton, atom, or whatever) is a particle, it must go through just one, and not both, of the two slits in the double-slit experiment. And if it goes through just one slit, it cannot form an interference pattern. But experiment shows that it does form an interference pattern.

Somehow, an electron is more than just a tiny particle during this experiment. As long as both slits are open, an electron acts like a spread-out wave right up until it actually hits the screen. Then when the flash appears on the screen, the electron "collapses" back into a tiny particle. An electron (or any other microscopic particle) seems to be a spread-out, wavy entity when it is not being observed and seems to collapse into a tiny particle every time it is observed.

To try to understand this confused situation, we turn again to experiment. We would like to look more closely at what is actually coming through each of the

All of modern physics is governed by that magnificent and thoroughly confusing discipline called quantum mechanics. . . . It has survived all tests and there is no reason to believe that there is any flaw in it. . . . We all know how to use it and how to apply it to problems; and so we have learned to live with the fact that nobody can understand it.

MURRAY GELL-MANN, PHYSICIST, IN *THE SEARCH FOR UNITY IN PARTICLE PHYSICS* two slits. So we set up a detector behind one or both slits, to see whether something comes through only one slit, or both. Experiments of this sort have been done. In every instance, the electron was observed to come through only one slit, not both!

At first, we might conclude that in the double-slit experiment, the electron really does come through only one slit. But this conclusion would be unwarranted, because in every instance in which we set up a detector of this sort, *the interference pattern is destroyed*, and we get just the simple pattern (c) of Figure 13.16. The mere act of observing the slits destroys the interference pattern! The electron seems to "know" that a detector is present and that it (the electron) should go through only one slit instead of both. Putting the detector behind the slits does not give us the information we seek concerning what comes through the slits in the interference experiment, because the mere presence of the detector destroys the interference pattern.

In summary, an electron behaves like a tiny particle whenever a detector is present that can determine through which slit the electron came, but it behaves like a spread-out wave that goes through both slits whenever there is no detector.

This throws into question a basic premise of science: Science has always assumed that the real world is essentially the same when we are not observing it as it is when we are observing it. But wave–particle duality seems to imply that the real world is essentially different when detectors are absent than when they are present.

■ **DIALOGUE 6**   List several differences between particles and waves. Can you think of any ways in which particles and waves are similar?

13.6 THE PSI FIELD: NATURE DOESN'T KNOW WHAT SHE WILL DO NEXT

A philosopher once said "It is necessary for the very existence of science that the same conditions always produce the same results." Well, they don't!

RICHARD FEYNMAN, PHYSICIST

The formation of an electron wave interference *pattern* is significant. In order for a pattern like that graphed in Figure 13.13(b) to emerge from tiny particles striking a screen, different particles must strike at different places. You might suppose that electrons hit at different places because they started out differently from the electron source. Could we then adjust the source so as to prepare every electron identically and make them all hit the same point on the screen? Experimentally, we find that no matter how finely we control the source, we get the same pattern. Even though electrons are prepared identically, they hit at different points.

What causes different electrons to hit at different places? The answer must be that nothing causes it, because all the electrons are identical and all were prepared identically. This general rule is found throughout the microscopic world. Despite our best efforts to control them, microscopic particles go their own way.

---

6. Some differences: A particle is limited in size, whereas waves are spread out and have no well-defined size; a particle has a well-defined mass, whereas we cannot define the mass of a wave; a particle is a thing or object, whereas a wave is a pattern; and a particle can exist by itself, whereas a wave can exist only in some medium. Some similarities: Both have a speed; both transport energy; and both can transport information.

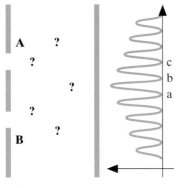

FIGURE 13.17
The pattern represents the statistics of the impacts—the number hitting between points a and b or b and c, for example.

Wave–particle duality implies irreducible uncertainties, called **quantum uncertainties**, in nature.

This is quite a break with deterministic Newtonian physics. A few physicists disagree with the notion that the future is fundamentally uncertain, arguing instead that our current understanding (quantum theory) is simply not deep enough to penetrate the true principles governing the microscopic world and that these true principles would restore predictability to nature. Einstein argued forcefully during the 1930s that "God does not play dice," citing detailed examples to try to show that an irreducible element of uncertainty would be absurd. But despite such arguments, quantum theory as formulated during the 1920s continues to have a perfect record of success in predicting observed phenomena. Predictions that Einstein and others believed to be absurd have now been tested and found actually to occur.

Despite the randomness of individual impacts in the double-slit experiment, the overall pattern *is* predictable. That is, we get the same interference pattern every time we do the experiment. The pattern represents the overall statistics of many impacts. In Figure 13.17, we might predict, for example, that 15% of the impacts will appear between points a and b, that 10% will appear between b and c, and so forth. *The overall statistics are predictable, even though individual impacts are not.*

In 1926, German physicist Max Born (Figure 13.18) was the first to conclude that the interference pattern must be a *probability pattern* for each electron. That is, Born concluded that data such as the graph in Figure 13.17 show the probabilities for a single electron to strike at various points on the screen. Following up on this idea, Born concluded that de Broglie's wave that accompanies each particle is a *wave of probability* or **probability wave** for the presence of the electron.

Probabilities were invented long before quantum theory and usually have nothing to do with quantum theory. Probabilities are useful whenever the outcome of a particular experiment is uncertain, but the overall statistics of many repetitions are predictable. A simple example, having nothing to do with quantum theory, is the flip of a coin. It is common to speak of a 50% probability of heads and a 50% probability of tails for a fairly flipped coin. What this statement means is that in a long series of tosses, roughly 50% will be heads. This probability, 50% or 0.5, can be regarded as a statistic, a number representing the pattern that emerges in many repeated trials of the experiment.

But there is a difference between the probabilities observed in macroscopic experiments such as the flip of a coin and the probabilities referred to in quantum theory. A coin flip obeys Newtonian physics to an excellent approximation, and so the outcome is predictable in principle. That is, with enough information regarding the tension in the flipper's thumb, the height of the coin above the table, the elastic properties of table and coin, and so forth, it is possible to use Newtonian physics to predict the outcome. Our uncertainty about a coin flip arises only from our ignorance of the precise details. But quantum events are not predictable even in principle. Quantum uncertainties don't arise from our ignorance of the details. Rather, the microscopic world is inherently unpredictable: Nature herself doesn't know what she will do next.

A wave of probability is a fairly abstract notion. Although each electron is described by a probability wave, that probability wave cannot be conclusively detected when just one electron goes through the double-slit apparatus, because

I think I can safely say that
nobody understands quantum
mechanics. . . . I am going to tell
you what nature behaves like. If
you will simply admit that maybe
she does behave like this, you will
find her a delightful, entrancing
thing. Do not keep saying to
yourself, if you can possibly avoid
it, "But how can it be like that?"
because you will get down the
drain, into a blind alley from which
nobody has yet escaped. Nobody
knows how it can be like that.

RICHARD FEYNMAN, IN *THE
CHARACTER OF PHYSICAL LAW*

Where did we get that [the
Schroedinger equation] from?
Nowhere. It is not possible to
derive it from anything. It came out
of the mind of Schroedinger.

RICHARD FEYNMAN

the pattern emerges only after many electrons have gone through. The probability wave for a single electron is a description of the possible places in which the single electron might be found if we decided to observe that electron's position. It is a wave of possible locations, a wave of possibilities.

Let's follow the probability wave for just one electron traveling through a double-slit apparatus. Figure 13.19 shows the probability wave at four different times, labeled (a), (b), (c), and (d). According to quantum theory, this single electron's probability wave undergoes the following development: (a) The wave leaves the source; (b) the wave spreads out so as to go through both slits; (c) the wave goes through both slits; and (d) the wave arrives at the screen in an interference pattern formed by the parts of the original wave going through each slit. The probability wave represents the likelihood of the electron's presence. For instance, at time (c), 50% of the probability wave is located just behind slit A, and 50% is located just behind B. The physical meaning of this is that if we observe the electron's position at time (c), we will have a 50% chance of finding the electron just behind A and a 50% chance of finding it just behind B.

The wave-interference pattern observed on the screen can be described by saying that the portions of the probability wave coming through each slit interfere with one another so as to reinforce one another at some places on the screen and cancel out one another at other places. Each individual electron must strike the screen at a point that agrees with this pattern.

The predictability of the statistical patterns shows that probability waves also are predictable. In 1926, Austrian physicist Erwin Schroedinger (Figure 13.20) invented a method of predicting probability waves. Schroedinger began with a well-known formula that had been used to describe waves in other situations not involving quantum theory. Into this wave formula, he inserted de Broglie's relation $\lambda = h/ms$, along with some judicious guesswork. The result was a formula, now called **Schroedinger's equation,*** that predicts the probability wave for electrons or any other material particles in a wide variety of situations. The quantity that does the waving in Schroedinger's equation is usually represented by the Greek letter Ψ (psi). We will call it the **psi field** of the particle to indicate that it is spread out over a region of space, like an electromagnetic field. A particle's psi field is nearly the same thing as its probability wave: The probability of finding a particle at some point is simply the square of the particle's psi field at that point. We often use *psi field* and *probability wave* interchangeably.[†]

Schroedinger's equation correctly describes such phenomena as the double-slit experiment with electrons (Figure 13.19), the single-slit experiment (Figure 13.11(b)), and the interference patterns observed when electrons pass through thin sheets of solid material. Most important historically, Schroedinger showed that his equation could be applied to an electron within an atom and that the predicted results agree with atomic experiments (Chapter 14).

---

*Here it is: $-(h^2/8\pi^2 m)\,\nabla^2\Psi + V\Psi = E\Psi$, in which Ψ (psi) represents the wave and is a so-called probability amplitude rather than a probability. The meaning of this is that actual probabilities are found by squaring Ψ. Schroedinger's equation is a *differential equation* involving a *second derivative*, $\nabla^2\Psi$. In addition, h is Planck's constant, m is the mass of the particle, and E is the numerical value of the particle's total energy. The symbol V represents the way that external forces act to alter the particle's energy. Because V is different in different physical situations, such as the double-slit experiment and within a hydrogen atom, Schroedinger's equation has a different form for each different situation.

[†]The psi field goes by many other names in other books: psi wave, matter wave, probability field, and Dirac field. Einstein called it a ghost field, and physicist/author Nick Herbert calls it quantum-stuff.

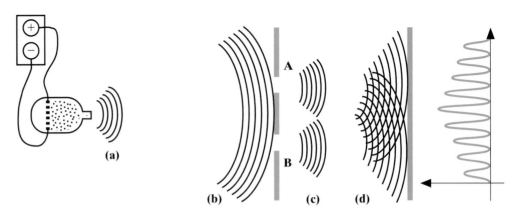

FIGURE 13.19
The probability wave for a single electron at four different instants: (a) as the wave (that is, the electron) leaves the source, (b) as it approaches the slits, (c) just after passing through the slits, and (d) just before arriving at the screen. The graph—of the probability wave at the position of the screen—shows the probabilities that the electron will hit at various positions.

For me, the main purpose of doing experiments is to show people how strange quantum physics is. Most physicists are very naive; most still believe in real waves or particles.
    ANTON ZEILINGER, PHYSICIST

Nature loves to hide.
    HERACLITUS OF EPHASUS,
    500 B.C.

FIGURE 13.20
Erwin Schroedinger. He invented the equation that predicts the statistical pattern, or *psi field*, in a wide variety of situations involving microscopic material particles.

Schroedinger's equation is the quantum equivalent of Newton's law of motion. Like Newton's law, it predicts the behavior of a material particle subjected to external forces. But whereas Newton's law predicts a particle's precise motion, Schroedinger's equation predicts only a particle's psi field and so the probabilities of various *possible* motions.

13.7 THE QUANTUM THEORY OF MATTER AND RADIATION

Let's summarize what we have learned so far:

THE QUANTUM THEORY OF MATTER AND RADIATION

Both matter and radiation are observed to occur as particles such as electrons, protons, and photons. But the behavior of an individual particle is unpredictable, while the statistical pattern formed by many particles, each of them acting separately and independently, is predictable. Thus such a spatially extended pattern must be associated with each individual particle, and the pattern can be interpreted as a probability pattern for a single particle. Such a predictable probability pattern is called a psi field. Since psi fields are wavelike, both matter and radiation have both particle and wave characteristics.

For material particles such as electrons and protons, the psi field can be predicted, in any particular physical situation, from Schroedinger's equation for that particle in that situation. The psi field's wavelength is related to the particle's mass and speed by means of de Broglie's relation, $\lambda = h/ms$, where h is Planck's constant.

For particles of radiation, or photons, the psi field's frequency is related to the photon's energy by Planck's relation, $E = hf$.

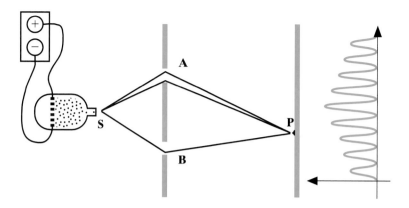

FIGURE 13.21
We must not imagine an electron moving along any particular path. Suppose the electron hits the screen at P. If we imagine the electron doing anything at all between S and P, we must imagine it moving along all the possible paths, such as the three shown here.

My own suspicion is that the universe is not only queerer than we suppose, but queerer than we can suppose. . . . I suspect that there are more things in heaven and Earth than are dreamed of, or can be dreamed of, in any philosophy. That is . . . why I have no philosophy myself, and must be my excuse for dreaming.

J. B. S. HALDANE, PHILOSOPHER AND BIOLOGIST

The discovery of quantum mechanics in the mid-1920s was the most profound revolution in physical theory since the birth of modern physics in the seventeenth century.

STEVEN WEINBERG, PHYSICIST, IN *DREAMS OF A FINAL THEORY*

We have always had a great deal of difficulty understanding the world view that quantum mechanics represents. At least I do, because I'm an old enough man that I haven't got to the point that this stuff is obvious to me. Okay, I still get nervous with it. . . . You know how it always is, every new idea, it takes a generation or two until it becomes obvious that there's no real problem. I cannot define the real problem, therefore I suspect there's no real problem, but I'm not sure there's no real problem.

RICHARD FEYNMAN IN 1982

Psi fields and Schroedinger's equation help clarify and predict nature's microscopic behavior, but they do not resolve the quantum paradoxes. For instance, Schroedinger's equation predicts the probability pattern on the screen, but it doesn't tell us what happens to an electron as it goes through a double-slit apparatus. Which path does it really follow?

To highlight this question, Figure 13.21 shows the double-slit interference experiment once again, with the probability pattern drawn at the position of the screen. Suppose we perform this experiment with just one electron. The electron travels from the source, marked S, through the apparatus to the screen. It may hit anywhere in the probability pattern except at the places of destructive interference. Suppose that it actually hits at the point marked P. By what path did it get from S to P? Any possible path must go through one or the other slit. Figure 13.21 shows three such possible paths. Surely, it seems, the electron followed *some* such path from S to P. And yet we have already seen (Section 13.4) that the commonsense assumption that the electron came either through slit A or slit B implies that there is no interference pattern. The conclusion is that we are not allowed even to *imagine* that the electron moves from S to P along any single path. We are not allowed to imagine such things as "the electron really went through only one of the slits, but since we didn't observe it, we described it with probabilities for going through both slits." If we imagine the electron doing anything at all between S and P, we must imagine it moving along all the possible paths, such as the three shown.

What is really going on at the microscopic level? We will probe this question more deeply in the next chapter.

■ **DIALOGUE 7** List at least two differences between Newtonian physics and quantum physics. Now list two similarities.

---

7. Differences: Newtonian physics is deterministic; it is objective; it assumes that matter is made of particles and that radiation is waves; and it is based on Newton's law of motion. Quantum physics is nondeterministic; it is dependent on the observer; it assumes that matter and radiation show both wavelike and particlelike behavior; and it is based on Schroedinger's equation. Similarities: conservation of energy, second law of thermodynamics, law of inertia, idea of universal natural laws.

"ACTUALLY I STARTED OUT IN QUANTUM MECHANICS, BUT SOMEWHERE ALONG THE WAY I TOOK A WRONG TURN."

SUMMARY OF IDEAS AND TERMS

Quantum theory The currently accepted physical theory of the microscopic world.

Photoelectric effect Light and other radiation shining on a metal surface can eject electrons from the surface, thereby providing evidence that radiation is made of photons.

Quantized Limited to specific allowed quantities.

Photons Particles of radiation. A photon moves at speed c, has zero rest mass, and carries a **quantum** (a specific allowed amount) of energy.

Particle theory of radiation Electromagnetic radiation is created by vibrating charged particles whose energy is quantized. A photon is created when a charged particle jumps instantaneously from one allowed energy to another lower energy. Each photon has energy hf, where f is the radiation's frequency and h is **Planck's constant**, 6.6×10^{-34} joule-seconds.

Wave theory of matter Every material particle is accompanied by a wave having wavelength h/ms, where m is the particle's mass and s is its speed.

Wave–particle duality of radiation and matter Both radiation and matter have wavelike and particlelike aspects.

Electron wave interference The wave interference patterns that are observed in experiments involving electrons. Such patterns occur for example when electrons pass through a thin metal foil, and when electrons pass through a pair of thin slits.

Double-slit interference experiment When photons or material particles are sent through two narrow slits, the distribution of many impacts on a screen forms a wave-interference pattern.

Quantum uncertainties Experiment shows that identical physical situations often lead to different outcomes. Thus, there are irreducible uncertainties in nature itself.

Probability An event's fractional number of occurrences in a long series of trials. Useful in situations involving uncertainties.

Probability wave A probability pattern that moves through space in a wavelike manner. According to quantum theory, every microscopic particle is represented by a probability wave.

Electron microscope Uses electron probability waves to form images of microscopic objects.

Psi field   The psi field for a particle is roughly the same as the particle's probability wave. The psi field is spread out in space. The square of its value at any particular point is the probability for finding the particle at that point. Thus, the particle is more likely to be found where the psi field is larger, and will not be found where the psi field is zero. Quantum theory predicts a particle's psi field.

Schroedinger's equation   Predicts the psi field for material particles.

The quantum theory of matter and radiation   Both matter and radiation occur as particles whose behavior is unpredictable but whose probabilities form predictable spatially extended psi fields, or wave patterns.

REVIEW QUESTIONS

QUANTIZATION

1. What is quantum theory?
2. Describe the photoelectric effect and its two troubling features. What does this effect tell us about radiation?
3. When a vibrating electron emits one photon, does the electron's energy increase, decrease, or remain unchanged? By how much?
4. Where do photons come from? What is their speed? Their rest mass?
5. Name a rather common device that utilizes the photoelectric effect.

RADIATION: WAVES OR PARTICLES?

6. What is a photon? What is the relation between its frequency and its energy?
7. How do we know that radiation is made of waves? How do we know that it is made of particles?
8. If we perform the experiment shown in Figure 13.5 using a light source that emits only one photon per second, what will we observe on the screen? What if we perform it with a million photons per second?
9. Why don't we normally notice that light is made of photons?

MATTER: PARTICLES OR WAVES?

10. Can a single electron have a wavelength? How can you calculate its wavelength?
11. How do we know that material particles are accompanied by waves?
12. What will be observed on the screen in Figure 13.9? In Figure 13.10?
13. If one electron per second is sent through the experiment in Figure 13.13(b), what will we observe? What if we send through a million electrons per second?

14. Which detects the smallest objects: a visible light microscope or an electron microscope? Why? List some differences between the two devices.
15. What experimental arrangement would produce each of the four patterns graphed in Figure 13.16?
16. Of the four patterns shown in Figure 13.16, which one would be created by bullets with only slit A open? By electrons with only slit A open? By bullets with both slits open? By electrons with both slits open but with a detector behind the slits capable of determining through which slit each electron will come? By electrons with both slits open and with no detector?
17. When a single electron travels through a double-slit apparatus, does anything go through both slits? How do we know?
18. What happens when we use a detector to tell us which slit each particle comes through in the double-slit experiment?

THE PSI FIELD

19. What is Schroedinger's equation?
20. What are quantum uncertainties? How do they differ from the ordinary uncertainties in a coin flip?
21. The impact point of each electron is unpredictable in the double-slit experiment. What *is* predictable?
22. Explain in detail the meaning of each part of Figure 13.19 as it applies to a single electron.
23. If an electron traveling through a double-slit apparatus strikes directly behind slit A, is it correct to say that the electron came through slit A?

HOME PROJECTS

1. Observe the metal heating element on an electric stove as it changes color while heating up. As it heats up, what can you say about the number and kind of photons it is emitting? Before it begins to glow, what kind of photons is it emitting?
2. Find the probability pattern for tossing two dice. Record the number of twos, threes, and so on in 100 tosses, and graph the results. According to your experimental results, what is the percentage probability of a seven? A two? Compare this with the theoretical results obtained by assuming that each possible combination of two faces (such as a three on die 1 and a four on die 2) has a probability of one-thirty-sixth. Do these probabilities have anything to do with quantum uncertainties?

FOR DISCUSSION

1. Discuss the similarities and the differences between matter and radiation.
2. Do electrons and other microscopic particles really exist? Are such particles more real than, say, the chair you are sitting on? Less real? Is the "redness" of a red lightbulb more real than the atoms that physicists tell us are emitting the red light? Less real?

EXERCISES

THE PHOTOELECTRIC EFFECT

1. What energy transformation (from what to what) occurs in the photoelectric effect?
2. How can the photoelectric effect be used to open a door when someone approaches?
3. Suppose that the photoelectric threshold frequency in a certain metal lies in the yellow region of the spectrum (Figure 9.7). Red, green, and violet light are all directed at the surface of this metal. Which, if any, of these colors will cause electrons to be ejected from the surface?
4. Suppose that the photoelectric threshold frequency in a certain metal lies in the red region of the spectrum (Figure 9.7). Yellow, green, and violet light are all directed at the surface of this metal. Which, if any, of these colors will cause electrons to be ejected from the surface?
5. In Exercise 3, would ultraviolet radiation cause electrons to be ejected? Would infrared radiation?
6. In Exercise 4, would ultraviolet radiation cause electrons to be ejected? Would infrared radiation?
7. A red lightbeam has a variable intensity, or brightness. As you increase the intensity, do the energies of the individual photons increase, decrease, or remain the same? Does the number of photons emitted each second increase, decrease, or remain the same?
8. A red lightbeam has a variable frequency. As you increase the frequency, does the color change? Do the energies of the individual photons increase, decrease, or remain the same? Do the photons' speeds change?

RADIATION: WAVES OR PARTICLES?

9. Which has greater energy: a radio photon or an infrared photon? Which has a longer wavelength? Larger frequency? Larger rest mass? Which moves faster?
10. What is meant by a red photon? A yellow photon? Which one has greater energy? Longer wavelength?
11. Explain why, in terms of photons, ultraviolet light can damage cells in your skin but visible light cannot.
12. Which has greater energy, a microwave photon or a visible photon?
13. When you charge a comb by running it through your hair, and then shake the comb back and forth, does it emit photons? What if the comb is not charged, and you shake it back and forth?
14. If you took a photograph using a shutter speed that was so fast that only 10 photons entered the lens, what would you see on the photograph?

MATTER: PARTICLES OR WAVES?

15. List the similarities and differences between a photon and an electron.
16. If electrons behaved only like particles and not like waves, what pattern would you observe in the double-slit experiment?
17. You don't notice the wave aspect of a pitched baseball. Is this because the wavelength is very long or because it is very short?
18. Which has the longer wavelength, an electron or a proton moving at the same speed?
19. Which has the longer wavelength, a slow proton or a fast proton?
20. If a "proton microscope" could be devised, how would you expect its wavelength to compare with the wavelength of an electron microscope?
21. Suppose that protons rather than electrons were used in the double-slit experiment. Would we still get an interference pattern on the screen? How would this pattern differ from the pattern for electrons?

THE PSI FIELD

22. What is the percentage probability of getting two heads in a row in fair coin tosses? Of getting five heads in a row? How could you experimentally test your prediction about getting five heads in a row?
23. Do quantum uncertainties have anything to do with the probabilities involved in a coin toss? Explain.
24. In the double-slit experiment with electrons, is the impact point predictable? Are there any points where we can predict that an electron will certainly not hit?
25. What is predictable in the double-slit experiment with electrons?
26. Would the answers to the preceding two questions be different if we were talking about photons instead of electrons?
27. Suppose that two electrons travel separately through a double-slit apparatus such as shown in Figure 13.19. The first electron hits at the upper end of the interference pattern, and the other hits at the lower end. Is it correct to say that the first electron came through slit A and the second came through slit B? Explain.

PROBLEMS

RADIATION: WAVES OR PARTICLES?

1. Which has greater energy, a microwave photon or a visible photon? About how many times greater (consult Figure 9.7)?

2. You charge an object by rubbing it, and then shake it at 1 Hz. Does it emit photons? How much energy does each photon carry?

3. How much energy does one photon of 10^{24} Hz gamma radiation carry?

4. How many visible photons would be needed to have enough energy to lift a 1 newton weight through 1 meter (consult Figure 9.7)?

5. Making estimates. About 10 visible photons are needed to cause a single photosynthesis reaction in living plants. About how much energy is carried by these 10 photons?

6. Making estimates. The human eye can detect as few as 10,000 photons per second entering the pupil. About how much energy is this, per second?

MATTER: PARTICLES OR WAVES?

7. If you double a hydrogen atom's speed, how does this affect its wavelength?

8. How would the wavelength of a proton compare with the wavelength of a *deuteron* (a proton and neutron held together by nuclear forces), assuming that both the proton and the deuteron have the same speed?

9. An electron and a proton are moving at the same speed. Which has the longer wavelength? How much longer? (Protons are about 1800 times more massive than electrons.)

10. Suppose we fire a high-velocity pellet gun, that accelerates 1-gram (10^{-3} kg) pellets to speeds of 1000 m/s (3 times the speed of sound). Find the wavelength of the pellet's probability wave.

11. Find the wavelength of an electron that strikes the back of a TV screen at a speed of 0.1c. The mass of an electron is 9.1×10^{-31} kg.

12. Individual electrons have been slowed down to speeds as low as several centimeters per second. The mass of an electron is 9.1×10^{-31} kg. What is the wavelength of a single electron moving at 0.1 m/s (10 cm/s)? Express your answer in millimeters.

13. In a recent experiment, sodium atoms were cooled until they were moving at only a few meters per second. The mass of a sodium atom is 38×10^{-27} kg. What is the wavelength of a single sodium atom moving at 2 m/s? Express your answer in millimeters.

CHAPTER 14

The Universe According to Quantum Theory

The previous chapter found the atomic world to be paradoxical, unpredictable, and quite non-Newtonian, all of which raised some deep questions about the meaning of quantum theory and even the meaning of the scientific enterprise. This chapter looks at some specific quantum principles that can help us understand these questions.

But first we devote two sections to one of the theory's many successful practical applications: the atom. We have already studied two theories of the atom: the Greek atom and the planetary atom. We saw that the failure of the Greek atom to explain electrical phenomena led to the planetary model. Similarly, certain failures of the planetary atom (Section 14.1) lead to the quantum atom. The quantum atom (Section 14.2) accounts for every atomic observation to date and, as far as we know, needs no further improvements. That is saying a lot because the observations have been many and they have been extraordinarily precise. For example, a certain radiation frequency for the hydrogen atom, known as the *Lamb shift,* is predicted by quantum theory to be 1057.860 ± 0.009 megahertz and has been measured as 1057.845 ± 0.009 megahertz. It seems miraculous that an abstract theory about unseen atoms can predict a seven-digit number that is then verified in the natural world. Quantum theory might be odd, but there is something right about it.

We then return to the question of the meaning of quantum theory. Section 14.3 presents the uncertainty principle, one of the theory's most useful ideas. Sections 14.4 and 14.5 present two topics that are closely related to the uncertainty principle and that could lay claim to being the oddest notions that have cropped up yet in physics or in all of science: quantum jumps and the nonlocality principle. We bolster these ideas with several recent experiments. Section 14.6 discusses the human significance of all of this. Section 14.7 returns to one of this book's themes: comparisons between Newtonian and post-Newtonian physics. Quantum theory is central to any such comparison and is perhaps an important guide toward finding a philosophical framework that can sustain humankind in the twenty-first century.

14.1 HOW DO WE KNOW? OBSERVING ATOMIC SPECTRA

The most accurate scientific measurements known are made with **spectroscopes**, devices that measure the frequencies or wavelengths present in radiation. Figure 14.1 shows how a spectroscope studies the visible radiation emitted by a light source such as a heated, glowing gas. Radiation from the source passes through a single thin slit and emerges as a narrow beam. This beam passes through a glass prism or other device that can separate the light beam's different frequencies (colors). Light beams bend when they pass from one medium into another, for instance, from air into glass. You might have noticed this effect in a pool of water, where partly submerged objects appear to bend at the water's surface. The reason that a prism separates a light beam's frequencies is that different frequencies bend by different amounts at each glass surface. This separation of frequencies is also seen in a rainbow, where each raindrop acts like a small prism for sunlight.

By the time the light exits the far side of the prism, it has separated into many beams, each with its own wavelength. A screen or photographic film intercepts all the light beams and displays their various colors. Each beam's frequency can be determined by measuring the position at which it strikes the screen. The set of frequencies measured in such an experiment is called the **spectrum** of the source that emitted the radiation.

Different kinds of spectroscopes operate in every part of the electromagnetic spectrum. For example, a radio receiver is a spectroscope for detecting the frequencies of the radio radiation present in a room.

Spectral measurements yield an enormous quantity and variety of information. Most of our data about the microscopic world come from spectral measurements. For instance, telescopes gather data about the mass, temperature, motion, chemical composition, and other properties of stars and galaxies by means of a spectroscope at the viewing end.

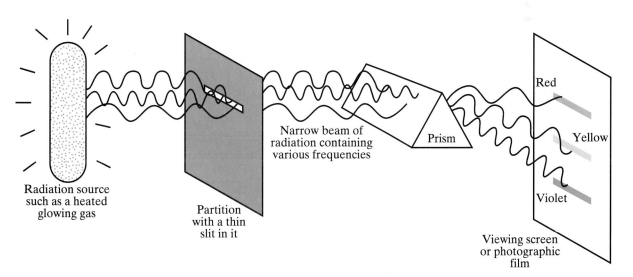

FIGURE 14.1
One type of spectroscope.

FIGURE 14.2
The continuous spectrum created by an incandescent bulb, and the line spectra produced by several different kinds of gases: sodium (Na), mercury (Hg), hydrogen (H), and helium (He). Frequency increases from left to right.

A glowing solid or liquid, such as a lightbulb's metal filament heated to around 3000°C, emits a **continuous spectrum**, one that contains an unbroken range of visible frequencies and is spread out in a continuous band of color, like a rainbow. But surprisingly, if a dilute gas is heated until it glows, it will emit a spectrum that is not continuous. Instead, it is quantized, or restricted to a limited number of precise frequencies, each frequency appearing on the screen as a narrow slit-shaped line (Figure 14.1). This collection of frequencies is called a **line spectrum**. Figure 14.2 shows a continuous spectrum and line spectra from several gases. As you can see, the line spectra for different gases are different. Because each gas has its own characteristic spectrum, it is possible for spectroscopy to identify different gases.

Heating is one way to **excite** a gas, in other words, to cause it to emit radiation. Most gases glow once they reach temperatures above about 2000°C. Flames are glowing gases of this sort, heated by combustion. The sun's light comes from hot gases on its visible surface, which has a temperature of 5500°C.

A second way to excite a gas is to send an electric current through it. This process, called **electric discharge**, creates the light seen in neon tubes, fluorescent bulbs, mercury or sodium vapor bulbs, sparks, and lightning strokes. Electric discharge tubes containing a dilute gas can be used to study the gas's spectrum (Figure 14.3).

How can we explain the observed spectra? As we know, when any substance is heated, the random kinetic energy of its atoms increases. The Greek model of the atom offers no obvious reason that this should cause materials to glow, but

Atoms are completely impossible from the classical [Newtonian] point of view.

RICHARD FEYNMAN

FIGURE 14.3
An electric discharge tube containing a dilute gas. With a large enough charge on the two *electrodes* at the ends of the tube, the electrodes *discharge* by forcing electrons off the negative electrode. These electrons then excite the gas by colliding with it as they move through the tube toward the positive electrode.

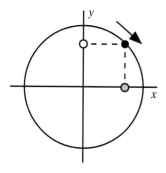

FIGURE 14.4
An orbiting electron (black circle) can be thought of as making two vibrational motions: When viewed from below, it appears to be vibrating along the x axis (blue circle), and when viewed from the side, it appears to be vibrating along the y axis (white circle).

the planetary model does: The subatomic parts of atoms are electrically charged; these charged particles move more energetically when the material is heated; and these vibrating and orbiting charged particles should send out electromagnetic radiation.

But why do gases emit line spectra rather than continuous spectra? Why are only some wavelengths emitted, rather than all wavelengths? What determines which wavelengths are emitted? The planetary atom does not answer these questions.

There is an even more glaring problem with the planetary atom. As explained in Figure 14.4, an orbiting electron can be thought of as vibrating along two directions at once. So the electromagnetic wave theory predicts that an orbiting electron should radiate electromagnetic energy all the time! But observations show that atoms do not radiate all the time. Worse yet, if an electron did radiate all the time, it would have to continually lose energy, and as it lost energy, it should spiral into the nucleus and cease orbiting. So when coupled with the electromagnetic theory of radiation, the planetary model is not self-consistent, because it predicts that atoms should collapse.

It is possible to imagine a universe in which Newtonian physics would be correct, even down to the smallest sizes. But that universe would be pretty boring. Atoms could not exist, so there would be no chemistry, so life would be impossible. The universe would in fact be a predictable machine.

■ **DIALOGUE 1**   You might have noticed that as you heat a metal hot plate, it first glows dark red and then becomes brighter and whiter. If you observe this process through a spectroscope, what will you see?

■ **DIALOGUE 2**   Instead of the thin slit in a spectroscope, could a round hole be used? How would a "line" spectrum then appear? Why is a slit better?

14.2 THE QUANTUM ATOM

We will examine the simplest atom, hydrogen, made of one proton and one electron. Because the electron is some 2000 times less massive than the proton, it does nearly all the moving, orbiting in the electromagnetic field of a nearly stationary proton. To a good approximation, we can ignore the proton's motion, treating it as a tiny material particle at rest.

The electron's behavior is described by its psi field, that is, by its probability wave. Imagine a hydrogen atom at rest and isolated. Since there is no reason for anything to be physically changing in such an atom, we would expect the electron's psi field to be stationary, unchanging.

Stationary wave patterns are observed in many phenomena in nature, phenomena usually having nothing to do with quantum theory. Such patterns are called **standing waves**. A vibrating violin string is an example. Figure 14.5 shows

1. At first, you will see a dim spectrum extending from the longest visible wavelengths (red) through a portion of the red part of the spectrum. As the plate gets hotter, the spectrum will get more intense, and it will extend through a larger portion of the visible spectrum.

2. Yes. The spectrum would appear as a series of circles instead of a series of lines. A slit works better because two neighboring lines on the screen are more distinctly separated than are two neighboring circles.

FIGURE 14.5
Four standing waves on a string. In the first state of vibration, the string vibrates in one segment; in the second state, it vibrates in two segments with one stationary point (which does not vibrate) in the middle; and so forth.

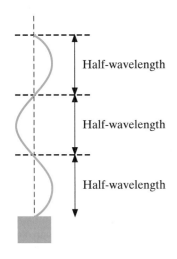

FIGURE 14.6
A "snapshot" of the third standing wave in Figure 14.5, at a single instant.

four different standing waves on a string. Each photo is a time exposure over the duration of a few vibrations.

As you can see, each standing wave is formed by vibrations of the entire string, but the resulting wave shape is not moving in either direction along the string. Nevertheless, the string has a wave shape at any particular instant. For example, Figure 14.6 shows what the third standing wave of Figure 14.5 would look like if it were photographed in a short-exposure snapshot. Each end of the string must be a point of nearly zero vibration. Within these fixed end points, the string is vibrating in one, two, three, and four loops in each of the successive photographs of Figure 14.5. Notice that only certain standing waves can fit onto the string, namely, those for which an integral number (1, 2, 3, 4, and so on) of loops fit onto the string's length.

Returning to the hydrogen atom, since the electron's psi field is expected to be stationary, we expect that it must form a standing wave. But this standing wave will not stretch along a straight line as in Figure 14.5; instead the hydrogen atom's spherical shape causes the psi-wave to form a circular standing wave, as shown in Figure 14.7(a). Just as the standing waves of Figure 14.5 must just fit onto the string, the standing probability wave must just fit around the nucleus. Figure 14.7(a) shows a probability wave that fits properly, and Figure 14.7(b) shows one that does not and so is not allowed.

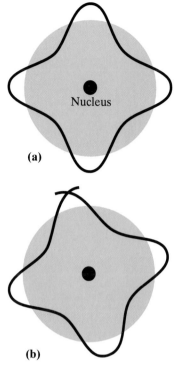

(a)

(b)

FIGURE 14.7
(a) A standing wave that just fits around the nucleus. (b) A standing wave that doesn't fit; its wavelength is a little longer than the wavelength in (a).

Nucleus

A detailed mathematical study of the Schroedinger equation for a hydrogen atom yields precise predictions as to the allowed standing wave patterns for the electron's psi field. Each one of these standing wave patterns is a possible psi field for the electron in a hydrogen atom. Figure 14.8 is one way of picturing a few of these **quantum states** of the hydrogen atom. Each of the 10 patterns shown is one of the allowed stationary probability patterns of the single electron in a hydrogen atom. To visualize the full three-dimensional probability patterns, imagine rotating the two-dimensional diagram around the vertical z axis shown in each diagram.

As we know, a single electron is always observed as a tiny particle, not a spread-out wave. The pattern shows only the probability that the electron will be observed to be at various places. The diagram represents this probability by darkening, and so the electron is more likely to be found at places where the diagram is darker. Physicists find it convenient to label these quantum states by means of three numbers called **quantum numbers**, indicated as N, L, and M in the figure.

For example, the state labeled $N = 1$, $L = 0$, $M = 0$ occupies a smaller volume than does any other state. In this state, the electron is highly likely to be found close to the nucleus and is equally likely to be found in any direction out from the nucleus (upward, downward, to the left, and so forth).

The state $N = 2$, $L = 0$, $M = 0$ is larger, so the electron is likely to be found farther from the nucleus than is an electron in the state $N = 1$, $L = 0$, $M = 0$. This state has an interesting gap partway out from the nucleus, representing a distance from the nucleus at which the electron will never be found. It is interesting that an electron in this state can be found inside or outside this distance but never at this distance. How can the electron get from the inside to the outside without sometimes being at this distance? Paradoxical questions like this are related to wave–particle duality: The electron is in some sense spread out on both sides of the gap, even though the electron is always observed at only one point.

The state $N = 3$, $L = 0$, $M = 0$ is larger still. The electron is likely to be found still farther from the nucleus. There are now two gaps where the electron will not be found. The electron's psi field is now spread out in three separated regions: inside the first gap, between the two gaps, and outside the second gap.

Unlike the states mentioned so far, states having $L = 1$ or $L = 2$ are not the same in every direction. The state $N = 2$, $L = 1$, $M = 1$ is shaped like a fat doughnut circling the z axis and is reminiscent of the planetary model of the atom. The state $N = 2$, $L = 1$, $M = 0$ is shaped like a dumbbell (two spheres) along the z axis. It is separated into two parts, between which the electron is not found.

Figure 14.8 shows all the states having N values equal to 1, 2, or 3. There is another set of states having $N = 4$, another set having $N = 5$, and so forth. There is no limit to the possible values of N, and so there is an infinite number of quantum states of hydrogen.

These diagrams represent the patterns of hydrogen, nature's simplest and most prevalent atom. Each pattern represents one way that a hydrogen atom can exist.* Atoms with more than one electron have more complex quantum states, but they all are found by solving the Schroedinger equation in the form appropriate to that particular atom and requiring that the psi field just fit around the nucleus.

Since each of these quantum states is a standing wave with one particular frequency and since the frequency of a wave is related to its energy, we expect each

---

*A hydrogen atom can also exist in a combination of two or more of these allowed quantum states.

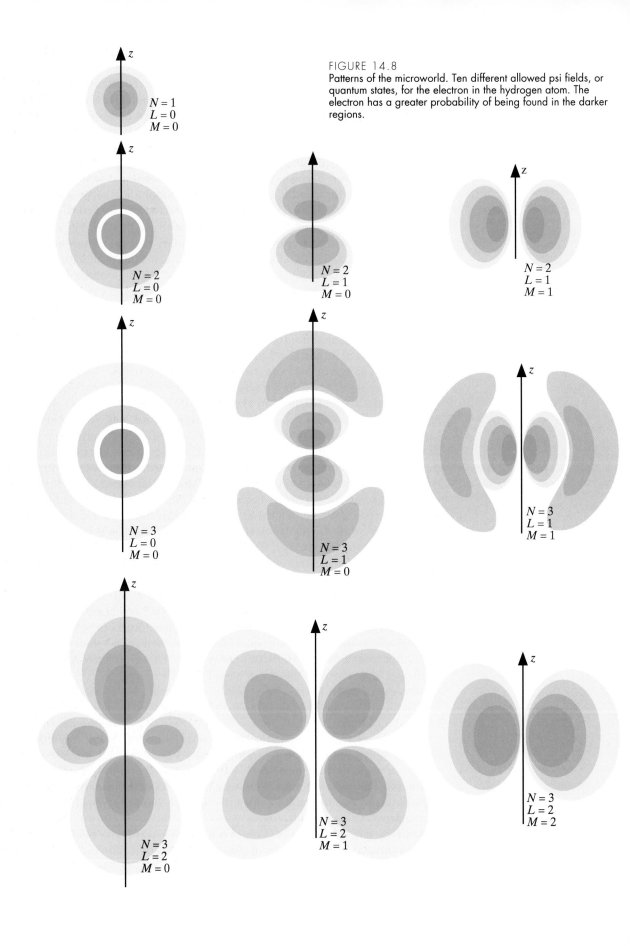

FIGURE 14.8
Patterns of the microworld. Ten different allowed psi fields, or quantum states, for the electron in the hydrogen atom. The electron has a greater probability of being found in the darker regions.

$N = 1$
$L = 0$
$M = 0$

$N = 2$
$L = 0$
$M = 0$

$N = 2$
$L = 1$
$M = 0$

$N = 2$
$L = 1$
$M = 1$

$N = 3$
$L = 0$
$M = 0$

$N = 3$
$L = 1$
$M = 0$

$N = 3$
$L = 1$
$M = 1$

$N = 3$
$L = 2$
$M = 0$

$N = 3$
$L = 2$
$M = 1$

$N = 3$
$L = 2$
$M = 2$

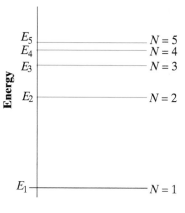

FIGURE 14.9
The lowest five energy levels for the electron in the hydrogen atom. When measured in joules, these atomic energy levels are quite small: The energy difference, $E_2 - E_1$, between the lowest two levels is only 1.6×10^{-18} joules.

You surely must understand, Bohr, that the whole idea of quantum jumps necessarily leads to nonsense. . . . If we are still going to have to put up with these damn quantum jumps, I am sorry that I ever had anything to do with quantum theory.

SCHROEDINGER, DURING A CONVERSATION WITH NIELS BOHR

state to have just one specific energy. That is, we don't expect the energy of any one of the states of Figure 14.8 to have any quantum uncertainty. From Figure 14.8, we can make some educated guesses about the amount of energy present in each state. Because the force by the proton on the electron is attractive, one would have to do work to pull an electron outward, away from the nucleus. So the electromagnetic energy of the electron goes higher as the electron travels farther from the nucleus. This is just like gravitational energy: Because Earth exerts an attractive gravitational force on a rock, the rock's gravitational energy increases as the rock travels farther from Earth.

So the smallest psi field, the one bunched most tightly around the nucleus, should have the lowest energy. Judging from Figure 14.8, this is the $N = 1$ state. Because of the gravitational analogy, this is called the **ground state**. It is the state in which the electron is as close to the nucleus as it can be. Other states are called **excited states** because they are more energetic. Because all the $N = 2$ states have about the same average distance from the nucleus, we expect them to have about the same energy. Similarly, it appears from Figure 14.8 that all the $N = 3$ states have about the same energy, an energy that is higher than that of the $N = 2$ states.

The precise energy of each quantum state can be calculated using Schroedinger's equation, which shows that the energies increase with increasing N, as depicted in Figure 14.9. The energy values E_1, E_2, and so on are labeled by their N values 1, 2, and so forth. A diagram like this, showing the energies of various quantum states, is called an **energy level diagram**.

Each of these quantum states represents an isolated hydrogen atom that isn't changing. What happens when something does change? What happens, for example, when a hydrogen atom emits radiation?

As we know, radiation comes in bundles of energy, photons. An atom must emit at least one photon whenever it radiates. In emitting a photon, an atom must lose energy, so it must be in an excited state to begin with, and it must change to a lower-energy state. This transition from one quantum state to another must be instantaneous, because the conservation of energy demands that as soon as the photon's energy is emitted, the atom's energy must drop by exactly this amount. Such an instantaneous transition from one quantum state to another is called a **quantum jump**.

Figure 14.10 shows a common way of representing quantum jumps. The transition is shown as an arrow on an energy-level diagram stretching from the initial to the final energy. The diagram represents an atom making a transition from the $N = 4$ to the $N = 2$ energy level and also indicates that a photon is emitted, carrying away the energy. In Figure 14.8, imagine that an $N = 4$ (not shown) pattern vanishes instantaneously, to be replaced by an $N = 2$ pattern. The hydrogen atom truly jumps from one pattern to another.

Now we can understand, in terms of quantum theory, how atoms emit radiation. Atoms emit radiation when they quantum-jump to a lower energy level, creating and emitting a photon in the process. Recall (Section 13.2) that a photon's energy is hf, where h is Planck's constant and f is the photon's frequency. Because of the conservation of energy, the energy hf of the emitted photon must equal the energy difference in the quantum jump; that is,

$$hf = \text{(energy of high-energy state)} - \text{(energy of low-energy state)}$$

So if you know the two energy levels, you can find the frequency of the photon emitted in a quantum jump between these two levels. Physicists can calcu-

FIGURE 14.10
A symbolic representation of a quantum jump from an N = 4 quantum state to an N = 2 quantum state. A photon, carrying energy $E_4 - E_2$, is given off at the instant the quantum jump occurs.

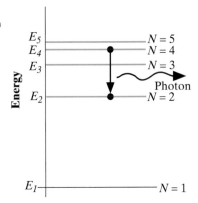

late the frequencies that hydrogen atoms can emit by finding the allowed energies of hydrogen, as calculated from the Schroedinger equation, and then finding the frequency of the photon emitted in each possible quantum jump between pairs of energy levels.

H O W   D O   W E   K N O W ?

Schroedinger's equation and the spectrum of hydrogen

Figure 14.11 shows, on an energy level diagram, the 10 downward quantum jumps that are possible between the lowest five energy levels for hydrogen. Since the photon's energy is equal to the atom's energy change, the length of the arrow representing each quantum jump is proportional to the frequency of the radiation emitted in that quantum jump. So a hydrogen atom can emit 10 different frequencies by quantum-jumping from the N = 2, 3, 4, or 5 levels downward into a lower level. Figure 14.12 shows these 10 frequencies quantitatively. Since

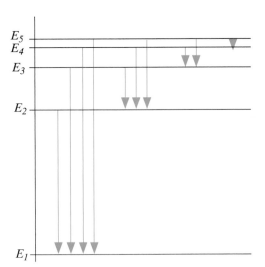

FIGURE 14.11
The possible downward quantum jumps between the lowest five energy levels of the hydrogen atom.

FIGURE 14.12
The frequencies of the photons that are given off during the quantum jumps shown in Figure 14.11.

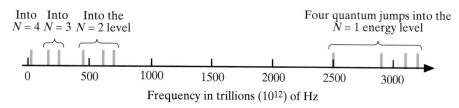

Into Into   Into the
$N = 4$ $N = 3$ $N = 2$ level

Four quantum jumps into the
$N = 1$ energy level

0   500   1000   1500   2000   2500   3000

Frequency in trillions (10^{12}) of Hz

Schroedinger's equation determines the energy levels (Figure 14.11), it also determines these frequencies.

When one uses a spectroscope (Section 14.1) to measure the spectrum of atomic hydrogen gas, the frequencies of the photons turn out to be precisely the frequencies indicated in Figure 14.12 and predicted by quantum theory. Schroedinger's equation first gained fame because Schroedinger was able to show that it correctly predicted these frequencies. Physicists knew that there must be something right about it if it could predict such precise numbers.

The absorption of radiation is similar to emission, only backward. A photon from the environment enters an atom. If the photon's energy just matches one of the atom's possible quantum jumps, the atom may quantum-jump upward to a higher energy level, absorbing the photon in the process.

> There is no part of chemistry that does not depend, in its fundamental theory, upon quantum principles.
>
> LINUS PAULING, CHEMIST

The quantum theory of more complex atoms and of molecules is similar. The carbon atom, the hydrogen molecule, and indeed a DNA molecule all could be studied by starting from the Schroedinger equation for this atom or molecule and by finding the allowed quantum states and their energies. This procedure gets difficult for more complex molecules, but the same quantum principles apply.

Throughout this section, we have studied how quantum theory explains and predicts the line spectra emitted by hydrogen gas and other gases. What about the other question posed in Section 14.1, the question of why the electron in an atom does not radiate all the time and hence spiral into the nucleus? Quantum theory deals very simply with this issue: Once it enters its lowest-energy state, $N = 1$ in Figure 14.8, an atom cannot emit radiation because it has no lower-energy state to quantum-jump to. The reason is that no smaller standing waves will fit, so this is the smallest the electron's psi field can become.

■ **DIALOGUE 3**   Describe the three-dimensional shapes of the three quantum states for hydrogen with $N = 3$ and $L = 2$.
■ **DIALOGUE 4**   Among the ten quantum jumps between the five energy levels of hydrogen shown in Figure 14.11, which one will create the photon with the highest frequency? Which will create the lowest frequency?
■ **DIALOGUE 5**   How many different frequencies can be created by quantum jumps among only the lowest six energy levels of hydrogen?

---

3. A small doughnut circling the nucleus, plus a dumbbell along the z axis. Two doughnuts circling the z axis, above and below the nucleus. A fat doughnut circling the nucleus.
4. $N = 5$ to $N = 1$. $N = 5$ to $N = 4$.
5. 15.

FIGURE 14.13
Werner Heisenberg. Using the Schroedinger equation, he derived the famous uncertainty principle, according to which every material particle has inherent and irreducible uncertainties in position and speed. Thus, in the microscopic world, the future is not entirely determined by the past.

I remember discussions with Bohr which went through many hours till very late at night and ended almost in despair; and when at the end of the discussion I went alone for a walk in the neighbouring park I repeated to myself again and again the question: Can nature possibly be as absurd as it seemed?

WERNER HEISENBERG

Would the universe, without consciousness, have remained a play before empty benches, not existing for anybody, thus quite properly not existing?

ERWIN SCHROEDINGER

14.3 THE UNCERTAINTY PRINCIPLE: A PARTICLE'S REALM OF POSSIBILITIES

Quantum theory's central idea is that everything is made of unpredictable particles whose statistical behavior follows a predictable wave pattern. In 1927, German physicist Werner Heisenberg (Figure 14.13) found that this wave–particle duality implies that the microworld has an inherent uncertainty that can be quantified.

To get a feel for Heisenberg's argument, let's consider just one electron moving through empty space, along a direction that we will call the x axis. Since the electron's position is unpredictable, its psi field must be spread out somewhat along the x axis, as indicated in Figure 14.14. The spread in possible positions, called the **uncertainty in position**, is indicated by the symbol Δx (delta x), the range of possible places where we might find the electron if we precisely measured its position. A traveling psi field like Figure 14.14, which extends over only a limited distance, is called a **wave packet**. It is a standard way of representing a moving particle whose position is somewhat uncertain.

Uncertainty is a common notion, usually having nothing to do with quantum theory. For example, suppose you hike along a straight road for 10 kilometers, give or take 1 kilometer. Then your uncertainty in your position (measured perhaps from your starting point) is $\Delta x = 2$ km.

Heisenberg showed that a wave packet like the one in Figure 14.14 represents an electron with a range of possible *speeds*, as well as a range of possible positions.* That is, a wave packet represents an electron whose position is uncertain and whose speed is *also* uncertain. A particle's range of possible speeds is called its **uncertainty in s**, abbreviated Δs.

Here is a simple example, again having nothing to do with quantum theory: If you glance at your car's speedometer and read it as "75 km/hr, give or take 2 km/hr," then your uncertainty in speed is $\Delta s = 4$ km/hr.

So every wave packet representing a moving electron has two kinds of uncertainties: Δx and Δs. Let's compare one wave packet A with another wave packet B that has been squeezed into half of A's length (Figure 14.15). As you can see, B's wavelengths are shorter. But de Broglie's formula (Section 13.4), $\lambda = h/ms$, tells us that shorter wavelengths correspond to higher speeds. So wave packet B represents a higher-speed electron than does packet A. But what about the uncertainty in speed, Δs—how does packet B's Δs compare with packet A's Δs? It seems plausible (and it can be shown[†]) that larger speeds mean a larger *uncertainty* in speed and that in fact the halving of Δx implies a doubling of Δs.

This illustrates a general feature of wave packets: Whenever Δx is squeezed by some amount, Δs expands by the same amount, and vice versa: squeezing Δs expands Δx. Quantitatively, this means that the *product* of the two uncertainties,

---

*The reason is that, according to the branch of mathematics known as "Fourier analysis," a wave packet is a superposition of many, different, infinitely long waves, each having a definite wavelength. But de Broglie's formula $\lambda = h/ms$ tells us that each different wavelength corresponds to different speeds. Thus, a wave packet has a range of possible speeds.

[†]Since B is squeezed to half of A's length, B's wavelengths are half as long as A's. So B's component speeds are twice as big as A's, because wavelength and speed are inversely proportional. So the range of speeds, Δs, is twice as big for B as for A.

FIGURE 14.14
The psi field representing a single particle whose uncertainty in position is Δx, moving along the x axis. A psi field like this, which is spread out over only a limited distance, is called a wave packet.

Direction of motion of wave packet ⟶

This again emphasizes a subjective element in the description of atomic events, since the measuring device has been constructed by the observer. . . . [W]e have to remember that what we observe is not nature in itself but nature exposed to our method of questioning.

WERNER HEISENBERG

The belief in an external world independent of the perceiving subject is the basis of all natural science.

EINSTEIN

In the experiments about atomic events we have to do with things and facts, with phenomena that are just as real as any phenomena in daily life. But the atoms and the elementary particles themselves are not as real; they form a world of potentialities or possibilities rather than one of things.

WERNER HEISENBERG

$\Delta x \cdot \Delta s$, remains unchanged. We refer to a particle's Δx and Δs, considered together, as the particle's **realm of possibilities**.*

Working through these ideas in detail, Heisenberg found that this rule holds for every particle (not just electrons) in every physical situation (not just when moving freely). This idea is called:

> HEISENBERG'S UNCERTAINTY PRINCIPLE
> Every material particle has inherent uncertainties in position and in speed. Although either one of these uncertainties can take on any value, the two uncertainties are related by the fact that their product must approximately equal h/m. In symbols,
>
> $$(\Delta x) \cdot (\Delta s) \approx h/m^{\dagger}$$
>
> where h is Planck's constant and m is the particle's rest mass.

The uncertainty principle tells us that any particle preserves its overall realm of possibilities in position and speed; any reduction in one uncertainty must expand the other by an equal amount.

We can visualize a particle's realm of possibilities in a diagram such as Figure 14.16(a), where a single point represents a precise position x and speed s. Newtonian physics assumes that every object has a precise x and s. For example, the location and motion of the center of a baseball can be described, according to Newtonian physics, by a particular x representing the center's location and a particular s representing the center's speed. Newton's law of motion is basically

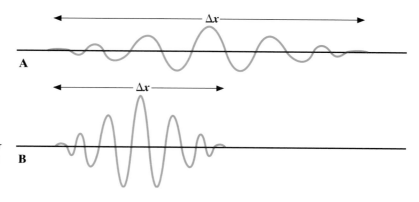

A

FIGURE 14.15
Two wave packets, having different Δx's. Packet B can be constructed by squeezing packet A to half its size. In this process, all of A's wavelengths get squeezed to half their original length, which means that the speeds and also the uncertainty in speed get doubled.

B

---

*This nice phrase is from Nick Herbert's wonderful nontechnical account of the meaning of quantum theory, *Quantum Reality* (New York: Doubleday/Anchor, 1985).
†More precisely, $(\Delta x) \cdot (\Delta s) \geq h/4\pi m$. The product can be greater than $h/4\pi m$, but not less.

(a) A single point on an *x*-versus-*s* diagram, such as the point shown here, represents a precise value of both *x* and *s*. Quantum theory does not allow such precise values.

(b) A realm of possibilities for a single particle, according to quantum theory. The total area of the shaded region, $(\Delta x)\cdot(\Delta s)$, must be roughly equal to h/m.

(c) If for any reason Δx is reduced, then Δs must expand to fill up the same overall realm of possibilities.

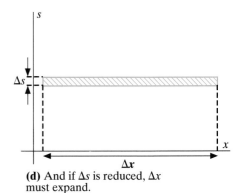

(d) And if Δs is reduced, Δx must expand.

FIGURE 14.16
Position and speed uncertainties.

> The probability wave . . . means a tendency for something. It's a quantitative version of the old concept of potentia in Aristotle's philosophy. It introduces something standing in the middle between the idea of an event and the actual event, a strange kind of physical reality just in the middle between possibility and reality.
>
> WERNER HEISENBERG

a method for predicting an object's future *x* and *s* from its present *x* and *s*. For example, given the position and speed of the center of a falling baseball at one time, we can predict its position and speed at any later time during the fall.

But quantum theory does not allow such precision. Quantum theory demands that an object's *x* and *s* have uncertainties Δx and Δs whose product is roughly h/m. In an *x*-versus-*s* diagram, this product is the area formed by the rectangle whose sides are Δx and Δs, as shown in Figure 14.16(b). If for any reason Δx is reduced, then Δs must expand to fill up the same overall realm of possibilities, as shown in (c). And if Δs is reduced, Δx must expand as in (d). Either *x* or *s* can be as highly predictable as you like, but if one is highly predictable, the other must be highly uncertain.

Since the uncertainty principle says that $(\Delta x) \cdot (\Delta s)$ is inversely proportional to *m*, more massive particles have smaller realms of possibilities. A proton's realm, for example, is 2000 times smaller than an electron's realm (Figure 14.17). Because *x* and *s* are what we need to predict an object's future behavior, a proton is much more predictable than an electron is.

A 1-kilogram baseball, one million trillion trillion times more massive than an electron, is so predictable that quantum uncertainties can be neglected. That is why Newtonian physics works just fine for large objects like baseballs. Even a grain of sand is so massive (it contains some 10^{18} atoms) that quantum uncertainties are negligible.

FIGURE 14.17
Because of its larger mass, a proton's uncertainties are much smaller than an electron's.

The world thus appears as a complicated tissue of events, in which connections of different kinds alternate or overlap or combine and thereby determine the texture of the whole.
WERNER HEISENBERG

The observer is never entirely replaced by instruments; for if he were, he could obviously obtain no knowledge whatsoever. . . . They must be read! The observer's senses have to step in eventually. The most careful record, when not inspected, tells us nothing.
SCHROEDINGER, IN *MIND AND MATTER*

Marvelous, what ideas the young people have these days. But I don't believe a word of it.
EINSTEIN, AFTER HEISENBERG'S 1927 LECTURE ENUNCIATING THE UNCERTAINTY PRINCIPLE

I cannot seriously believe in [the quantum theory] because it cannot be reconciled with the idea that physics should represent a reality in time and space, free from spooky actions at a distance.
EINSTEIN

Suppose a particle's Δx has been squeezed into a very small uncertainty. This particle must then have a large Δs. But you can't have a large Δs without at the same time having a large s; for instance, if Δs were 1000 km/s, the lowest (slowest) realm of possibilities for s would be 0 to 1000 km/s, so the average s must be at least 500 km/s. So when Δs is large, s must be large too. This means that a highly confined particle (Δx small) must move fast. The smaller the confinement, the faster the motion. The uncertainty principle will not permit the microscopic world to sit still!

For example, the protons and neutrons in a nucleus are confined to a region 10,000 times smaller than the atom itself, by the strong nuclear forces that hold the nucleus together. So these protons and neutrons must move rapidly, which is why nuclear energies must be large.

The uncertainty principle prevents atoms from collapsing. As we have seen, the planetary atom has the grave defect that it should collapse. The uncertainty principle prevents this: As an electron radiates energy and falls inward, it is confined to a smaller region. This process of losing energy cannot continue indefinitely because the particle would eventually not be able to fill out its realm of possibilities.

Quantum uncertainties lie at the heart of the nuclear phenomenon known as radioactive decay (Chapter 15) and cause this process to be fundamentally unpredictable.

When a child is conceived, the DNA molecules of each parent are randomly combined in a process in which the quantum features of the DNA's chemical bonds play a role. So quantum uncertainties operating at the microscopic level play a role in our genetic inheritance.

In fact, according to the so-called inflationary theory of the manner in which the big-bang creation of our universe occurred, it is possible that our entire universe originated as a tiny fluctuation due to quantum uncertainties, and then expanded to become our present universe. See Section 18.6 for further details.

In these and other ways, we are in the hands of the god who plays dice.

■ **DIALOGUE 6** Arrange these objects in order, beginning with the object having the largest realm of possibilities and ending with the one having the smallest: proton, glucose molecule $C_6H_{12}O_6$, helium atom, baseball, electron, grain of dust, water molecule, automobile.

14.4 QUANTUM JUMPS

Quantum theory continues to surprise and perplex. Einstein was among those who found the theory too counterintuitive to believe. He and two other physicists showed in 1935 that because of quantum uncertainties, quantum theory predicts some phenomena that are, as he put it, so "spooky" that "no reasonable definition of reality could be expected to permit this." Einstein and others took these predictions as evidence that a correct theory would not contain quantum uncertainties. However, Einstein did not suggest a way to put quantum theory's spooky predictions to an experimental test.

Because quantum theory proved so gloriously successful in practice, few physicists worried much about such untested objections. Among those who did

6. Electron, proton, helium atom, water molecule, glucose molecule, grain of dust, baseball, automobile.

worry were David Bohm and John Bell (Figure 14.18). Bohm began publishing his theoretical analysis of quantum theory during the 1950s. Working from these ideas, Bell showed in 1964 that some of quantum theory's spooky predictions are experimentally testable. John Clauser (Figure 14.18) and four collaborators* carried out the first such test in 1972 and found that contrary to the expectations of Einstein and others, the spooky phenomena actually occur! In 1982, Alain Aspect (Figure 14.18) refined Clauser's test so as to leave little doubt that the real world is stranger than Einstein and others had thought.

All the spooky predictions are related to the psi field, especially its sudden changes. Suppose, for example, that a freely moving electron's psi field is a wave packet (Figure 14.14), with a position uncertainty Δx. Suppose we then measure the electron's position to within an accuracy better than Δx, perhaps $\Delta x/2$. Then just after this measurement, the electron must be represented by a new wave packet having position uncertainty $\Delta x/2$. The measurement caused the electron's psi field suddenly to change. Such sudden changes in a psi field are called **quantum jumps**.

For another example, we return to the double-slit experiment using electrons.[†] In Chapter 13, Figure 13.19(d) shows an electron's psi field just before the electron hits the screen, along with the possible impact points and their probabilities. But once the particle actually hits the screen, it is suddenly "localized" at the impact point. So its psi field jumps instantaneously, from the spread-out pattern in Figure 13.19(d) (this is what is graphed on the screen in the figure) to a small region at the impact point.

Quantum jumps can affect quite large regions. For example, it happens that the psi field for each photon from any very distant star is spread out over many kilometers by the time it reaches Earth. British physicist Robert Hanbury Brown confirmed this prediction in 1965 by measuring, for the light from an individual star, interference patterns that were over 100 meters in diameter. Despite the psi field's large size, it instantaneously contracted to a point when the photon hit a detector.

For another example of quantum jumps, activation of a particle detector behind one slit in the double-slit experiment destroys the interference pattern (Chapter 13). Merely switching on the detector causes the psi field to jump from the interference pattern of Figure 14.19(a) to the noninterference pattern of Figure 14.19(b). As a normal or Newtonian explanation of this strange effect, perhaps the detector interacts with each particle as it comes through a slit, forcing the particle to follow pattern (b) rather than pattern (a). To check this, experimenters placed the detector far behind the slits, near the screen, as in Figure 14.20. With the detector far from the slits, it is implausible that the detector could exert forces on the particle as it comes through the slits. And yet when the detector is on, we again get pattern (b) instead of (a).

The detector can even be switched on after the particle is well past the slits, and the interference pattern still will jump to the noninterference pattern at the instant the switch is thrown! It is as though the particle comes through both slits

---

*Michael Horne, Abner Shimony, and Richard Holt collaborated on the design of the experiment, and Stuart Freedman collaborated on its execution.

[†]Most of the following experiments have been done using photons rather than material particles such as electrons, because the experiments are easier with photons. But the quantum predictions are the same for electrons or other material particles, and physicists have little doubt that the results would be the same for electrons.

FIGURE 14.18
Four explorers of quantum theory:
Clockwise from upper left: David
Bohm, John Bell, Alain Aspect talking
with Bell (r.) and physicist Albert
Messiah, John Clauser.

FIGURE 14.19
Merely switching on a particle detector at
a point such as D causes the psi field to
jump from the interference pattern (a) to
the noninterference pattern (b).

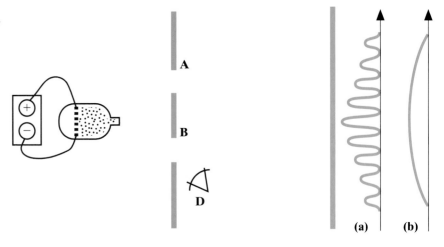

(at least, its psi field comes through both slits) right up until the experimenter decides to switch on the detector; then the particle "changes its mind" and decides that it actually came (in the past!) through only one slit.

We see here hints of an effect known as *nonlocality*. The psi field changes instantaneously, all over the screen, the moment that the detector is switched on. The detector seems intimately connected with the particle's psi field, instantaneously, everywhere on the screen. Such an instantaneous connection, across a distance, is called a **nonlocal effect**.

H O W   D O   W E   K N O W ?

Mandel's experiment on the nonlocal effect of detectors

To what extent does the mere detection of an event cause changes in related events some distance away? In 1991 Leonard Mandel and coworkers conducted a striking test of quantum theory's predictions about the effect that detectors have on distant events.

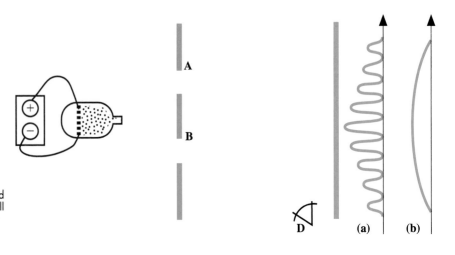

FIGURE 14.20
Even if the detector is placed far behind
the slits, near the screen, the pattern still
jumps from pattern (a) to (b) whenever
the detector is activated.

Figure 14.21 shows Mandel's experimental arrangement. A source emitted a single particle whose psi field moved along two paths labeled path A and path B. This single particle was then converted into *two* particles, 1 and 2, each moving along two possible paths, labeled 1A, 2A, 1B, 2B. Particle 1 moved along either path 1A or path 1B through a double-slit apparatus having slits A and B, and particle 2 moved along either path 2A or path 2B toward a detector. As long as the experimenters had no way of knowing which slit (A or B) particle 1 came through, the psi field for particle 1 came through both slits and so the impact points on the screen formed the interference pattern (a), just as in the double-slit experiment with electrons (Figure 13.13(b)).

The unique feature of this arrangement is that the two particles must be *either* on paths 1A and 2A, *or* on paths 1B and 2B. That is, the two particles' psi fields were connected with each other or "entangled" (see the following section) in such a manner that, if particle 1 was on path 1A, then particle 2 was necessarily on path 2A rather than 2B; and if particle 1 was on path 1B, then particle 2 was necessarily on path 2B.

Now suppose that a barrier is placed in the way of path 2A, as shown, and that a single particle is emitted from the source. Then if we see a flash on the screen and also detect a particle hitting the detector, we can conclude that particle 2 came along path 2B, so the flash on the screen was caused by a particle moving along path 1B. And if we see a flash on the screen but we do *not* detect a particle hitting the detector, we can conclude that particle 2 came along the blocked path 2A, so particle 1 must have moved along path 1A.

In other words, with the barrier in place, the detector placed along one of the paths of particle 2 can determine which slit particle 1 came through. On the other hand, if the barrier is removed, then the detector in the path of particle 2 no longer gives any information about particle 1. Note that the barrier and detector interact only with particle 2. Thus, the detector and barrier constitute an

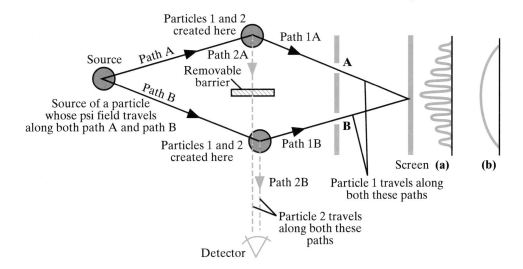

FIGURE 14.21

Mandel's experiment. A barrier placed in path 2A causes the pattern on the screen to quantum-jump from pattern (a) to pattern (b), even though the barrier does not directly disturb the particle that impacts the screen, and even though the barrier and detector can be quite distant from the screen.

extremely remote and noninterfering detection scheme for particle 1. It is hard to believe that the placement or removal of the barrier along path 2A could affect what happens to particle 1 at the screen.

Yet quantum theory predicts, and the experiment confirms, that when the barrier is placed in path 2A, the interference pattern (a) quantum-jumps to the noninterference pattern (b). Furthermore, the effect persists even when the detector is removed—the mere blocking of path 2A destroys the interference between paths 1A and 1B! Apparently the mere possibility that an observer *could* insert a detector and thus determine whether path 1A or 1B was taken causes the interference pattern to switch to noninterference.

We conclude that detectors that can provide information about distant events can change those events even without directly physically interfering with those events.

14.5 THE INTERCONNECTEDNESS PRINCIPLE: SPOOKY ACTION AT A DISTANCE

Except for Mandel's experiment, the quantum uncertainties discussed so far have involved only one particle at a time. For example, in the double-slit experiment with electrons, the essential uncertainty concerned whether a single electron goes through slit A or slit B.

We now consider the possibility of quantum uncertainties involving two or more particles at a time. If two particles are involved in the same microscopic event, for example, if they are created together in a single event or if they physically interact with each other, their psi fields usually become intimately connected to each other. The two particles then become a single quantum system so that there is no psi field for either particle separately but only a single inseparable psi field that the two particles share. In a sense, the particles are not separate at all but are one object in two different places. The particles are then said to be **entangled**. Figure 14.22 is an attempt to picture the entanglement of two particles that start out separately and then interact to become entangled.

H O W   D O   W E   K N O W ?

The position-entanglement experiment*

This experiment begins with the creation of two entangled particles that move directly away from each other, as diagrammed in Figure 14.23. Each particle then passes through a double-slit apparatus and is reflected by a mirroring device that directs it toward a screen. Because of their opposite directions, if parti-

*The experiment was suggested in 1986 by Michael Horne and Anton Zeilinger and performed in 1990 by John Rarity and Paul Tapster; a similar experiment was performed by Z. Y. Ou and Leonard Mandel in 1989. The description presented here simplifies some of the experimental details without changing the physical essentials.

FIGURE 14.22
When two particles interact and then separate, their psi fields usually become entangled.

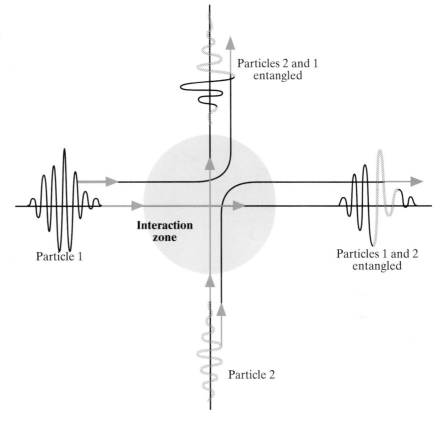

Particles 2 and 1
entangled

Interaction
zone

Particle 1

Particles 1 and 2
entangled

Particle 2

cle 1 goes through slit A, particle 2 must go through slit B (the dashed double arrow in the diagram), and vice versa (solid arrow). There is a separate viewing screen (or other detection device) for each particle.

As in the ordinary double-slit experiment (see Figure 13.19, for instance), each particle's psi field goes through both slit A and slit B. The mirrors are arranged to bring the psi fields from the two slits back together to create an interference pattern on each particle's screen. The experiment could be described as a "duplicate double-slit experiment" with two entangled particles. If the two particles were not entangled, each half of the experiment (the left side and the right side of the figure) would yield the usual double-slit interference results discussed previously.

In this experiment, the entangled particles must separate in opposite directions. Because of this, there are so-called *statistical correlations* between the impact point of particle 1 on its screen and the impact point of the particle 2 on its screen.

Statistical correlations are common in situations involving uncertainties. For a typical example—having nothing to do with quantum theory—suppose your friend tells you that he has sealed a gold coin and a silver coin in separate envelopes and mailed one to you in Tokyo and one to Betty in Paris. Having no information about which coin was mailed to Tokyo, it would be reasonable for you to assess your probabilities of receiving either coin at 50% for the gold coin and

No elementary phenomenon is a real phenomenon until it is an observed phenomenon.

JOHN WHEELER

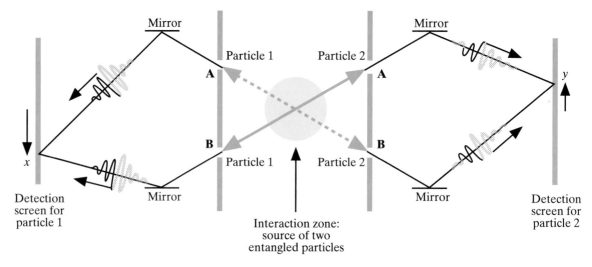

FIGURE 14.23
The position-entanglement experiment. Because of their entanglement, particles 1 and 2 coordinate their impact points x and y instantaneously, regardless of the distance between them.

> If we take quantum theory seriously as a picture of what's really going on, each measurement does more than disturb: it profoundly reshapes the very fabric of reality.
>
> NICK HERBERT, PHYSICIST, IN QUANTUM REALITY: BEYOND THE NEW PHYSICS

50% for the silver coin. However, you do have information connecting your outcome with Betty's outcome: If Betty informs you that she received a gold coin, then your probabilities jump to 0% for gold and 100% for silver; and if Betty tells you that she received a silver coin, then your probabilities jump to 100% for gold and 0% for silver. Such a connection between the probabilities of different events is called a **statistical correlation**. Two different events are statistically correlated if the outcome of one affects the probabilities associated with the other. Note that in this example, the correlation is not the result of any actual physical interaction between the two coins in the two cities; that is, neither coin causes the other actually to change from gold to silver. The correlation is due only to the prior information that your friend put a gold coin in one envelope and a silver coin in the other.

In the position-entanglement experiment, the two particles separate in opposite directions, so we expect some correlation. Referring to Figure 14.23, if particle 1 impacts at some distance x below the midpoint of its screen, then particle 2 should impact at approximately the same distance (shown as y in the diagram) above the midpoint of its screen; that is, y should be approximately equal to x. Such a correlation is similar to the gold and silver coin correlations and is due simply to the prior opposite directions of the two particles.

The interesting correlations occur when x and y differ from each other, which they can because of quantum uncertainties in the precise impact point of each particle. Given that particle 1 impacted at distance x, how likely is it that particle 2 will impact at some different distance y? Quantum theory predicts an interference pattern for these correlations: If particle 1 impacts at x, then quantum theory predicts positions y where particle 2 is particularly likely to hit, and other positions where particle 2 will certainly not hit. The second case is the most interesting: Specifically, particle 2 never hits at points y for which x and y *differ* by certain fixed amounts, such as 0.5, 1.5, or 2.5 mm. The two particles somehow "know" that they are not supposed to impact at points x and y that differ by these particular amounts. *How can they "know" that?*

Each particle's interference pattern, for a fixed impact position of the other particle, is dependent on where the other particle impacts its screen. As another way of putting this: Each particle's psi field quantum-jumps into a new pattern when the other particle impacts at a particular point—the two particles instantaneously adjust their psi fields to each other's impact point! The two particles are truly entangled with each other—anything that happens to one of them affects the two-particle psi field and hence simultaneously affects the other particle.

The two screens can be as widely separated as you like; in one recent experiment of this sort they were separated by 11 kilometers. They could be in different galaxies, yet quantum theory predicts the same results. Each entangled particle "knows" instantaneously what the other is doing. They coordinate their impact points so as not to hit when the difference between x and y is, say, 0.5 mm. If particle 1 happens to strike at $x = 0.3$ mm, particle 2 will instantaneously "know" that it must not hit at $y = 0.8$ mm, 1.8 mm, 2.8 mm, and so on, and if particle 1 happens to hit at $x = 0.4$ mm, particle 2 will not hit at $y = 0.9$ mm, 1.9 mm, 2.9 mm, and so forth. How can they instantaneously cooperate this way when they are far apart?

When you don't look, the world is made of waves. And then when you look at the world, it turns into real particles. The act of looking is what turns it into particles. This is the way the theory works. Whenever you don't look, you have to describe the world as this half-real waviness. . . . If you don't treat it this way, you get the wrong answers.

NICK HERBERT

Maybe this cooperation is not spooky. Maybe it is merely of the "gold-and-silver-coin" variety, due entirely to the prior connection. John Bell analyzed this question in 1964 and proved that the correlations are not of this ordinary variety. In other words, Bell proved that this cooperation is due to a real, instantaneous, physical contact between the particles—each particle really does "know" what the other particle is doing from one instant to the next. The cooperation occurs because the two particles behave as a single entity; they are in a single quantum state so that they both must quantum-jump at the same time. We summarize Bell's idea as follows:

Non-locality means that we cannot discuss the different parts of space independently.

JOHN BELL

> **BELL'S INTERCONNECTEDNESS PRINCIPLE**
> Quantum theory predicts that entangled particles exhibit correlations that can be explained only by the existence of real **nonlocal** (that is, instantaneous and distant) connections between the particles.*

For me it's a dilemma. I think it's a deep dilemma, and the resolution of it will not be trivial; it will require a substantial change in the way we look at things.

JOHN BELL, REFERRING TO THE IMPLICATIONS OF ASPECT'S EXPERIMENT

Bell was able to discover quantum predictions involving correlations of entangled particles that required a nonlocal explanation and that could be experimentally tested. As just mentioned, Clauser and then Aspect were the first to carry out such tests. Aspect's experiment included a refinement that showed that the connections really were instantaneous, or at least faster than light. The position-entanglement experiment is an example of such nonlocal connections between

---

*The proof of this is surprisingly nontechnical. See, for instance Nick Herbert, *Quantum Reality* (New York: Doubleday/Anchor, 1985), pp. 215–223.

FIGURE 14.24
Niels Bohr's engagement photograph, taken in 1911. Two years later he developed the first accurate theory of atomic structure.

There is no quantum world. There is only an abstract physical description. It is wrong to think that the task of physics is to find out how nature is. Physics concerns what we can say about nature.
NIELS BOHR

entangled particles, whose results fully confirm the "spooky" quantum predictions.

Such a conclusion might seem to contradict relativity theory's prohibition on faster-than-light motion. But relativity says only that energy (in the form of material objects or of radiation) cannot travel faster than light. The statistical correlations referred to in Bell's principle cannot transfer energy, so Bell's principle does not contradict relativity.

■ **DIALOGUE 7**   Can two particles be entangled even when they are not exerting forces on each other?

14.6 WHAT DOES IT MEAN?
QUANTUM THEORY AND REALITY

It isn't easy to say what all of this means. Danish physicist Niels Bohr (Figure 14.24) was the leading contributor to our understanding of quantum theory. In 1913, he invented a partially quantized version of the planetary atom, a useful stepping-stone to the later full quantum theory of the atom (Sections 14.1 and 14.2). During the 1920s and 1930s, Bohr, along with Heisenberg, Born, and others, developed a philosophy of quantum theory, a view of what the theory means, that has withstood the test of time and new experiments such as those showing nonlocal connections. This view is generally, but not universally, accepted among physicists today. The remainder of this chapter presents an updated version of this now standard **Copenhagen interpretation**.

Perhaps the most characteristic feature of quantum theory is its uncertainties. The essence of quantum uncertainty is that identical physical situations lead to different outcomes. The Copenhagen interpretation concludes that microscopic events really are unpredictable, that God really does play dice. Furthermore, when we say that the position of a microscopic particle is uncertain, we do not mean simply that we lack knowledge of its position. Rather, we mean that the particle actually *has* no definite position. A particle described by a wave packet (Figure 14.14), for example, has no definite position and no definite speed. The particle is in some sense all over its entire psi field; it fills up its entire realm of possibilities $\Delta x \cdot \Delta s$. As Heisenberg put it, a particle's psi field "introduces something standing in the middle between the idea of an event and the actual event, a strange kind of physical reality just in the middle between possibility and reality."

Now consider the effect of a precise position measurement on a particle described by a wave packet. Such a measurement finds the particle to be at some specific position x. It would be a mistake, however, to conclude that the particle was at or even near x just before the measurement. Instead, we must visualize that the particle resides all over the range Δx before the measurement and that the measurement actually *creates* the particle's position rather than simply discovering it. Measurements to some extent create the properties they detect. A

7. Yes. Once entangled particles have separated sufficiently widely, they exert no forces on each other.

FIGURE 14.25
The effect of a position measurement or of a speed measurement is to create a position and speed for the measured particle.

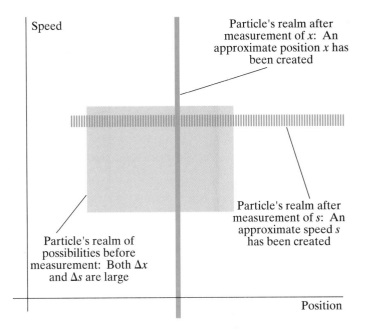

position measurement creates a position, and a speed measurement creates a speed (Figure 14.25).

The uncertainty principle prevents the simultaneous existence of both a precise x and a precise s. It is as though a baseball could be either white or spherical but not both at once. Because the existence of either property precludes the existence of the other, these two properties are said to be *complementary* to each other. When Bohr was knighted as an acknowledgment of his achievements in science and his contributions to Danish culture, he chose as a suitable motif for his coat of arms the Chinese symbol representing a similar complementary relationship of the archetypal opposites yin and yang (Figure 14.26).

The meaning of quantum theory is bound up with observation. Observations can instantaneously and radically change the observed system. But "observation" must be understood broadly. Observations are possible without physical disturbances and without human observers. For instance, the flash created by an electron hitting a screen is an observation, even though no human eye or camera or other recording device ever records it. If a macroscopic detection device *could* record the event, it is considered an observation. Indeed, we have seen in Mandel's experiment (Figure 14.21) that the mere threat of possible detection can cause a quantum jump.

The properties of particles have meaning only in the context of the entire experimental environment, because that environment helps determine a particle's psi field. An electron's position, for example, refers to the position as it would be defined by some particular position-measuring device. It is improper to think of the electron as having a position in the absence of such a measurement. As Bohr often said, attributes of microscopic particles do not belong to the particle itself but reside in "the entire measurement situation." This view that a particle's properties are created by the experimental context is called **contextual reality**, or **observation-created reality**.

There are two sorts of truth: trivialities, where opposites are obviously absurd, and profound truths, recognized by the fact that the opposite is also a profound truth.

NIELS BOHR

When it comes to atoms, language can be used only as in poetry. The poet, too, is not nearly so concerned with describing facts as with creating images.

NIELS BOHR

FIGURE 14.26
Niels Bohr's coat of arms. The inscription above the yin–yang symbol reads "opposites are complementary." In this way quantum theory's most profound philosopher acknowledged the harmony between seeming opposites, a harmony contemplated in the ancient Eastern teaching that the essence of all natural and human phenomena lies in the dynamic interplay of complementary opposites.

To what appear to be the simplest questions, we will tend to give either no answer or an answer which will at first sight be reminiscent more of a strange catechism than of the straightforward affirmatives of physical science. If we ask, for instance, whether the position of the electron remains the same, we must say "no"; if we ask whether the electron's position changes with time, we must say "no"; if we ask whether it is in motion, we must say "no." The Buddha has given such answers when interrogated as to the conditions of a man's self after his death; but they are not the familiar answers for the tradition of seventeenth and eighteenth century science.

J. ROBERT OPPENHEIMER

It moves.
It moves not.
It is far,
and it is near.
It is within all this,
And it is outside of all this.

THE UPANISHADS, HINDU
SACRED SCRIPT, CA.
1000 B.C.

Microscopic particles do not have the same reality status as, say, a penny does. A macroscopic object like a penny is well described by Newtonian physics, so it makes sense to speak of its position and speed in the ordinary sense. But to the extent that an object's behavior must be described by quantum theory, the object has, as Heisenberg put it, "a strange kind of physical reality just in the middle between possibility and reality." But microscopic particles are by no means subjective or only in the mind. For example, the flash of an electron on a screen is real. The microscopic world is real, but its reality status is not what we are used to.

Entanglement represents an extreme form of contextual reality. When two particles are entangled, each particle's nature is bound up with the other particle. Each particle becomes the context for the other. The two particles form a single object, only in two different places. Just as we must not think of an electron in a wave packet as really being at one point x and we must not think of an electron in the double-slit experiment as really coming through one slit, we must not think of entangled particles as separate. Any attempt to do so will run into contradictions with experiment.

All of this inverts the conventional relationship between the microscopic and macroscopic worlds. In the usual view, macroscopic objects such as tables are made of atoms. But quantum theory provides a sense in which tables are more fundamental than atoms, because macroscopic detectors define the conditions of existence for atoms.

Again quoting Heisenberg:

Some physicists would prefer to come back to the idea of an objective real world whose smallest parts exist objectively in the same sense as stones or trees exist independently of whether we observe them. This however is impossible. . . . Materialism rested upon the illusion that the direct "actuality" of the world around us can be extrapolated into the atomic range. This extrapolation, however, is impossible—atoms are not things.

14.7 TOWARD A POST-NEWTONIAN WORLDVIEW

Scientifically, humankind is now one century into the post-Newtonian age, but it is unclear where we are philosophically. What is our metaphysical direction today? Let us recall four key features of the Newtonian worldview (Section 5.5):

Atomism.   Atoms form the fundamental reality. Newton called them "solid, massy, hard, impenetrable particles" that "never wear or break in pieces." Tastes, colors, and the like are mere names. For example, a napkin's redness is due to motions of atoms in the napkin and in the observer. Things are not really red in themselves.

Objectivity.   Galileo wanted to remove all human influence from his experiments, in order to study nature itself without human influences. It was assumed that objectivity actually was possible, at least in principle. If everything is caused by atoms and if atomic behavior is determined by physical laws, then what is real does not depend on humans. So it should be possible to observe the universe, uninfluenced by humans.

Predictability.   Every physical system is entirely predictable. Once started, the clockwork universe was required to do precisely what it has done and what it will do.

Analysis.   Ever since Galileo, science has progressed by separating phenomena into their simplest components and studying those components. This led to a focus on the simplest and smallest components: atoms.

Today's physics denies all four of these Newtonian principles:

Atomism.   Atomism was first contradicted by the electromagnetic field, which is physically real but not made of atoms. Today, the microscopic world, including atoms, appears to be made of psi fields, but a psi field is certainly not made of atoms. Atoms themselves are made of energy. Far from being solid, hard, and impenetrable, they are empty. They are made of fields, and their rest mass, their matter, is a consequence of the energy of these fields. Far from never wearing or breaking, atoms can be annihilated. Although energy is indestructible, matter can be destroyed and created.

Atoms were once viewed as tiny things in the traditional sense. For example, when Ernest Rutherford (Chapter 8) was asked whether atomic nuclei really existed, he replied, "Not exist—not exist! Why I can see the little beggars there in front of me as plainly as I can see that spoon!" But atoms are not things in the same way that a spoon is a thing. An atom's nature is dependent on its surroundings. Descartes argued that secondary qualities such as a napkin's redness are dependent on primary realities such as atoms. Turning this order around, quantum theory views macroscopic objects as the experimental context for the atoms.

Objectivity.   The idea of objectivity was crucial to scientific progress in Galileo's and Newton's time. But over time, objectivity took on a mythical status as scientists came to believe that natural phenomena could always be divorced from their surroundings.

[W]e shall always be able to imagine other [false] theories—like the boring world of particles governed by Newtonian mechanics.

STEVEN WEINBERG, PHYSICIST, IN *DREAMS OF A FINAL THEORY*

Without being aware of it . . . we exclude the Subject of Cognizance from the domain of nature that we endeavor to understand. We step with our own person back into the part of an onlooker who does not belong to the world, which by this very procedure becomes an objective world. [Our] science . . . is based on objectivation, whereby it has cut itself off from an adequate understanding . . . of the mind. But I do believe that this is precisely the point where our present way of thinking does need to be amended, perhaps by a bit of blood-transfusion from Eastern thought.

ERWIN SCHROEDINGER, FROM *MIND AND MATTER*

In order to understand atomic structure, we must accept the idea that the future is uncertain. It is uncertain to the extent that the future is actually created in every part of the world by every atom and every living being. This point of view, which is the complete opposite of machinelike determinism, is something that I believe should be realized by everyone.

EDWARD TELLER, PHYSICIST

I don't think there's one unique real universe. . . . Even the laws of physics themselves may be somewhat observer dependent.

STEPHEN HAWKING

It would be a poor thing to be an atom in a universe without physicists. . . . A physicist is an atom's way of knowing about atoms.

GEORGE WALD, BIOLOGIST

Their [Bohr's and Heisenberg's] theoretical structure did not extend down and anchor itself on fundamental microscopic space–time realities. Instead it turned back and anchored itself in the concrete sense realities that form the basis of social life.

HENRY STAPP, PHYSICIST

We have found a strange foot-print on the shores of the unknown. We have devised profound theories, one after another, to account for its origin. At last, we have succeeded in reconstructing the creature that made the foot-print. And Lo! it is our own.

SIR ARTHUR EDDINGTON, ENGLISH ASTRONOMER AND PHYSICIST

Beginning at least with Einstein's relativity in 1905, the measurement process intruded on this belief in the possibility of complete objectivity. It became essential to specify exactly how quantities like time and space can be measured. In quantum theory the entire experimental context becomes essential to the defining properties of the objects under study. Does an electron have a position? The answer depends on the experimental context. This does not make physics subjective, because every observer using the same observational equipment still sees the same result, but it does make reality dependent on the experimental surroundings.

Predictability.   Identical causes no longer lead to identical effects. Individual events are not predictable, even in perfectly controlled experiments. A single radioactive decay, the flash of a photon, and individual chemical reactions such as those that determine a person's genetic inheritance are unpredictable quantum events. The universe is not like a clock. But statistical patterns are predictable, even though single events are not. Chance rules individual events, but nature determines the odds.

Analysis.   The analytic process assumes that it is possible to divide a phenomenon into parts without changing it. Quantum theory contradicts this notion. The analytic process works quite well for macroscopic systems. For instance, it is useful to separate the solar system into the sun, planets, and so forth and to consider the ways that each part interacts with each other part. But quantum theory says that we cannot necessarily consider a microscopic system as made of separable parts. The electron double-slit experiment cannot be broken down into an electron, on the one hand, plus the double-slit apparatus, on the other, because the electron is in part defined by the apparatus. Whereas Mars would still be Mars even if Venus did not exist, an electron changes its nature when its environment changes. Quantum entanglement is the most striking example. Two entangled particles are so closely connected that it is not possible even to think of them as independent particles. There is a microscopic wholeness that is not obvious to our macroscopic eyes.

Despite nearly a century of post-Newtonian physics, a post-Newtonian worldview is still not in sight. What grand metaphors will we eventually adopt to replace the Newtonian metaphors? Or do we even need grand metaphors? The Newtonian age certainly had its grand metaphor, one to which nearly all physical scientists subscribed, from Kepler through Maxwell, namely, the clockwork or mechanical universe. Today this metaphor continues to deeply influence our culture's view of physical reality.

Perhaps a biological rather than mechanical analogy would be more fitting. It appears from quantum physics that natural systems are closely tied to their surroundings, that they really possess the macroscopic contextual properties such as redness or hotness that they appear to possess, that they are partly undetermined, and that they cannot be separated without essentially changing their nature. All of this seems similar to our perceptions of biological organisms in interaction with their environments. Perhaps we live in an "organic" universe rather than a "mechanical" universe.

Will we construct a scientifically accurate and humane worldview that can sustain us in the post-Newtonian age and regain our philosophical roots? Humankind has barely scratched the surface of this task. It might be the critical issue for the modern world.

"I THINK YOU SHOULD BE MORE EXPLICIT HERE IN STEP TWO."

SUMMARY OF IDEAS AND TERMS

Spectroscope   A device that measures the **spectrum**, or set of frequencies, emitted by a radiation source.

A **continuous spectrum**   covers a range of frequencies; a **line spectrum** contains only precise frequencies.

Standing wave   A wave in which the medium vibrates in a wave pattern but the pattern does not move.

The **planetary model of the atom,** in which electrons move in planetlike orbits around the nucleus, cannot explain line spectra, and predicts that atoms will lose energy until they collapse.

The **quantum model of the atom,** in which the psi fields of an atom's electrons are standing waves surrounding the nucleus, agrees with all experiments so far.

The **quantum states** of the hydrogen atom are the various possible psi fields for its electron. Each quantum state is a standing-wave pattern that obeys Schroedinger's equation and represents a possible probability pattern for the atom's electron.

Quantum numbers   The numbering system used to label an atom's quantum states. The states of hydrogen are labeled by three quantum numbers, usually called N, L, and M.

The **energy level** of a quantum state is the precise, predictable energy the atom has when it is in that state. An **energy-level diagram** shows the collection of energy levels for an atom.

The **ground state** of an atom is its quantum state of lowest possible energy. Higher-energy states are **excited states.** A gas can be **excited** by heating and by an **electric discharge,** in other words an electric current flowing through it.

Quantum jump An instantaneous change of an entire psi field. When an atom quantum-jumps, the psi field for its electron changes to a new quantum state.

Emission of radiation occurs when an excited atom quantum-jumps into a lower energy level, emitting a photon. Since the photon's frequency is determined by the atom's energy levels, an atom's spectrum can be predicted from Schroedinger's equation.

Heisenberg's uncertainty principle Every material particle has an inherent **uncertainty in position (Δx)** and **uncertainty in speed (Δs)**. Although either one of these uncertainties can take on any value, the two uncertainties are related through the fact that their product must approximately equal h/m.

Realm of possibilities The range of possible positions and speeds that a microscopic particle can have at any particular time, as determined by the particle's psi field. This range cannot be smaller than is permitted by the uncertainty principle.

Wave packet A psi field moving through space and spread out over only a limited distance Δx.

Nonlocal effect An instantaneous connection between two particles that are separated in space.

Mandel's experiment Demonstrates that a detector can cause quantum jumps even if the detector is remote from the particle being detected.

Two particles are **entangled** when they interact and then separate in such a way that they form a single inseparable psi field.

Two events are **statistically correlated** if the outcome of one affects the probabilities of the other.

The **position-entanglement experiment** demonstrates nonlocal connections between two entangled particles, each of which goes through a double-slit apparatus. The statistics of the impact points of the two particles are instantaneously correlated across a distance.

Bell's interconnectedness principle Entangled particles cooperate in a way that can be explained only by the existence of real nonlocal connections. Entanglement of this sort is predicted by quantum theory and has been confirmed by experiments.

Copenhagen interpretation The consensus interpretation of quantum theory. Uncertainties, complementarity, and nonlocality are inherent in nature; observations help create the observed properties; microscopic reality is a **contextual** or **observer-created reality**.

REVIEW QUESTIONS

OBSERVING ATOMIC SPECTRA

1. The theoretical value of the Lamb shift (in the chapter's introduction) is 1057.860 ± 0.009, and the measured value is 1057.845 ± 0.009 megahertz. Do these two results agree?
2. What is the purpose of the prism in a spectroscope? What is the purpose of the thin slit?
3. Exactly what is measured by a spectroscope?
4. Describe two ways to excite a gas. What does *excited* mean, in microscopic terms?
5. Describe one way in which the planetary model disagrees with observations of atomic spectra. Describe one problem with the concept of the planetary atom.

THE QUANTUM ATOM

6. What are standing waves? How are they related to the quantum theory of the atom?
7. Describe the three-dimensional shapes of some of the states in Figure 14.8. Exactly what does one of these states represent?
8. Which state(s) in Figure 14.8 has the lowest energy? The highest? Which is a ground state? An excited state?
9. Consider any one of the states in Figure 14.8. In this state, does the electron have a predictable energy? A predictable position? A predictable speed?
10. Describe the process by which atoms create radiation.
11. What is meant by a quantum jump in an atom?
12. How many different radiation frequencies are emitted in the quantum jumps shown in Figure 14.11?

THE UNCERTAINTY PRINCIPLE

13. Does the particle represented by the psi field of Figure 14.14 have a precise position? A precise speed?
14. Which wave packet of Figure 14.15 has the more precise position? The more precise speed?
15. Does the uncertainty principle say that a particle must have a Δx that is larger than some prescribed value? What does it say?
16. Does a baseball have large quantum uncertainties or small ones? Why?

QUANTUM JUMPS

17. What happens to a particle's wave packet when a new and more accurate position measurement is performed?
18. Is it possible for a single microscopic particle's psi field to be spread out over macroscopic dimensions, such as several meters or larger? Give an example.
19. Give an example in which the switching on of a detector causes a particle's psi field to quantum-jump.
20. What is meant by a nonlocal effect?
21. What does Mandel's experiment demonstrate?

THE NONLOCALITY PRINCIPLE

22. What is a statistical correlation? Give a common example of statistically correlated events.
23. What is entanglement, and what does Bell's principle tell us about entangled particles?
24. What does the position-entanglement experiment demonstrate?

25. Can the nonlocal connections described by Bell's principle transfer energy instantaneously?

QUANTUM THEORY AND REALITY
26. According to the standard interpretation of quantum theory, which of the following are actually inherent in nature: uncertainties, wave–particle complementarity, nonlocal connections, contextual reality?
27. Describe at least two key ideas of the Newtonian worldview that are contradicted by quantum theory. Are some aspects of the Newtonian worldview contradicted by Einstein's relativity theory?

HOME PROJECTS

1. Burn different substances and observe the colors of their flames. Try scattering a little salt into a candle flame. Give a microscopic explanation, based on the quantum theory of the atom, of what you are seeing.
2. Use your radio as a spectroscope to determine the frequencies of the radio radiation present in your room. Record the frequencies and each frequency's relative intensity (don't touch the volume knob) on a scale of 1 to 10 as your ear perceives it. How long are the wavelengths your radio is detecting (consult Figure 9.7)?
3. Observe a streetlight containing mercury vapor (a gas made of mercury atoms), and a streetlight containing sodium vapor. What color dominates mercury's spectrum? The sodium atom's spectrum? What are you seeing, according to the quantum theory of the atom?

FOR DISCUSSION

1. Is the moon still there when you are not looking at it? How do you know? How could you experimentally determine whether or not it is still there when nobody is looking at it?
2. Does social materialism (the desire to possess TV sets, cars, and so forth) have anything to do with philosophical materialism (the belief that reality consists only of matter in motion)?
3. Are there senses in which contemporary U.S. culture might be said to be Newtonian? Pre-Newtonian? Post-Newtonian?

EXERCISES

OBSERVING ATOMIC SPECTRA
1. In what ways is your radio a type of spectroscope? In what ways does it differ from the spectroscope described in the text?
2. Why, when different materials burn, do they often create flames of different colors? How might the chemical composition of a burning substance be determined?
3. If you compared the spectra from two sodium vapor lightbulbs, would they be the same? What if you compared a sodium vapor bulb with a mercury vapor bulb?

THE QUANTUM ATOM
4. Explain, in terms of inertia, why the electron does nearly all the moving in a hydrogen atom.
5. Assuming that the string's length in Figure 14.5 is 2 m, could the string vibrate in a standing wave whose wavelength is 2.1 m? 1.9 m? 0.5 m? Explain.
6. Describe the three-dimensional shapes of the two quantum states shown in Figure 14.8 having $N = 3$ and $L = 1$.

7. If a very accurate measurement of an atom's mass could be made in an excited state and in its ground state, would any difference be found? (*Hint:* Remember $E = mc^2$.)
8. What happens to an atom's mass when it emits a photon? (*Hint:* Remember $E = mc^2$.)
9. In Figure 14.11, which quantum jump creates the higher frequency photon, $N = 4$ to $N = 3$ or $N = 4$ to $N = 2$? Which of the two photons has the longer wavelength?
10. In Figure 14.11, which quantum jump creates the highest frequency, $N = 5$ to 4, 4 to 3, 3 to 2, or 2 to 1? Which creates the longest wavelength?
11. The four spectral lines of hydrogen photographed in Figure 14.2 have wavelengths and frequencies that agree precisely with the four lowest-energy transitions into hydrogen's $N = 2$ energy level. Which three of these four lines are graphed in Figure 14.12? Give the initial and final N values for all four lines in the photo.

THE UNCERTAINTY PRINCIPLE

12. If Planck's constant were smaller than it is, how would the uncertainty principle be affected? What if Planck's constant were zero?
13. How would it affect you if Planck's constant were 1 J-s instead of 6.6×10^{-34} J-s?
14. If Planck's constant were smaller than it is, would this affect the sizes of atoms? How? What would become of atoms if Planck's constant were zero?
15. Think of a few common situations, unrelated to quantum theory, in which observation changes reality. Would public opinion polls be an example? Would looking at the moon be an example?
16. One everyday example in which a measurement disturbs the measured object is the measurement of the temperature of a pan of water using a thermometer. How does this disturb the temperature? Is this a quantum effect?

QUANTUM JUMPS

17. What would happen to the wave packet of Figure 14.14 if an accurate speed measurement were performed? How would the measurement affect Δx and Δs?
18. Your friend flips a coin but covers it up so that neither of you can tell whether it is heads or tails. What odds (probability) would be fair to put on heads? Suppose he uncovers it and you see that it is tails. What odds should you now assign to heads? Does this sudden shift in the probabilities have anything to do with quantum theory?
19. The graph behind the screen in Figure 13.12(b) shows the pattern formed by the psi-wave on the screen in a single-slit experiment. Is this a graph of a single electron's psi field just before or just after the electron hits the screen? What happens to the psi field when the electron hits the screen?

THE INTERCONNECTEDNESS PRINCIPLE

20. Suppose two dice are physically connected in such a way that an "even" outcome (2, 4, or 6) on either die always occurs with an even outcome on the other, and odd outcomes also always occur together. Could we then say that the outcome on one die entirely determines the outcome on the other die? Could we say that the two dice are statistically correlated? If the first die comes up 2, what are the probabilities for each of the six sides of the other die?
21. Is the statistical correlation described in the preceding exercise a quantum effect?
22. What experiment demonstrates statistical correlations that are not explainable in terms of Newtonian concepts but that are predicted by quantum theory?

QUANTUM THEORY AND REALITY

23. List several general ways in which nature is non-Newtonian and several specific phenomena (such as radioactive decay) that are non-Newtonian. In what ways is nature Newtonian? List several specific phenomena (such as the fall of a rock) that are, to a very good approximation, Newtonian.

PROBLEMS

THE QUANTUM ATOM

1. For the hydrogen atom, the energy difference $E_2 - E_1$ between the lowest two levels (Figure 14.9) is 16×10^{-19} J. Find the frequency of the photon emitted when a hydrogen atom quantum jumps from state 2 to state 1. In which region of the spectrum is this (Figure 9.7)?
2. For the hydrogen atom, the energy difference $E_3 - E_2$ between the second and third levels (Figure 14.9) is 3×10^{-19} J. Find the frequency of the photon emitted when a hydrogen atom quantum jumps from state 3 to state 2. In which region of the spectrum is this (Figure 9.7)?
3. From the information given in the preceding two problems, find the energy difference $E_3 - E_1$ between the third and first energy levels of the hydrogen atom. Find the frequency of the photon emitted when a hydrogen atom quantum jumps from state 3 to state 1. In which region of the spectrum is this?
4. According to Section 14.2, quantum states of the hydrogen atom occupy larger volumes as the quantum number N gets larger. Quantitatively, the electron's psi field occupies a more-or-less spherical region, centered on the nucleus, whose radius is proportional to N^2. For a ground-state ($N = 1$) hydrogen atom, the diameter of this psi field is about 10^{-10} m. Find the diameter of a hydrogen atom in the state $N = 3$. Find the diameter of a state having $N = 50$, expressing your answer in mm. How do the *volumes* of the $N = 50$ and $N = 1$ states compare? Scientists have created such giant hydrogen atoms.
5. It is difficult for laboratory scientists to create the giant hydrogen atoms described in the preceding problem, because these atoms' electrons are so weakly attached that the slightest disturbance destroys the high-N atomic states. But hydrogen atoms in space are sufficiently isolated that giant hydrogen atoms occur frequently enough to be observed from Earth (by viewing their spectra). Atoms having N higher than 100 have been observed. Calculate the diameter of an atom having $N = 100$, expressing your answer in millimeters. How many ground-state hydrogen atoms could fit into the volume occupied by one such giant atom?

THE UNCERTAINTY PRINCIPLE

6. Your friend tells you that it is 250 miles to Kansas City, give or take 10 miles. What is the uncertainty in your friend's estimate? Is this related to quantum uncertainties?

7. An electron (mass = 9.1×10^{-31} kg) has a speed uncertainty $\Delta s = 1$ *m/s*. How large must its position uncertainty be? Express your answer in millimeters.

8. A proton (mass = 1.7×10^{-27} kg) has a speed uncertainty $\Delta s = 1$ *m/s*. How large must its position uncertainty be? Express your answer in millimeters. If you worked the preceding problem, then compare the two answers.

9. Making estimates. The electron in a ground-state hydrogen atom is confined, by electric forces, to remain within a sphere measuring roughly 10^{-10} meters across. An electron's mass is about 10^{-30} kilograms. Use this data along with Heisenberg's uncertainty principle to estimate the speed of this electron. (*Hint:* In the ground state, the electron's speed should be roughly equal to the *uncertainty* in its speed, in other words $\Delta s = s$. See the discussion at the end of Section 14.3.) What fraction of light-speed is this?

PART FIVE

WITHIN
THE ATOM
▼
fire of the
nucleus, fire
of the sun

The Nucleus and Radioactivity

CHAPTER 15

a new force—

One way that science has expanded human awareness is by expanding the "distance scales" that humans comprehend. Prescientific cultures were aware of what they could see, down to the smallest dust particles (10^{-5} m) and up to the distance a person might see on Earth (100 km, or 10^5 m). Today, telescopes have extended our awareness to other galaxies at over 10^{21} m and to the edge of the observed universe at some 10^{26} m. At the other end of the scale, microscopes of various sorts, including the giant accelerators that investigate subatomic objects, have extended our awareness down to atoms (10^{-10} m in size), the nucleus (10^{-14} m), and subnuclear particles (10^{-19} m and still counting).

The nucleus is the star of this and the next chapter. Although the nucleus might seem remote from human concerns, technology has exploited it as a source of great power. Power always has both dangers and opportunities. It is up to all of us to see that when used, it is used wisely. So we study nuclear physics not only for its intellectual significance but also because you and I must figure out how to use this powerful knowledge beneficially and nondestructively.

Quantum principles are important to any object as small as the nucleus. And we have seen that processes within such a small region of space must occur at high energies, so high that relativistic principles are important. Thus, nuclear physics is an offspring of both of the post-Newtonian physical theories.

This chapter explores the structure of the nucleus and the nuclear reaction known as radioactive decay. Chapter 16 then studies the other two major types of nuclear reactions, fusion and fission. Section 15.1 of this chapter discusses the forces acting within the nucleus, and Section 15.2 explores the energetics of the nucleus, a topic crucial to our further discussions of nuclear physics. Then we present the physics of radioactive decay (Section 15.3), including the concept of half-life (Section 15.4). The next two sections explore two radioactivity-related cultural and societal topics: radioactive dating and its implications for Earth's history (Section 15.5), and human exposure to radioactivity (Section 15.6). Section 15.7 expands on Section 15.6 by studying technological risks more generally.

■ **DIALOGUE 1**   Where should nuclear physics be placed in Figure 5.17?

---

1. In "quantum + special relativity."

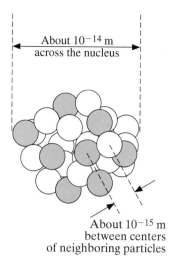

About 10^{-14} m
across the nucleus

About 10^{-15} m
between centers
of neighboring particles

FIGURE 15.1
A schematic representation of a nucleus. The colored spheres represent protons, and the white spheres represent neutrons. Diagrams like this, showing subatomic particles as if they were ordinary Newtonian objects, are useful aids to thought, but they are highly simplified. Nature at this level is quantized, having both wave and particle properties, and cannot be properly represented with any picture.

15.1 NUCLEAR FORCES: THE THIRD GLUE

Scientists are like detectives, forming hypotheses and deductions from observed clues. Let's do some nuclear detective work.

The main clue is Rutherford's evidence that the atom is mostly empty space. An atom's protons and neutrons are tightly bunched in a tiny region 10^{-14} m across (Figure 15.1).* Since protons are positively charged and neutrons are not charged, there must be an attractive force between these nuclear particles to bind them to each other, because otherwise the repulsion between the protons would blow the nucleus apart.

This attractive force must be strong enough to overcome the great repulsion that exists between two protons as close as 10^{-15} m. Gravity is always attractive. Could it hold the nucleus together? As you might guess, gravitational forces between nuclear particles are insignificant compared with electric and other forces, so gravity can't do the job. Now recall that the forces experienced in your daily life can be explained in terms of only two fundamental forces: gravity and electricity. Since the force that holds the nucleus together cannot be either of these, we deduce that there must be a third fundamental force that holds the nucleus together.

This third force must strongly attract protons to one another when they are separated by about 10^{-15} m. It must also attract neutrons to one another and to protons, since otherwise neutrons would fall out of the nucleus. But despite its great strength at short separations, this force cannot extend very far. It certainly cannot extend from one nucleus to the next in solid matter (about 10^{-10} m), because if it did, all the nuclei would clump together. In fact, we could hypothesize (remember, this is a detective story) that this force extended only far enough to hold the largest nuclei together, because if it extended much farther, still larger nuclei would be possible and would be observed in nature.

This force, which is strongly attractive between protons and neutrons at around 10^{-15} m and negligible at much larger distances, is called the **strong force**. It holds the nucleus together.

We can distinguish among three kinds of "glue" that bind things together. At the subatomic level, the strong force holds the nucleus together. At the atomic level, the electric force binds orbital electrons to their nuclei, binds atoms into molecules, and holds solids and liquids together. And at the astronomical level, the gravitational force holds planets, stars, solar systems, and galaxies together and reaches across clustered galaxies to determine the shape of the universe.

So far as we know, every force in nature can be reduced to the action of these three forces plus just one other: the **weak force** (Section 15.3). Together, these **four fundamental forces** determine the structure of our universe. We don't know why they have the properties that they do, properties that make the universe such an interesting and varied place. What if the properties of any one of them were different? For example, what if the strong force had a longer range, or what if it were weaker? What if the electric or gravitational force were stronger or weaker than it actually is? What if there were three kinds of electric charge or only one? What if gravity were repulsive instead of attractive? If any one of the

*Nuclear diameters range from about 3×10^{-15} (hydrogen) to 20×10^{-15} m (uranium).

four fundamental forces had properties that differed much from the properties that they actually have, the universe as we know it could not exist and we would not be here.

15.2 NUCLEAR ENERGY AND NUCLEAR STRUCTURE

The strong force is by far the strongest of the four fundamental forces. Since this force is so strong, the energy changes accompanying individual nuclear processes are large, compared, for example, with chemical processes (dominated by the electric force). Nuclear weapons explosions and nuclear reactors yield huge energies per kilogram of material, and gamma radiation from the nucleus is the most powerful form of electromagnetic radiation.

There is a fundamental quantum reason that nuclear energies must be large. Because the strong force has a short range, any nuclear particle is trapped in a small region of space. The uncertainty principle tells us that any such well-localized particle must have a highly uncertain speed and that the average speed itself must therefore be large (Section 14.3). Since nuclear particles must move quickly, they must have high energies. When this argument is worked out quantitatively, the conclusion is that any nuclear particle trapped in a region as small as 10^{-14} m must have an average speed of at least 3×10^7 m/s, or about 10% of lightspeed. Relativistic phenomena are important at such high speeds, and so nuclear processes necessarily involve both quantum and relativistic physics. This is indicated in Figure 15.2, which is a more detailed version of Figure 5.17. The x at the edge of the region that is forbidden by the quantum uncertainty principle indicates a typical nuclear size and speed.

A similar argument applies to entire atoms: Since the psi field for an electron in an atom (Figure 14.8) is about 10^{-10} m across, the uncertainty principle implies that atomic electrons must move at about 0.5% of lightspeed. This is indicated in Figure 15.2 by the point y.

Since nuclei are made of parts, we suspect that it is possible to alter their structure. Any process that does so is called a **nuclear reaction**. Most nuclear reactions fall into one of three types: radioactive decay, fusion, and fission.

A favorite enterprise of the medieval scientist-magicians known as *alchemists* was to try to transform one chemical element into another, especially lead into gold. Neither their chemistry nor their magic produced any gold, but the dream has turned out to be achievable. Although magic doesn't help, chemical reactions are on the right track. The problem is that ordinary atomic and molecular chemical energies are far too low to do the job. Today, *nuclear* chemists routinely turn one element into another. It is even possible to convert lead to gold, but you can't make money this way: The needed energy costs much more than the gold is worth.

Nuclear reactions are a little like chemical reactions. The chemical properties of an atom are determined by its number of orbital electrons, that is, by the **element** to which the atom belongs, specified by its atomic number. The elements and their atomic numbers are listed in the periodic table (inside back cover). The atomic number is important to nuclear reactions, too, because it is the number of protons in the nucleus. But neutrons are also significant in determining

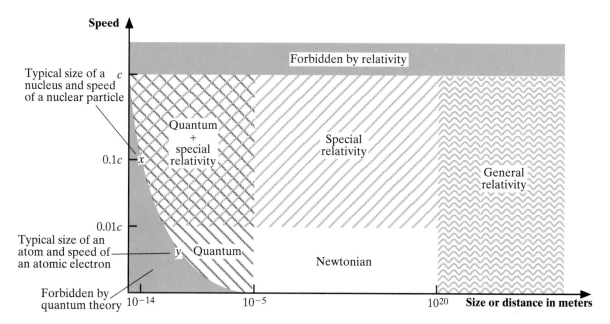

FIGURE 15.2
The domains of Newtonian, relativistic, and quantum physics. The numbers and boundaries are only representative; there are no definite borders between the domains of the theories. As shown on the diagram, atomic physics falls into the quantum domain, while nuclear physics falls into the quantum relativistic domain.

nuclear behavior. Nuclei with identical numbers of both protons and neutrons are said to belong to the same **isotope**. Just as atoms of a particular element have identical properties in chemical reactions, nuclei of a particular isotope have identical properties in nuclear reactions.

The numbering system for isotopes is only slightly more complicated than that for elements. An isotope is numbered by its **atomic number**, the number of protons it contains, and also by its **mass number**, the total number of protons and neutrons it contains. The mass number is simply the total number of particles in the nucleus. Because protons and neutrons have about the same mass, the mass of the entire nucleus is nearly proportional to the number of nuclear particles, that is, to the mass number. That's why it's called the mass number. For example, a nucleus with mass number 8 would be about twice as massive as a nucleus with mass number 4.

As an example of the numbering system for isotopes, the element carbon has atomic number 6, so every carbon nucleus has 6 protons. Because some carbon nuclei contain 6 neutrons, others contain 7, and still others contain 8, there are three different isotopes of carbon. These three isotopes have the same chemical properties but different nuclear properties. Their mass numbers are 12, 13, and 14.

We indicate specific isotopes by their chemical symbol preceded by their atomic number as a subscript and their mass number as a superscript. For example, the three isotopes of carbon are written $^{12}_{6}C$, $^{13}_{6}C$, $^{14}_{6}C$. But we often drop the atomic number because the chemical symbol actually specifies it and simply write ^{12}C, ^{13}C, ^{14}C, pronounced "carbon-12, carbon-13, carbon-14."

■ **DIALOGUE 2** Compare and contrast nuclear reactions with chemical reactions: (a) Which of the four fundamental forces is involved in each? (b) What is changed, or rearranged, by each?

■ **DIALOGUE 3** How does the mass of a ^{14}C nucleus compare with the mass of ^{12}C nucleus? How do their electric charges compare?

■ **DIALOGUE 4** How many protons and how many neutrons are in a nucleus of each of the following isotopes: $^{1}_{1}$H, $^{2}_{1}$H, $^{3}_{1}$H, $^{235}_{92}$U, $^{238}_{92}$U?

■ **DIALOGUE 5** In what way are ^{3}H and ^{3}He similar? In what way are they different?

15.3 RADIOACTIVE DECAY: SPONTANEOUS NUCLEAR DISINTEGRATION

In 1896, French physicist Henri Becquerel finished his research for the week and stored a certain uranium compound away in a drawer for the weekend. By chance, an unexposed photograph plate was stored in the same drawer. When he returned the following week, Becquerel found to his surprise that the film had been exposed, despite having been kept in a dark drawer. A lesser scientist might have shrugged his shoulders and tossed out the ruined film, but Becquerel suspected a connection between the uranium and the exposure. He discovered that he could reproduce the effect whenever he placed the uranium near photographic film. Apparently, the uranium radiated something that could expose a photographic plate. The process was called **radioactivity**. When Becquerel subjected the uranium to various chemical treatments, they produced no change in the effect. So radioactivity had little to do with chemistry. Science had had its first brush with the nucleus, even though the nucleus had not yet been discovered.

Two years later, French physicist Marie Curie (Figure 15.3) and her husband Pierre detected radioactivity in pitchblende, a tarry black substance. They were not surprised at the radioactivity because pitchblende is a known ore of uranium, but they were surprised to find that the radiation was more intense than the radiation from pure uranium, even though the pitchblende contained only low concentrations of uranium. Apparently some other substance, much more radioactive than uranium, was present in pitchblende. The Curies then took on the monumental task of chemically separating this substance from 8 tonnes of pitchblende. This involved some real detective work, because the new substance and its chemical properties were unknown. They managed to get only a bare powdery pinch, 0.01 grams, of the stuff. Like uranium, it radiated spontaneously, but it gave off rays at a much higher rate than did an equal mass of

---

2. (a) Chemical reactions involve electric forces, whereas nuclear reactions involve nuclear forces (and electric forces also, because protons are charged). (b) Chemical reactions alter the structure of molecules (by altering the electron orbits), whereas nuclear reactions alter the structure of nuclei.

3. One-sixth larger mass (the mass ratio is $14/12 = 7/6$). Same charge.

4. 1 and 0, 1 and 1, 1 and 2, 92 and 143, 92 and 146.

5. They both have 3 nuclear particles, and they have about the same mass. ^{3}H has 1 proton and 2 neutrons, while ^{3}He has 2 protons and 1 neutron.

FIGURE 15.3
Marie Curie shared, with Pierre Curie and Henri
Becquerel, the 1903 Nobel Prize for physics. Pierre
died in a traffic accident in 1906. Marie Curie
continued her research, confident that she could
succeed despite the widespread prejudice against
women in physical science. She did succeed: She
was the first woman to teach at the Sorbonne, she
was the first person to receive two Nobel prizes,
and element number 96 is named in her honor. Her
daughter Irène became a nuclear scientist, winning
the 1935 Nobel Prize in chemistry. In 1934 Marie
Curie developed cataracts and lesions on her
fingers and died of radiation-induced leukemia.
Irène Curie also died of leukemia.

Marie Curie is, of all celebrated
beings, the only one whom fame
has not corrupted.

ALBERT EINSTEIN'S EULOGY
UPON THE DEATH OF MARIE
CURIE IN 1934

uranium. Because of its powerful radiation, they named the new element
radium.

Since the discovery of uranium and radium, scientists have found that there
are a lot of radioactive substances. Every isotope heavier than bismuth (atomic
number 83, just heavier than lead in the periodic table) is radioactive, and there
are many **radioactive isotopes** of the lighter elements. For example, among the
three isotopes of carbon, ^{14}C is radioactive, but ^{12}C and ^{13}C are not.

Experiments (Figure 15.4) show that radioactive materials emit three distinct
types of rays, known as **alpha**, **beta**, and **gamma rays**. The way that these rays respond to electric or magnetic fields shows that alpha rays are positively charged,
beta rays are negatively charged, and gamma rays are not charged. These rays
come from the nuclei of various isotopes.

Closer examination reveals these rays are created in one of two types* of
spontaneous nuclear processes, known as **radioactive decay** processes. In **alpha
decay**, a radioactive nucleus spits out a particle called an **alpha particle** that is
identical with the nucleus of helium: $^{4}_{2}He$, two protons and two neutrons bonded
together by the strong nuclear force (Figure 15.5). If the alpha particle gets into
the air, collisions with air molecules will soon slow it down, after which it will
pick up two electrons from nearby atoms to become a normal atom of helium.

In **beta decay**, a radioactive nucleus spits out an electron (Figure 15.6). This is
surprising because there are no electrons in the nucleus! We'll straighten out
this mystery later. If it gets into the air, the electron collides with air molecules,
slows down, and is captured by some nearby atom to become an ordinary orbital

---

*In addition to the two types discussed here, there are other less common types of radioactive decay
processes, such as positron decay.

FIGURE 15.4
An experiment showing that radioactive materials can emit three different types of rays, alpha, beta, and gamma. The two metal plates create an electric field in the space between the plates. The experiment shows that because they are attracted to the negative plate and are repelled by the positive plate, alpha rays carry a positive electric charge. Similarly, beta rays are negatively charged, and gamma rays are not charged.

Daughter nucleus after radioactive decay

Alpha particle breaks away from the nucleus

FIGURE 15.5
Alpha decay.

Beta particle is created in the nucleus and immediately ejected

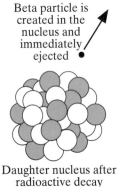

Daughter nucleus after radioactive decay

FIGURE 15.6
Beta decay.

electron. But when it is emitted by a radioactive nucleus, we call it a **beta particle** to indicate the process that created it.

Most radioactive isotopes decay by only one of these two processes. For example, uranium and radium are alpha emitters, and ^{14}C is a beta emitter. Because alpha and beta decay violently disturb the nucleus, both cause the nucleus to emit electromagnetic radiation, so each decay is normally accompanied by a gamma-ray photon. Although alpha rays and beta rays are often called radiation because they radiate outward from the nucleus, they are not electromagnetic radiations. They are, instead, streams of material particles. Gamma rays, however, are a form of electromagnetic radiation.

Radioactivity occurs because some nuclei do not hold together very well and eventually fall apart spontaneously. Such nuclei are said to be **unstable**. In a **stable nucleus**, on the other hand, the balance of nuclear forces holds the nucleus together forever, or at least until some outside influence breaks it apart. One source of instability is simply large size: A very large nucleus has a hard time sticking together because each proton tends to be pushed out of the nucleus by the repulsive force of all the many other protons. This is why all isotopes heavier than bismuth (atomic number 83) are radioactive.

In alpha decay, a small part of the nucleus is pushed out (Figure 15.7). Because the alpha particle, $^4_2$He, is one of nature's most stable structures, it is this combination of protons and neutrons that breaks away. Once separated from the nucleus, an alpha particle is pushed strongly away by the electric force from the protons in the remaining nucleus, called the **daughter nucleus**.

Whereas alpha decay is the sort of falling apart that one might expect in an unstable nucleus, beta decay is more surprising. It is caused by the weak force, which actually causes a neutron to transform spontaneously into a proton while

Alpha particle about
to break away

FIGURE 15.7
During alpha decay, an
alpha particle becomes
separated from the rest of the
nucleus and is then pushed
rapidly away by the repulsion
from the protons in the
daughter nucleus.

Before

After

FIGURE 15.8
The details of beta decay: A neutron
is transformed into a proton and an
electron, and the electron moves
rapidly away. Compare this with
Figure 15.6.

Now, the special interest of radium
is in the intensity of its rays, which
is several million times greater than
the uranium rays. And the effects
of the rays make the radium so
important. . . . These effects may
be used for the cure of several
diseases. . . . What is considered
particularly important is the
treatment of cancer. . . . Radium
is more than a hundred thousand
times dearer than gold.

MARIE CURIE, IN A TALK AT
VASSAR COLLEGE, 1922

simultaneously creating a high-energy electron (Figure 15.8). When this process occurs within a nucleus, the nucleus loses a neutron and gains a proton. At the same time, the electron created in the process is spewed out of the nucleus because it is created at high energy. This is the source of beta particles.

The nucleus itself transforms into a different isotope during radioactive decay. It is the alchemist's dream, and it has been occurring spontaneously all the time! For example, when $^{14}_{6}C$ beta decays, it loses a neutron and gains a proton, so its atomic number increases by 1 and its mass number remains unchanged. The daughter nucleus, with atomic number 7 and mass number 14, is $^{14}_{7}N$. We represent nuclear reactions the way we represent other processes: with an arrow from the initial situation to the final situation. For example, the beta decay of ^{14}C is

$$^{14}_{6}C \rightarrow ^{14}_{7}N + beta$$

As another example, $^{238}_{92}U$ is an alpha emitter. Alpha decay reduces the atomic number by 2, from 92 to 90 (thorium), and reduces the mass number by 4, so this nuclear reaction is represented by

$$^{238}_{92}U \rightarrow ^{234}_{90}Th + alpha$$

It is always useful to view processes in energy terms. What energy transformation occurs in radioactive decay? Alpha and beta particles carry microscopic kinetic energy (thermal energy) into the environment around the nucleus, and gamma photons carry radiant energy. These energies came from the nuclear structure of the radioactive material. So the universe loses nuclear energy and gains thermal and radiant energies, and the energy transformation is

$$nuclear\ energy \rightarrow thermal\ energy + radiant\ energy$$

Radioactive decay is like a landslide, but caused by nuclear forces instead of gravitational forces. In a landslide, gravity pulls part of a hill downward into a more stable, compact configuration. The slide transforms the gravitational energy of the elevated land into kinetic energy during the slide and finally into thermal energy. In radioactive decay, the forces in the nucleus cause a similar spontaneous "sliding" of an unstable nucleus into a more stable configuration, causing the nucleus to fall apart instead of downhill. In the process, nuclear energy (instead of gravitational energy) is converted to thermal and other forms.

Radioactive isotopes exist—or can be manufactured artificially in nuclear reactors—for nearly every chemical element. Because they have chemical properties identical to the chemical properties of the stable isotopes of the same element, radioactive isotopes have many medical and other uses, as well as many dangers for human health (Section 15.6). Another significant application of radioactive isotopes is radioactive dating (Section 15.5). But first we need to discuss the concept of *half-life*.

■ **DIALOGUE 6**   What is the daughter nucleus in the beta decay of radioactive iodine, $^{131}_{53}I$? In the alpha decay of radium, $^{226}_{88}Ra$?

6. $^{131}_{54}Xe$. $^{222}_{86}Rn$.

15.4 HALF-LIFE: WHEN DOES A NUCLEUS DECAY?

The central feature of quantum theory is that individual microscopic events are unpredictable, but the statistics of these events are predictable. It's like flipping coins: You cannot predict the outcome of one coin flip, but you can predict that the statistics of many flips will be approximately 50% heads. The difference between a macroscopic process like a coin flip and an unpredictable microscopic process is that "nature knows" how the coin will fall. You could use Newtonian physics to predict the outcome of a coin flip, although the measurements and calculations would be very complex. But microscopic processes are non-Newtonian and inherently unpredictable.

Given the submicroscopic nature of the nucleus, it is not surprising to learn that radioactive decay is an inherently unpredictable event. It is one of nature's most prominent examples of quantum uncertainties.

To be specific, consider the alpha decay of one uranium nucleus. Because of nature's preference for alpha particles, you can think of the surface of a big nucleus like uranium as a collection of alpha particles that drift near the surface, attracted to the nucleus by the strong force but also repelled by the nucleus's protons (Figure 15.9). The alpha particle is governed by a quantum probability wave of the same sort that is responsible for the wave properties of electrons. So there is a certain chance that any particular alpha particle will find itself outside the nucleus, far enough outside that the strong force (which, remember, has a short range) can no longer be felt. Such an alpha particle will be immediately pushed out of the nucleus by the long-range electric forces of all the protons.

Because of quantum uncertainties, nature knows only the chances that an alpha particle will be ejected, not when it will be ejected. A particular uranium nucleus might decay during the next second, in 5 minutes, or in 20 billion years.

But quantum theory does predict the statistics of radioactive decay. For example, given a large number of uranium nuclei, it is possible to predict roughly what fraction of them will decay in any particular period of time, even though it is impossible to predict exactly which nuclei will decay and which ones won't. Coin tosses are statistically predictable in this way, too: If we toss a large number of coins a single time, we can predict that roughly 50% will come up heads, but we can't predict which particular coins will be heads and which will be tails.

For any particular radioactive isotope, the most important statistic of this sort is the **half-life**, the time during which 50% of a large collection of those nuclei will decay. Because some radioactive nuclei are highly unstable and others are much more stable, different isotopes have widely different half-lives. A highly unstable nucleus is like a landslide waiting to happen: It will probably decay soon, just as the waiting landslide will probably slide soon. A more stable nucleus will probably take longer to decay. Table 15.1 gives the half-lives of several radioactive isotopes.

For example, suppose you are given 1 gram of ^{14}C, an isotope with a 6000-year half-life that beta-decays to ^{14}N, the stable form of nitrogen. After 6000 years, 50% of the ^{14}C nuclei will have decayed, and you (or rather, your descendants) will have just half a gram of ^{14}C. The other half-gram will have been transformed into nitrogen.

I hope that no one discovers how to release the intrinsic energy of radium until man has learned to live at peace with his neighbour.

ERNEST RUTHERFORD, IN 1915

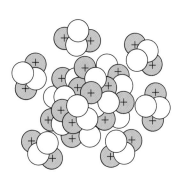

FIGURE 15.9
Think of the surface of the nucleus as a collection of alpha particles that drift near the surface. When, by chance, one of these finds itself so far from the nucleus that it no longer feels the inward pull of the short-range strong force by its neighbors, it is immediately pushed away by the long-range electric forces of all of the protons in the entire nucleus.

TABLE 15.1

Half-life and decay process of several radioactive isotopes

Isotope	Name of element	Decay process	Half-life (approx.)
$^{14}_{6}C$	carbon	beta	6000 yr
$^{90}_{38}Sr$	strontium	beta	30 yr
$^{131}_{53}I$	iodine	beta	8 days
$^{214}_{84}Po$	polonium	alpha	0.00016 s
$^{222}_{86}Rn$	radon	alpha	4 days
$^{235}_{92}U$	uranium	alpha	0.7×10^9 yr
$^{238}_{92}U$	uranium	alpha	4.5×10^9 yr
$^{239}_{94}Pu$	plutonium	alpha	24,000 yr

How much ^{14}C is left after another 6000 years? The answer is half of the half-gram, or 0.25 gram, because the half-life statistic still applies. Six thousand years is the time during which 50% of *any* macroscopic amount of ^{14}C decays. So there will be 0.50 grams of ^{14}C after 6000 years, 0.25 grams after 12,000 years, 0.125 grams after 18,000 years, and so forth. These values are graphed in Figure 15.10 and connected with a smooth curve showing the amount remaining at any time. This is the **decay curve** for ^{14}C. The decay curve for any radioactive isotope is identical to that for ^{14}C, only with a different half-life. Thus Figure 15.10, with the time measured in half-lives instead of in years (bottom line along the horizontal axis), may be used for any radioactive isotope.

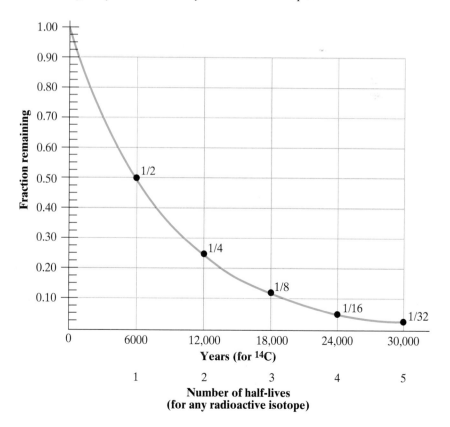

FIGURE 15.10
Radioactive decay curve for ^{14}C and for any other radioactive isotope.

■ **DIALOGUE 7 MAKING ESTIMATES** Starting with 16 pennies, suppose you toss all of them and remove the ones that come up tails. You toss the remaining coins and again remove the tails. If you continue the process, about how many tosses must you make before you get down to a single remaining penny? (Try it!) What if you started with 100 pennies? What is the half-life of pennies in this game?

■ **DIALOGUE 8** If you start with a gram of pure ^{14}C, how much will remain after 36,000 years? Approximately how much will remain after 10,000 years?

15.5 RADIOACTIVE DATING: WHEN DID WE COME FROM?

Radioactive decay is a kind of clock. If you know how much of a substance has decayed, you can read the elapsed time from the decay curve (Figure 15.10). For instance, if a radioactive substance has decayed away to 25% of its original amount, you will know that it has been two half-lives since it started decaying.

Carbon dating is one example. Essentially all of Earth's carbon is one of the two stable carbon isotopes, ^{12}C and ^{13}C. But in Earth's atmosphere, about 1 carbon atom in a trillion is the radioactive form, ^{14}C. Since the half-life of ^{14}C is only 6000 years and Earth is far older than 6000 years, you might wonder how any ^{14}C could still be in the atmosphere. The answer is that **cosmic rays**, high-energy particles that travel through outer space, continually replenish it. They occasionally enter the atmosphere, collide with atmospheric nitrogen, and transform the nitrogen nucleus into ^{14}C.

Because the carbon in all biological organisms comes ultimately from the atmosphere, ^{14}C is distributed throughout the biological world at 1 ^{14}C atom per trillion carbon atoms, about the same as the atmospheric ratio. This ratio of radioactive to stable carbon is maintained until a living organism dies. Then the ^{14}C gradually decays. The time elapsed since death can be determined by measuring the amount of ^{14}C remaining compared with the amount of stable carbon. For instance, if an old ax handle has only a quarter of its normal amount of ^{14}C, the tree from which the ax handle was made must have died two half-lives ago, or 12,000 years.

Such a measurement can determine the time elapsed since death if one knows how much ^{14}C was present when the organism died. The usual assumption is that the fraction of ^{14}C in the atmosphere in the past was nearly the same as it is today, so the fraction in a long-dead tree when it died would have been 1 ^{14}C per trillion carbon atoms, the same as today. But is this assumption correct? We will discuss this in a moment.

Many different radioactive decay processes are used for a variety of dating methods. Each method compares a radioactive isotope with a second nucleus to determine what fraction of the radioactive isotope has decayed and how long it has been decaying.

Radioactive as well as nonradioactive methods have taught us "when we came from" (Table 15.2). To put these dates in perspective, the table's third col-

In all of us there is a hunger, marrow deep, to know our heritage, to know who we are, where we have come from. Without this enriching knowledge, there is a hollow yearning. No matter what our attainments in life, there is a vacuum, an emptiness, and a most disquieting loneliness.

ALEX HALEY, IN *ROOTS*

We shall not cease from
   exploration
And the end of all our exploring
Will be to arrive where we started
And know the place for the first
   time.

T. S. ELIOT, IN *FOUR QUARTETS*

---

7. After 4 tosses, you will (on the average) be down to 1 penny. If you start with 100 pennies, you will be down to about 1 after 6 to 7 tosses. The half-life of pennies is 1 toss.

8. 1/64 of a gram, or 0.016 gram. Figure 15.10 shows that after 10,000 years, which is 1.67 half-lives, about 30%, or 0.30 gram, will remain.

TABLE 15.2

When we came from: some approximate dates relevant to our species

Event	Years before present	On a 12-hour clock
Creation of Earth	4.6 billion	noon
Life		
Earliest fossils (single cell)	3.3 billion	3:00 P.M.
First vertebrates	500 million	10:45 P.M.
First reptiles	300 million	11:15 P.M.
First mammals	200 million	11:30 P.M.
First primates	70 million	11:50 P.M.
Humans		Seconds before midnight
Earliest hominids[a]	4 million	40 s
Use of stone tools	2 million	20 s
Genus *Homo*	1 million	10 s
Homo sapiens	100,000	1 s
Culture		
Invention of agriculture	10,000	0.1 s
First cities, earliest writing	5,000	0.05 s
Scientific age (Copernicus)	500	0.005 s
Industrial age	250	0.002 s
Twentieth century	100	0.001 s

[a]Humanlike creatures that were direct human forebears.

umn compresses Earth's history into 12 hours. This perspective can be an eye-opener (Figure 15.11). Throughout most of Earth's history, the dominant life-forms have been simple organisms such as algae. On the 12-hour clock, complex animals appear only late in the evening; the earliest humans evolve at less than a minute before midnight; and our species (*Homo sapiens*) appears at 1 second before midnight. All of human culture spans far less than 1 second. Great human movements that changed the face of our planet, events such as the

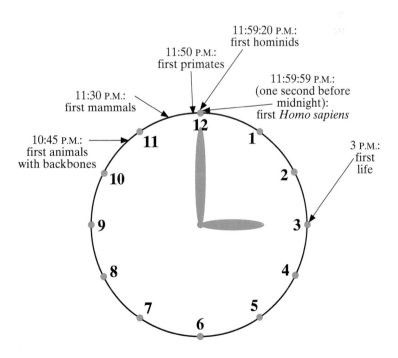

FIGURE 15.11
The "cosmic clock" compresses Earth's history into 12 hours. *Homo sapiens* does not appear until one second before midnight. A lot happens during this final second: agriculture is invented at 0.1 s before midnight, followed by cities, the scientific revolution, and the industrial revolution.

Science can purify religion from error and superstition; religion can purify science from idolatry and false absolutes. Each can draw the other into a wider world, a world in which both can flourish. . . . Such bridging ministries must be nurtured and encouraged.

POPE JOHN PAUL II

About 43% of Americans, for example, doubted that humans are descended from other animal species and 45% said that early humans lived at the same time as dinosaurs.

NATIONAL SCIENCE BOARD, SCIENCE AND ENGINEERING INDICATORS, 1989

spread of agriculture, the human population explosion, the Industrial Revolution, and the information revolution, have occupied only a moment on the world stage. As Norman Mailer put it (Section 1.1), the itch has been to accelerate.

H O W   D O   W E   K N O W ?

Cross-checking radioactive dating

How much confidence should we have in the dates given in Table 15.2? Scientists continually seek to check such results by verifying them in a variety of independent ways. If several different methods, perhaps some of them based on radioactive decay and some based on independent nonradioactive processes, all agree, our confidence in all of them is increased. It would be amazing if they were all wrong *in exactly the same way*; scientists tend not to believe in such amazing coincidences. Because of this habit of cross-checking, science has been called a *web of consistency*.

There is reason to expect errors in carbon-inferred dates: If the amount of cosmic rays hitting the atmosphere were different in the past than it is today, the ^{14}C content in the biological world would have been different then than now. For this reason, scientists have cross-checked the carbon-based dates and have found that carbon dating actually is in error, giving dates that are 10 to 20% too young.

One method of cross-checking is by studying ^{14}C in the annual growth rings of trees. We can date living trees (up to 5000 years old) by simply counting these growth rings, and we can date dead wood of unknown age by matching the rings' growth patterns (variations in ring thicknesses, caused by varying climate) with the growth patterns in dated specimens. Tree specimens up to 10,000 years old can be dated in this way. There is little reason to doubt these dates, for they are

FIGURE 15.12
Checking carbon dating by using the tree-ring method (crosses) and a uranium method (dots). Apparently carbon dating gives ages that are a few thousand years too young for objects that are around 10,000 to 20,000 years old. The greatest discrepancy between carbon dating and the other two methods is 15%.

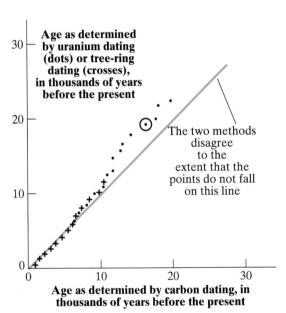

This is of course a drastic suggestion—orthodox geologists indeed reject it out of hand. However, there is no other alternative. If the Bible is the Word of God—and it is—and if Jesus Christ is the infallible and omniscient Creator—and He is—then it must be firmly believed that the world and all things in it were created in six natural days and that the long geological ages of evolutionary history never really took place at all.

HENRY M. MORRIS, DIRECTOR OF THE INSTITUTE FOR CREATION RESEARCH, IN *SCIENTIFIC CREATIONISM*, CREATION-LIFE PUBLISHERS, SAN DIEGO, 1974

We do not look upon the Bible as an authority for science or history. We see truth in the Bible as not to be reduced solely to literal truth, but also to include salvation truths expressed in varied literary forms.

PASTORAL STATEMENT FOR CATHOLICS ON BIBLICAL FUNDAMENTALISM, NATIONAL CONFERENCE OF CATHOLIC BISHOPS, 1987

It is false, however, to think that the theory of evolution represents an irreconcilable conflict between religion and science. A great many religious leaders and scientists accept evolution on scientific grounds without relinquishing their belief in religious principles.

SCIENCE AND CREATIONISM, NATIONAL ACADEMY OF SCIENCES, COMMITTEE ON SCIENCE AND CREATIONISM

The Bible clearly teaches that God created the heavens and the earth and all that is in them in six days of approximately 24 hours. . . . Not only is the Bible clear about creation, but science itself, properly done, will never deny the Creator. It is men's conclusions and beliefs about the created world which challenge the Biblical account of creation, not the facts. Thousands of good scientists today, many of them famous, are creationists.

OUR MIRACULOUS WORLD, BIBLE-SCIENCE ASSOCIATION, MINNEAPOLIS, 1987

obtained by simple counting. The results show that the older rings are as much as 10% older than is indicated by their carbon dates.

Another method of cross-checking studies the ^{14}C content in coral reefs. These reefs can be dated by the carbon method and also by a uranium method that is thought to be reliable to within 1%. Figure 15.12 shows the results of this cross-check (dots on the graph) and also the tree-ring cross-check (crosses). The dots show the uranium age (along the vertical axis) at which a given carbon-inferred age (horizontal axis) occurs. If both dating methods were accurate, these dots would lie on the straight line shown, along which the uranium age equals the carbon age. For example, the data point circled shows that a carbon age of 16,000 years corresponds to a uranium age of 19,000 years, a difference of about 15%.

Tree-ring dating is entirely independent of radioactive methods, and uranium dating is unrelated to cosmic rays striking the atmosphere. It is hard to believe that all three could be far wrong and yet that they all would arrive at about the same answer.

Many such cross-checks are possible, using a variety of radioactive and nonradioactive methods. Together, they produce a strong scientific consensus in support of such results as Table 15.2.

In the past, many scientists and nonscientists believed that Earth was only a few thousand years old. For example, Kepler suggested (Chapter 1) that "God waited six thousand years" for an observer such as Tycho Brahe. Such estimates were based on counting backward through the generations of the Old Testament and so calculating a date of about 6000 years ago for Adam and Eve.

Creationism (see also Sections 7.4 and 12.7) is the belief that the Old Testament, including Genesis, can be read as science and that Earth and the major types of biological organisms, including humans, all were created separately (not connected by evolution) and at roughly the same time, just a few thousand years ago.* Many Americans continue to hold creationist beliefs. Questions about human origins remain controversial because of their religious significance, tending to divide Christians into those who interpret the Bible "liberally," as a spiritual and moral guide but not as an authority for science or history, and those who interpret it "literally," as an authority for science and history.

There is a debate in the United States today about the teaching in public schools of topics such as Earth's age and the age and origin of different species, including humans. This question is perhaps the most important of the pseudoscience issues (astrology, extrasensory perception, unwarranted faith in extraterrestrial visitations, and so forth). These debates are really about scientific methodology, that is, about the validity of science itself.

Today, the hypothesis that Earth is only a few thousand years old conflicts with data and principles from astronomy, physics, chemistry, geology, biology, paleontology, archaeology, and history. There is a broad scientific consensus in favor of an Earth that is far older than a few thousand years. Scientific results are always open to question, and the estimates in Table 15.2 will doubtless be revised as new results come in, but the general picture of an Earth measured in bil-

*This is essentially the definition of creationism given by one of its major proponents, Henry M. Morris, in his book *Scientific Creationism* (San Diego: Creation-Life Publishers, 1974).

lions of years, with a long-term evolution toward more complex organisms and human forebears measured in millions of years, appears to be both durable and fruitful.

■ **DIALOGUE 9**   A future archaeologist digs up a body and measures an average of only about 1 ^{14}C decay per minute in 1 gram of the body's carbon. This is about 10% of the decay rate when the person was alive. About how long has this person been dead?

■ **DIALOGUE 10 MAKING ESTIMATES**   How radioactive are you? The human body is 18% carbon by weight. The biological world contains some 50 billion ^{14}C atoms per gram of carbon, of which some 10 atoms decay every minute. Estimate the number of radioactive carbon atoms in your body and the number inside you that decay in 1 minute.

15.6 HUMAN EXPOSURE TO IONIZING RADIATION

The walls of your room, the air you breathe, and even your body are radioactive (Dialogue 10). What are the effects? Are the effects different today than they were before the nuclear age? What are the risks? Can or should anything be done?

The alpha, beta, and gamma particles emitted by radioactive materials damage biological cells. The damage is done when these high-energy particles pass, like little bullets, through a cell, ionizing (knocking electrons out of) some of the cell's molecules. Because they can do this, alpha, beta, and gamma rays are called **ionizing radiations**. Since X rays also have enough energy to ionize biological material, they, too, are classified as ionizing radiation, although they do not come from the nucleus.

Ionization changes a molecule's chemistry. Because nature has spent several billion years perfecting the biological cell, any change is likely to be for the worse. Any of the cell's functions can be altered, depending on which molecules are ionized. The most important effects result from damage to a cell's DNA molecules, because DNA carries the biological information inherited by other cells.

The biological damage done in humans by ionizing radiation is measured in a unit called the **rem** (for Roentgen Equivalent in Man). If a person receives ionizing radiation, the number of rems received is a direct measure of the number of damaged cells. A quantitative feel for the rem is best obtained by looking at examples.* For instance, the amount of radiation that an average person in the United States receives every year from all sources is about 0.3 rems, or 300 **millirems**. At the other extreme, a sudden dose of 1000 rems causes death within 30 days.

---

9. The decay curve, in Figure 15.10, is down to 10% after about 19,000 years.

10. If your body's mass is 60 kg (130 pounds), you have about 12 kg, or 12,000 gm, of C. The number of ^{14}C atoms in your body is about $(50 \times 10^9) \times (12,000) = 600 \times 10^{12}$, or 600 trillion. The number of atoms that decay in 1 minute is $10 \times 12,000 = 120,000$. Your body's total radioactivity is actually twice this large, owing to other radioactive isotopes in your body.

*Here is the precise definition of the rem: For gamma rays and X rays, the rem is the amount of radiation that would produce 0.01 joule of absorbed ionizing energy in 1 kilogram of biological material. The definition is a little more complicated for alpha and beta rays.

FIGURE 15.13
A victim of radiation sickness, 23 days after the Hiroshima bombing. The spots on his face are hemorrhages beneath the skin, caused by a weakness of the blood vessels and blood-clotting defects. He died a few days later. These acute, short-term effects are quite different from the long-term effects, primarily cancer, caused by low levels of ionizing radiation.

There are three main types of biological damage to humans. The most immediately obvious is **radiation sickness**, damage to the red blood–forming cells of the bone marrow and to the cells that line the intestinal wall. A sudden dose of 25 to 100 rems to the whole body causes short-term changes in the blood that the person might not notice. At 100 to 300 rems, the effects on the blood and the intestines produce typical symptoms of radiation sickness: fever, vomiting, damaged red blood cells, reduced white blood cells and platelets, loss of hair, spontaneous internal and external bleeding from weakened blood vessels, and small hemorrhages beneath the skin. A dose of 500 rems produces 50% fatalities; 1000 rems cause death within 30 days; and 10,000 rems cause death within hours.

Fortunately, humans have seldom experienced doses large enough to produce acute effects. The most severe examples have been the nuclear bombs dropped by the United States in 1945 on the Japanese cities of Hiroshima and Nagasaki (Figure 15.13) and the 1986 Chernobyl (in Ukraine) nuclear reactor accident. Most of our knowledge of radiation damage to humans has come from these events, and from the medical uses of radiation.

The second form of radiation damage is **mutation**, an inheritable alteration of the genetic material in a sperm or egg cell. Radiation-caused mutations have been observed in experiments with fruit flies and other species. Mutations can produce successive generations of altered offspring. They are almost always harmful, but occasionally a mutation is advantageous, and this effect is in fact essential to the process of biological evolution. To date, studies of radiation survivors and their descendants at Hiroshima and Nagasaki have found no observable inherited mutations, but this does not guarantee that no mutations have occurred. Since recessive mutations can lie dormant for several generations before becoming apparent, it will be some time before mutations can be completely ruled out among descendants of nuclear bomb survivors.

The third form of damage is **cancer** in ordinary body cells. The rate of some cancers observed in nuclear bomb survivors is far above normal. Among the Hiroshima survivors, the leukemia rate between 1950 and 1985 was four times the normal rate for an unexposed population. Many other forms of cancer were double the normal rate. Based on the available statistics, the National Academy of Sciences estimates that a sudden radiation dose of 50 rems produces about a 4% probability of eventual death by radiation-caused cancer. This means that out of every 100 people exposed to 50 rem, an average of 4 will ultimately die of radiation-caused cancer.

Although ionizing radiation can cause cancer, it can also be used to treat it. Cancer cells, because they grow rapidly, are especially susceptible to destruction by radiation. To minimize damage to normal cells, a narrow beam of gamma rays or X rays is directed at the tumor. To keep the dose as low as possible to all other parts of the body, the radiation source is rotated while remaining focused on the tumor. The radiation may come from a radioactive isotope such as ^{60}Co, or from an X ray machine (Figure 9.9). In some cases, a tiny radioactive source is inserted directly into the tumor. To treat thyroid cancer, the radioactive isotope ^{131}I is injected into the blood. Because the thyroid gland concentrates any iodine present in the bloodstream, the ^{131}I becomes concentrated in the thyroid, particularly in cells that are growing abnormally, killing the defective cells.

Radioactive isotopes are also put to beneficial use as tracers in medicine, agriculture, and industry. A given compound is chemically synthesized using a radioactive isotope such as ^{14}C or ^{3}H. These "tagged" molecules can then be followed by radiation detectors as the compound moves through the human body, through a living plant, or through an industrial process. For example, the details of how food is digested or how a medical drug distributes itself in the body can be traced in this way.

The principal open question about radiation and cancer concerns the effects from lower exposures, below 50 rem. Here, statistical evidence gives uncertain conclusions. The working hypothesis is that smaller exposures do have effects and that these effects are proportional to the exposure level. In other words, since 50 rems cause an estimated 4% probability of cancer, 25 rems is assumed to cause a 2% probability, 12.5 rems is assumed to cause a 1% probability, and so forth.

As an example, one cancer-causing isotope produced by nuclear weapons explosions and nuclear reactors is radioactive strontium, $^{90}_{38}$Sr. After a nuclear explosion, ^{90}Sr and other radioactive isotopes attach themselves to atmospheric dust particles that eventually fall to Earth as **radioactive fallout**. As you can see from the periodic table, strontium is chemically similar to calcium. If you breathe or eat ^{90}Sr, it will do what calcium does, namely, migrate to your bone marrow. Since the bone marrow is where red blood cells are created, the connection between radioactivity and leukemia (a blood condition) is not surprising. Isotopes such as ^{90}Sr that are both radioactive and chemically active in the human body are especially dangerous, because once inside the body they combine with other chemicals and stay there, rather than being quickly released. Because of isotopes like ^{90}Sr, radioactivity is much more dangerous inside the body than outside it.

There is a lot of disagreement about the long-term effects of low radiation doses, comparable to the amounts people receive every day. Because the effects are difficult to observe among human populations, predictions are difficult to check. A prudent working hypothesis is that even a single microscopic decay process has a small probability of causing a cancer or mutation and that at these low levels the probability of harm is proportional to the amount of radiation received. This hypothesis is considered prudent or conservative in the sense that the facts might be less severe than this hypothesis predicts.

Table 15.3 shows what kinds of radiation most Americans receive. As you can see, people generally receive about one-third of a rem during each year of life. In addition to the scientific uncertainty about the damage caused by such low doses, the doses in Table 15.3 are themselves uncertain estimates.

Figure 15.14 shows the relative significance of each source of radiation. Most of it is natural, from sources that have been present in nature for thousands of years. Two widely discussed artificial sources, nuclear power and nuclear weapons tests, contribute only a tiny fraction. By far the largest artificial source is medical.

Most of the average dose is thought to come from a single source, **radon**, which comes from the alpha decay of radium in the ground. Because radon is a gas, it escapes into the atmosphere shortly after it is created underground. It is not really radon itself that is dangerous, because it is chemically inert and is breathed in and right back out before it can decay (its half-life is 4 days). But radon's decay products are both radioactive and chemically active, and they attach themselves to microscopic airborne particles, are breathed in, become lodged in the lungs, and can lead to lung cancer. Because radon collects inside closed houses, it is five times more concentrated in the average U.S. home than in outdoor air. Most of the radon in homes enters through the substructure.

TABLE 15.3

Ionizing radiation received by the U.S. population. Estimated annual effective dose received in one year from various sources, averaged (per person) across the U.S. population.

Source	Annual dose per person (millirem)
Natural	
Radon from the ground	200
Cosmic rays	27
Rocks and soil	28
Internal consumption	40
Subtotal	295
Artificial	
Medical and dental X rays	39
Nuclear medicine	14
Consumer products	10
All other[a]	1
Subtotal	64
Grand total	359

[a]Occupational, nuclear weapons, nuclear power and its fuel cycle, and miscellaneous.

SOURCE: National Research Council's fifth committee on the biological effects of ionizing radiation (BEIR V), 1990

Radon levels vary widely among different houses, with some not much above outdoor levels and others a hundred or a thousand times greater.

You can check the radon level in your home by buying a small device called a track detector at a hardware store. Alpha particles leave a track as they pass through the detector. After the detector has been in your house for a few months, you can mail it to a laboratory where the tracks are counted and used to calculate your home's radon level. If your home has a high level, you might want to seek professional assistance in reducing it. Sealing cracks doesn't help much, because air pressure differences suck air from the ground through the substructure and up into the house, even in well-sealed houses. Simple ventilation can help, though. The most effective solution is a blower inserted between the house's concrete base and the gravel on which the concrete is laid to change the pressure system beneath the house so that the house will no longer suck air from the ground.

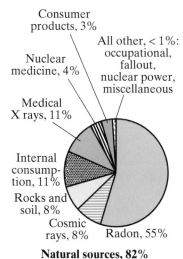

Nonnatural sources, 18%

Consumer products, 3%

All other, < 1%: occupational, fallout, nuclear power, miscellaneous

Nuclear medicine, 4%

Medical X rays, 11%

Internal consumption, 11%

Rocks and soil, 8%

Cosmic rays, 8%   Radon, 55%

Natural sources, 82%

FIGURE 15.14
Sources of radioactive risk: the relative contributions from various radiation sources that the average U.S. citizen commonly receives.

15.7 RISK ASSESSMENT: DEALING WITH RISK IN A TECHNOLOGICAL SOCIETY

To live is to risk: There is risk in drinking a cup of coffee, there is risk in going out of the house, and there is risk in not going out of the house (such as breathing in radon).

The nuclear accident at **Chernobyl** provides a good starting point for thinking about risk. In 1986, an explosion and fire at the former Soviet Union's Chernobyl nuclear power plant threw massive amounts of radioactive isotopes into the air. It was history's worst nuclear power accident. The immediate toll was

237 cases of radiation sickness, with 31 fatalities within a month, all among the reactor's personnel or emergency response personnel such as firefighters.

It is possible to estimate the number of long-term cancer deaths that may be caused by Chernobyl during the succeeding several decades. Radioactive fallout from the accident spread over much of Europe, producing an irregular fallout pattern (Figure 15.15). Many Europeans received total individual doses ranging from 10 millirem (equivalent to one diagnostic X ray) to over 1 rem. The fallout doses were much larger closer to Chernobyl. Over 100,000 people were evacuated from a 30-kilometer zone around the accident, after receiving some 20 rems per person within a few days. Significant effects persist today in Ukraine and Belarus in the former Soviet Union.

The two most harmful isotopes were radioactive iodine and cesium, ^{131}I and ^{137}Cs. Both are biologically active. Although ^{131}I has a half-life of only 8 days and so soon decayed to harmless levels, it was inhaled during the first 8 days after the accident, and it also fell onto grass that cows ate, so it entered milk supplies where much of it was consumed during the first 8 days (see Figure 15.15, which shows only ^{131}I exposures). ^{137}Cs has a half-life of 30 years and so will remain dangerous in the soil for decades. Both isotopes are expected to cause cancer deaths.

Some 17,000 excess cancer deaths might result from Chernobyl, with about half of those in the former Soviet Union nations of Russia, Belarus and Ukraine, and the remainder in Europe. In the same population, one might normally expect about 125 million cancer deaths. So, although the number of deaths is cer-

FIGURE 15.15
The spread of radiation following the Chernobyl nuclear power plant accident. The three shaded regions represent three different levels of exposure to ^{131}I. During the first four days following the accident, persons in the three gray areas received the iodine radiation exposures indicated.

tainly large, they represent an increase in the cancer death rate of only 17×10^3 / 125×10^6, or about 0.01%. This tiny percentage change will be undetectable in the overall cancer statistics, because it is far smaller than the unpredictable natural variations, or "noise," in the total cancer rate.

However, some of the more specific effects are or will be detectable. For example, the childhood thyroid cancer rate among children living within 500 kilometers of Chernobyl has risen many-fold above its pre-1986 rate, an effect that stands out in the data. Of the 1 million children exposed to radiation in Belarus, it is thought that a few percent (i.e., a few tens of thousands) will eventually contract the disease—a significant and detectable effect.

One way to make sense of such numbers is by making numerical comparisons with other risks. For example, the 17,000 long-term worldwide deaths from Chernobyl are fewer than half the number of people killed by automobiles in 1 year in the United States alone (nearly 50,000—a number that I like to remember for comparison with other statistics that I come across in the news) and about the same as the number of murders in the United States in 1 year (20,000). Comparisons such as this can help put abstract numbers into perspective.

For another example, radon exposure causes between 5000 and 20,000 (the number is highly uncertain) U.S. lung cancer deaths per year. Since this is comparable to (although somewhat less than) the annual numbers of U.S. car deaths and murders—and these are generally considered serious problems—radon should be considered serious, too.

One general way to comprehend such risks is to look at probabilities of death for one person. For example, fallout from Chernobyl might kill 17,000 out of a total population in Europe and the former Soviet Union of about 750 million, representing a fraction of 17×10^3 / 750×10^6, or about 2×10^{-5}, or $2 / 10^5$. One way to look at this is that each of these people has on the average about 2 chances in 100,000 (10^5) of dying from Chernobyl-caused cancer. That is, in a randomly chosen group of 100,000, about 2 will die of this cause.

Quantitative risk estimates such as these give us a rational way to compare risks. Table 15.4 lists several activities, all of which carry the same quantitative risk of death, 1 in 1 million. All these risks are equivalent in the sense that if everybody in a population of 1 million were to perform any one of the indicated activities, one of them (on the average) would eventually die from it. From an individual person's point of view, every time he or she does one of these activities, there is 1 chance in 1 million (still on the average) that it will eventually kill that person. It's a little unnerving.

To say that everything has its risks is really another way of saying that we all must die. In the long run, the total risk of death is 100%. It is sad but true, and one might as well try to deal with it sensibly. Many risks, such as radon exposure, are impossible to reduce to negligible proportions, and it is usually impossible to reduce any particular risk to absolutely zero. Furthermore, attempts to reduce risk can sometimes be counterproductive. For example, if you don't get that chest X ray, you might die from tuberculosis. You could significantly reduce your radon risk by living outdoors, but that would expose you to other risks. If you really wanted to reduce your radon risk, you could live outdoors and breathe only once every hour. But then you would suffocate. You could decide to do nothing that is nonessential if it carried any risk. But then you would probably die of boredom.

Radiation safety specialists have thought a lot about radiation risk, and they offer two general pieces of wisdom: First, every radiation dose may carry some

TABLE 15.4

The risks of daily life: Activities carrying an average risk of death of one part per million

Activity	Cause(s) of death
Ionizing radiation	
One chest X ray at a good hospital	Cancer from ionizing radiation
Traveling cross-country once by jet	Cancer from cosmic ionizing radiation
Living 1 week in a building	Cancer from indoor radon
Living 5 weeks outdoors	Cancer from outdoor radon
Living 2 months in Denver	Cancer from cosmic ionizing radiation
Living 5 years next to nuclear power plant	Cancer from ionizing radiation
Living 50 yrs within 5 mi of nuclear power plant	Accident
Internal consumption	
Smoking 1.4 cigarettes	Cancer and heart disease
Living 2 months with a cigarette smoker	Cancer and heart disease
Drinking 0.5 liters of wine	Cirrhosis of the liver
Normal consumption of tap water for 1 year	Cancer from chloroform
Drinking 30 12-ounce cans of diet soda	Cancer from saccharin
Eating 40 tablespoons of peanut butter	Liver cancer from aflatoxin B
Eating 100 charcoal broiled steaks	Cancer from benzopyrene
Travel	
3 miles by motorcycle	Accident
10 miles by bicycle	Accident
30 miles by car	Accident
800 miles by train	Accident
1000 miles by commercial airplane	Accident
6000 miles (cross-country) by jet	Cancer from cosmic ionizing radiation
Work	
Spending 1 hour in a coal mine	Black lung disease
Spending 3 hours in a coal mine	Accident
Other	
Living 2 days in New York or Boston	Air pollution

SOURCE: Richard Wilson, "Comparing Risks," *Physics and Society*, October 1990, pp. 3–5.

risk, so no exposure is permissible unless it carries a compensating benefit. Second, the dose to any person should be kept as low as is reasonably possible, taking into consideration all other factors (social, economic, and so forth). In other words, be aware, and balance the risks against the benefits. Similar principles apply to all the risks of living.

■ **DIALOGUE 11**   Given that ^{131}I and ^{137}Cs have half-lives of 8 days and 30 years, respectively, how much time must pass after the Chernobyl accident before the radiation from each isotope will have decreased to 1% of its original level?

■ **DIALOGUE 12 MAKING ESTIMATES**   Your doctor recommends a diagnostic X-ray exposure of 1000 millirem (1 rem). The lifetime risk of death from a single 1 rem dose is 0.1%. In a population of 1 million, how many would be killed by this exposure? Suppose the disease is detected, thanks to the X rays, and then treated. If the disease had a 10% chance of killing you, did you reduce your risk by undergoing the X-ray diagnosis?

---

11. After 6 to 7 half-lives, the decay curve is down to about 1%. For ^{131}I, 6 half-lives is 48 days, and for ^{137}Cs, 6 half-lives is 180 years.

12. Because 0.1% is 1 part per 1000, or 1000 parts in 1,000,000, about 1000 persons would be killed. You reduced your risk from 10% to 0.1% (a risk-reduction factor of 100).

SUMMARY OF IDEAS AND TERMS

Strong force   Holds the nucleus together, acts between nuclear particles (protons, neutrons), is strongly attractive at separations of around 10^{-15} m, and is negligible at larger distances.

The four fundamental forces   Gravitational, electromagnetic, strong, and weak.

Nuclear reaction   Any process that alters the structure of a nucleus.

Element   Same as *chemical element* (Chapter 2). A substance that cannot be chemically decomposed. An atom whose nucleus has a specific **atomic number** (number of protons), regardless of how many neutrons it might have.

Isotope   A particular type of nucleus. Specified by its atomic number (number of protons) and **mass number** (total protons and neutrons). A symbol like $^{14}_{6}C$ represents an isotope with the atomic number 6 and the mass number 14.

Radioactive decay   The spontaneous emission, by a nucleus, of an **alpha particle** (a helium nucleus that breaks off a larger nucleus) or a **beta particle** (an electron created in the nucleus). These two processes are called **alpha decay** and **beta decay**, respectively. At the time of decay, a **gamma particle**, or high-energy gamma-ray photon, is also emitted. The streams of particles that are emitted by a macroscopic sample of radioactive material are called **alpha rays, beta rays,** and **gamma rays**.

Daughter nucleus   The nucleus that remains after a radioactive decay has occurred.

A **radioactive** or **unstable nucleus** is a nucleus that can undergo radioactive decay.

Radioactive isotope   An isotope that is radioactive. An isotope that is not radioactive is **stable**.

Half-life   of a radioactive isotope. The time during which half of a macroscopic amount of the isotope will decay.

Decay curve   A graph of the amount of a radioactive material remaining, versus time.

Radioactive dating   Determining the ages of old objects by using radioactive methods. In **carbon dating**, radioactive ^{14}C in a dead organism is measured as a fraction of the total carbon in order to determine how long the ^{14}C has been decaying and when the organism died.

Cosmic rays   High-energy particles that travel through outer space.

Ages   Some approximate ages determined by several radioactive and other methods are Earth, 5 billion years; life, 3 billion years; humans (hominids), 4 million years.

Creationism The belief that the biblical book of Genesis can be read as scientific truth and that Earth and the major biological organisms all were created only a few thousand years ago.

Ionizing radiation Any atomic or subatomic emissions that are capable of ionizing biological molecules. Alpha particles, beta particles, gamma rays, and X rays are the main types. The biological damage is measured in a unit called a **rem**. A **millirem** is one-thousandth of a rem. The main types of damage are **radiation sickness** (damage to the red blood-forming cells of the bone marrow) which can appear quickly following exposure, and two longer-term effects: **mutation** (an inheritable alteration of the genetic material in a sperm or egg cell) and **cancer**.

Radioactive fallout Dust that falls to the ground carrying radioactive isotopes from a nuclear explosion or nuclear accident.

Natural radiation is ionizing radiation from natural sources, and **artificial radiation** is from human-made sources. Natural sources include **radon gas**, cosmic rays, the ground, and internal consumption. The main artificial source is medicine.

Chernobyl Site of history's worst nuclear power accident, in 1986. A few tens were killed from short-term effects, and thousands more are expected to die of long-term cancers.

Quantitative risk estimates The quantitative evaluation of human-made and natural risks, especially for the purpose of comparing different risks.

REVIEW QUESTIONS

NUCLEAR FORCES

1. In what ways is the nucleus non-Newtonian?
2. Name the four fundamental forces, and describe the main function of each of the three that act as "glues" to hold things together.
3. Which of the four forces holds the nucleus together, and which tends to push it apart?
4. What property of the strong force causes nuclei to be so small? Why can't much larger nuclei exist?

NUCLEAR ENERGY AND NUCLEAR STRUCTURE

5. Explain why nuclear particles must move rapidly and so must have high energy.
6. How are nuclear reactions similar to alchemy?
7. What is an isotope, and what quantities need to be known in order to specify a particular isotope?
8. What are the differences among $^{12}_{6}C$, $^{14}_{6}C$, and $^{14}_{7}N$? Compare their numbers of protons, numbers of neutrons, chemistry, nuclear chemistry (behavior in nuclear reactions), numbers of orbital electrons in the neutral atom, and mass.

RADIOACTIVE DECAY

9. What kinds of rays, or particles, are emitted by radioactive nuclei? Describe each kind.
10. What happens to the atomic number and mass number during alpha decay?
11. What happens to the atomic number and mass number during beta decay?
12. What energy transformation occurs during radioactive decay?

HALF-LIFE

13. In radioactive decay, what quantity cannot be predicted? Of what basic principle is this an example?
14. We know that ^{131}I has an 8-day half-life. If you start with 100 grams, how much will remain after 24 days?

15. If you started with 100 grams of ^{14}C (6000-year half-life) and only 3 grams remain, about how much time has elapsed?

RADIOACTIVE DATING

16. Explain how carbon dating works. What kinds of objects does it work on? Where does radioactive carbon come from?
17. An ax handle has a C^{14}/C ratio that is only 1/32 of the ratio found in living organisms. About how old is this ax handle?
18. According to science, about how old is Earth: a few thousand years, a few million years, a few hundred million years, a few billion years, or a few trillion years? How about the human race (hominids)?
19. Describe at least one way to cross-check a particular dating method such as carbon dating.

HUMAN EXPOSURE TO IONIZING RADIATION

20. Which of the following are ionizing radiations: radio, ultraviolet, gamma, alpha rays, X rays?
21. What is a rem?
22. Name and describe the three main types of biological damage caused by ionizing radiation.
23. List two natural sources of ionizing radiation and one artificial source.
24. List some useful applications of radioactive isotopes.

RISK ASSESSMENT

25. What will be the long-term health effects of the Chernobyl accident?
26. What is fallout, and where can it come from?
27. What is the meaning of a risk (of cancer, for example) of 3/1,000,000 per person?
28. Where do we find radon in our normal living environment? How does it get there?

during the twentieth century and are still far from a fundamental scientific understanding of it.

Fusion and fission, the topic of this chapter, are nuclear reactions in which the strong nuclear force comes fully into play. The most useful way for us to think about the physics of these reactions is in terms of energy, one of the themes of this book. Sections 16.1 and 16.2 continue the discussion of nuclear energy that was begun in Section 15.2. Section 16.1 presents the energetics of hydrogen fusion and why the stars shine. Section 16.2 is a general discussion of the nuclear energy present in the various nuclei found in nature. Section 16.3 applies these ideas to tell one of nature's most fascinating tales: how the atoms in your body and elsewhere were forged in stars. Section 16.4 presents the physics of fission, and Section 16.5 discusses the method, called a chain reaction, by which fission is used to get useful energy from the nucleus. Unfortunately, the first major application of humankind's nuclear knowledge was the nuclear bomb. Because the bomb continues to be a threat to the world, and because of its many lessons for our scientific age, Sections 16.4, 16.5, and 16.6 discuss the fission bomb in historical context. Finally, Section 16.7 presents another unfortunate product of our nuclear knowledge, the fusion bomb.

16.1 FUSION: THE FIRE IN THE SUN

Radioactive decay (Chapter 15) is one way to convert nuclear energy into other forms. There are two other ways, called *fusion* and *fission*. These are the subjects of this chapter. Briefly, in **nuclear fusion**, two nuclei combine or *fuse* to form a single larger nucleus, and in **nuclear fission**, a single nucleus splits or *fissions* roughly in half to form two smaller nuclei.

Continuing the nuclear detective story begun in Section 15.1, let's see what we can deduce about the energetics of the simplest multiparticle nucleus, the hydrogen isotope ^2_1H, made of one proton and one neutron. Suppose you tried to separate this nucleus into an isolated proton and neutron. Because the proton and neutron are held together by the strong nuclear force, this would not be easy (Figure 16.1).

■ **DIALOGUE 1** Which situation has more *energy*, the ^2_1H nucleus before separation or the separated proton and neutron? In which situation is the nuclear *force*, acting between the proton and neutron, greater?

The separated proton and neutron has more energy than the ^2_1H nucleus. This excess energy is a form of **nuclear energy**, since it is due to nuclear forces. The energy relationship here is like that of two objects, such as Earth and a rock, that are attracted by gravity: The gravitational energy is larger when the rock is held higher above Earth's surface, in other words, when the two objects have a larger separation (Figure 16.2).

FIGURE 16.1
You would have to do work to pull apart the two particles in the nucleus of ^2H.

1. Because you must do work on the nucleus to separate it, the work-energy principle says that the system's energy increases. So the separate proton and neutron have more energy than the ^2_1H nucleus. But the nuclear force is larger when the proton and neutron are closer together, before separation.

FIGURE 16.2
Nuclear energy is similar to gravitational energy: Greater separation means greater energy.

(a) In which case does the system have greater gravitational energy: when the rock is at a lower or a higher altitude?

(b) Which has greater nuclear energy: a proton and neutron that are closer together or farther apart?

Now imagine putting a $^2_1$H nucleus together, instead of pulling it apart. This process could be represented as

$$n + p \rightarrow\,^2_1\text{H},$$

where n and p stand for neutron and proton. As we have seen, the left side of this process has more nuclear energy than the right side. And, since energy is always conserved, this energy must show up in some other form after this process. In other words, this process is one way to transform nuclear energy into other forms. It is, in fact, one form of nuclear fusion. It occurs whenever neutrons and protons (or $^1_1$H atoms) are in sufficiently close proximity to come within range of each other's strong force. The strong force then pulls the two tightly together—like a stretched rubber band snapping together because of elastic forces. When a large mass of neutrons and protons fuse in this manner, this snapping together causes heating and the emission of high-energy gamma radiation. So the energy transformation is

$$\text{NuclE} \rightarrow \text{RadE} + \text{ThermE}$$

Neutron-proton fusion was especially important during the first few minutes after the big-bang creation of the universe (Section 16.3) and formed much of the $^2_1$H that is still present today. This reaction occurs today during supernova explosions of giant stars (see Section 5.4), and contributes to the radiant and thermal energy output of these explosions.

A more significant fusion reaction in today's universe is the fusion of two hydrogen nuclei, either $^1_1$H (a proton) or $^2_1$H. Again, nuclear energy is converted to other forms when the two nuclei come within range of each other's nuclear force and snap together. In fact, stars are made mostly of hydrogen, and the fusing of $^1_1$H with $^2_1$H is responsible for much of the energy of the stars, including the sun.

■ **DIALOGUE 2** What nucleus is created by the reaction $^1_1$H + $^2_1$H? Write out the complete reaction formula. What is the energy transformation?

But there is an interesting catch in the ^{1}H + ^{2}H fusion process. Unlike the $n + p$ reaction, the electric force is involved in the ^{1}H + ^{2}H reaction because both nuclei are charged. The two nuclei repel each other electrically at relatively great

---

2. The helium isotope $^3_2$He is created. $^1_1$H + $^2_1$H $\rightarrow\,^3_2$He. NuclE $\rightarrow$ RadE + ThermE.

distances, much greater than the nuclear distances at which their strong forces attract each other. So it's not easy to get the two nuclei sufficiently close for them to feel each other's strong force. Stars solve this problem by first heating up to millions of degrees at their centers during the star-formation process (see Section 5.3). At such temperatures, the electrons are stripped off the fast-moving atoms to create an ionized gas of rapidly moving hydrogen nuclei and unattached electrons. The nuclei are moving so rapidly that they are able to move through one another's region of repulsive electric force and get close enough to fuse.

So thermal energy is required if fusion is to occur in a collection of hydrogen nuclei. Once the hydrogen is hot enough to fuse, the fusion reaction itself creates radiant and thermal energy. We can summarize these energetics as follows:

$$\text{ThermE}_{\text{in}} + \text{NuclE} \rightarrow \text{ThermE}_{\text{out}} + \text{RadE}$$

Quantitatively, the thermal energy output is far larger than the thermal energy input needed to make the reaction occur. So the reaction, once started, is self-sustaining. Energetically, this is like a fire. It takes thermal energy, for instance a burning match, to start paper burning; but once started, the combustion reaction creates more thermal energy than is needed to make the reaction occur, so it sustains itself. The difference between fusion and a fire is that fusion involves nuclear forces and nuclear energy, while a fire involves chemical forces (electromagnetic forces between atoms) and chemical energy. Because of the central role of thermal energy, a self-sustaining fusion reaction is called a **thermonuclear reaction.**

H O W   D O   W E   K N O W ?

Checking $E = mc^2$

Nuclear reactions are good examples of mass–energy equivalence (Section 11.4), because the energies involved are large enough to produce significant mass changes. As an example, let's consider the ^{1}H + ^{2}H reaction. Since ^{3}He has less nuclear energy than the separated ^{1}H and ^{2}H nuclei, the principle of mass-energy equivalence predicts that the helium nucleus should also have less mass (Figure 16.3).

The masses of all three nuclei are known:

$$\text{mass of } ^1_1\text{H} = 1.6727 \times 10^{-27} \text{ kg,}$$

$$\text{mass of } ^2_1\text{H} = 3.3437 \times 10^{-27} \text{ kg,}$$

$$\text{mass of } ^3_2\text{H} = 5.0066 \times 10^{-27} \text{ kg.}$$

The first two masses add up to 5.0164×10^{-27} kg, which is 0.0098×10^{-27} kg more than the mass of the helium nucleus. Just as Einstein predicted, the mass becomes less when the system loses energy.

To validate mass–energy equivalence quantitatively, the energy loss during the reaction (the energy that is transformed into radiant and thermal energy) must be directly measured. This can be done by using the transformed energy to

FIGURE 16.3
The mass of the whole does not equal the sum of the masses of its parts: When ^{1}H and ^{2}H fuse to form ^{3}He, the ^{3}He has less total energy when it is at rest, so it must have less mass.

warm up water and measuring the resulting temperature change of the water. The measured energy *released* (in other words, transformed) per individual fusion reaction turns out to be 8.815×10^{-13} joules. Let's see if this does equal the known mass difference times the square of the speed of light:

$$\begin{aligned}
0.0098 \times 10^{-27} \times c^2 &= 0.0098 \times 10^{-27} \times 9 \times 10^{16} \\
&= (0.0098 \times 9) \times (10^{-27+16}) \\
&= 0.08815 \times 10^{-11} \\
&= 8.815 \times 10^{-13} \text{ joules}
\end{aligned}$$

It checks.

■ **DIALOGUE 3**   The sun's total power output of 400 trillion trillion watts (400×10^{24} joules per second) comes from three types of fusion reactions whose net effect is to convert the sun's hydrogen into $^4_2$He. How much mass does the sun lose every second? Despite this enormous mass loss, the sun will end its life 5 billion years from now with only 1% less mass than it had when it turned on, 5 billion years ago!

16.2 THE NUCLEAR ENERGY CURVE

Many other nuclei, in addition to ^{1}H and ^{2}H, can be fused together to form larger nuclei. Here are two examples:

■ **DIALOGUE 4**   What nucleus is created by the fusion of two $^4_2$He nuclei? Would this process transform nuclear energy into other energy forms, or would it transform other energy forms into nuclear energy? Write down the energy transformation in as much detail as possible.
■ **DIALOGUE 5**   What nucleus is created when $^4_2$He fuses with $^8_2$Be? Write down the energy transformation.

As we see from these examples, fusion builds heavier nuclei such as $^8_4$Be and $^{12}_6$C out of lighter nuclei such as $^4_2$He. This is, in fact, how the heavier nuclei were created in the universe (see Section 16.3). The energy transformation for these processes is

$$\text{ThermE}_{in} + \text{NuclE} \rightarrow \text{ThermE}_{out} + \text{RadE}.$$

But there is an important difference in the energetics of these different fusion processes. As an example, let's compare the fusion of ^{1}H and ^{2}H with the fusion of ^{4}He and ^{4}He. In the ^{4}He + ^{4}He process, the electric repulsion between the nu-

---

3. mass = E/c^2 = 400×10^{24} / 9×10^{16} = $(400/9) \times 10^{24-16}$ = 44×10^8 kg = 4.4×10^9 kg, or 4.4 million tonnes every second!

4. $^8_4$Be. Nuclear energy transforms into other forms. Like the fusion of ^{1}H and ^{2}H (Section 16.1), this reaction uses thermal energy and also creates thermal energy: $\text{ThermE}_{in} + \text{NuclE} \rightarrow \text{ThermE}_{out} + \text{RadE}$.

5. $^{12}_6$C. $\text{ThermE}_{in} + \text{NuclE} \rightarrow \text{ThermE}_{out} + \text{RadE}$.

clei is stronger than it is in the ^{1}H + ^{2}H process, because each He nucleus contains two protons while each H nucleus contains only one proton. So a higher temperature is needed for the He + He process than for the H + H process, requiring a larger thermal energy input. As you can see from this example, the fusing together of heavier and heavier nuclei requires higher and higher temperatures and thus greater and greater amounts of thermal energy input.

It turns out that, as we fuse together heavier and heavier nuclei, we eventually come to a point where the enormous thermal energy input required to sustain the reaction becomes larger than the thermal energy output, so the reaction cannot be self-sustaining. It happens that this limit is reached at atomic number 26 (iron). That is, it is possible to create nuclei that are lighter (lower in atomic number) than iron by reactions in which more thermal energy is created than is used in each reaction. But this favorable energy balance turns around for elements heavier than iron.

The **nuclear energy curve**, Figure 16.4, shows this point graphically. It graphs nuclear energy (more precisely, the nuclear energy per nuclear particle) versus the mass number (the number of nuclear particles), for all of the isotopes. The curve shows the two features that we have emphasized: First, the entire curve lies below the straight line representing the energy of the separated nuclear particles. This must be so, because it takes work to separate any nucleus into its component protons and neutrons (Section 16.1). Second, the curve *decreases* until we get to about mass number 55, corresponding to iron (atomic number 26), after which it *increases*. The turn-around point occurs because of the greater electric repulsion between heavier nuclei, as described in the preceding paragraph. The decreasing portion (mass numbers 1 to 55) tells us that, for nuclei lighter than iron, fusion converts nuclear energy to other forms such as thermal energy and thus can be self-sustaining in energy. But the increasing portion (above 55) tells us that for nuclei heavier than iron fusion results in an *increase* in nuclear energy and thus a decrease in other forms such as thermal energy. Such a reaction can be maintained only if there is an outside source to provide the required thermal energy.

FIGURE 16.4
The nuclear energy curve. The solid curve shows the energy per nuclear particle for nuclei of various mass numbers. Iron, with a mass number of about 55 and an atomic number of 26, has the least nuclear energy per particle, so it is the most stable nucleus.

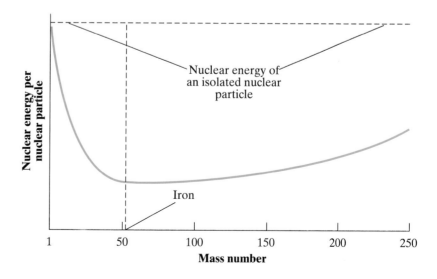

Although nuclear energy cannot be released (converted to other forms) by means of the fusion of heavy nuclei, the shape of the nuclear energy curve suggests another possibility: If, instead of fusing heavy nuclei, we began with a heavy nucleus and *fissioned* (split) it, nuclear energy would be released. Such a process could be self-sustaining in energy.

Let's summarize this important point: *Nuclear energy can be released (converted to other forms) not only by the fusion of very light nuclei but also by the fission of very heavy nuclei.*

16.3 CREATION OF THE UNIVERSE AND THE ATOMS: WE ARE STAR DUST

The scientific consensus, since at least 1965, has been that our universe began in a single event some 12 billion years ago, a violent explosion that created space and time themselves, and that created the different forms of energy and matter. The creation event is called the **big bang** (Section 11.7).

HOW DO WE KNOW?

Evidence for the big bang

Four independent lines of evidence support the big-bang theory:

1. Astronomers proposed the theory in 1929 because they discovered evidence that all the galaxies throughout the universe are receding from one another just as if they had been driven apart by an explosion. There is now considerable evidence, based on observation of the radiation from distant galaxies, confirming this expansion of the universe. Judging from the separation speeds and distances that we see today, the galaxies should have been together at the same place some 12 billion years ago.

2. In 1964, radio astronomers detected the *cosmic background radiation*, the faint remnant of the hot initial explosion, that still fills the universe. The radiation has now cooled all the way down to −270°C.* This cooled radiation has too little energy to be visible and is now in the radio region of the spectrum. Its observed characteristics, such as its temperature, agree with the big-bang theory's predictions.

3. In 1992 an observing satellite mapped the high-frequency radio radiation from all directions in space. The results showed that the background radiation contains large but very subtle "ripples" of just the sort expected if the initial big bang did indeed develop into the structured universe that we see today (Figure 16.5).

4. The fourth line of evidence concerns nuclear fusion and creation of the elements. During the first moments following the big bang, the only kinds of

If you're religious, it's like looking at God.

GEORGE SMOOT, LEADER OF THE TEAM THAT ANNOUNCED IN 1992 THE DISCOVERY OF THE STRUCTURE OF THE COSMIC BACKGROUND RADIATION THAT REMAINS FROM THE CREATION OF THE UNIVERSE

*This is just 3 degrees above absolute zero, the lowest possible temperature, the temperature at which all microscopic motion is the least it can be without violating the uncertainty principle.

FIGURE 16.5
First light. A 12-billion-year-old "fossil": map of the temperature differences in the background radiation emitted by the big bang. The light regions are slightly warmer than the dark regions. The radiation responsible for this map was emitted just 100,000 years after the initial creation event. Before 100,000 years, the universe was so hot that protons and electrons could not unite to form hydrogen, and the resulting mix of charged particles absorbed any radiation present. At 100,000 years, the universe had cooled enough for electrons to combine with protons to form neutral hydrogen atoms and radiation was able to propagate through the universe for the first time. Although it has now cooled into the microwave region of the spectrum, you are looking at the pattern of the universe's first light.

TABLE 16.1

Predicted nuclear composition of universe at 4 minutes after start of big bang. Current spectroscopic observations of the oldest material in the universe agree well with these predictions.

Isotope	Relative concentration by mass
$^1_1$H	75%
$^2_1$H	5–10 parts in 100,000
$^3_2$He	2–5 parts in 100,000
$^4_2$He	25%
$^7_3$Li	2–5 parts in 10 billion

ordinary matter that could exist were unattached protons, neutrons, and electrons. The universe was too hot for protons and neutrons to stick together and form nuclei or for electrons to go into orbit around the protons. It was not until 3 minutes after the beginning that the universe had cooled enough for protons and neutrons to attach themselves to one another. According to well-developed theories of nuclear physics, conditions during just the next 1 minute were right for the nuclei to fuse. After this single minute, the universe was too cool and too dilute for fusion. Nuclear physics predicts that during this fourth minute of the universe, about 25% of the protons ($^1_1$H) and neutrons fused into just four different isotopes: $^2_1$H, $^3_2$He, $^4_2$He, and $^7_3$Li. The universe was then made of just five isotopes, in the proportions stated in Table 16.1.

Astronomers have made spectroscopic measurements of the light from the oldest stars, stars that presumably formed from the original material created in the big bang and that have changed little since that time. These measurements show relative amounts of $^1_1$H, $^2_1$H, $^3_2$He, $^4_2$He, and $^7_3$Li that are in excellent agreement with the theoretically predicted amounts (Table 16.1). This quantitative agreement with the big-bang theory's predictions for so many different isotopes is the most important evidence in support of the theory.

The universe is still made mostly of hydrogen and helium, although heavier elements created since the big bang now contribute a small percentage. Nearly all the hydrogen and helium can be traced back to the big bang. Although our bodies contain no helium, the hydrogen made in the big bang is one of the four common elements (C, O, H, N) present in living organisms.

Once matter began gravitating together, nuclear fusion lit up the stars and began transforming lighter nuclei into heavier ones. Stars such as the sun convert hydrogen to helium during most of their history. But a star cannot burn (fuse, actually) forever. After a few billion years, a star such as the sun uses up the available hydrogen at its center. Then the fusion ceases; the star collapses further; this collapse causes an increase in temperature; and the higher temperature ignites new fusion processes such as

$$ {}^{4}_{2}\text{He} + {}^{4}_{2}\text{He} \rightarrow {}^{8}_{4}\text{Be} $$

$$ {}^{4}_{2}\text{He} + {}^{8}_{4}\text{Be} \rightarrow {}^{12}_{6}\text{C} $$

These hotter thermonuclear reactions then cause the star to expand. It becomes a *red giant*, far larger than the original star, destroying any life that might have existed on its planets. The star converts its helium to heavier elements such as carbon. Eventually, all possible fusion material is used up; fusion ceases; and gravity reasserts itself. The star collapses once again, but there is now nothing to stop it until it has shrunk all the way down to become a white dwarf (Section 5.3). The elements, such as carbon and oxygen, created by such a star remain bound up in the white dwarf.

Where, then, did all the various elements that we see on Earth come from? The answer is that they came from stars that exploded. Most stars settle down without too much fuss to become white dwarfs. But a few, namely, those that are much more massive than the sun, go through violent death throes. Because high-mass stars are pulled strongly inward by their own gravity, they become extremely hot during the collapse that follows the consumption of the star's usable hydrogen. The high temperatures ignite thermonuclear processes that create heavier and heavier elements, all the way up through iron, with atomic number 26. But the nuclear energy curve (Figure 16.4) tells us that thermonuclear reactions cannot create elements heavier than iron, because such reactions would consume thermal energy, cooling the star and ending the reaction. Iron continues forming at the star's core until the core becomes so massive that it cannot hold itself up against its own gravity. The solid iron core then suddenly collapses, reducing the core's radius from 1000 km to 20 km in just 1 second, an unimaginably violent process. The energy sent outward by this collapse heats and lights the star dramatically. All of this literally rips apart the star in a **supernova explosion** and throws much of its material out into space. The remaining core material collapses further, becoming either a spinning, superdense neutron star or a black hole (Section 5.4).

What about the elements heavier than iron? The nuclear energy curve tells us that no self-sustaining process can create these elements because they require a net increase of nuclear energy and this energy must come from some outside source. Supernova explosions provide just such an outside source. Some of the energy of the explosion itself is transformed into the nuclear energy needed to fuse together the elements heavier than iron.

Supernova explosions are fairly rare events. In our galaxy, only five have been visible to the naked eye during the past 1000 years. In addition, one was visible in a neighboring galaxy in 1987 (Figure 16.6). From Earth, a supernova explosion looks like a new (*nova*) star because of the star's sudden brightening; hence its name. The 1987 supernova occurred in a neighboring galaxy, the Large Magellanic Cloud, 160,000 light-years away from Earth.

Without supernova explosions, only the material listed in Table 16.1 could be spread throughout the universe; all stars and planets would be formed from that material; and nobody would be here to tell this story. Because our sun is a "second-generation" star that formed several billion years after our galaxy first formed, it incorporated heavier elements from supernovae that exploded before the sun was born. We can thank our stars for the rare, ancient, and distant supernova explosions that sent forth the elements heavier than helium to places such as this one, where a new star and a new planetary system eventually formed and where a tiny fraction of the fused star stuff has been borrowed, for a few years, by our bodies.

■ **DIALOGUE 6** Because the supernova shown in Figure 16.6 occurred in a galaxy 160,000 light-years away, when did it actually occur?

■ **DIALOGUE 7** During a supernova explosion, if an iron nucleus captured a neutron and then the nucleus beta-decayed, into what element would it be transformed? What if the iron nucleus captured three neutrons and then beta-decayed twice? This is the type of process that created the elements heavier than iron.

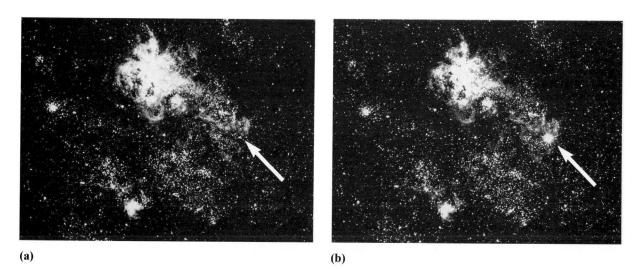

(a) **(b)**

FIGURE 16.6
The supernova of 1987, the brightest supernova in 400 years. Its light reached Earth on February 23, 1987. "Before" (a) and "after" (b) photos show the star as it looked before and shortly after the explosion. The supernova is the bright star on the right in figure (b).

---

6. 160,000 years ago.

7. Atomic number 27: cobalt. Atomic number 28: nickel.

The unleashed power of the atom
has changed everything save our
modes of thinking, and we thus
drift toward unparalleled disaster.

EINSTEIN

16.4 THE DISCOVERY OF FISSION: PASSAGE TO A NEW AGE*

It has been said that those who cannot learn from history are condemned to repeat it. The saga of the discovery and first use of nuclear fission is a powerful case study in science and society, a story that we must assimilate if we want to use science and technology without, in one way or another, destroying ourselves. It is a heroic story of a job that needed to be done in the face of a world-threatening enemy. Once these events were set in motion in the early days of World War II, the story's conclusion at Hiroshima and Nagasaki may have become, like all high tragedy, inevitable.

Scientists discovered radioactive decay, the first known nuclear reaction, in 1896. During the next several decades, they studied the new phenomenon intensely. Using a simple technique now common in high-energy physics, scientists bombarded tiny things such as nuclei by throwing other tiny things at them and observing what happened.

In 1933, Irène Joliot-Curie (daughter of Marie and Pierre) and her husband, Frédéric Joliot (Figure 16.7), bombarded a thin aluminum foil with alpha particles. This created a previously unknown isotope of phosphorus. Although natural phosphorus is stable, the new isotope was radioactive. It was the first creation of a radioactive isotope and the first artificial release of nuclear energy in radioactive decay. Joliot foresaw the potential consequences: "We are entitled to think that scientists, building up or shattering elements at will, will be able to bring about transmutations of an explosive type. . . . If such transmutations do succeed in spreading in matter, the enormous liberation of useful energy can be imagined."[†]

At about the same time, scientists bombarded beryllium nuclei with alpha particles and detected a previously unknown particle that was ejected from the beryllium during the collision. The new particle was electrically neutral. They had discovered the neutron.

The subtle new particle played a key role in the application of nuclear energy. Being uncharged, neutrons can sneak into a nucleus without having to overcome the electrical repulsion that protons feel when they approach the positively charged nucleus. And unlike the electron, which does not feel the strong force, the neutron interacts strongly once it sneaks inside.

Hungarian physicist Leo Szilard (Figure 16.8) was a visionary and a lifelong admirer of another visionary, the author H. G. Wells. Wells's 1914 novel, *The World Set Free*, predicted nuclear energy, nuclear bombs, nuclear war, and world government. Szilard was extremely inventive and quick to grasp the possibilities presented by the neutron: Perhaps some energetically favorable nuclear reaction in some material would emit neutrons; perhaps these neutrons could then bombard other nuclei in the same material and create a whole series of similar reactions in a large mass of material. In this way, neutrons might be the key

FIGURE 16.7
Irène and Frédéric Joliot-Curie at the Radium Institute in Paris, in about 1935. They created previously unknown isotopes by bombarding the nuclei of the elements with fast-moving alpha particles. The institute was founded by Irène's mother, Marie Curie.

---

*I am indebted to Richard Rhodes's *The Making of the Atomic Bomb* (New York: Simon & Schuster, 1986) for most of the historical details and quotations in Sections 16.4, 16.5, and 16.6. Rhodes's book is the definitive work on the topic, a beautiful and intelligent account.

[†]Rutherford and others had pursued the release of such energies since the beginning of the century.

FIGURE 16.8
Leo Szilard. This extremely inventive physicist was the first to understand that the neutron could be the key to unlocking large amounts of nuclear energy by means of a chain reaction. He found experimentally that the neutrons released in the fission of ^{235}U would be sufficient to sustain such a reaction. He was an activist both in promoting U.S. development of nuclear weapons during the Second World War, and then in promoting world peace and banning the use of nuclear weapons after the war.

FIGURE 16.9
Enrico Fermi in 1930. In 1934, he bombarded nuclei using the newly discovered neutron instead of with alpha particles as the Joliot-Curies had been doing. In 1938, he received the Nobel Prize for the discovery of many new radioactive isotopes created by this process. He then immigrated from Mussolini's fascist Italy to the United States where, concerned that the Axis powers would beat the Allied powers to the development of nuclear weapons, he played a key role in the Manhattan Project. Element 100 is named in his honor.

to releasing useful amounts of nuclear energy. But Szilard also realized that such a vision, if realized, would be a two-edged sword, both hopeful and fearful.*

In 1934, Enrico Fermi (Figure 16.9) tried bombarding nuclei using the newly discovered particle, the neutron, instead of using alpha particles as the Joliot-Curies had been doing. He reasoned that a neutron, being uncharged, should be able to find its way into the nucleus much more easily than could an alpha particle. He worked his way through the elements to uranium, creating 40 new radioactive isotopes in the process. One element that Fermi bombarded was uranium, element number 92 and the heaviest natural element. The results were ambiguous.

German chemist Ida Noddack published an interpretation: "One could assume . . . that when neutrons are used to produce nuclear disintegrations, some distinctly new nuclear reactions take place which have not been observed previously. . . . [Perhaps] when heavy nuclei are bombarded by neutrons . . . the nu-

*For a comprehensive biography of the life and times of this highly creative individual, read *Genius in the Shadows: A Biography of Leo Szilard, the Man Behind the Bomb*, by William Lanouette with Bela Silard (Chicago: University of Chicago Press, 1994). Leo Szilard's prophetic political testament and science fiction fable *The Voice of the Dolphins* (Stanford, CA: Standford University Press, 1992) is a worthwhile and entertaining read.

FIGURE 16.10
Lise Meitner, about 1930. In Berlin she did nuclear research on uranium until 1938, when Germany's anti-Semitism forced her to flee to Stockholm, Sweden. There, she did the theoretical calculations that clarified the work that her previous group in Berlin had been doing, showing that the group had, in fact, discovered an entirely new nuclear process. Taking a cue from microbiology, she named it *nuclear fission*. Element 109 is named in her honor.

The vision she never lost, that filled her life, was of physics as a battle for final truth.

OTTO FRISCH, SPEAKING OF HIS AUNT AND COLLEAGUE, LISE MEITNER

cleus breaks up into several large fragments." In 1934, her prophetic words were universally ignored.

In 1938, Irène Curie reported that the slow neutron bombardment of uranium had produced a mysterious element that she could chemically separate from the uranium target by using lanthanum (atomic number 57) as a "carrier" to pick up the new element and carry it away. But the mysterious element was chemically inseparable from lanthanum. Surely it could not actually *be* lanthanum. How could the bombardment of uranium, element 92, create an element that was 35 places away from uranium in the periodic table? Otto Hahn, who led a uranium research group in Berlin, was skeptical of such unusual results and accordingly directed his research toward what he believed were needed corrections in Curie's work.

1938 was a fateful year. Austria was annexed by Hitler's Germany; the Munich Agreement allowed Hitler to occupy part of Czechoslovakia; and anti-Jewish mobs in Germany torched synagogues and beat Jewish families in the streets. Lise Meitner (Figure 16.10), an Austrian Jew, had worked with Hahn since 1907. She had been protected from German anti-Semitism by her Austrian passport, but the annexation of Austria made her a German citizen. In July, with Hahn's help, daringly traveling on her now invalid Austrian passport, she fled from Germany to Stockholm, Sweden.*

As the clouds of war gathered across Europe, Hahn and his coworkers in a laboratory in a peaceful suburb of Berlin bombarded uranium with neutrons, trying to find the error in Irène Curie's work. The mystery only increased. Not only lanthanum but also a second element, barium (element 56), made a good carrier for some of the radioactive isotopes created by the bombardment. Once again, it proved impossible to separate chemically the mysterious isotope from its barium carrier. Hahn communicated the results to Meitner in Stockholm. Far from exposing any error in Curie's earlier work, Hahn's results were similar to Curie's.

Meitner enjoyed frequent 10-mile hikes "to keep me young and alert." On Christmas Eve 1938, she and her nephew Otto Frisch took a long walk on cross-country skis through the Swedish countryside. Frisch was a colleague, a physicist, and together they pondered Hahn's data. Niels Bohr had suggested that a nucleus could be viewed as a liquid drop. With this picture in their minds, they debated whether a neutron added to a uranium nucleus might cause the nucleus to oscillate and elongate. Electric forces would then push the two ends away from each other, and two smaller nuclei would appear where the one had been before (Figure 16.11). It was like a drop of water that elongates and splits in two. One of the smaller nuclei might be barium or lanthanum. Taking a cue from microbiology, they named the hypothetical new process **nuclear fission**.

As the nuclear energy curve, Figure 16.4, shows, this process converts nuclear energy to other forms. The nuclear energy that is released transforms into the microscopic kinetic energy seen in Figure 16.11(d) and the energy of the radiation created by this process. Using the liquid drop model, Meitner calculated that the energy with which the two **fission fragments** should be pushed apart by electric forces should be about 3×10^{-11} joules. She then recalculated the energy in an entirely different way, by applying Einstein's $E = mc^2$ to the experimentally known mass difference between uranium and the fragments (Figure

*For an excellent biography of her life and times, see *Lise Meitner: A Life in Physics,* by Ruth Lewin Sime (Berkeley: University of California Press, 1996).

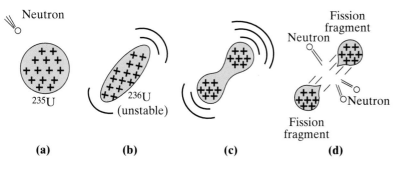

FIGURE 16.11
The fission of a U nucleus after being struck by a neutron.

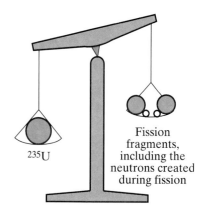

FIGURE 16.12
The uranium nucleus is more massive than the pieces into which it splits (see the nuclear energy curve). According to Einstein, the mass difference multiplied by the square of lightspeed should equal the energy released during fission.

16.12). That mass difference was known to be about one-fifth of a proton's mass. Multiplying this mass by c^2, she got about 3×10^{-11} joules. It checked.

Meitner's calculations were solid evidence of fission. The nuclear age had begun.

■ **DIALOGUE 8**   When the Joliot-Curies bombarded $^{27}_{13}$Al with alpha particles, each aluminum nucleus absorbed an alpha particle and emitted one neutron. Use the periodic table to determine the isotope created by this process.

16.5 THE CHAIN REACTION: UNLOCKING THE STRONG FORCE

By 1939, leading scientists such as Einstein, Fermi, and Szilard had fled to the United States from Hitler's Europe. When Szilard heard that uranium could absorb a neutron and then break into two parts, he foresaw a way to realize H. G. Wells's dream of nuclear energy. Because neutrons, being uncharged, are an especially good nuclear "glue," heavier elements contain many more neutrons than protons. The lighter fragments formed when uranium splits should have many more neutrons than is normal for their atomic number. Consequently, individual neutrons should split off during the reaction, which could then fission other uranium nuclei. As this **chain reaction** proceeded from one uranium nucleus to the next, a large mass of uranium might be fissioned (Figure 16.13). The number of neutrons would multiply quickly as the chain reaction spread, causing several kilograms of uranium to fission in a few millionths of a second.

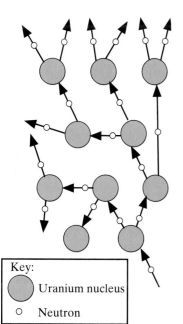

Key:
⬤ Uranium nucleus
○ Neutron

FIGURE 16.13
A chain reaction. At the lower right, a single neutron strikes a uranium nucleus. When the uranium nucleus fissions, it emits two neutrons that then fission two other uranium nuclei, and so forth. In this way, several kilograms of uranium can be quickly fissioned.

8. $^{30}_{15}$P.

Szilard thought that if neutrons were in fact emitted during fission, this fact should be kept secret from the Germans.

Within a week of the announcement of fission, Fermi and others had independently hit on the idea of a chain reaction using neutrons and were making estimates of the energy that might be released. Once, standing at his Columbia University office window overlooking the bustling streets of New York City, Fermi cupped his hands as if he were holding an orange. "A little bomb like that," he mused, "and it would all disappear."

Szilard devised a simple experiment to detect directly the neutrons that he suspected were released when a neutron fissioned a uranium nucleus. From the data, he estimated the average number of neutrons released per fission. A number larger than one could be enough to allow a chain reaction to build up quickly and fission a large mass of uranium. The number turned out to be about two. Szilard immediately telephoned a fellow Hungarian physicist now living in the United States, Edward Teller. Szilard said only one thing: "I have found the neutrons." That night there was little doubt in Szilard's mind that the world was headed for grief.

Hitler ordered the invasion of Poland on September 1, 1939, starting World War II. German scientists were aware of the weapons potential in fission. Accordingly, the German government banned the sale of uranium and in 1939 started a secret nuclear weapons program. It was the beginning of the international nuclear arms race.

Physicists such as Szilard and Teller understood the possibilities. Fission bombs could be winning weapons for Hitler if he were allowed to build them sooner than the United States could. They discussed their fears with Einstein, who agreed to lend his prestige to an effort by scientists to alert the U.S. government to the problem. Together they drafted a letter from Einstein to President Franklin D. Roosevelt that they delivered in October 1939. Einstein suggested that the U.S. government stay informed of further developments and financially support fission research. The final paragraph noted, "Germany has actually stopped the sale of uranium from the Czechoslovakian mines which she has taken over," and "in Berlin . . . some of the American work on uranium is now being repeated."

As it turned out, Germany did try to develop nuclear weapons throughout the war but never got close to its goal. A U.S. project, on the other hand, was successful, although not in time for use against the enemy whom Einstein had feared.

Einstein's letter had little effect. There was a meeting and a committee report. Nothing more. America did not enter the war until December 1941, and in 1939 only Szilard, Einstein, and other knowledgeable physicists took nuclear dangers seriously.

In 1940, Germany invaded and occupied most of Europe and bombed Britain in preparation for an invasion. Both sides began bombing cities, and massive civilian casualties became a reality of modern warfare.

Although the U.S. government declined an active role, U.S. uranium research proceeded between 1939 and 1941. It gradually became clear that there was an important difference between the two uranium isotopes, $^{235}_{92}$U and $^{238}_{92}$U. When a neutron strikes uranium, only ^{235}U has much chance of fissioning; ^{238}U just absorbs the neutron to become a new radioactive isotope, ^{239}U. This means that a nuclear bomb requires nearly pure ^{235}U to sustain the rapid chain reaction

I was shocked and depressed beyond measure. The thought of the unspeakable misery of countless innocent women and children was something that I could scarcely bear…. By the end of a long evening of discussion, attempts at explanation, and self-reproaches, I was so agitated that others became seriously concerned on my behalf.

OTTO HAHN, CODISCOVERER OF FISSION, WHILE INTERNED WITH OTHER GERMAN NUCLEAR SCIENTISTS IN ENGLAND AFTER THE WAR IN 1945

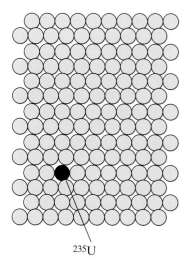

^{235}U

FIGURE 16.14
In natural uranium, only 1 atom in 140 is ^{235}U. The others are ^{238}U.

Some recent work by Fermi and Szilard leads me to expect that the element uranium may be turned into a new and important source of energy.... It may become possible to set up a nuclear chain reaction in a large mass of uranium....This new phenomenon would also lead to the construction of bombs, and it is conceivable, though much less certain, that extremely powerful bombs of a new type may thus be constructed.

EINSTEIN, IN HIS LETTER TO
PRESIDENT ROOSEVELT,
AUGUST 1939

You who are scientists may have been told that you are in part responsible for the debacle (W W II) of today . . . but I assure you that it is not the scientists of the world who are responsible. . . . What has come about has been caused solely by those who would use, and are using, the progress that you have made along lines of peace in an entirely different cause.

PRESIDENT ROOSEVELT,
SPEAKING TO A MEETING OF
SCIENTISTS, 1940

needed to fission a large mass of uranium. If much ^{238}U is present, it will absorb most of the neutrons, and the bomb will fizzle.

Natural uranium has very little ^{235}U, less than 1%. To make a bomb, this 1% must be separated from the 99% that is ^{238}U (Figure 16.14). To many scientists, this appeared essentially impossible. The problem was that two isotopes of the same element must behave identically in every chemical reaction, so chemistry cannot separate them. The difficulties of extracting enough ^{235}U to build a bomb seemed so great that Niels Bohr insisted that "it can never be done unless you turn the United States into one huge factory." Bohr believed that it would therefore not be done. His words proved prophetic, although not in the way he imagined.

In 1940, scientists created the first nonnatural chemical element, that is, one not found naturally on Earth. They found evidence that when $^{238}_{92}U$ is bombarded with neutrons, it absorbs a neutron to become $^{239}_{92}U$, which then quickly emits a beta particle to become element number 93. Its discoverers named the first element beyond uranium *neptunium*, for Neptune, the planet lying just beyond Uranus, the planet for which uranium was named.

Like all elements heavier than lead, neptunium is radioactive. Because it is a beta emitter, it decays to a higher atomic number, creating yet another nonnatural element, number 94 (Figure 16.15). Early in 1941, scientists detected the new element, which turned out to have an important property: Like ^{235}U, element 94 fissions readily when struck by a neutron. Furthermore, the new element can be chemically separated from the uranium in which it was created, thereby avoiding the difficulties of separating two isotopes of the same element. It was not until 1942 that its discoverers proposed a name for this new element that fissions like ^{235}U but that can be chemically separated from uranium. They called it *plutonium*, for the outermost planet, which had in turn been named for Pluto, the Greek god of the underworld.

In October 1941, scientists convinced President Roosevelt that fission weapons could work.

The Japanese attacked the United States at Pearl Harbor on December 7, 1941. In 1942 a U.S. nuclear weapons program, code-named Manhattan Engineer District and known as the **Manhattan Project**, began in earnest.

■ **DIALOGUE 9** When a ^{235}U nucleus is struck by a neutron, it splits into two large fragments and emits two to four neutrons. Although the fragments are different in different instances, one typical fragment is $^{142}_{56}Ba$. Use the periodic table to deduce the chemical name of the other fragment in this particular fission process.

■ **DIALOGUE 10** In the preceding dialogue, suppose the second fragment is $^{91}_{36}Kr$. Write a *reaction equation* showing the isotopes and other particles that go into and come out of this reaction.

■ **DIALOGUE 11** By roughly what percentage do the masses of one atom of each of the two uranium isotopes differ?

9. The other fragment must have atomic number $92 - 56 = 36$. This is krypton.

10. $142 + 91 = 233$. Two additional neutrons ($235 - 233$) must have been released. The overall reaction equation, including the neutron that struck the ^{235}U nucleus, is

$$^{235}_{92}U + \text{neutron} \rightarrow \,^{142}_{56}Ba + \,^{91}_{36}Kr + 3 \text{ neutrons}$$

11. They differ by 3 parts ($238 - 235$) in 238, or a little more than 1%.

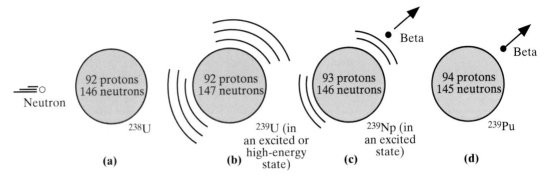

FIGURE 16.15
The creation of plutonium, element 94. After ^{238}U absorbs a neutron, it beta-decays twice to become ^{239}Pu.

The fateful question of the human species seems to me to be whether and to what extent the cultural processes developed in it will succeed in mastering the derangements of communal life caused by . . . aggression and self-destruction. In this connection, perhaps, the phase through which we are at this moment passing deserves special interest. Men have brought their powers of subduing the forces of nature to such a pitch that by using them they could now very easily exterminate one another to the last man. They know this; hence arises a great part of their current unrest, their dejection, their mood of apprehension.

SIGMUND FREUD

Fundamentally, and in the long run, the problem which is posed by the release of atomic energy is a problem of the ability of the human race to govern itself without war.

FROM A REPORT OF THE
U.S. SECRETARY OF STATE,
JANUARY 1953

16.6 THE MANHATTAN PROJECT AND FISSION WEAPONS

Constructed on a monumental scale, the U.S. project to build a **fission bomb** or *A-bomb** was a measure of U.S. fears about a German fission bomb. In December 1942, a research group under Fermi at the University of Chicago created the first self-sustaining chain reaction (Figure 16.16). To achieve this, Fermi's group constructed the world's first **nuclear reactor**, a device to transform, in a controlled way, nuclear energy into other energy forms. We discuss nuclear reactors in Chapter 17.

The world's first reactor illustrates several facts about a possible chain reaction of a different kind: **proliferation of nuclear weapons** to countries around the world. Just as Fermi's reactor was a stepping-stone to a U.S. fission bomb, countries seeking to build their own nuclear weapons are likely to begin by building a reactor.† Although the primary purpose of nuclear reactors today is to provide peaceful electric power, reactors can also provide research relevant to nuclear weapons. An apparently peaceful nuclear power program can be used as a cover for a secret weapons program, as was demonstrated by South Africa, Iraq, North Korea, and other nations.

Another problem is that reactors can produce militarily useful plutonium by converting ^{238}U to ^{239}Pu, providing one possible path to a fission bomb. Since uranium is common in Earth's crust, the ^{238}U is easy to obtain. A reactor can create enough plutonium for a fission bomb during just a few weeks of operation. Near Hanford, Washington, the Manhattan Project engineers built three large natural-uranium reactors for making plutonium.

^{235}U is the key to the nuclear era. Among all the naturally occurring isotopes, it is essentially the only one that will chain react. The only other chain-reacting isotope is ^{239}Pu, and chain reactions in ^{235}U are essential to produce it. There is no obvious reason that there should be exactly one naturally occurring chain-reacting nucleus rather than some other number, such as zero. In a sense, this

---

*Atomic bomb or *A-bomb* is an inappropriate name, because only the nucleus, rather than the entire atom, is the energy source. The term *nuclear weapon* is appropriate, but it applies equally to fission and fusion weapons. We use *fission bomb* and *fusion bomb*.

†However, it is possible for a country to proceed to a bomb by the enriched-uranium route (discussed next), without building a reactor.

FIGURE 16.16
A painting of the opening ceremony, in 1942, of the world's first nuclear reactor beneath the football field (now removed) at the University of Chicago. Photographs were not allowed because of wartime security. Note the "suicide squad" of three young physicists in the back; they were holding jugs of neutron-absorbing liquid to pour into the reactor in case something went wrong.

It was not very clear that the job of separating large amounts of uranium 235 was one that could be taken seriously.

ENRICO FERMI, 1939

isotope just barely exists on Earth: Uranium is the last, heaviest, natural element in the periodic table, and less than 1% of it is ^{235}U. It seems ironic, like some kind of test for the human race, that nature gave us this powerful isotope.

Both uranium and plutonium offer a path to fission bombs. The Manhattan Project pursued both paths. Along the uranium path, the key problem was separating natural uranium to isolate the 1% that is ^{235}U from the 99% that is ^{238}U. Because these isotopes are identical chemically, any **isotope separation** method must be based only on the tiny mass difference between the two.

A technique used during the Manhattan Project, and still widely used, is **gaseous diffusion**. Recall (Chapter 2) that gas molecules that are initially concentrated in one region will spread out as a result of their thermal motion, a phenomenon known as **diffusion**. Lighter molecules, which have less inertia, have more thermal motion than heavier molecules, so lighter molecules diffuse faster than heavier ones do. If uranium is put into a gaseous form, ^{235}U will diffuse faster than ^{238}U, and this difference can be exploited to separate gaseous uranium into a part that is slightly **enriched** in ^{235}U and a part that is depleted in ^{235}U. After many such enrichment steps, **highly enriched uranium** is obtained. The material is suitable for nuclear weapons once the enrichment reaches 90%.

Physicists at Columbia University studied gaseous diffusion between 1941 and 1943. In 1943, in Oak Ridge, Tennessee, army engineers began constructing a gaseous diffusion isotope separation plant. It was of monumental size (Figure 16.17), to house the thousands of diffusion containers needed.

FIGURE 16.17
The gaseous diffusion plant at Oak Ridge, Tennessee. The original Manhattan Project building, in the center, it is half a mile long with 43 acres under its roof. Built during World War II, it produced a total of about 50 kg of highly enriched uranium during the war—enough for the Hiroshima bomb.

This war, in contradistinction to all previous wars, is a war in which pure and applied science plays a conspicuous part.

SIR WILLIAM RAMSAY IN 1915, COMMENTING ON WORLD WAR I

The nuclear proliferation lesson is that it is not easy to enrich uranium. Acquiring the tens of kilograms of highly enriched uranium needed for a bomb is the key obstacle to building a uranium bomb.

In October 1942, the U.S. Army selected physicist J. Robert Oppenheimer (Figure 16.18) to direct the laboratory that would design and build the world's

(b)

FIGURE 16.18
J. Robert Oppenheimer was one of the United States's most brilliant physicists, politically left-wing and controversial, a wide-ranging intellectual and sensitive man given to literary pursuits, and scientific director of the laboratory that designed and built the world's first nuclear weapons. (a) In the 1930s. (b) During the Manhattan Project.

(a)

If the scientists in the free countries will not make weapons to defend the freedom of their countries, then freedom will be lost.

PRESIDENT ROOSEVELT, SPEAKING TO A MEETING OF SCIENTISTS, 1940

Taken as a story of human achievement, and human blindness, the discoveries in the sciences are among the great epics.

J. ROBERT OPPENHEIMER

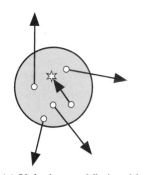

(a) If the lump of fissionable material is too small, most neutrons emitted during fission will escape through the surface without striking any nuclei.

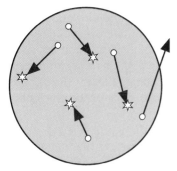

(b) But if the lump of material is large enough, most neutrons will strike other nuclei before they escape through the surface.

FIGURE 16.19
The concept of critical mass.

first nuclear weapons. It proved to be an inspired choice. Oppenheimer chose a high, isolated desert mesa in New Mexico as the site of the new laboratory. They named the place after the boys' school that had been on the mesa: Los Alamos.

A certain minimum amount of fissionable material is needed to sustain a chain reaction. The reason is that, for very small amounts, too many of the neutrons created during fission will simply pass right out through the surface without hitting anything, and a chain reaction cannot be sustained (Figure 16.19). One of the first things that the Los Alamos scientists did was calculate this **critical mass** needed to sustain a chain reaction. For ^{235}U, the critical mass is a grapefruit-sized 15 kilograms. For ^{239}Pu, it is 5 kilograms—the size of an orange, as Fermi had guessed.

The design chosen for the world's first uranium bomb was so straightforward that there was no need to test it before use. High explosive simply slammed two subcritical hunks of ^{235}U together to make a critical mass; simultaneously, the critical mass was showered with neutrons to start the chain reaction. There is a proliferation lesson here: A country need not test a uranium bomb in order to have one. If enough highly enriched uranium can be made, bought, or stolen, building a crude but dependable bomb is not terribly difficult.

But the simple design chosen for the uranium bomb was inefficient: It needed 42 precious kilograms of ^{235}U, nearly three times the minimal critical mass. For the plutonium bomb, a more efficient but more subtle spherical design was chosen (Figure 16.20).* At its center is a small subcritical sphere of plutonium. Although subcritical, it contains enough plutonium to become critical if squeezed into a much smaller volume, because at the smaller volume the nuclei are so close together that neutrons created by each fissioning nucleus are highly likely to fission other nuclei rather than escape out the sides of the plutonium sphere.

High explosive is arranged around the plutonium sphere, which is surrounded by metal to briefly confine and thus enhance the reaction. The high explosive detonates from the outside, and a wave of exploding material moves inward, or *implodes*, squeezing the plutonium sphere into a small enough volume to become critical. At this instant, a neutron source releases neutrons at the center. Although this design is technologically demanding, plutonium is not as hard to obtain as is the highly enriched uranium needed to construct a uranium bomb.

Highly enriched uranium began arriving at Los Alamos in late 1944, and by the following summer there was enough for one uranium bomb. By May 1945, enough plutonium had arrived to allow final critical-mass experiments for the plutonium bomb.

In May 1945, Germany surrendered. The German bomb had never approached completion. Nonetheless, the United States did not halt work on the Manhattan Project, even though the feared German bomb had been its initial driving force. There are lessons here. Once we find that something can be done, there is often a drive, a **technological imperative**, to do it. And once a project is started, it often develops a self-justifying **technological momentum**. It seems that

---

*Simply slamming two subcritical hunks together won't work for plutonium, for the following reason: Plutonium begins to spontaneously chain-react very quickly once a critical mass is assembled, so quickly that spontaneous fission begins as soon as the two subcritical hunks approach each other, "spoiling" the explosion.

FIGURE 16.20
The implosion design of the Nagasaki bomb.

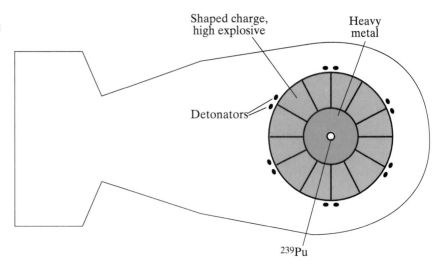

Shaped charge, high explosive

Heavy metal

Detonators

^{239}Pu

what can be done must be done, and what is being done should be done. On the other hand, there were good military arguments for continuing the project: Nuclear weapons might shorten the war that still continued against Japan.

Oppenheimer helped choose the test site for the plutonium bomb, in a barren landscape south of Los Alamos. He code-named the site Trinity, referring to a line by poet John Donne: "Batter my heart, three person'd God." Just before dawn on July 16, 1945, a burst of brilliant purple never seen before lit up the desert, and a small part of Earth was brought to temperatures that were unprecedented within the solar system, save at the center of the sun.

The energy released was the same as would be released by 18,000 tons, or 18 *kilotons*, of a chemical explosive such as TNT. A **kiloton** is a unit of energy,* the energy that would be released by 1000 tons of exploding TNT. For comparison, typical chemical bombs carry perhaps one-quarter ton of high explosive.

The untested uranium bomb was loaded onto a B-29 bomber on August 5. Most Japanese cities had been firebombed to ashes by this time. But Hiroshima, an industrial city with an army depot, an ocean port, and 400,000 people, was still untouched. The B-29 arrived over Hiroshima by 9 A.M. on August 6. As the world's first combat nuclear weapon fell toward its target, the airplane quickly turned and dove away to escape the blast. By the time its crew looked back, the city was hidden by an awful cloud (Figure 16.21). The bomb released 12 kilotons of nuclear energy; 140,000 lay dead with an equal number wounded; and the city, 5 kilometers (3 miles) across, lay in ruins (Figure 16.22). The dying continued: By 1950, the total had reached 200,000, 50% of the city's population.

And the war continued. A debate raged among Japanese leaders about whether it was time to surrender. The Soviet Union, no longer fighting against Germany, was poised to go to war against Japan.

On August 9, just three days after the bombing of Hiroshima, the United States dropped its plutonium bomb. The target was Nagasaki, a city somewhat smaller than Hiroshima. The bomb released 22 kilotons of energy and killed 70,000 outright and 140,000 all together by 1950, again a death rate of 50%.

Upon hearing Roosevelt's words, my mind was made up [to participate in weapons research] and it has not changed since.

EDWARD TELLER, UPON HEARING ROOSEVELT'S SPEECH IN 1940

There floated through my mind a line from the Bhagavad-Gita in which the Krishna is trying to persuade the Prince that he should do his duty: "I am become death, the shatterer of worlds." I think we all had that feeling more or less.

J. ROBERT OPPENHEIMER, AT THE TEST OF THE FIRST ATOMIC BOMB

I just could not understand why our surroundings had changed so greatly in one instant. . . . I thought it might have been . . . the collapse of the Earth which it was said would take place at the end of the world.

YOKO OTA, JAPANESE WRITER AND HIROSHIMA SURVIVOR

---

*One kiloton = 4.2×10^{12} joules.

FIGURE 16.21
By the time the B-29's crew looked back, Hiroshima
was hidden by an awful cloud.

FIGURE 16.22
Hiroshima, August 1945. The bomb
released 12 kilotons of nuclear
energy; 140,000 people lay dead,
with an equal number wounded; and
the entire city, 5 kilometers (3 miles)
across, lay in ruins. The dying
continued: By 1950, the total had
reached 200,000, 50% of the
population.

We turned back to look at Hiroshima. The city was hidden by that awful cloud—boiling up, mushrooming, terrible and incredibly tall.

COLONEL PAUL TIBBETS, PILOT OF THE *ENOLA GAY*, THE B-29 THAT BOMBED HIROSHIMA

If I live a hundred years, I'll never quite get these few minutes out of my mind.

CAPTAIN ROBERT LEWIS, COPILOT OF THE *ENOLA GAY*

Those scientists who invented the atomic bomb, what did they think would happen if they dropped it?

WRITTEN BY A YOUNG WOMAN WHO WAS A FOURTH-GRADE STUDENT IN HIROSHIMA

In some sort of crude sense which no vulgarity, no humor, no overstatement can quite extinguish, the physicists have known sin; and this is a knowledge which they cannot lose.

J. ROBERT OPPENHEIMER, 1947

When you see something that is technically sweet, you go ahead and do it and you argue about what to do about it only after you have had your technical success.

J. ROBERT OPPENHEIMER

How well we meant.

TITLE OF 1983 SPEECH BY PHYSICIST I. I. RABI, AT THE LOS ALAMOS WEAPONS LABORATORY ON THE FORTIETH ANNIVERSARY OF ITS FOUNDING

One may even say that the arms race is a global demonstration of the mechanistic hope for a technological fix for social problems. Here the proof of the futility of power has been reached at too great a cost for all parties concerned.

PETER KAPITZA, PHYSICIST AND PRESIDENT OF THE PHYSICAL SOCIETY OF THE FORMER SOVIET UNION

On August 14, 1945, Japan surrendered.

■ **DIALOGUE 12** About how many 500-pound chemical bombs must be dropped to release the energy of the plutonium bomb? How many large highway trucks (30-ton freight capacity) would be needed to carry this much high explosive? How many World War II heavy bombers (5-ton bomb capacity) would be needed to carry this much high explosive?

■ **DIALOGUE 13** It was discovered during the 1980s that North Korea had built two large nuclear reactors and a chemical processing plant for nuclear fuel. Which, if either, of the two kinds of fission bombs might North Korea have been developing?

16.7 FUSION WEAPONS: STAR FIRE ON EARTH

Following World War II, the U.S.–Soviet cold war rivalry commenced. Like the rest of the scientific world, Soviet physicists learned of fission in early 1939, and by 1940 they had concluded that a chain reaction could be established in uranium. Fifty to one hundred Soviet scientists had worked on isotope separation and uranium and plutonium bomb designs during the war, and the bombings of Hiroshima and Nagasaki only caused them to redouble their efforts. The Soviet Union achieved a chain reaction in a reactor in 1946 and tested a fission bomb in 1949.*

In 1941, Edward Teller began calculations about a bomb based on the energy released when hydrogen fuses to form helium. Whereas the research during World War II focused on a fission bomb, Teller concentrated his efforts on fusion. His fusion group expanded after the war, and by 1949 it appeared that they might eventually succeed. But the United States had not yet decided to build a fusion weapon.

The Soviet fission bomb test, coming as cold war tensions were building, alarmed the United States. President Harry Truman and a small circle of officials and scientists debated whether to proceed with an all-out U.S. effort to build a hydrogen bomb, or **fusion bomb**. Their fear was similar to earlier fears about Germany: The Soviets might build the winning weapon first. Believing that this would be intolerable, Truman decided in 1950 to develop a fusion bomb.

In 1950, the U.S. fusion bomb project was, according to Oppenheimer, "a tortured thing that you could well argue did not make a great deal of technical sense." But during 1951, Teller and mathematician Stanislaw Ulam devised a clever design that solved many of the project's problems. In a revealing comment on the allure of high technology, Oppenheimer stated, "By 1951 the program was technically so sweet that you could not argue about that."

---

12. 500 pounds = 0.25 tons. 22,000/0.25 = 88,000 bombs. 22,000/30 = 730 trucks. 22,000/5 = 4,400 bombers.

13. Plutonium, since an enrichment plant is not mentioned.

*Physicist Klaus Fuchs, a researcher at Los Alamos between 1942 and 1949, later confessed to being a Soviet spy. His information might have shortened the Soviet fission bomb project by 1 or 2 years. Fuchs also passed on information about U.S. fusion bomb research, but this information probably actually hindered the Soviet fusion project because U.S. fusion research was headed along a wrong path up until 1951.

In October 1952, the United States exploded the world's first thermonuclear fusion device. As occurs at the center of the sun, two hydrogen nuclei were fused to form helium. In this and all other fusion weapons, the hydrogen isotopes used were ^{2}H and the rare and radioactive isotope ^{3}H, because these isotopes fuse at a lower, more attainable, temperature than do the ^{1}H and ^{2}H that are fused in the sun. You might guess then that this reaction would create ^{5}He, but ^{5}He is such an unstable isotope that the reaction results instead in ^{4}He plus a free neutron:

$$^2\text{H} + {}^3\text{H} \rightarrow {}^4\text{He} + n$$

The Dark Ages may return on the gleaming wings of Science.
WINSTON CHURCHILL, PRIME MINISTER OF GREAT BRITAIN DURING WORLD WAR II

Concern for man himself and his fate must always form the chief interest of all technical endeavors . . . in order that the creations of our mind shall be a blessing and not a curse to mankind. Never forget this in the midst of your diagrams and equations.
EINSTEIN

Figure 16.23 shows a typical design. To bring the hydrogen fuel to the multi-million-degree temperatures needed for fusion, an implosion-type fission bomb is used as a "trigger." Inexpensive natural uranium is used at several places in the design. Under the conditions that exist while the hydrogen is fusing, even natural uranium can be made to fission, yielding a larger blast that is also higher in radioactive fallout.

Although the critical mass limits the size of fission weapons, there is no natural upper limit to a fusion weapon. The yield of the world's first fusion explosion was 10,000 kilotons, or 10 **megatons**, equivalent to 10 million tons of TNT. A single fusion weapon such as this releases a thousand times more energy than did the Hiroshima bomb, or roughly twice the total explosive energy released by all combatants during all of World War II.

The success of the Soviet fission bomb test and reports that the U.S. government was considering a fusion program stimulated the Soviets to move ahead rapidly with their own fusion program. In November 1955, the Soviet Union also tested a thermonuclear fusion weapon.

We see here, once more, the action–reaction cycle. This dangerous and expensive process is hard to control as long as nations feel threatened, and so similar arms race cycles, involving nuclear and other sorts of weapons, are likely to continue among many nations.

The U.S.–Soviet nuclear arms race was, by any measure, extreme. By the mid-1980s each side possessed about 25,000 nuclear weapons, enough to destroy

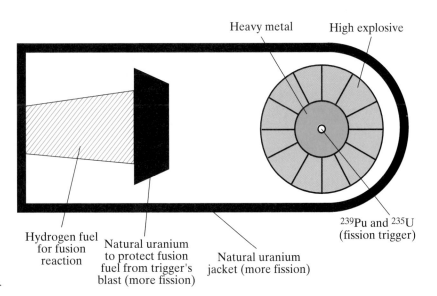

FIGURE 16.23
The design of a fusion bomb.

Heavy metal   High explosive

Hydrogen fuel for fusion reaction

Natural uranium to protect fusion fuel from trigger's blast (more fission)

Natural uranium jacket (more fission)

^{239}Pu and ^{235}U (fission trigger)

the other side as a functioning society many times over. Each side held its weapons out of fear of what the other side might do if it had a significant advantage. Each side's arsenal was designed to "deter" the other side from initiating an attack. Any significant increase in either stockpile always produced a compensating, or overcompensating, increase in the other. Each side's nuclear arsenal contained about 10,000 strategic fusion weapons, each one of them capable of destroying the center of a large city such as New York City or Chicago. Thousands of weapons like this still exist.

■ **DIALOGUE 14 MAKING ESTIMATES: HOW BIG IS 1 MEGATON?**   Estimate the number of large highway trucks (capacity 30 tons) needed to haul 1 megaton of TNT and the length of a single line of trucks carrying such a load on the highway. What about 10 megatons?

TRYING TO DESCRIBE THE SIZE OF THE BIG BANG

---

14. 1,000,000/30 = 33,000 trucks. A large highway truck is about 15 m long, so the length of 33,000 trucks is 500,000 m, or 500 km, with no spacing. If trucks are separated by one truck length, the line is 1000 km long. For 10 megatons, it is 10,000 km, or twice across the United States.

SUMMARY OF IDEAS AND TERMS

Nuclear fusion   A nuclear reaction in which two nuclei combine to form a single larger nucleus. Nuclear energy is released (converted to other forms) whenever two light nuclei are fused to create a nucleus that is lighter than iron (atomic number 26).

Nuclear fission   A nuclear reaction in which a large single nucleus splits into two roughly equal smaller nuclei. Nuclear energy is released whenever a heavy nucleus is fissioned into two nuclei that are both heavier than iron.

Nuclear energy   The energy resulting from the structure of a material's nuclei.

Nuclear energy curve   A graph that shows the energies of the different nuclei versus their mass number (Figure 16.4). The graph shows a lowest point at mass number 52 (iron), indicating that nuclear energy can be released by the fusion of light nuclei and also by the fission of heavy nuclei.

Thermonuclear reaction   A fusion reaction that creates the thermal energy needed to sustain itself.

Big bang   The event about 12 billion years ago that created our universe.

Supernova explosion   The explosion of a giant star. Supernova explosions spread the chemical elements into space and so are the source of the elements heavier than helium in our solar system.

Mass–energy equivalence and the nucleus   When a nucleus releases nuclear energy, the total rest mass of the participants must decrease. The energy release is related to the mass decrease by $E = mc^2$.

Fission fragment   One of the two pieces that results from the fissioning of a nucleus.

Chain reaction   A series of neutron-induced fission reactions that proceed from one nucleus to the next by means of the neutrons released during each fission reaction.

Action–reaction cycle   A mutually reinforcing cycle of increased armaments by two or more hostile nations.

Manhattan Project   The U.S. project during World War II to build fission bombs.

Hiroshima and **Nagasaki**   The Japanese cities that were fission-bombed near the end of World War II.

Isotope separation, or **enrichment**, of uranium. Any process that increases the percentage of ^{235}U relative to ^{238}U. One method is based on **gaseous diffusion**. Weapons-grade uranium is **highly enriched** to about 90% ^{235}U.

Critical mass   The minimum amount of fissionable material that will sustain a chain reaction.

Nuclear weapons proliferation   The spread of nuclear weapons to more and more nations.

Nuclear reactor   A device that controllably transforms nuclear energy into other energy forms.

Fission bomb, or atomic bomb (A-bomb). A bomb that gets its energy from a fission chain reaction. The fuel can be either the uranium isotope ^{235}U, or plutonium. In the ^{235}U bomb, the design can be as simple as bringing two subcritical masses together to form a critical mass. If the bomb contains Pu, a subcritical mass is made critical by squeezing it to high density.

Fusion bomb, or hydrogen bomb (H-bomb). A bomb that gets its energy from the fusion of hydrogen, triggered by a fission bomb.

Technological imperative, **technological momentum**   The tendency to build whatever technology is possible, and to continue a technological project once it is started.

Kiloton, megaton   The amount of energy that would be released in the explosion of 10^3 tons and 10^6 tons, respectively, of TNT.

REVIEW QUESTIONS

FUSION AND THE NUCLEAR ENERGY CURVE

1. Give the reaction formula for one fusion reaction.
2. What isotope is created when ^{12}C fuses with ^{4}He? When ^{14}N fuses with ^{2}H? (Use the periodic table.)
3. What energy transformation occurs in radioactive decay? In nuclear fusion? Nuclear fission?
4. Of the four fundamental forces, which are important inside the nucleus?
5. Which has more nuclear energy, a separated proton and neutron or an ^{2}H nucleus? How do you know? Which has greater mass? How do you know?
6. If you separate a nucleus into its individual protons and neutrons, does its nuclear energy increase or decrease, or does the answer depend on which nucleus you started with?

7. Suppose you separate a nucleus into two equally massive parts. Does the system's nuclear energy increase or decrease, or does the answer depend on which nucleus you started with?
8. Sketch the nuclear energy curve. What does this curve tell us about the possibilities of getting useful energy from fusion or fission?

CREATION OF THE ELEMENTS

9. List two pieces of evidence for the big bang.
10. List two isotopes created in the big bang.
11. Where did your body's hydrogen nuclei originate? What about your body's oxygen nuclei?
12. Where do stars get their energy?
13. What is a *supernova*?

14. Where did most of the chemical elements originate? How did they get to our solar system?

FISSION AND THE CHAIN REACTION
15. Why do neutrons work better than protons do in causing fission? Why do neutrons work better than alpha particles do? Why do neutrons work better than electrons do?
16. Describe what happens during a chain reaction.
17. When $^{238}_{92}$U absorbs a neutron, what isotope does it become? What new element is then created by a single beta decay of this isotope?
18. Where does plutonium come from?
19. What is meant by *uranium enrichment*?

20. What would happen if a fission chain-reaction were attempted in a piece of ^{235}U whose mass was below the critical mass? Explain why this would happen.

NUCLEAR WEAPONS
21. Can natural uranium fuel a fission weapon? Why?
22. Why must uranium be enriched before it can be used as a weapon material?
23. Name the two chain-reacting materials that can be used in fission weapons.
24. Explain how an implosion fission bomb works.
25. How is a fusion bomb heated to fusion temperatures?
26. What is a kiloton, in reference to a weapon? A megaton?

HOME PROJECT

Ask each class member to get three sheets of paper and bunch up each sheet tightly. Gather the class closely together in the center of the room. Ask them to toss their three bunched sheets in the air when they are struck with one sheet. Toss one bunched sheet into the crowd, and watch the chain reaction. Add a few "control rods"—people who throw no bunched sheets into the air. Can you see the difference? How many control rods are needed to stop the reaction? How many control rods are needed to make the reaction "barely critical," neither expanding nor dying out?

FOR DISCUSSION

1. Science and the search for truth. Is it always better to know than not to know? And is it always better to know sooner rather than later? The contrasting opinions of scientists Pierre Curie and Ernest Rutherford are of interest here: "Humanity will obtain more good than evil from future discoveries" (Curie, in 1903); "I hope that no one ... release(s) the intrinsic energy of radium until man has learned to live at peace" (Rutherford, in 1915). Consider three examples: the 1930 discovery of chlorofluorocarbons (Chapter 9), the 1938 discovery of fission, and the 1951 discovery of a way to initiate a fusion bomb explosion. Would we be better off if any of these had not been discovered or had been discovered later? If scientists had discovered fission not in 1938 but in 1934, as they could have done by following Ida Noddack's suggestion, would the world have been better off?
2. Science and democracy. Science-related decisions that affect citizens can be difficult in a democracy, because the questions are sometimes highly technical and understood only by specialists, yet they affect all of us. Examples include radioactive waste, nuclear weapons, and global warming. How should democracies deal with such questions? If all such questions were left to specialists such as scientific experts, would it still be a democracy? What role should be played by specialists? By political leaders? By citizens?
3. Science and secrecy. Science-related decisions are especially difficult when they also involve secrecy. For example, the general public was excluded from the decision in 1941 to build a fission bomb and the decision in 1950 to

build a fusion bomb. Does the public's right to know extend to such cases? If a few officials and experts make these decisions, is this significantly nondemocratic? Who should make such decisions?
4. Science and secrecy, an extreme example. In 1942, Oppenheimer gathered a few scientists to discuss nuclear weapons. One speaker suggested that a fusion bomb might inadvertently trigger fusion reactions in the atmosphere and destroy Earth. Other scientists made calculations to see whether this was possible. Their calculations showed that even under worst-case assumptions, the temperature needed to ignite the atmosphere exceeded by a factor of 100 the calculated temperature of the fusion bomb. So the scientists felt certain that fusion bombs would not ignite the atmosphere. Should the public have been told about this possibility and been allowed to come to its own conclusions, before the first fusion test in 1952?
5. International effects of technological decisions. Technological decisions by one nation can affect the entire planet. Should all people of the world have a hand in such decisions, since they are affected? For example, should all people have a say about the United States's decisions concerning chlorofluorocarbons or carbon-dioxide emissions? About Chinese decisions concerning birth control? About Russian decisions concerning nuclear power safety?
6. Science and war. What might have been some of the dilemmas faced by scientists as they considered whether to work on the Manhattan Project? What would have

been your decision, if you had been one of those scientists? What might have been some of the moral, political, or military dilemmas faced by U.S. military and political leaders as they considered using fission bombs against Japan? What would have been your decision, if you had been one of those leaders?

EXERCISES

FUSION AND THE NUCLEAR ENERGY CURVE

1. How does a nucleus's mass compare with the sum of the masses of its protons and neutrons? Does the answer depend on which nucleus you are considering?
2. Can matter be destroyed? If so, give an example. Can energy be destroyed? If so, give an example.
3. An alpha particle is removed from $^{16}_{8}O$. Is this an example of fusion or fission? Write down the nuclear reaction formula for this reaction. Which side of this reaction formula, the left or the right, represents more nuclear energy?
4. An alpha particle is fused with $^{16}_{8}O$. Write down the nuclear reaction formula for this reaction. Which side of this reaction formula, the left or the right, represents more nuclear energy?
5. Would fusion, or fission, or neither, release nuclear energy from carbon? From gold? From iron?
6. In which of the following processes is there a change in rest mass due to the mass-energy equivalence principle? In which is the change in rest mass actually measurable: warming a cup of coffee; explosion of a fusion bomb; lifting a book; matter-antimatter annihilation (Section 11.4)?
7. In which of the following processes is there a change in rest mass due to the mass-energy equivalence principle? In which is the change in rest mass actually measurable: operation of a nuclear reactor; explosion of TNT; explosion of a fission bomb; throwing a baseball?
8. Which of the four fundamental forces holds the nucleus together? Which one of the four forces tends to push the nucleus apart, that is, to separate it into pieces?
9. One of the forces operating inside the nucleus is the electromagnetic force. Does this force tend to assist nuclear fusion, or oppose it?

CREATION OF THE ELEMENTS

10. The fusion of helium nuclei into heavier nuclei occurs only during the later stages of a star's history, when the star has reached a higher temperature due to collapse. Why would you expect helium fusion to occur only at higher temperatures than those needed for hydrogen fusion?
11. In what sense have we always been sustained by the energy from nuclear fusion?
12. Before it exploded, the 1987 supernova fused many elements. In one reaction, ^{12}C fused with ^4He. What nucleus did this create? Did this reaction release (convert to other forms) nuclear energy?
13. Another reaction occurring in the 1987 supernova, prior to explosion (see the preceding exercise), was the fusion of two ^{12}C nuclei. What nucleus did this create? Did this reaction release (convert to other forms) nuclear energy?
14. During the 1987 supernova explosion, one reaction that occurred was the fusion of iron with helium. What element did this create? Did this reaction release (convert to other forms) nuclear energy? Where then did the needed energy come from?
15. Since the sun is powered by nuclear fusion, why do you suppose Earth doesn't receive a large amount of gamma radiation from the sun?

FISSION AND THE CHAIN REACTION

16. A neutron strikes a ^{235}U nucleus and creates lanthanum. What other element is created?
17. A neutron strikes a ^{239}Pu nucleus and creates strontium. What other element is created?
18. List at least one similarity between combustion and fission. List at least one difference.
19. List similarities and differences between a nuclear chain reaction and igniting and burning a sheet of paper.
20. One of the forces operating inside the nucleus is the electromagnetic force. Does this force tend to assist fission, or oppose it?
21. Why does the chain reaction work in ^{235}U but not in ^{238}U?
22. Give two reasons why it is easier to fission a macroscopic amount of ^{235}U than to fission a macroscopic amount of ^{14}N.
23. In order to obtain the most efficient chain reaction with the minimum amount of material, what shape should the critical mass of fissionable material take: cylinder or sphere? Why?
24. Would you expect that a chain reaction would be able to proceed in natural uranium ore? Why or why not?

NUCLEAR WEAPONS

25. Is plutonium radioactive? Defend your answer.
26. Can natural uranium metal spontaneously explode? Why?
27. Is a chain reaction possible in a substance that emits no neutrons when it fissions?
28. The hydrogen bomb's fusion energy comes from the fusion of ^2H with ^3H. One of the reaction products is an alpha particle. Write out the complete nuclear reaction formula.

PROBLEMS

FISSION AND THE CHAIN REACTION

1. When ^{235}U fissions, it loses about 1% of its rest-mass. Suppose that 10 kg of ^{235}U fissions. How much rest-mass is lost? How much nuclear energy is released?

2. Making estimates. A large nuclear power plant supplies energy at a rate of about 1000 MW, or 10^9 joules/second. The energy comes from ^{235}U. About how much rest-mass vanishes in one day?

3. In the preceding question, suppose that the energy came from a coal plant rather than a nuclear power plant, still at a rate of 1000 MW. Would this change the answer? Explain.

FISSION WEAPONS, FUSION WEAPONS

4. Since 1 kiloton = 4.2×10^{12} J, how much rest-mass vanishes when a 15-kiloton atomic bomb explodes? How much rest-mass vanishes when a 15-megaton hydrogen bomb explodes?

5. How high could the energy of a 15-kiloton atomic bomb (preceding problem) lift the U.S. population, assuming a population of 250 million with an average weight of 600 N (mass of 60 kg) per person. How high could a 15-megaton hydrogen bomb lift the U.S. population?

6. Suppose that all the energy released in a 1 megaton (4.2×10^{15} joules) fusion bomb were used to lift people to a height of 1600 meters (about 1 mile—the elevation of Denver, Colorado). About how many people could be lifted? What fraction of Earth's population is this? (Recall from Chapter 6 that GravE = weight $\times$ height. You will need to estimate the weight, in newtons, of an average person).

17

The Energy Future

If you want to physically change anything, you must exert a force through a distance—you must do work. So energy—the ability to do work—is the ability to physically change things. It isn't surprising then that energy is basic to society, especially today when humans are physically changing so many things.

Because the impacts of energy on society are so all-pervasive, we tend not to notice them, at least until they cause problems or are taken away from us. The differences between stone-age culture and modern culture, for example, have everything to do with different ways of organizing energy to do work. Today, energy-related developments such as electrification, modern agriculture, industrialization, the automobile, and global warming have changed our world and our individual lives in ways that are hard to appreciate or imagine.

The energy future is not predictable. This is because the energy future will be determined by human actions, and human actions are not predictable. Even though our energy use is all-pervasive and not often noticed by most of us, we can *decide* to notice it, to determine ourselves what its impacts should be, and to change our energy use in the ways that we choose. We have the power to choose an energy path that will promote peace, health, and prosperity, or the kind that will promote strife, environmental exhaustion, and poverty. And so, because we do have the power to choose, we have a responsibility to study the many possible futures of energy.

Section 17.1 takes two views of the history of energy use: long term and short term. Section 17.2 looks at where we are now and outlines our future options. Because some of our traditional energy resources are either running out or destroying the environment, the present era is a watershed in the history of energy. Perhaps the most important issue is the trade-offs between traditional nonrenewable sources—natural gas, oil, coal, and nuclear—and the renewable options, especially solar energy and energy efficiency. Sections 17.3, 17.4, and 17.5 look at the traditional energy sources, especially nuclear and coal sources. Section 17.3 studies how nuclear power works, and Section 17.4 makes a comparative assessment of nuclear and coal power. Section 17.5 considers some of the issues for nuclear power. Section 17.6 explores the physics and a little of the economics of renewables, especially the solar-related energy sources. Finally, Section 17.7 looks at the fascinating economics and physics of energy conservation.

A personal note: Most people who have spent much time studying society's energy problems have formed opinions about the topic. I am no exception. However, I have made every effort to restrain my own opinions and to present fairly all of the reasonable options together with some of their pros and cons. Do not, for example, assume that the emphasis given to nuclear power means that I am either especially friendly, or especially hostile, to nuclear power. Rather, I have tried to emphasize those topics that are important for our energy future and that have a strong physics component.

17.1 A BRIEF HISTORY OF ENERGY

Before deciding where you want to go, it's a good idea to know where you've been. Figure 17.1 presents an approximate long-term view of our energy past. The earliest humans used only their own bodily energy, obtained by respiration of food and oxygen. A typical person puts some 2000 Calories, or 8 million joules (8 MJ), of energy to work every day, roughly equivalent to a single 100-watt bulb burning continuously.

During the agricultural age, beginning some 10,000 years ago, humans used the chemical energy of farm animals to do work and the chemical energy released in fires for cooking and warmth. This might have increased the energy put to work by one person about fivefold, to 40 MJ per day, equivalent to five 100-watt bulbs burning continuously.

Heat engines fueled by coal, the first fossil fuel, ushered in the industrial age. Fossil fuels store the chemical energy created by millions of years of accumulating layers of energy-rich plant and animal remains. Time and pressure trans-

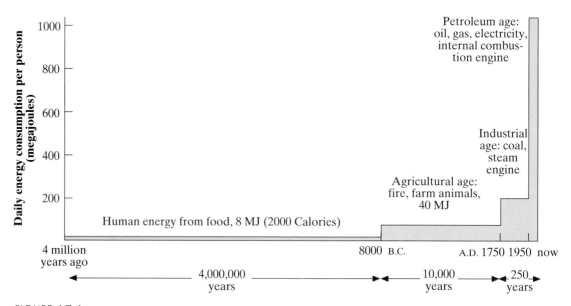

FIGURE 17.1

A brief and approximate history of the use of energy resources: individual daily consumption versus the approximate date. Note that the time axis is not drawn to scale: The preagricultural period is 400 times as long as the agricultural period, which is in turn 40 times as long as the industrial period.

formed these remains into great beds of coal, pools of oil, and pockets of gas. In only 250 years of fossil fuel burning, we have sent much of this carbon—removed from the atmosphere over millions of years—back into the atmosphere. This enabled people in industrialized countries to increase their personal energy use by another factor of 5. From about 1750 to 1950, each person in the industrialized world put perhaps 200 MJ of energy to work every day, equivalent to twenty-five 100-watt bulbs.

During the past 50 years, the intense use of fossil-fueled heat engines (mostly for transportation and electricity) has pushed industrialized nations' energy use upward by yet another factor of 5, to 1000 MJ per person per day, equivalent to one hundred twenty-five 100-watt bulbs burning all day long.

Figure 17.2 takes a closer look at the recent past in the United States. The numbers plotted along the vertical axis are in units of 10^{18} joules, or *exajoules*. Each of the six portions marked Wood, Hydro, Coal, Oil, Gas, and Nuclear represents the energy provided by one particular resource, and the upper boundary represents the total energy provided by all six resources. Note the dominance and growth of fossil fuels since 1900, the growth of oil and gas since World War II (about 1940), and the large overall rise in energy consumption. During the 150 years shown, the nation's *per capita* energy consumption was rising fivefold (Figure 17.1) while the nation's population was rising tenfold, for a fiftyfold overall increase in energy use.

■ **DIALOGUE 1** According to Figure 17.2, in approximately what year did the United States begin getting as much, or more, energy from fossil fuels as from wood? In approximately what year did the combined energy from oil and natural gas begin to exceed that from coal?

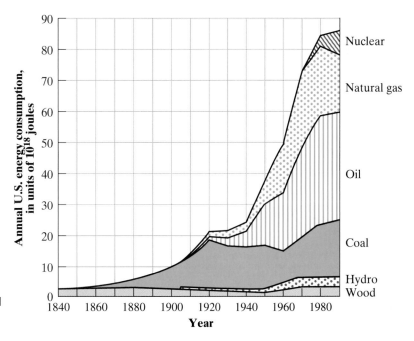

FIGURE 17.2
History of U.S. energy use since 1840: total annual U.S. energy resource consumption from various sources.

1. About 1880. About 1940 or 1950.

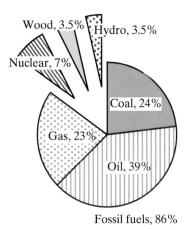

Wood, 3.5%   Hydro, 3.5%

Nuclear, 7%

Coal, 24%

Gas, 23%

Oil, 39%

Fossil fuels, 86%

FIGURE 17.3
The U.S. energy mix today.

■ **DIALOGUE 2**   How much total energy did the United States use in each of the years 1880, 1900, 1920, 1940, 1960, and 1980? By about what factor did the energy consumed in 1980 exceed that consumed 100 years earlier?

■ **DIALOGUE 3 MAKING ESTIMATES**   Use Figure 17.2 to document roughly the estimate shown in Figure 17.1 for the period from 1950 to the present. The U.S. population is about 250 million.

17.2 ENERGY USE TODAY AND FUTURE POSSIBILITIES

We have noted that the United States presently consumes about 85 exajoules annually (Figure 17.2). Figure 17.3 shows how this is divided among the six primary resources; Fossil fuels are responsible for nearly all of it, with smaller contributions from nuclear, wood, and hydroelectric resources. Figure 17.4 shows how the 85 exajoules flow through the U.S. economy, beginning with the six primary resources. More than one-third goes into the generation of electricity that is then used in the industrial, residential, and transportation sectors of the economy, while the other two-thirds goes directly to these three sectors. Hydroelec-

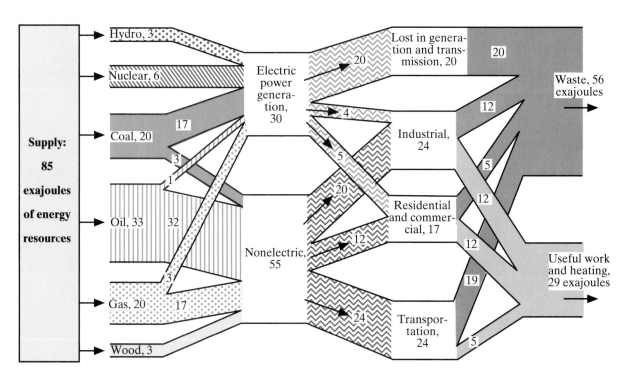

FIGURE 17.4
Approximate energy flows in the U.S. economy in exajoules (10^{18} J).

2. 5, 10, 21, 25, 50, and 85 exajoules. 85/5 = 17.

3. In 1980 the total annual energy consumption was 85×10^{18} J, so the annual consumption per person was $85 \times 10^{18}/250 \times 10^6 = 340 \times 10^9$ J, and the daily consumption per person was about 340×10^9 J/365 ≈ 10^9 J = 1000 MJ.

tric, nuclear, and coal resources go mainly to electric energy production, while oil, natural gas, and wood go mainly to nonelectric uses.

Figure 17.4 reflects the two great principles of energy. We see the law of conservation of energy in the fact that the energy flows (the pipe widths and the numbers) match: Energy in always equals energy out. We see the second law's effect in the transformation of energy from 85 exajoules of high-quality chemical, nuclear, and gravitational energy into 56 exajoules of waste thermal energy and 29 exajoules of useful energy, some of which again produces thermal energy for heating. The second law's rigorous restrictions on heat engine efficiencies shows up in the lower efficiency of electric power generation and transportation, both powered mainly by heat engines.

As we will see, present energy use patterns might change considerably during the next few decades. Environmental problems and resource depletion might force large reductions in fossil fuel use. If so, what might be our energy future?

One possibility is an expansion of nuclear power. This could take the form of new and safer versions of today's uranium-fueled fission reactors (Section 17.3), or of a new type of fission reactor called a *breeder reactor* (Section 17.5), which produces and uses plutonium. A more futuristic possibility is the **fusion reactor**, based on the same reaction that fuels a fusion bomb:

$$^2\text{H} + {}^3\text{H} \rightarrow {}^4\text{He} + \text{neutron}$$

In a fusion reactor this reaction would take place nonexplosively, yielding a continuous source of thermal energy that could boil water to make power-plant steam. The ^2H would be extracted by isotope separation from ordinary water, and the ^3H, a radioactive isotope of hydrogen that is not found naturally, would be made in nuclear reactors.

It's not easy to make a fusion reactor. To start a self-sustaining thermonuclear reaction, one must heat the fuel to millions of degrees (Chapter 16). In one design, this is done by passing an electric current through the gaseous hydrogen fuel. To get a net energy gain, three conditions must be met: The fuel must be sufficiently compressed, the temperature must be sufficiently high, and these two conditions must be sustained for a sufficiently long time (about a tenth of a second). It's hard to maintain the required compression and temperature for this long. For instance, if the fuel touches the walls of its container, the walls will immediately cool the fuel far below the fusion temperature.

Despite steady progress during the decades of intense research, the conditions for creating usable energy from fusion have not yet been met, and a commercial fusion reactor could not be ready until at least the middle of the twenty-first century. But if fusion reactors do become operational, they might revolutionize the energy picture. Despite its possible future importance, we will not discuss fusion further, simply because it's too far off and there are so many nearer-term decisions to be made.

Renewable energy resources will be more important in the future than they are today. An energy resource is renewable if it is continuously available, like sunlight, or if it can be replaced within a human lifetime, like wood. Other resources, such as fossil fuels (which are really forms of solar energy stored over long periods) and uranium, are **nonrenewable**.

Sunlight energized the agricultural revolution thousands of years ago, and renewable wood helped energize the beginning of the Industrial Revolution a few centuries ago. Industrial societies are now energized mainly by nonrenewable fossil fuels and uranium, with renewable wood and hydroelectric resources playing a small but significant role (7% of the total).

Future renewable energy could take many forms. Hydroelectric could expand a little. Biomass—wood, waste paper, garbage, agricultural wastes, sugar crops, and grains—can be burned for thermal energy or transformed into liquid or gaseous fuels such as alcohol and methane. Wind can generate electricity. The sun can provide direct heating, and it can generate electric power in photovoltaic cells or by heating water to make steam to make electricity (Section 17.6). All of these renewables are related to the sun. A nonsolar option is the geothermal energy of Earth's hot steam, hot water, and hot rock. Because there is so much geothermal energy, we classify it as renewable, even though strictly speaking, it is replaced over time periods much longer than a human life. We will discuss the renewables further in Section 17.6.

Energy conservation—getting along with less energy—will also be more important in the future. Although not really an energy resource, conservation acts in many ways like an energy resource. **Energy conservation** refers to all those measures that reduce our energy consumption so that fewer resources are needed. Conservation includes **energy efficiency** measures, such as home insulation, energy-efficient lighting, and energy-efficient automobiles, that reduce energy consumption without reducing the services provided by that energy. Conservation also includes saving energy by lifestyle changes such as greater use of mass transit, living in smaller homes, and building more compact communities. Many studies show that energy consumption can be reduced dramatically by efficiency alone, with no need for controversial changes. We will discuss conservation further in Section 17.7.

Table 17.1 summarizes the main future possibilities, grouped into four categories: fossil, nuclear, renewables, and conservation.

U.S. energy patterns changed beginning in 1973. Figure 17.5 documents this important point. This graph is similar to Figure 17.2, but with more detail. The curve marked "Total annual energy" traces the total consumption of natural energy resources. Fossil fuels dominate, with a small long-term contribution from renewable wood and hydro resources and an expanding contribution from nuclear resources beginning in the early 1970s. The curve marked "U.S. annual inflation-adjusted GNP" traces the nation's annual gross national product (GNP), its annual output of goods and services. The dollar GNP amounts are not shown. Instead, the scale of the GNP graph is chosen to make the GNP and energy graphs coincide in 1950.

Since GNP measures the services that the nation actually receives from the energy it consumes, it is not surprising that U.S. energy consumption rose almost exactly in step with GNP between 1950 and 1973: As energy services increased, so did energy consumption. The first international oil crisis, in 1973, changed this. Energy prices increased rapidly in response to decreased foreign oil supplies, and the nation looked for ways to save money by conserving energy. The effect of this crisis, and of the second crisis in 1979, is evident in the graph. Although GNP fell briefly following each crisis, it soon resumed its upward trend. But total energy use declined after each crisis and did not resume its upward trend. Apparently the conservation measures, promoted by increased prices, became permanent. By using energy more efficiently, the nation has obtained a

TABLE 17.1

Primary energy options

Fossil fuels
  Coal
  Oil
  Natural gas
Nuclear power
  Uranium-fueled reactors
  Plutonium-fueled and breeder
    reactors
  Hydrogen-fueled fusion reactors
Renewable resources
  Hydroelectric
  Biomass: wood, trash, agriwaste,
    crops
  Geothermal
  Wind
  Solar-thermal electricity
  Photovoltaic (solar) cells
  Active solar heating
  Passive solar heating
Conservation
  Energy efficiency with no
    change in energy services
  Lifestyle changes to reduce energy
    use

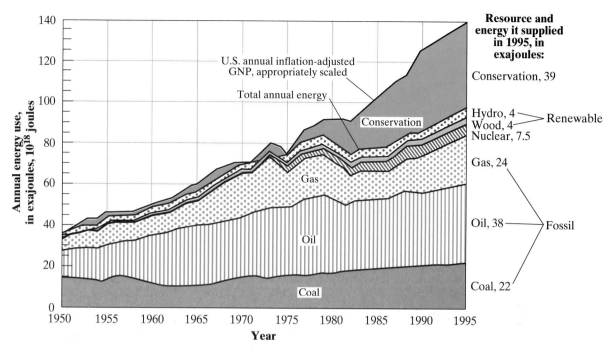

Resource and
energy it supplied
in 1995, in
exajoules:

Conservation, 39

Hydro, 4 ⟩ Renewable
Wood, 4
Nuclear, 7.5

Gas, 24

Oil, 38 ⟩ Fossil

Coal, 22

FIGURE 17.5
A more detailed view of the history of U.S. energy consumption since 1950. The upper boundary traces the GNP, one measure of the goods and services provided by energy. The gap that opens between the total energy consumed and the GNP, beginning at the time of the 1973 energy crisis, is a measure of the effect of conservation since 1973.

steadily increasing level of goods and services out of a roughly constant supply of energy since 1974.

Without conservation, it is likely that the United States' total energy consumption would have stayed in step with its GNP. So the difference between the two graphs is appropriately labeled "Conservation." By 1995, conservation contributed about 40 exajoules to the nation's energy mix, more than did any other energy resource.

■ **DIALOGUE 4** From Figure 17.4, find the overall energy efficiency of each of the following: electric power generation, the industrial sector, the residential–commercial sector, the transportation sector.

■ **DIALOGUE 5 MAKING ESTIMATES** In Dialogue 3, we used the 85 exajoules of annual U.S. energy consumption to verify that the daily energy consumption per person is 1000 MJ (Figure 17.1). Now use this figure, 1000 MJ, to estimate the average person's power consumption, in watts. Express this in kilowatts. If all this energy were in the form of electricity, how many 100-watt bulbs could it light up?

---

4. Electric power 9/29, or 31%; industrial 50%; residential–commercial 71%; and transportation 21%.

5. Power consumed is energy consumed per second (Chapter 6). Accordingly, divide the 1,000 MJ of daily energy consumption by the number of seconds in a day: $1,000 \times 10^6$ J/$(60 \times 60 \times 24$ s$) \approx 12,000$ watts, or 12 kilowatts. This is the equivalent of one hundred twenty 100-watt bulbs.

17.3 NUCLEAR POWER: HOW IT WORKS

Nuclear power comes mostly from uranium chain reactions that provide energy
for steam-generated electricity. Most U.S. power reactors are of the type shown
in Figure 17.6. By comparing this with Figure 7.19, you can see that nuclear
power plants operate pretty much like every other steam-electric generating

FIGURE 17.6
Schematic diagram of a steam-electric generating plant powered by a light-water reactor. Compare this
with Figure 7.19.

plant: Thermal energy from the primary energy source (nuclear fission, in this case) heats water to make high-temperature steam that turns a turbine that turns a generator that makes electricity.

All power reactors have four essential ingredients: **fuel** to provide energy; **coolant** to transfer thermal energy away from the fuel; neutron-absorbing **control rods** that can be moved in or out to slow down, stop, or speed up the chain reaction; and a **moderator** to slow down the neutrons to make it more likely that they will cause fission.

The coolant—natural water in most U.S. reactors—circulates through the center of the reactor and removes the thermal energy created by the chain reaction. The same water acts as the reactor's moderator, by slowing down the neutrons to their most efficient speeds when these neutrons collide with the hydrogen nuclei in the water.

The fuel in light-water reactors is uranium, slightly enriched to 3%. The energy comes only from the ^{235}U nuclei (3%), and not from the nonfissionable ^{238}U nuclei (97%). Natural uranium, made of roughly 99% ^{238}U, won't work in water-cooled reactors, because both ^{238}U and ^1_1H can absorb neutrons and spoil the reaction. Since power reactors use slightly enriched or natural uranium rather than highly enriched (90%) bomb-grade uranium, it is impossible for a power reactor to explode the same way that a full-fledged nuclear weapon would. And it is impossible to use the uranium directly in a bomb.

The uranium fuel is shaped into small cylindrical pellets and stacked in long thin metal tubes. Some 40,000 of these fuel rods, 100 tonnes of fuel, form the power-producing core of the reactor. The core is enclosed in a heavy steel *reactor vessel* (Figure 17.7) designed to withstand high pressures, to shield against radiation, and to absorb the many neutrons that escape from the core.

Water circulates through two physically separated loops (Figure 17.6). The first loop circulates through the core, removing heat from the fuel rods. Although this water is heated to above 300°C, the high pressure maintained inside the reactor vessel keeps it from boiling. This water circulates around to a *steam generator* where it heats the water in the second loop. This thermal energy is exchanged by placing the first loop's pipes in contact with the second loop's pipes, so that thermal energy flows from the hot loop to the cooler one without allowing their water to mix.

Because the reactor's operation causes the core, the reactor vessel, and the first water loop to become highly radioactive, these elements are encased in thick concrete. For further shielding and protection against accidents, a large airtight **containment dome**—made of steel-reinforced concrete about a meter thick and built to absorb an impact as great as that of a jetliner crash—surrounds the radioactive elements (Figure 17.8).

Since the secondary loop is not kept at high pressure, its water immediately turns to steam in the steam generator. The rest of the operation is identical to every other steam-electric power plant (Figure 17.6).

■ **DIALOGUE 6** Uranium is one possible fission-reactor fuel. Name another. How could this second fuel be obtained?

FIGURE 17.7
Looking into the reactor vessel, or core, of a nuclear reactor as the first load of fuel is placed inside. The core is seen here below many meters of water.

6. Plutonium, obtainable from nuclear reactors themselves. As we saw in Chapter 16, plutonium is produced from the nonfissionable isotope ^{238}U, an isotope that is present in the reactor fuel.

FIGURE 17.8
Nuclear power plant under construction. This view shows a cooling tower and a reactor containment dome under construction.

We nuclear people have made a Faustian bargain with society. On the one hand, we offer an inexhaustible source of energy. But the price that we demand of society for this magical energy source is both a vigilance and a longevity of our social institutions that we are quite unaccustomed to.

ALVIN WEINBERG, FORMER DIRECTOR OF OAK RIDGE NATIONAL LABORATORY, 1972

To supply future energy demand, every variety of energy source will be called upon. . . . The nonfossil sources will undoubtedly be stimulated by global climate concerns . . . [but] might at best provide about one-third of the input. Hydro and nuclear will undoubtedly expand, but the bulk of the input will continue to come from fossil sources. By the middle of the next century, only coal will remain the inexpensive and abundant fossil fuel. Thus, clean coal technology, improved coal power plants, and coal conversion to liquid fuel, should continue to be a principal thrust of national global energy policy.

CHAUNCEY STARR, PRESIDENT EMERITUS OF THE ELECTRIC POWER RESEARCH INSTITUTE AND FORMER PRESIDENT OF THE NUCLEAR DIVISION OF ROCKWELL INTERNATIONAL

17.4 TECHNOLOGY ASSESSMENT: ASSESSING THE NONRENEWABLES

From energy to ozone, science-related social issues are part of the landscape of our age. For our society to remain both industrialized and democratic, it is essential for all citizens to learn something of technology and science in order to evaluate such questions as our future energy options and what to do about the ozone layer. Societies cannot turn these issues over to the experts and still remain democratic, for these issues affect our environment, our economy, and ultimately our values.

In fact, citizens do constantly decide these issues as they pursue their daily lives. Millions of individual decisions about which product to buy, whether to insulate one's home, modes of transportation, recycling, and so forth, collectively determine our uses and misuses of technology and of science. Should everybody become experts, then, on all science-related social issues? Obviously that is impossible. But at the other extreme, thoughtlessly following our beliefs and our traditions is likely to be a prescription for disaster.

Individuals and groups of all sorts can clarify these issues through a process of **technology assessment** that encourages rational, informed decisions. The idea of technology assessment is to clarify the issues first, before debating about values. To clarify the issues, all "stakeholders" should draw up a single, objective balance-sheet showing all the costs and benefits of each plausible option. Only then, after a complete balance sheet is drawn up so that every stakeholder understands the implications of each option, should the debate over the human valuation of these costs and benefits begin. Technology assessment is far from a complete prescription for solving technology-related problems, but it is an essential first step.

As an instructive example, let's briefly assess two of the more plausible large-scale energy options that are already in wide use: coal and nuclear power. Between them, they supply 76% of the United States's electricity, with coal power providing 56% and nuclear power providing 20%. Do we want either or both of these to be major energy sources in the future? The way to find out is to stack up the costs and benefits and evaluate them.

The decline of [the nuclear] industry began well before Chernobyl; the accident was not a one-shot catastrophe but a galvanizing of doubts that had persisted at least since the . . . accident at Three Mile Island. . . . In fact, the nuclear industry has been beset by problems since its inception . . . problems now believed to be endemic to the technology. Equally noteworthy is that as nuclear power has faltered . . . the renewable sources—wind and solar—have boomed. . . . The challenge now is to shift the investment priorities of governments, which have heavily subsidized nuclear power.

N. LENSSEN AND C. FLAVIN, ENERGY SPECIALISTS

The new generation of nuclear power plants . . . will not only provide electricity at a lower cost than coal-generated electric power, but they will also be a thousand times safer than plants of the present generation.

BERNARD L. COHEN, HEALTH PHYSICIST, UNIVERSITY OF PITTSBURGH

Global climate change . . . poses dangers which may be very large. The problem can be ameliorated by reducing fossil fuel consumption through conservation and expanded use of nuclear and solar power. . . . The most obvious role for nuclear power is to replace coal in electricity generation. It could also replace all the oil and much of the natural gas, leaving some use of natural gas for peaking purposes. . . . A program of this sort could be implemented over three or four decades, as existing fossil fuel plants are retired.

DAVID BODANSKY, PHYSICIST AND ENERGY SPECIALIST

■ **DIALOGUE 7** Draw a line down the center of a single sheet of paper. Write "coal power" at the top of one side and "nuclear power" on the other side. List as many possible problems (costs) and strengths (benefits) as you can think of under each heading. If you intuitively favor one of the options, compensate for this possible bias by looking especially hard for the costs of that option and the benefits of the other option. After each entry on either side, consider whether there should be a comparable entry on the other side. For example, nuclear power plant accidents might be listed opposite coal mine accidents.

Table 17.2 lists costs and benefits of coal and nuclear power. Because both resources are used to generate electricity, the two lists are similar. In fact, there are parallel entries for all but two problems: global warming caused by coal and nuclear weapons proliferation caused by nuclear power. But the similarity of problems does not necessarily mean that the evaluations will be similar. For example, the amount of land degraded is greater for coal than for nuclear power, because the volume of the fuel is so much greater. For another example, terrorists are more of a threat to nuclear power than to coal power, because they could release radiation or steal materials for nuclear weapons.

One fundamental issue is resource availability. Table 17.3 shows the resource picture for coal and uranium along with oil and natural gas. Only 2% of U.S. coal has been consumed, whereas oil and natural gas are about half gone, and uranium stands between these extremes. A similar picture emerges when one looks at the number of years before each resource would be used up if the current annual rate of U.S. production of that resource remained unchanged (second column).* Apparently the United States has a lot of coal and not so much oil or gas, with uranium lying somewhere in the middle.

Air pollution and solid waste are serious problems for coal, but coal's biggest drawback is undoubtedly the looming issue of global warming (Chapter 9). Car-

TABLE 17.2

Costs and benefits for coal and nuclear power

Coal power	Nuclear power
Benefit: electricity	Benefit: electricity
Resource: enough for many centuries	Resource: enough for many decades
Land degradation from mining	Land degradation from mining
Air pollution: acid rain, SO_2, NO_2, ash	Radiation in environment
Workers' health: black lung disease	Workers' health: radiation in mines and plants
Accidents: mine explosions and collapses	Power plant accidents
Heating of lakes and rivers	Heating of lakes and rivers
Solid waste: ash, sludge	Used fuel rods, low-level radioactive waste
Cost of plant, fuel, operations	Cost of plant, fuel, plant closing, waste disposal
Global warming due to CO_2 released	—
—	Proliferation of nuclear weapons to nations, terrorists
Sabotage: shut off electricity	Sabotage, terrorists: release radiation, build bomb

7. See Table 17.2 for suggestions.

*But of course the rate will not remain unchanged. As oil and natural gas approach depletion, they will become more expensive to produce, and so production will decline. Oil is already past the high point of the bell-shaped curve (Figure 7.25) that is typical of nonrenewable resource consumption. Despite these qualifications, the years remaining at current rates of use (Table 17.3) is a useful measure of a resource's remaining lifetime.

TABLE 17.3

Percentage consumed and years remaining at present rates of use for four nonrenewable energy resources in the United States

	Percentage of total resource consumed by 1992[a]	Years until total resource is gone at present rate of use
Coal	2	3600
Oil	70	25
Natural gas	40	69
Uranium (without breeder reactors)	8	200

[a]The "total resource" means the amount estimated to be recoverable in the United States by known technical means, at reasonable prices.

bon-dioxide emissions cause 55% of the global warming problem, and essentially all of this comes from the three fossil fuels. Switching from coal to natural gas can help, because natural gas produces only about half as much carbon dioxide (per joule of energy) as coal does. But because of global warming and also limited gas supplies, this can be only a temporary solution. Global warming is causing energy analysts and governments to look more seriously at the nonfossil alternatives: nuclear power, renewable resources, and conservation.

■ **DIALOGUE 8 MAKING ESTIMATES**   U.S. coal plants collect 80 million tons of unburnable ash every year, most of which is disposed of in ponds, where it settles to the bottom. How much is 80 million tons? For instance, how many large highway trucks would it fill, at a 30-ton load per truck? If these trucks formed a bumper-to-bumper caravan across the United States, how long would the line be? (Highway trucks are some 15 m long.)

17.5 ISSUES FOR NUCLEAR POWER

There are many science-and-society lessons in the history and problems of nuclear power in the United States.

The first commercial nuclear plant opened in 1957. By 1975, with 54 plants providing 10% of the nation's electricity, the nuclear age seemed firmly established. But then the nation hesitated. The last firm order for a new plant was placed in 1974. Despite the growing number of operating plants, it is now clear that the nuclear power era ended, for now at least, during the 1970s (Figure 17.9).

Nuclear power has always attracted strong proponents and opponents. Proponents argue that global warming, fossil-fuel shortages, fossil-fuel pollution, and the prospect of a safer generation of nuclear reactors call for a revival of nuclear power, the only nonfossil resource that has proved capable of producing large amounts of energy. Opponents contend that nuclear power is unacceptable because of safety concerns, waste disposal problems, economics, and nuclear

8. $80 \times 10^6/30 = 2.7 \times 10^6$: nearly 3 million trucks. $(15 \text{ m}) \times (2.7 \times 10^6) = 40 \times 10^6$ m, or 40,000 km, or about eight times across the United States.

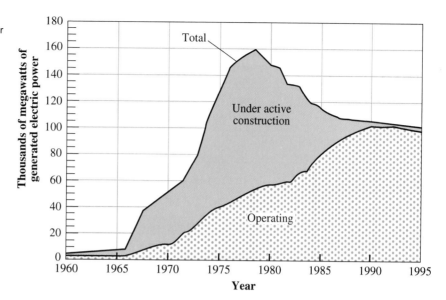

FIGURE 17.9
The United States's commitment to nuclear power, 1960–1995: total megawatts installed and under construction.

Nuclear power must play a significant role in a sustainable future, but it must be carefully administered. Safety standards, design features, waste storage, and the entire fuel cycle must be under the jurisdiction of an international regulatory agency. Only then can the full potential of nuclear power be realized. In view of the worldwide population growth and its accompanying problems, I sincerely doubt that turning backward will allow us to master the future. Instead we have to go forward.

WOLF HAFELE, PHYSICIST, FORMER SCIENTIFIC ADVISER TO THE WEST GERMAN GOVERNMENT

The "electricity-specific" uses that need, and economically justify, electricity's high quality are only 8 percent of U.S. end-use needs. Positing huge new markets for coal or uranium is thus like pushing on a string—imaginative but ultimately unsatisfying.

AMORY LOVINS, PHYSICIST, ENERGY CONSULTANT, AND DIRECTOR OF RESEARCH AT THE ROCKY MOUNTAIN INSTITUTE AT SNOWMASS, COLORADO

proliferation, that conservation plus natural gas can gain time for a few more decades, and that renewable resources and conservation will then be able to meet our needs. There is, however, agreement on both sides that it is time to start disengaging from fossil fuels.

Nuclear waste, reactor accidents, uranium resources, and nuclear weapons proliferation have been central and interrelated issues for nuclear power. We will discuss these here.

Although lower-level forms of nuclear waste pose some problems, the intensely radioactive **high-level nuclear waste** that comes from used fuel rods is the most serious problem. Used fuel rods are currently stored on site in large pools of water inside the containment domes of the country's 109 nuclear power plants. Because they must be stored safely for thousands of years, a long-term solution is needed.

Used fuel rods contain the fission fragments and other radioactive isotopes created during the chain reaction. Shorter-lived isotopes dominate the radioactivity for the first few centuries, so that the waste's radiation level decreases by a factor of 300 during the first 10 years, and by a factor of 100,000 during the first 1000 years. Nevertheless, some isotopes remain potentially toxic for tens of thousands to millions of years, so planning must take an extremely long-term view. In the United States, the official time frame for high-level waste planning extends to 10,000 years in the future—a time longer than recorded history—and a study panel within the National Academy of Sciences has recommended that planning be extended to 1 million years. Experts believe that such long-term burial can be achieved by incorporating the wastes into glassy materials, putting these materials into steel canisters with a lifetime of 10,000 years, and storing the canisters in human-made caverns in geologically stable sites several hundred meters below ground. Various sites have been proposed, most recently at Yucca Mountain in Nevada, but there is opposition to each one.

The accident in 1979 at **Three Mile Island** illustrates how things can go wrong in complex technologies. It began with a routine event: A pump failed in the water loop that carries steam to the turbine (Figure 17.6). Control rods immedi-

No scientist or engineer can give an absolute guarantee that radioactive waste will not someday leak in dangerous quantities from even the best of repositories. . . . Should repository construction be authorized? Or should it be delayed . . . ? These are not technical questions, but questions for the man in the street and his elected representatives.

KONRAD KRAUSKOPF,
GEOLOGIST, STANFORD
UNIVERSITY

The proposed storage of high-level radioactive wastes in the Nevada desert poses no apparent risk of groundwater contamination.

FROM A NEWSPAPER REPORT
ON THE FINDINGS OF A
NATIONAL RESEARCH COUNCIL
STUDY ON THE SUITABILITY OF
YUCCA MOUNTAIN, IN
NEVADA, AS A HIGH-LEVEL
NUCLEAR-WASTE DISPOSAL
DEPOSITORY

In the end, it might not be economics, but spent fuel storage and disposal that determine the viability of nuclear power in the United States.

IVAN SELIN, CHAIR, NUCLEAR
REGULATORY COMMISSION

No event is more antinuclear than an accident.

PETER BRADFORD, FORMER
MEMBER OF THE NUCLEAR
REGULATORY COMMISSION

Radioactive waste . . . is a rather trivial technological problem.

BERNARD L. COHEN

ately and automatically dropped into the reactor, and fission ceased. But stopping fission in a reactor doesn't stop all the heating, because some 5% of a reactor's power comes not from fission but from radioactivity. So water must be kept moving through the reactor to prevent overheating.

So a backup pump took over. But a valve in this backup pump had been left closed by mistake, and a warning light that should have alerted operators to the closed valve was obscured by a tag, so this water did not get into the reactor for another 8 minutes. During this time, the temperature and pressure rose, forcing a pressure-relief valve to open. Unfortunately, this valve stuck open and, for 2 hours, was not discovered to be open. The open valve allowed water to escape from the reactor, flooding the containment building floor with radioactive water. Some of this water was automatically pumped to an adjoining building, releasing radioactivity.

Water in the core evaporated and escaped. Because of the threat of overheating, every reactor contains a tank of emergency cooling water for use in an accident. But unfortunately, again, the operators interpreted their control room dials to mean that there was too much water in the core, rather than too little. So they shut off the emergency cooling water. The water level dropped below the top of the fuel rods, and the fuel heated until much of the core had melted, an event known as a **melt-down** because it causes the fuel to slump downward. This permanently destroyed the reactor. Only the dwindling water still in the bottom of the core prevented the entire reactor vessel from melting and spilling molten fuel into the containment building.

Unusual chemical reactions in the hot reactor created hydrogen gas that remained in the reactor vessel for several days, causing concern that a hydrogen explosion might rip open the reactor vessel and release a large amount of radioactivity. The explosion did not happen because of a lack of oxygen in the reactor vessel.

There were no immediate deaths, although the excess radiation might cause one extra long-term cancer death. The total radioactivity released was small. But the accident cost the utility company $1 billion to clean up, plus the loss of the reactor.

History's worst nuclear power accident happened at **Chernobyl**. Unlike the current U.S. power reactors, the Chernobyl reactor was moderated by graphite rather than water, and it lacked a containment dome.* Although low-enriched reactor fuel cannot explode in the same way that a full-fledged nuclear weapon can, graphite-moderated reactors can suffer a runaway chain reaction that releases thermal energy rapidly enough to create a low-grade nuclear explosion, similar in its energy release to that of a conventional chemical bomb. Such a runaway chain reaction occurred at Chernobyl, causing a melt-down, contaminating a large area, and causing many short-term and long-term casualties (Section 15.6). The fuel melted through the containment vessel and partly penetrated the plant floor. If it had melted entirely through the floor, it could have contacted groundwater and set off a massive radioactive steam explosion.

Although these accidents led to new safety procedures, experts estimate that with 400 power reactors operating worldwide today, the chances of another core melt during the next 10 years is between 4 and 40%. If there is to be a revival of nuclear power, it will probably be based on new, safer reactors.

---

*However, U.S. military reactors of this type operated until 1987.

Even now, after more than 50 years of effort and the expenditure of vast sums of money, the United States does not have a technically sound or socially acceptable policy for dealing with its high-level nuclear wastes. Indeed, its current attempt to solve the problem by creating a national repository at Yucca Mountain, Nevada, is mired in conflict and destined for failure.

J. FLYNN, SOCIAL SCIENTIST; R. E. KASPERSON, POLITICAL SCIENTIST; H. KUNREUTHER, PUBLIC POLICY EXPERT; P. SLOVIC, SOCIAL SCIENTIST

Citizens of the distant future will look upon [plutonium] as one of God's greatest gifts to humanity. . . . The antinuclear movement has devoted so much energy to trying to convince the public that [plutonium] is an important health hazard. Those with scientific background among them must realize it is a phony issue.

BERNARD L. COHEN

The current system would not be adequate to the task of separating nuclear power from nuclear weapons technologies, if nuclear power were developed on a large scale. . . . Reconstructing nuclear power therefore requires that commercial reprocessing and plutonium recycling be halted and deferred indefinitely, and that emphasis instead be given to nuclear technologies that do not involve those activities.

ROBERT WILLIAMS, PHYSICIST, AND HAROLD FEIVESON, CENTER FOR ENERGY AND ENVIRONMENTAL STUDIES, PRINCETON UNIVERSITY

Those who advocate a large future role for nuclear energy envision something like a fivefold expansion in nuclear power plants during the next few decades. But under such a scenario, uranium resources would run out in about 40 years. The solution to this problem is the **breeder reactor**. Recall (Section 16.5) that, while ^{235}U is fissioning to provide energy inside a reactor, the much more numerous ^{238}U nuclei are absorbing neutrons and transforming into plutonium. As we know, plutonium is a second chain-reacting nucleus that can be used in a nuclear reactor. A breeder reactor is a reactor designed to efficiently produce plutonium for just this purpose. It is actually possible to create more than one new Pu nucleus for every ^{235}U nucleus fissioned in a reactor because uranium contains so much more ^{238}U than ^{235}U. Thus, breeder reactors can convert much of the ^{238}U in any country's uranium resources to fissionable plutonium, greatly extending these resources.

Although breeder reactors can solve the resource problem, they make the **nuclear weapons proliferation** problem worse. As we saw in Chapter 16, there are several connections between nuclear power and nuclear weapons. As one example, uranium-fueled power reactors can produce militarily useful plutonium, although the plutonium must first be processed to separate it from the remaining uranium and from the highly radioactive isotopes in the used fuel rods. India followed this route to a fission bomb when it utilized the used fuel rods from a nuclear power reactor, along with a supposedly peaceful processing plant, to obtain plutonium. The ongoing nuclear arms race between Pakistan and India is intertwined with nuclear power issues.

Breeder reactors exacerbate such proliferation problems because reactor-grade plutonium can fuel a nuclear weapon while reactor-grade uranium cannot. Widespread breeder reactors and reprocessing will create large amounts of already-separated plutonium to be stored, shipped to other reactors, and used. An expanded nuclear power industry could put more than a million kilograms of plutonium into global commerce per year. Because fewer than 10 kilograms are needed for a fission bomb, it might be difficult to keep significant amounts from going into weapons.

■ **DIALOGUE 9** Does a coal-burning generating plant have combustion products that are analogous to a reactor's fission products? What is done with them? Is this harmful?

17.6 FUTURE ENERGY OPTIONS: RENEWABLES

Among the renewable resource options (Table 17.1), only hydroelectric and biomass contribute significantly to today's U.S. energy budget. However, other renewables are expanding, and their future possibilities are significant.

Hydroelectric and biomass resources each contribute 4% of the energy budget (Figure 17.5). **Hydroelectric energy**, the gravitational energy of water that the sun's warmth has raised by evaporation, goes entirely to centralized elec-

---

9. The main product of combustion is carbon dioxide, which goes into the atmosphere where it contributes to global warming.

Responsible scientists have argued strongly that trade in plutonium poses risks of both proliferation and instances of diversion which might further undermine public confidence in nuclear power. And physicists might well read and learn from the substantial literature about both the history and the psychology of the relationship between nuclear power and nuclear weapons.

ROBERT GOBLE, PROFESSOR OF ENVIRONMENT, TECHNOLOGY, AND SOCIETY AND ADJUNCT PROFESSOR OF PHYSICS, CLARK UNIVERSITY

tricity. Although more energy could be squeezed out, it is close to its practical limit.

Biomass energy, the chemical energy of organic substances such as wood, sugar crops, grains, and trash, is available from many sources and can be transformed into many forms. It can be burned for heating or to make power-plant steam for electricity—its main uses today. Bacteria can decompose biomass and convert it to alcohol for transportation and to methane gas for heating. Biomass can also be chemically processed to create gaseous or liquid fuels. All together, an estimated 20% of current annual energy consumption could be supplied in these ways. Biomass makes no long-term net contribution to global warming, provided that the trees or crops grown for energy are replanted.

The remaining renewables in Table 17.1 contribute very small amounts today, but each has substantial future potential. Of these renewables, geothermal, wind, solar–thermal, and photovoltaic cells are used to make electricity. The remaining two, active and passive solar, are used for direct heating.

Geothermal energy is the thermal energy that radioactivity and pressure create underground. The easiest way to tap it is by drilling to directly remove hot water or steam. But hot water and steam are very limited resources. The large, essentially unlimited geothermal resource is **hot dry rock** lying several kilometers underground and amounting to many times more energy than is contained in coal deposits. The drilling and underground fracturing technology to recover this energy is highly demanding and is as yet unproved (Figure 17.10).

Wind energy is the kinetic energy of air set into motion when the sun warms the daylight side of Earth. Since antiquity, it has driven sailing ships and turned windmills for grinding and pumping. Today, it also generates electricity. The world's largest installation is at Altamont Pass in California, where small electric generators attached to 7500 **wind turbines** have a capacity equivalent to two large power plants (Figure 17.11). This is a proven technology: The Altamont installation is over 50 years old. In regions where wind turbines are feasible, wind power is already financially competitive with coal for large-scale electric power generation. New designs should reduce costs further, potentially making 12 U.S. Great Plains states major wind-power producers. Widely spaced wind turbines placed over the windiest 10% of the country—largely agricultural land—could in principle provide most of U.S. electricity.

Solar–thermal electricity is generated from thermal energy created by the sun. This resource generates the equivalent of one large power plant today, using two types of technologies. In the first, reflective solar collectors track the sun and focus it on a liquid that is then piped to a central location where it produces steam to drive an electric generator (Figure 17.12). The second uses sun-tracking mirrors to reflect solar energy to a central boiler that produces steam (Figure 17.13).

Photovoltaic cells use the photoelectric effect (Section 13.2) to transform solar radiation directly into electric current. The cells are made of **semiconductors**, such as silicon, that have electrical properties lying midway between conductors and insulators. Normally, semiconductors behave like insulators: It is difficult to make their electrons flow. But if some of a semiconductor's electrons are given a small amount of energy, these electrons can flow easily, as in a conductor. Because of these properties, semiconductors are the basis for modern electronic technology.

FIGURE 17.10
Thermal energy might someday be extracted from dry hot rocks by circulating water through large cracks created by hydraulic fracturing.

FIGURE 17.11
Wind generators, each one producing 100 kilowatts of electricity (enough for about 100 households), in northern California.

In photovoltaic cells, light provides the energy that puts electrons into the conducting state. A typical cell is made of two thin layers of silicon constructed differently so as to have different electrical properties, called *n-type* (negative) and *p-type* (positive) (Figure 17.14). The difference is that nonsilicon "impurity" atoms of different types are introduced into the two different layers. N-type semiconductors are made of silicon containing impurity atoms that have more semiconducting electrons (per atom) than does silicon. P-type semiconductors contain impurities that have fewer semiconducting electrons than does silicon.

Suppose the two layers are simply placed together, without any external electrical contacts. Because of the impurities, microscopic forces acting within each layer cause electrons to quickly flow from the n-type (where there were more electrons to begin with) to the p-type (where there were fewer). Within a fraction of a second, this process builds up a sufficiently strong electric field at the p–n boundary (the junction between the two layers) to stop the flow of electrons.

If light now shines on the two layers—still with no external electrical contacts—the photoelectric effect will energize some of the electrons, causing those to cross back across the p–n boundary from the p side to the n side. At this

FIGURE 17.12
Solar collectors for a solar–thermal power plant. The mirrored faces focus sunlight onto the pipe running the length of the collectors. The pipes transfer thermal energy to a steam turbine coupled with an electric generator.

FIGURE 17.13
A 10-megawatt solar–thermal test facility in Barstow, California. The mirrors reflect sunlight onto a boiler on top of the tower, and the boiler makes steam for electricity generation.

point, the n side is like the negative terminal of a battery that is not connected to anything, and the p side is like the positive terminal. Now suppose we electrically connect this tiny "battery" to an external device, as shown in Figure 17.14. Then the energized electrons that were pushed to the n side will flow through the wires back around to the p side. As electrons arrive at the p side, light continues to energize electrons to move across the junction from the p side to the n side, creating a complete loop of flowing electrons. The p–n junction acts like a battery energized by light.

Arrays of such cells can provide electricity for centralized electric power, or they can power individual buildings or appliances, especially in remote locations that are difficult to reach with centralized power (Figure 17.15). This resource is under intense development and should become increasingly competitive for large-scale electric power.

Finally, solar energy can be used directly for **solar heating**. Figure 17.16 shows a rooftop "flat-plate" collector where a pumped liquid is heated and then is circulated back indoors to be used for space or water heating. Collectors using pumps are **active** forms of solar heating. Solar energy use can also be **passive**, based on natural flow. The methods are older than humankind (animals and

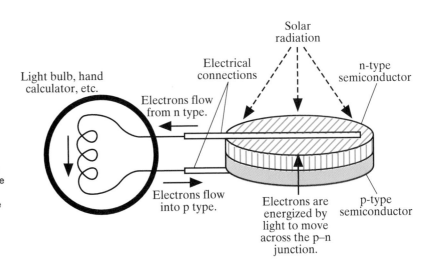

FIGURE 17.14
How a photovoltaic cell works. The cell is almost as thick as a dime (1 mm). Because of the differing microscopic properties of the two semiconductors, electrons that are energized by light cross the p–n junction from the p to the n side. These electrons can then flow through an external circuit. The junction acts like a battery.

FIGURE 17.15
A series of photovoltaic cells, attached together, can provide a large electric current.

Substantial increases in importation of oil . . . could restore control of the market to OPEC by the year 2000 or before. Unless there are economic incentives for improving energy efficiency, conservation, and substitution, and for increasing domestic supplies, the United States will become increasingly vulnerable to a great, multicomponent energy crisis.

PHILIP ABELSON, PHYSICIST, PAST EDITOR OF *SCIENCE*

Black metal absorber plate with liquid tubes built in

Rear cover and insulation

Heated liquid out

Double-glass cover

Cool liquid in

FIGURE 17.16
A flat-plate solar collector that uses forced circulation of a liquid to collect and transfer solar thermal energy.

plants seek the sun), as simple as the backyard clothesline, and as new as the latest triple-pane argon-filled multilayer-coated windows. Figure 17.17 illustrates several passive concepts. Even in cold climates, these methods can reduce heating needs by 60 to 80%, so it is surprising that during the 1980s fewer than 10% of new houses incorporated passive solar features.

The economics of renewable resources are in rapid flux. Figure 17.18 graphs the past and expected future costs of electricity from three renewable sources and compares them with electricity from new coal-fired "base-load" plants that are operated continuously to provide most of the electricity consumed, and with natural gas-fired "peaking-load" plants that are operated only intermittently to provide electricity at times of high demand. Peak-load electricity costs more than twice that of base-load electricity. Wind and solar thermal electricity are competitive for peaking power today, and in some locales wind is competitive for base-load power. Early in the twenty-first century, all three might compete with coal for base-load electricity.

Many energy analysts have noted that if a market economy is to preserve its environment and resources, it must incorporate "externalities" such as pollution and future resource availability into the price of goods. According to this view, coal, for example, is being subsidized by the environment because the price of coal does not reflect its full pollution costs, and this hidden subsidy distorts the market, leading to the overuse of coal because it is underpriced. Many analysts have suggested that such externalities be incorporated into the economics of fossil fuels. Several methods have been suggested, the simplest being a tax on the carbon content of all fuels, with the proceeds used to clean up the environment or to support development of less harmful energy resources.

■ **DIALOGUE 10** Which of the following are heat engines; that is, which ones convert thermal energy into other forms: hydroelectric power plant, geothermal power plant, wind turbines, solar–thermal electricity generation, photovoltaic cells?

---

10. Geothermal and solar–thermal.

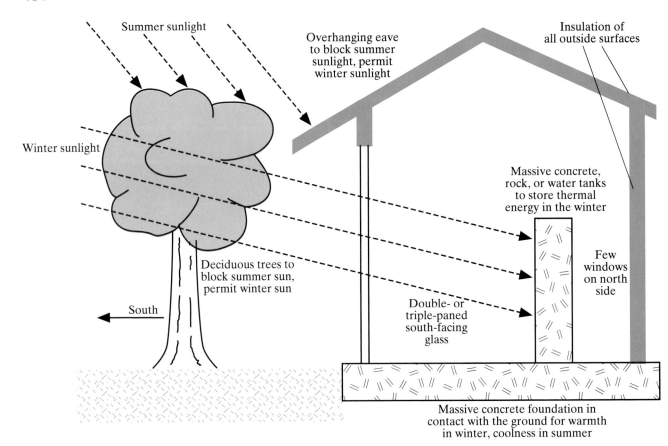

FIGURE 17.17
Several passive solar energy concepts for keeping houses warm in cold weather and cool in hot weather.

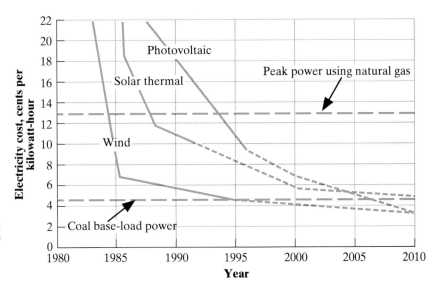

FIGURE 17.18
Solar electricity costs compared with fossil-fueled electricity. The electricity costs of renewables (vertical axis) are based on actual (solid lines) or projected (broken lines) costs for centralized power generation.

17.7 FUTURE ENERGY OPTIONS: CONSERVATION

Conservation contributes more to the nation's energy services than any of the actual energy resources (Figure 17.5) but with the bonus of not causing the pollution and resource depletion that accompanies actual energy use. Without the efficiency measures taken since the 1973 oil embargo, the United States would be using about 40% more energy than it is using. And the saved energy has reduced U.S. energy costs by over $200 billion annually.

Are further savings possible? Figure 17.19 suggests that the answer is "yes." The figure compares the energy consumed per person in nine nations. Germany, Great Britain, and Japan, three nations whose energy services (as measured by the gross national product per person) are similar to those of the United States, have per capita energy consumptions that are about half that of the United States. This suggests that U.S. energy use might be reduced by 50% without reducing energy services.

As one example of the savings that are possible when energy efficiency is taken seriously, we will look at recent advances in efficient lighting.

Traditional incandescent bulbs create light by simply heating a thin wire until it glows. This naturally produces a lot more heat than light. **Fluorescent bulbs** operate on a more energy-efficient principle, one that does not involve heating. The bulb's glass tube is filled with dilute mercury vapor or some other gas. An electric current passes through the tube, ionizing some of the mercury atoms by collision and putting many atoms into excited quantum states (Sections 14.1 and 14.2). The excited mercury atoms radiate invisible ultraviolet photons that are absorbed by a powdery material called the *phosphor* that coats the inside of the glass tube, causing the phosphor to radiate visible light. This process is five times more efficient than incandescent bulbs, so the same amount of lighting can be obtained for one-fifth as much energy.

Fluorescent bulbs have attained even higher efficiencies recently because of the development of high-frequency *ballasts*. A fluorescent bulb's ballast is the device that controls the current flowing through the tube. In standard electrical circuits in homes, electrons don't keep flowing in one direction but instead vibrate back and forth 60 times per second. Older fluorescent bulbs operate at this 60 Hz frequency. But both bulb quality and efficiencies would be higher if the frequency within the bulb were much higher than 60 Hz. The reason is that at 60 Hz, electrons and ions have a full one-sixtieth of a second for each back-and-forth motion—a long time in the world of electrons and ions. This is time enough for large numbers of electrons and ions to run into the ends of the tube, creating heat instead of light. This inefficiency is also unsightly because it causes the light to dim near the end of the tubes.

New electronically controlled ballasts cause the current to oscillate at up to 50,000 Hz. This reduces energy losses at the ends of the tube by restricting the electrons and ions to much shorter distances of vibration. Along with improving the efficiency and solving the end-dimming problem, high-frequency ballasts make possible shorter, compact fluorescent bulbs that can be screwed into ordinary incandescent light sockets (Figure 17.20).

The consequences of this single efficiency improvement are surprising. One compact fluorescent bulb, purchased at $18 and used over its normal 10,000-hour lifetime, saves the consumer about $40 (see Dialogue 11). It is estimated that in a few decades, when the market is saturated, the electricity savings to the

FIGURE 17.19
Per capita energy resource consumption in several nations in 1994.

FIGURE 17.20
Compact fluorescent bulbs.

Least risk leads to the same conclusion as least cost. Energy companies can afford only options that are small, fast, cheap, and modular. Efficiency, renewables, and some natural-gas technologies meet these tests. Coal and nuclear generally don't. . . . Coal will retain some niche markets for metallurgy and perhaps for process heat. Yet coal's central-station market will slowly die of an incurable attack of market forces. The competitor [for coal] to beat isn't nuclear power; it's efficiency costing less than coal at $10 a ton, effectively burned in a free, zero-emission plant.

AMORY LOVINS

Growth [in energy consumption] has declined primarily because conservation turned out to be far less expensive than new supply— 3 to 5 times cheaper than new power plants.

ARTHUR H. ROSENFELD, PHYSICIST WHO PROVIDED THE ENERGY SERVICES OF 55 LARGE GENERATING PLANTS BY DEVELOPING HIGH-FREQUENCY BALLASTS FOR HIGH-EFFICIENCY FLUORESCENT BULBS

United States from compact fluorescents will equal the output of 55 large power plants, cutting home electricity consumption by 8%, producing net savings (after paying for the bulbs) of $5.5 billion annually, and reducing carbon-dioxide emissions by 55 million tons per year.

Energy efficiency and economics are closely entwined. Energy efficiency usually saves money, and when it does it is possible for governments or companies to stimulate energy-saving programs that reward consumers financially. For example, some electric companies and some governments are making large-volume low-cost purchases of compact fluorescent bulbs and selling them to consumers at below cost, with the difference recovered in electricity charges to those consumers. The consumers save money because of reduced electricity consumption, which more than compensates for the extra charges paid for the bulbs. Everybody wins: the company, the consumer, and the environment.

Because energy efficiency has been overlooked for so long, small investments can result in large savings. For example, a $7.5 million compact-fluorescent lamp factory saves as much electricity as a $1 billion power plant generates, while also avoiding the power plant's fuel cost and pollution. A $10 million "superglass" factory making windows that block heat but permit light can produce the comfort that would be provided by the air conditioners run by $2 billion worth of generating stations.

The price of energy strongly affects the amount and types of energy consumed. It is instructive to note that the lock-step link between growth in GNP and in energy was broken only when the 1973 oil embargo raised energy prices. Beginning in 1973, as energy efficiency began to take hold, GNP grew but energy use did not (Figure 17.5). As a result of the higher energy prices resulting from the oil embargo, an enormous amount of energy has been saved.

Many studies suggest energy taxes as an incentive for conservation. Energy taxes could be made "revenue neutral" by lowering other taxes in compensation, to discourage energy use while not increasing overall taxes. Incentives could encourage energy-conscious shopping. For example, the U.S. Office of Technology Assessment suggests imposing gas-guzzler taxes on inefficient automobiles and gas-sipper rebates on efficient ones. An underlying view is that global warming, resource depletion, and other environmental concerns must be incorporated into the costs of doing business and consuming goods if a market economy is to address such concerns.

■ **DIALOGUE 11** A typical 18-watt compact fluorescent bulb costs $18, lasts 10,000 hours, and provides as much light as does a 75-watt incandescent bulb that costs 75¢ and lasts 750 hours. How much does it cost for each type of bulb to provide 10,000 hours of light? Do not forget to include energy costs, at about 9¢ per kilowatt-hour.

■ **DIALOGUE 12** In Dialogue 11, how many joules of electrical energy are saved by using the compact fluorescent rather than the incandescent bulb for 10,000 hours? (Recall that 1 watt is 1 joule per second.)

---

11. Compact fluorescent: One bulb is needed, at $18. The cost of the energy is $0.09 \times 0.018 \times 10,000 = $16.20. Total cost = $18 + $16.20 = $34.20. Incandescent bulb: The number of bulbs needed is 10,000/750 = 13.3 (14 will be needed). Cost of bulbs = $0.75 \times 14 = $10.50. The cost of the energy is $0.09 \times 0.075 \times 10,000 = $67.50. Total cost = $10.50 + $67.50 = $78.

12. Each type of bulb burns for $3,600 \times 10,000 = 36 \times 10^6$ seconds. The 18-watt bulb consumes $18 \times 36 \times 10^6 = 650$ million joules. The 75-watt bulb consumes $75 \times 36 \times 10^6 = 2,700$ million joules. Energy savings = about 2,000 million joules.

"THE ONLY OTHER SOLUTION IS THAT WE MAY EVOLVE INTO A SPECIES IMMUNE TO ALL THIS JUNK."

SUMMARY OF IDEAS AND TERMS

Energy resources Natural resources containing useful energy. The major U.S. resources today are fossil (coal, oil, natural gas), nuclear (uranium), hydroelectric, and wood. A resource is **renewable** if it can be replaced within a human lifetime; otherwise it is **nonrenewable**.

Energy conservation Measures to reduce energy consumption, including both **efficiency** measures to save energy while providing the same services, and changes in the quality of services.

Nuclear power reactor A device in which chain-reacting nuclei transform nuclear energy into thermal energy for electric power. Its main components are **fuel** to provide energy, a **moderator** to slow the neutrons so that they will split the nuclei more efficiently, neutron-absorbing **control rods** to control the reaction, and a **coolant** to transfer thermal energy from the fuel. Most reactors are enclosed in a thick concrete **containment dome** to prevent the escape of radioactivity into the environment.

Technology assessment The process of clarifying the scientific and technological issues, and objectively listing the costs and benefits of the available options, as a prelude to making a technology-related decision.

Breeder reactor A reactor that creates more than one ^{239}Pu nucleus (from ^{238}U) for each ^{235}U nucleus it fissions and so creates more fuel than it uses.

Fusion reactor A nuclear reactor that obtains its thermal energy from fusion rather than fission. Now under development, it might be commercially viable by the middle of the twenty-first century.

Global warming Human-caused greenhouse-effect warming of Earth's atmosphere, due mainly to fossil fuels.

High-level nuclear waste Used reactor fuel rods containing highly radioactive fission products.

Three Mile Island Site of the most significant U.S. nuclear power plant accident. Little radioactivity escaped even though the fuel suffered a **melt-down**—the fuel melted together into a solid radioactive mass that slumped downward inside the reactor.

Chernobyl Site of world's worst nuclear power accident. The fuel melted down; the reactor suffered a "slow nuclear explosion," comparable to a chemical bomb; and a large amount of radioactivity escaped.

Nuclear weapons proliferation The spread of nuclear weapons. The use of plutonium as reactor fuel, and breeder reactors, could worsen this problem.

Renewable energy resource A natural source of energy that is continuously available, or that can be replaced within a human lifetime. Examples include the following:

- **Hydroelectric**, the gravitational energy of raised water.
- **Biomass**, the chemical energy of organic substances.

- **Geothermal**, the thermal energy of hot underground steam, water, or rock.
- **Wind**, the kinetic energy of moving air.
- **Solar–thermal electricity**, generated from thermal energy created by the sun.
- **Photovoltaic electricity**, generated from solar radiation using the photoelectric effect. Two thin layers of silicon, constructed somewhat differently, cause electrons to move from one layer to the other when illuminated with light. The device acts like a battery. Silicon is a typical **semiconductor**—a material having properties lying midway between insulators and conductors. Normally, it is difficult to make electrons flow through a semiconductor. But if some of a semiconductor's electrons are given a small amount of energy, these electrons can flow easily, as in a conductor.
- **Active solar**, in which a solar-heated liquid or gas is pumped elsewhere for space or water heating.
- **Passive solar**, in which solar radiation, natural air flows, energy storage, and insulation provide direct **solar heating** and cooling.

Fluorescent bulb An electric current flows through a gas that fills the bulb, exciting the gas, which emits ultraviolet radiation, which is absorbed by the phosphor coating, which emits light. New **high-frequency ballasts** have recently been developed for increased efficiency.

REVIEW QUESTIONS

ENERGY HISTORY AND FUTURE

1. Is it correct to say that this chapter studies the *predicted energy future*? Suggest another, more accurate, phrase.
2. List the six main U.S. energy resources today. Which three are the most used?
3. Which of the six main U.S. energy resources are renewable?
4. Which of the six main U.S. energy resources are primarily used to generate electricity?
5. In Figure 17.4, why is the residential sector of the economy so much more energy-efficient than the transportation sector?
6. List the four broad categories of options for resolving the energy problem.
7. Which yields more usable energy in the United States today: fossil fuels, nuclear power, or renewable energy resources?
8. List the fossil fuels. Why are they called *fossil fuels*?
9. Explain why the gap between the GNP graph and the total energy graph in Figure 17.5 is labeled "conservation."
10. Explain why, in Figure 17.5, GNP and total energy coincide for many years, and why they eventually separate.

NUCLEAR POWER

11. What is a nuclear reactor?
12. Which of the following is needed in a coal-fired power plant but not in a nuclear power plant: condenser, cooling tower or lake, turbine, stack, containment dome?
13. Which of the following is needed in a nuclear power plant but not in a coal-fired power plant: condenser, cooling tower or lake, turbine, stack, containment dome?
14. List the four essential components of a nuclear power reactor, and explain the function of each.
15. Is the enrichment of U.S. reactor fuel closest to 3%, 10%, 25%, 50%, or 90%?
16. Is the enrichment of bomb-grade uranium fuel closest to 3%, 10%, 25%, 50%, or 90%?
17. Where might a fusion reactor get its fuel?

ASSESSING NUCLEAR AND COAL POWER

18. List five problems with nuclear power and five problems with coal power.
19. Of the four most-used nonrenewable energy resources, which one(s) does the United States possess in greatest abundance (most years remaining at current rate of use)? In least abundance?

20. Do most energy experts feel that coal use will continue rising over the long term? Why?
21. What are high-level nuclear wastes, and where are they stored today? What is the most likely long-term solution to this problem?
22. Describe in general terms what happened at Three Mile Island.
23. Describe in general terms what happened at Chernobyl.
24. What is a breeder reactor? How could breeder reactors affect the uranium resource problem?
25. How could breeder reactors affect the nuclear proliferation problem?

RENEWABLES AND CONSERVATION

26. Which two renewable energy resources are in widest use today?
27. List three renewable resources that could produce significant amounts of electricity in the future, and describe briefly how each works.

28. What are semiconductors, and what role do they play in photovoltaic cells?
29. What is the difference between active and passive solar heating? Describe some of the techniques used in each.
30. What is the difference between base-load power and peaking power?
31. How does the per capita energy consumption of other nations compare with that of the United States? What does this suggest about the potential of further energy efficiencies in the United States?
32. How does a fluorescent bulb work?
33. Why are fluorescent bulbs more energy efficient than incandescent bulbs?
34. Describe a recent improvement in fluorescent bulbs that has made them even more energy efficient.

HOME PROJECTS

1. What color absorbs solar energy best? Get several sheets of construction paper of different colors, place an ice cube on each, and put them in the sun or under a heat lamp. Which melts first? Do you get the same result if the paper is on top of the ice cube? What color should a house roof be in a hot climate? In a cold climate?
2. Measure the *solar angle* (the angle that the sun's noon-time rays make with the vertical) on several different dates during this semester. What do your results tell you about the proper design of a passive solar house (Figure 17.17)?
3. Make a tour of your home, listing all the electrical devices. Include lightbulbs and their power output. Use the labels on these devices, and also the figures given in Table 6.1, to estimate the total number of kilowatt-hours (kW·h) of electrical energy your home uses each day. Check your estimate against the number of kW·h stated on your monthly electric bill. The average U.S. home consumes electrical energy at an average rate of 1 kW, or 24 kW·h every day. Does your home consume more, or less, than this? What measures might be taken to reduce

your home's electrical consumption to half of this amount, that is, to 12 kW·h every day?
4. Write to your U.S. senator or congressperson about one of the science-related social concerns discussed in this course. Your opinion makes a difference—if you express it! Write to

        Senator _____
        U.S. Senate
        Washington, DC 20510

or to

        Congressman (or Congresswoman) _____
        U.S. House of Representatives
        Washington, DC 20515

Include your name and address, recipient's name and address, and date. Discuss only a single topic. Express your own opinion—your representatives want to know what you think. Back up your opinion with reasons. Keep it to the point—and on one page. Be specific. Try to refer to a specific bill before Congress. Feel free to disagree, but keep it friendly.

FOR DISCUSSION

1. Make a list of all the energy-consuming devices you used this week that were not available a hundred years ago. How would the world and your life be different without these devices?
2. List several ways by which you and the nation could reduce energy consumption without seriously harming your (or its) overall quality of life. When considering alternatives such as mass transit or reduced lighting, you

will have to decide whether such measures would actually harm the quality of life.
3. Outline two different scenarios for the nation's energy future, both of them plausible. Give a reasoned argument for your preferred scenario. Now give a reasoned argument for the other scenario. Graph each of your scenarios by extending (on a blank sheet of paper placed next to the graph) the graph of Figure 17.5 to the year 2050.

EXERCISES

ENERGY HISTORY AND FUTURE
1. Which are renewables: wood, uranium, trash (as fuel), coal, wind, natural gas, hydroelectric?
2. Which options listed in Table 17.1 contribute to global warming?
3. Which resources listed in Table 17.1 do *not* come ultimately from the sun?
4. Use Figure 17.4 to calculate the energy efficiency of the U.S. residential–commercial sector (answer: 71%). From this percentage, what can you plausibly conclude regarding the use of heat engines by this sector?
5. Use Figure 17.2 to estimate the amount of energy the United States got from coal in each of the following years: 1900, 1920, 1940, 1960, 1980.
6. Use Figure 17.2 to estimate the amount of energy the United States got from oil in each of the following years: 1900, 1920, 1940, 1960, 1980.
7. Which energy resource was dominant in the United States in 1860? 1880? 1900? 1920?
8. Which energy resource was dominant in the United States in 1940? 1960? 1980?

NUCLEAR POWER
9. Every heat engine has a thermal energy input, a work output, and a thermal energy output. At what places in Figure 17.6 does each of these occur?
10. Figure 17.8 shows a tall tower. Is it analogous to a coal-fired plant's stack? What is the function of this tower? Might coal plants have a tower like this?
11. Does a nuclear power plant have a "smoke" stack? Why or why not?
12. Figure 17.6 shows that a nuclear power plant has a condenser. What is the function of this device? Do coal plants also have condensers?
13. Why is it important that a nuclear power plant have two separate water or steam loops to get heat to the turbine? What could happen if there were only one loop?
14. *Thermal pollution* is thermal energy put into a cooling lake or stream. Consider a coal-burning power plant and a nuclear power plant, both operating at an energy efficiency of 40%. Which would create more thermal pollution, per joule of electric power output?
15. Neutrons, created when particles from outer space hit the atmosphere, often hit Earth's surface. Could these neutrons start a fission chain reaction in natural uranium in the ground? Why?
16. Suppose the main water pipe breaks in a nuclear power plant, shutting off the water flow. If the control rods fall immediately into place, stopping the chain reaction, is there still a problem? Why or why not? What safety feature should be used in this case?

ISSUES FOR NUCLEAR AND COAL POWER
17. List two advantages that coal has over nuclear power.
18. List two advantages that nuclear power has over coal.
19. Since power plants fueled by natural gas create carbon dioxide, why would it reduce global warming to switch from coal to natural gas power plants?
20. Are radioactive wastes hot? Why?
21. How can a breeder reactor create more fuel than it consumes? Why doesn't this violate the law of conservation of energy?
22. How might a country's nuclear power industry contribute to developing a uranium bomb?
23. How might a country's nuclear power industry contribute to developing a plutonium bomb?

RENEWABLES
24. Describe how hydroelectric energy is renewed.
25. List one possible disadvantage of each of the eight renewable energy resources listed in Table 17.1.
26. What are some ways that solar energy is routinely used around the home?
27. Is renewable energy used today for any form of transportation? Explain.
28. What physical energy transformation occurs for each of these ways of making electricity: coal, uranium, hydroelectric, biomass?
29. What physical energy transformation occurs for each of these ways of making electricity: geothermal, wind, solar thermal, photovoltaic?
30. What energy transformation occurs when you hang your wash outdoors to dry in the sun?
31. Suppose that a carbon tax caused coal base-load power to be priced at 10¢/kW·h. According to Figure 17.18, when would wind, solar–thermal, and photovoltaics then compete economically with coal for base loads?

CONSERVATION
32. In what way was the oil crisis, caused by the 1973 Mideast oil embargo, a good thing for the United States?
33. Are there any industrialized nations in which the energy use per person is half, or less than half, what it is in the United States?

PROBLEMS

ENERGY HISTORY AND FUTURE

1. Use Figure 17.2 to estimate the *percentage* of U.S. energy resource consumption that came from coal in each of the following years: 1900, 1920, 1940, 1960, 1980.
2. Use Figure 17.2 to estimate the *percentage* of U.S. energy resource consumption that came from oil in each of the following years: 1900, 1920, 1940, 1960, 1980.
3. Total U.S. energy use was roughly exponential (Section 7.8) during 1880 to 1920. Use Figure 17.2 to estimate its doubling time. Find its annual percentage increase (see Section 7.8).
4. From Figure 17.4, find the approximate energy efficiency of the overall U.S. economy. Of the three major sectors, namely industrial, residential/commercial, and transportation, which one is least efficient? Why?
5. According to Figure 17.4, what fraction of U.S. energy resources goes into the generation of electricity?
6. According to Figure 17.4, what fraction of coal energy goes into the generation of electricity? What fraction of oil? Of natural gas? Of nuclear resources?
7. From Figure 17.5, estimate the equivalent amount of energy services (in exajoules) provided by post-1973 conservation measures, in each of the following years: 1975, 1980, 1985, 1990, 1995. Now estimate the *percentage* of the nation's total energy services provided by conservation, in each of these years.
8. Use Figure 17.5 to estimate the percentage increase in annual U.S. energy use (actual energy—not energy services) during 1950–1960. Repeat for 1960–1970, for 1970–1980, and for 1980–1990. Do you notice any general trends here? Speculate on the reasons for any trend that you notice.

COAL AND NUCLEAR POWER

9. *Thermal pollution* is thermal energy put into a cooling lake or stream. Consider a coal-burning power plant and a nuclear power plant, both operating at an energy efficiency of 40%. How many joules of thermal pollution does each plant create, per joule of electrical energy?

10. A nuclear power plant's reactor heats steam to 450°C and exhausts hot water into a lake at 80°C. Using the formula

$$\text{efficiency} = (T_{in} - T_{out})/T_{in},$$

calculate the theoretical, or maximum possible, energy efficiency of this power plant. Suppose that the actual energy efficiency is 40%. For every joule of electrical energy created by this plant, how many joules of nuclear energy must be used?

RENEWABLES

11. According to Figure 17.18, when will photovoltaics and solar thermal become competitive with coal for large-scale electricity generation? Suppose that a carbon tax were placed on coal due to the threat of global warming, causing the cost of coal-generated electricity to rise to 8¢/kWh. In this case, when would photovoltaics and solar thermal become competitive?
12. Making estimates. If you covered a football field with photovoltaic cells, for about how many households could it provide electricity? Use the following information: Solar energy hits each square meter of Earth's surface at an average rate (averaged over day and night) of 200 watts, an average household consumes electricity at a rate of one kilowatt, and photovoltaic cells are 20% efficient.
13. Making estimates. Use the information in the preceding problem to show that about 10,000 square kilometers of land area (3% of a sunny state such as Arizona) would need to be covered with photovoltaic cells to provide all of the United States's electricity. The United States consumes electricity at an average rate of about 400 billion watts. What complications might arise if the nation tried to provide all its electricity this way?
14. Use Figure 17.19 to compare (by finding the ratio) the per capita energy consumption of the United States with Great Britain. Compare Great Britain with India.

CHAPTER 18

Quantum Fields
relativity meets the quantum

The unexpected and the incredible belong in this world. Only then is life whole.

CARL GUSTAV JUNG

What happens when relativity meets quantum theory?

The two theories extend Newtonian physics in different ways. One extends it up to lightspeed, and the other extends it down to the smallest dimensions yet measured, 10^{-19} meters, 10,000 times smaller than an atomic nucleus. But separate relativity and quantum theories cannot be the whole story: relativity theory doesn't contain the quantum principles so it doesn't work at small sizes, and quantum theory doesn't contain the relativity principles so it doesn't work at high speeds. Is it possible to combine the two into one theory covering all sizes and all speeds?

The answer is yes. The combination is called **quantum field theory**. One part of this theory, quantum electrodynamics, is the most accurate scientific theory ever invented. Like most post-Newtonian physics, quantum field theory is simple but takes some getting used to. An underlying idea of this theory, and an enduring theme of post-Newtonian physics, is the **field view of reality** already discussed in connection with Einstein's mass-energy relation (Section 11.5): the view that the universe is made of fields, subject to the rules of special relativity and quantum theory. This is the central idea of this chapter.

Section 18.1 presents the general idea of quantum field theory, and the remaining sections apply this idea to each of the four known fundamental forces: electromagnetic (Sections 18.2 and 18.3), weak (Section 18.4), strong (Section 18.5), and gravitational (Section 18.6).

18.1 QUANTIZED FIELDS: THE REASON THERE ARE PARTICLES

Recall (Chapter 9) that a force field, or simply a **field**, is spread out over a region of space. This region needn't contain any matter or any "thing" at all; it can be empty space, a vacuum, containing no matter. A field is a kind of stress in space and can be described as the possibility of a force. For example, an electric field is the possibility of an electric force. Electric fields are created by charged objects and are spread out in space wherever charged objects would feel a force. Magnetic fields and gravitational fields are other familiar examples of force fields.

—although we shall always be able to imagine other theories that are totally different from the true final theory—like the boring world of particles governed by Newtonian mechanics.

NOBEL PRIZE–WINNING PHYSICIST STEVEN WEINBERG'S COMMENT ON A UNIVERSE RULED AT THE MICROSCOPIC LEVEL BY NEWTONIAN MECHANICS, IN *DREAMS OF A FINAL THEORY*

The one part of today's physics that seems to me likely to survive unchanged in a final theory is quantum mechanics.

STEVEN WEINBERG

The notion that reality is a set of fields that give the probabilities for finding their associated quanta is the most important consequence of relativistic quantum field theory. It is the central concept for the picture of reality. Not only did the idea of matter disappear into the field concept, but the field specified the probability for finding quanta.

HEINZ PAGELS, PHYSICIST

At the core of quantum field theory is a view that we have discussed before in connection with mass–energy equivalence. It is the idea that everything is made only of fields, that every thing is "empty," that matter is really just a collection of force fields. Fields have mass because they have energy. Material particles such as electrons and protons are just fields, and the position of a particle is just the center of its field. As poet Gertude Stein put it, there is no "there" there. The table on which this book rests is simply a configuration of quivering force fields, and so is the book. The book doesn't fall through the table, however, because the electron force fields in the table repel the other electron force fields in the book. And your eye (which is also just fields) sees the book because the book's force fields emit radiation.

There is nothing: no thing. Only fields.

Quantum field theory began in 1926 when the quantum theory of the electromagnetic field was invented by Max Born, Werner Heisenberg, and Pascual Jordan. Its basic idea is to treat the electromagnetic field itself as the fundamental system that is subjected to both quantum and relativity principles, and to develop a mathematical description of this system (the field) that is analogous to the description that Schroedinger developed for material particles. In other words, this theory presents a sort of Schroedinger equation for the motion of the electromagnetic field itself, rather than for the motion of any particle such as an electron or proton or photon.

The Schroedinger equation for, say, an electron predicts that electron's psi field; that is, it predicts the probability of finding the electron at each point in space. In the same way, the quantum theory of the electromagnetic field predicts the electromagnetic field's psi field; that is, it predicts the probability of finding the electromagnetic field to have various strengths at each point in space. Just as the Schroedinger equation for a particle predicts energy quantization for that particle, the quantum theory of the electromagnetic field predicts that the field will be quantized into bundles of energy of various frequencies. These bundles are photons! And the theory predicts that the energy of one bundle will be $E = hf$, just as Planck and Einstein had discovered around 1900. But now, instead of having to *assume*, for no known reason, that radiation comes in the form of photons having energy hf, the quantum theory of the electromagnetic field *predicts* that electromagnetic fields will exhibit themselves as photons having this energy. The new theory explains why photons exist.

This is typical of all quantized field theories: One starts with a field, quantizes it, then out pop the particles associated with that field. These particles are the **quanta of the field**, the discrete bundles of energy associated with that field.

In 1929, Heisenberg and Wolfgang Pauli took a new step: They applied the quantized field idea to matter by quantizing the psi field of the electron. They treated this "electron field" as a real physical system that should itself obey the principles of quantum theory and relativity theory. Note that an electron field (the psi field of an electron) is different from an electric field: The electron field is a matter field, not a radiation field. Heisenberg and Pauli then discovered that this quantized electron field must exhibit itself as tiny bundles of energy, the quanta of the electron field. But the quanta of the electron field are not photons—they are electrons! So this theory predicts the existence of electrons.

The basic idea of quantum field theory is to quantize all of the force fields and all of the psi fields. We quantize (that is, we subject to quantum principles) the photon's psi field (the electromagnetic field), the electron's psi field (the electron field), the proton's psi field, the neutron's psi field, and so forth. All of the particles of nature are quanta of their associated fields, just as photons are quanta of the electromagnetic field. So quantum field theory explains why nature exhibits itself as particles. Nature is made of particles because nature is made of fields, and when a field is subjected to the principles of quantum theory, the field's energy turns out to come in tiny bundles—particles. The list of the fundamental ingredients of the world no longer needs to include any particles at all—it needs to include instead a few kinds of fields and the principles of quantum theory.

By viewing nature as made of quantized fields, physicists can combine quantum theory and relativity and put matter and radiation on an equal footing: Both material particles and radiation particles are just bundles of field energy. These particles are subject to the usual quantum uncertainties. The probability of finding a particle at some particular point is determined by the corresponding field's psi field at that point. All the different fields interact with one another, and these interactions determine everything that happens in nature. To summarize:

> **THE QUANTUM THEORY OF FIELDS**
> The essential reality is a set of fields. Everything that happens in nature is a result of the interactions and motions of these fields. All fields obey special relativity and quantum theory. Quantum theory requires these fields to exhibit themselves as tiny bundles or quanta of field energy. All of nature's particles of matter and radiation are quanta of this sort. The intensity of a field at a point determines the probability for finding the associated particle.

Quantum field theory resolves—as nearly as it can be resolved—the old wave-versus-particle dilemma. The resolution is that nature is made of fields (waves) but that these fields are subject to quantum principles and so must exhibit themselves as particles of field energy. Fields and particles are the yin and the yang of relativistic quantum theory.

18.2 QUANTUM ELECTRODYNAMICS: THE STRANGE THEORY OF ELECTRONS AND LIGHT

Quantum field theory emerged during the 1930s as the world was marching toward war. Although nuclear physics flourished (Chapter 16), fundamental quantum theory had to wait. After the war, in 1947, two New Yorkers in their 20s, Richard Feynman and Julian Schwinger (Figure 18.1), independently completed quantum field theories of the electromagnetic force. The two theories, known as **quantum electrodynamics**, say the same thing but they look different, perhaps owing to their authors' differing personalities: Whereas Schwinger preferred to work in Einstein-like solitude, Feynman was known to jolt his mind out of a rut by working at a back table in a nightclub, inspired by the blare of the sound system. We present Feynman's more intuitive version.

Quantum electrodynamics is about photons, the quanta of the electromagnetic field, and electrons, the particle that experiences the electromagnetic force

(a)

(b)

FIGURE 18.1
(a) Richard Feynman in about 1960. Feynman loved to play the drums at parties and while out in the woods. He studied African drumming from experts and kept a set of bongo drums in his office. (b) Julian Schwinger, who independently invented a theory of quantum electrodynamics.

FIGURE 18.2
A schematic diagram showing a single quantum interaction between two electrons. Diagrams like this are known as "Feynman diagrams."

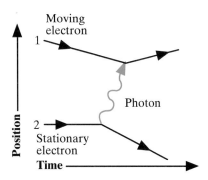

in its purest form. Sounds simple. But the requirement that electrons and photons obey both relativity and quantum theory leads to astonishing results.

Feynman's theory replaces the continuous force between two electrons with a "package" transfer in the form of a photon. Figure 18.2 pictures this idea. The diagram graphs the positions of two electrons at various times. It shows a single quantum interaction between the two electrons. Initially, electron 2 is at rest (its position isn't changing as time proceeds), and electron 1 is moving downward. Then electron 2 radiates a photon that travels through space to electron 1, and electron 1 absorbs this photon. When electron 2 emits the photon, electron 2 veers downward, and when electron 1 absorbs the photon, electron 1 veers upward. The electrons repel each other by means of **photon exchange**, much as basketball players interact by passing a basketball back and forth. Surprisingly, however, quantum electrodynamics allows two oppositely charged particles, such as a proton and electron, to veer toward each other when a photon is exchanged.

Every quantum event has quantum uncertainties. In Figure 18.2, the emission and the absorption of the photon are uncertain: That is, it is uncertain whether the emission and absorption will occur in the first place, and if they do, it is uncertain where and when they will occur. Quantum field theory replaces the deterministic electric force law with a formula giving the probability of emission and absorption of a photon. In this theory, for a particle to be **electrically charged** means that it has the ability to emit and absorb photons.

This theory replaces the smooth deterministic Newtonian paths with jerky nondeterministic paths (Figure 18.3). If the individual photons have low energy, quantum theory predicts a fairly smooth, nearly Newtonian, path (Figure 18.3(a)). At higher energies, the quantum predictions are decidedly non-Newtonian (Figure 18.3(b)).

So far, this is the kind of thing we might have expected from quantizing the electric interaction: quantized force packages, and randomness. But something quite unexpected also comes out of this theory. In order for the theory to obey the special theory of relativity, a new type of material particle must exist in nature.

The argument that leads to this prediction is an interesting one and is typical of the reasoning used in modern physics. It is based on symmetry. In order to obey special relativity, quantum field theory must be "symmetric under time reversal." In other words, if we imagine a universe precisely like ours, only with time running the other way, the laws of quantum field theory must be valid in

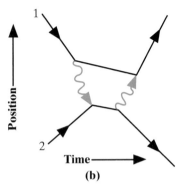

FIGURE 18.3
(a) A Feynman diagram for a series of interactions between two electrons at lower energy. The electrons' paths approach smooth Newtonian paths. (b) At higher energies, the paths deviate considerably from smooth paths. Individual quantum events and quantum uncertainties become more important at higher energies, and Newtonian physics is no longer a good approximation.

that universe.* Feynman found that an electron that is imagined to move backward in time would have precisely the same observable effects as would another particle just like the electron, only carrying a positive charge and moving forward in time. In order for the laws of physics to be properly symmetric under time reversal, this positive electron, or **positron**, had to exist.

The positron was first predicted much earlier, in 1928, by the British theorist Paul Dirac (Figure 18.4). Dirac did not use a field-theory approach but instead searched for and found a version of the Schroedinger equation for particles (not fields) that, unlike Schroedinger's original equation, obeyed special relativity. Dirac's new equation turned out to predict a new particle: the positron. It was the first major new prediction to come out of the marriage of quantum theory to special relativity.

H o w   D o   W e   K n o w ?

Discovery of the positron and the muon

The path of a subatomic particle can be made visible in a device known as a **cloud chamber**. A container is filled with dust-free air containing saturated water vapor, or gaseous H_2O that is just at the point of converting to droplets of liquid water but that lacks the tiny solid "nucleation centers" such as the dust particles that normally start droplets forming from water vapor. When a charged subatomic particle, such as an electron, speeds through the chamber filled with this vapor, it nudges many air molecules along its path. About 30 times in each centimeter of its path, these collisions are strong enough to ionize an air molecule. These ions make excellent nucleation centers for water droplets, so a droplet forms around each ion, and a trail of droplets reveals the particle's path. Jet planes form similar vapor trails in the atmosphere, revealing the plane's

FIGURE 18.4
Paul Dirac (left) and Werner Heisenberg in 1933. Dirac was an extremely reserved Englishman, and his colleague and good friend Heisenberg was a much more outgoing German.

*This raises the question of why the forward direction in time is actually different from the backward direction if our most basic physical theory is symmetric in time. For example, why aren't as many people growing younger as are growing older? The answer to this very deep question is not understood, but it is connected to the second law of thermodynamics (Chapter 7).

path. The cloud chamber was the workhorse of subatomic physics between 1930 and 1960. Its successor is the **bubble chamber**, based on the formation of tiny bubbles in a liquid. Its inventor, Donald Glaser, is said to have come up with this innovation in a bar in Ann Arbor, Michigan, while watching the bubbles in a glass of beer.

In 1932, Carl Anderson of the California Institute of Technology generated a strong magnetic field in a cloud chamber. Recall that magnetic fields exert sideways forces on moving charged particles. This sideways force makes electrons curve as they move through magnetic fields. Anderson generated a magnetic field strong enough to make this curvature measurable. A moving particle's speed and mass can be assessed from this curvature, because faster particles have straighter paths and because if two particles move at the same speed, the more massive one will have the straighter path. A particle's speed can also be independently assessed from the thickness of its path, because faster particles ionize fewer air molecules per centimeter and so create fewer bubbles and thinner paths.

In 1932, the only high-energy particles available for experiments came from high-energy particles from space. Anderson allowed these "cosmic rays" to pass through his cloud chamber, and his photographs show a surprising number of thin and fairly straight paths. Electrons and protons were the only charged particles known at that time. The paths indicated a mass too small for the proton. Rather, the paths appeared to be made by fast-moving electrons, but the direction of their curvature was the reverse of what was expected, indicating that the particles carried a positive charge. Anderson's first hypothesis was that these paths were made by electrons that were somehow moving upward through the cloud chamber, despite the expectation that cosmic rays should move downward. He clinched the matter by inserting a thin lead plate across the middle of the chamber. Although the fast-moving particles passed easily through the lead, they slowed down in the process, and so the path's curvature increased after passing through. In Figure 18.5—the photograph that won Anderson a Nobel Prize—the particle is clearly moving from top to bottom, so its curvature shows that it carries a positive charge. Anderson had discovered Dirac's predicted positron.

In order to observe cosmic rays before they interact with much air, Anderson in 1936 built a new magnetic cloud chamber on Pike's Peak in the Colorado Rockies. He found curious tracks that did not seem to fit the behavior of protons or of electrons, even positive ones. The paths were too curved for protons, yet the particles passed easily through lead plates that should have stopped any particle whose mass was as small as the electron's. Anderson found that this new particle was identical to the electron only 200 times more massive, and about one-tenth as massive as a proton. There had been no Dirac to predict this "fat electron," and it was a real surprise. As Columbia University physicist I. I. Rabi put it, "Who ordered that?" Today, we still do not know. This particle is called a **muon**.

FIGURE 18.5
The photo that won a Nobel Prize. This photo alone established the existence of a positive electron.

18.3 ANTIMATTER

The positron was science's first encounter with **antiparticles**. Relativity's requirement that quantum theory be symmetric under time reversal implies that for every existing type of particle, there must be an antiparticle. A particle and its antiparticle are identical except that their electromagnetic properties are mirror images of each other. For example, the electron's antiparticle is the positively charged positron. Similarly, the proton's antiparticle is the negatively charged **antiproton**, and the neutron's antiparticle is the uncharged **antineutron**. Although it carries no overall charge, the antineutron does have magnetic properties that are the mirror image of those of the neutron.

One of the most profound consequences of high-energy physics is the prediction and observation of the **creation and annihilation of matter**. Recall that quantum field theory states that fields come in tiny bundles or quanta of field energy. Photons, for example, are the quanta of the electromagnetic field. Quantum electrodynamics gives the probabilities for this field to exist in the form of various numbers of photons, and it gives the probability that additional photons will be created (increase in number) or annihilated (decrease in number). Such random increases and decreases are called *fluctuations*. Similarly, quantum electrodynamics states that the electron field can exist in the form of various numbers of electrons and allows for the possibility of the creation and annihilation of electrons. However, energy must be conserved, and so electrons must be created from some other form of energy, such as photons. Furthermore, one of the microworld's rules is that the total electric charge must remain unchanged, so it is always **electron–positron pairs** that are created or annihilated.

This is a very non-Newtonian development. As Heisenberg commented:

> I believe that the discovery of particles and antiparticles . . . has changed our whole outlook on atomic physics. . . . As soon as one knows that one can create pairs, then one has to consider an elementary particle as a compound system; because virtually it could be this particle plus a pair of particles plus two pairs and so on, and so all of a sudden the whole idea of elementary particles has changed. Up to that time I think every physicist had thought of the elementary particles along the lines of the philosophy of Democritus [Chapter 2], namely by considering [them] as unchangeable units which are just given in nature and are always the same thing, they never change, they never can be transmuted into anything else. They are not dynamical systems, they just exist in themselves. After Dirac's discovery everything looked different, because one could ask, why should a photon not sometimes be a photon plus a pair of electron and positron and so on? . . . Thereby the problem of dividing matter had come into a different light.

Quantum electrodynamics predicts that a photon has a certain probability of actually being observed as an electron–positron pair, or as more than one pair, and that such a pair has a certain probability of being observed as one or more photons. Figure 18.6, a Feynman diagram for part of a photon's life history, con-

> Just as it is possible for a particle to be in a quantum state in which it is neither definitely here nor there . . . so also it is possible to have a particle in a state in which it is neither definitely an electron nor definitely a neutrino until we measure some property that would distinguish the two, like the electric charge.
>
> STEVEN WEINBERG

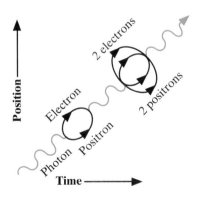

FIGURE 18.6
A few moments in the life history of a photon (which is not a photon but is one electron–positron pair during part of this history and two pairs during another part).

veys this notion that a photon can be an electron–positron pair, or two pairs. And conversely, if a particle and its antiparticle are close to each other, they can instead be photons. When this happens, the particle–antiparticle pair are said to "annihilate" each other.

This development implies the possibility of **antimatter**, similar to normal matter but made of antiprotons, antineutrons, and positrons, instead of the protons, neutrons, and electrons of normal matter. The first human-made atoms of antimatter were created in 1996 in a high-energy physics lab, when antiprotons were brought together with positrons to form nine atoms of antihydrogen. Large collections of antimatter, such as antigalaxies, are possible but are thought not to exist, because if they did, we would observe high-energy radiation from annihilation processes when a galaxy and antigalaxy collided. Although we are able to observe many colliding galaxies, we never observe such annihilation processes. The universe is believed to consist almost entirely of matter and very little antimatter. But symmetry seems to suggest that the universe should be made of equal amounts of both. Why so much matter and so little antimatter?

Russian physicist Andrei Sakharov suggested in 1967 that the big bang may have created equal amounts of matter and antimatter and that certain processes during the first second gave rise to a slight excess—less than a part in a billion—of matter, and then the rest of the matter and antimatter annihilated so that the tiny excess formed all the matter in the universe today. It's a good thing for life in the universe, including us, that things worked out this way. If it weren't for that slight excess, the universe would be made nearly entirely of radiation, and we wouldn't be here to think about antimatter!

H O W   D O   W E   K N O W ?

Observing the subatomic world

Matter is routinely created and annihilated in the world's high-energy physics labs when a high-energy particle enters a bubble chamber and collides with the particles of liquid (Figure 18.7). This creates a shower of new particles, including high-energy photons and particle–antiparticle pairs. Carl Anderson got his high-energy incoming particles from naturally occurring cosmic rays. Today the incoming particles are first accelerated to high energies by electromagnetic forces in **particle accelerators** (Figure 18.8).

We have sought for firm ground and found none. The deeper we penetrate, the more restless becomes the universe; all is rushing about and vibrating in a wild dance.

MAX BORN

No point is more central than this, that empty space is not empty. It is the seat of the most violent physics.

JOHN WHEELER

Among the strange phenomena of the microworld, one of the most fascinating is our new view of a **vacuum**—an empty space containing no matter. As we know, fields exist even in regions that are devoid of material objects, even in regions as empty as the spaces between the galaxies. But if fields do exist, then at any point in a vacuum there is a certain probability that a photon, or a particle–antiparticle pair of any type, will pop into and out of existence. Quantum uncertainties allow the energy present at any point in space to fluctuate around its long-term average value. The smaller the region of space is, the larger these **energy fluctuations** can be. Large fluctuations, however, are highly likely to subside in a very short time. So even in empty space there is always some probabil-

(a)   **(b)**

FIGURE 18.7

(a) A bubble-chamber photograph of electron–positron pair creations, caused by gamma-ray photons. In the event at the top, a photon has struck an atomic electron and knocked it out of its atom (long curving line), and it simultaneously created an electron–positron pair (tightly curling spirals). Toward the bottom, a different photon creates an electron–positron pair. How can you tell that each pair has two particles of opposite charge? Of the two pairs, which pair has the highest energy and speed? Why can't you see the path of the photons? (b) A high-energy particle striking a particle in a bubble chamber creates a "spray" of particles of various sorts.

ity of even very high-energy events occurring in small regions during short times. Empty space is not the quiet, uninteresting place we had imagined. Microscopically, a vacuum is a seething soup of creation and annihilation.

It seems that in nothingness, much is possible.

HOW DO WE KNOW?

The Lamb shift

It is ironic how physics turned out in this century. The 19th and early 20th century was characterized by a materialistic outlook which maintained a sharp distinction between what actually was in the world and what wasn't. Today that distinction still exists, but its meaning has altered. . . . Nothingness contains all of being.

HEINZ PAGELS, PHYSICIST

One consequence of energy fluctuations in a vacuum is a tiny effect on the hydrogen atom's energy levels (Section 14.2). In Schroedinger's nonrelativistic treatment of the hydrogen atom, the energies of the quantum states labeled $N = 2$, $L = 0$ and $N = 2$, $L = 1$ (see Figure 14.8) are identical. But when quantum field theory is applied to the hydrogen atom, it is found that vacuum energy fluctuations cause the orbiting electron to jiggle a little and that the energy of this jiggling is slightly different for the $N = 2$, $L = 0$ state than for the $N = 2$, $L = 1$ state. This difference was first noticed experimentally in careful measurements of the hydrogen spectrum by Willis Lamb in 1947. After the experimental discovery of this **Lamb shift**, quantum field theorists calculated it. The theoretically predicted frequency of the radiation absorbed or emitted when a hydrogen atom shifts between these two closely spaced levels is 1057.860 $\pm$ 0.009

FIGURE 18.8
A brief visual history of particle accelerators. The earliest accelerators were electron discharge tubes, similar to today's TV tubes and fluorescent lightbulbs. Compare these figures with Figure 1.2: Whether viewing outer space, or the inner subatomic space, large instruments are required at the limits.

(a) Ernest Lawrence invented the cyclotron that uses electromagnetic fields to bend charged particles into circular paths while accelerating them. He is shown here holding an early cyclotron in 1930, the year he first proposed the cyclotron idea.

(b) Lawrence, next to a 27-inch cyclotron in 1934. In 1937, this machine created the first artificial element, technetium. Although this element has atomic number 43, which puts it in the middle of the periodic table, all of its isotopes are radioactive with short half-lives, so it is not found naturally on Earth.

(c) This accelerator, at the University of California at Berkeley, was the first to have enough energy to create proton–antiproton pairs.

(d) The Stanford Linear Accelerator (SLAC) at Stanford University in California accelerates electrons down a 3-kilometer straight track. These high-speed electrons would radiate much of their energy away if they moved in a circle.

(e) Fermilab near Chicago. The main ring is 6 kilometers around. The small circle in front of the 16-story main lab building is a smaller "booster" accelerator that injects high-energy protons into the main ring.

megahertz. The measured value is 1057.845 ± 0.009 megahertz. This uncanny agreement is testimony to both the accuracy of the theory and the precision of spectral measurements.

Quantum electrodynamics describes not only electrons and positrons but also the electronlike muons (discussed earlier in this section) along with antimuons. Furthermore, a third type of electron, along with its antiparticle, was discovered in 1976. Called the **tau**, it is much heavier than the muon, weighing in at 3500 electron masses, or nearly twice as massive as the proton. Again, nobody knows "who ordered that." These three "generations" of electronlike particles appear today to be among the most fundamental constituents of matter. All three, along with their antiparticles, interact by exchanging photons, and all of their interactions are correctly described by quantum electrodynamics.

The muon and tau are "unstable"; in other words, they decay spontaneously into lower-energy entities. Today's universe contains muons and taus only when fleeting pairs of them are created by vacuum fluctuations or in high-energy interactions. However, these two heavy electrons might have played a crucial role during the big bang. Sakharov's process, mentioned earlier, for creating a slight excess of matter over antimatter requires all three generations. Although they seem esoteric and irrelevant today, we might owe our existence to the existence of muons and taus during the first second of the universe.

Are there more generations of still heavier electrons? As we will see, theory combined with astronomical observations indicate that the answer is no.

■ **DIALOGUE 1**   Would you like to travel to an antigalaxy? Why?
■ **DIALOGUE 2**   A certain gamma-ray source emits photons that have, according to quantum theory, a 20% chance of being found as an electron–positron pair. The source emits 400 photons. How many pairs will be found? Is this an exact figure or only an approximate one? Why?

18.4 ELECTROWEAK UNIFICATION: NEUTRINOS

In 1930, Wolfgang Pauli suggested that during radioactive beta decay, the nucleus emitted, in addition to a beta particle, another particle of an entirely new type. The hypothesized new particle was dubbed the **neutrino**, or "little neutral one." We discussed the history of this hypothesis, and its connection with energy conservation, in Section 6.5. Although neutrinos would not be discovered experimentally for another 25 years, Enrico Fermi took them seriously and argued that they indicated that a new fundamental force was at work. The new force became known as the *weak force*. Fermi was aware of the work in progress on the quantum theory of the electric force, and he quickly adapted these ideas to the new force. Fermi's quantum theory of the weak force succeeded in predicting the half-lives of radioactive nuclei and the spectrum of energies with which beta particles emerged from the nucleus during beta decay.

1. No. You would be annihilated, for one thing.
2. 20% of 400 is 80 pairs. This is only approximate, because of quantum uncertainties.

FIGURE 18.9
Abdus Salam, coinventor of the
electroweak force, was born in
Pakistan. He is one of the most
prominent scientists of the Islamic
faith. He donated his share of the
Nobel Prize to the institute with which
he is associated in Trieste, Italy,
which encourages scientists from
developing countries.

FIGURE 18.10
U.S. physicist Steven Weinberg,
coinventor of the electroweak force
that unified the electromagnetic and
weak forces into a single entity. He
has written a book for nonscientists
about the fundamental forces and
other topics, entitled *Dreams of a
Final Theory*.

The weak force is the most obscure of nature's four fundamental forces. Gravity and electromagnetism show up all the time in our macroscopic world because they can act over long distances. The strong force holds the nucleus together, is the major actor in nuclear power and nuclear weapons, and is responsible for alpha decay. The most noticeable example we have of the weak force is radioactive beta decay. The weak force is elusive because on a microscopic scale, it is both short ranged and weak.

A neutrino barely exists at all. That is, it has almost no properties: It has no charge and perhaps no rest mass (we're still not sure), and it feels neither the electric nor the strong force. Moving at or near lightspeed and feeling only the weak force (and gravity), this "little neutral one" is the most elusive known particle and one of the most fantastic.

Because neutrinos have only weak interactions, they travel through matter without "feeling" it. It would take 8 light-years of lead to stop half the neutrinos emitted during beta decay (8 light-years is twice the distance to the nearest other star)! No wonder the physicists studying beta decay had so much trouble trapping this thing in their calorimeters (Chapter 6). There are millions of neutrinos from space passing in all directions through your body at any instant, yet it will probably be years before even one of them has even a single quantum interaction within your body. The neutrinos now passing downward through you will exit Earth's far side in less than a tenth of a second and will within 2 seconds be beyond the orbit of the moon.

In 1967 Pakistani physicist Abdus Salam (Figure 18.9) and U.S. physicist Steven Weinberg (Figure 18.10), working independently, uncovered a close connection between the weak force and the electric force. They proposed a new quantum field theory that incorporated both forces into a single **electroweak** force. This unification was comparable to the unification, by Maxwell and others during the nineteenth century, of the electric and magnetic forces into a unified electromagnetic force.

Recall that quantum electrodynamics describes the electric interactions of electrons and positrons and that this interaction occurs via the photons that are exchanged between the charged particles. The Weinberg–Salam theory is a broader version of this picture. It describes the electric and weak interactions of electrons, positrons, neutrinos, and antineutrinos and states that this interaction occurs via the exchange of certain other particles. These **exchange particles** include the photon and three other particles as well. The three new exchange particles are quite unlike the photon, the main difference being that all three have mass—in fact, rather large masses for subatomic particles: Each of them is about 100 times more massive than a proton. They are labeled W^+, W^-, and Z and can be thought of as photons that have, for reasons unknown, acquired a mass. The electroweak theory argues that the weak and electric interactions are really the same thing but only appear to differ because the weak exchange particles (W^+, W^-, Z) have mass and the electric exchange particle (photon) does not. Another difference from the photon is that the two W's are charged, positively and negatively. The Z is, like the photon, not charged. The existence of a massive, charge-neutral exchange particle was a striking new prediction, and its experimental detection 6 years later in 1973 was a key confirmation of the Weinberg–Salam theory. The theory also correctly predicts the masses of all three exchange particles, although it does not explain why these particles have mass in the first place.

CHAPTER 18 QUANTUM FIELDS · 484

After all, the photon performs its exchange function perfectly well without having any mass, so why should the W and Z need it?

Recall that, besides the electron, there are two other generations of heavier, electronlike particles, the muon and tau. Since the electroweak force binds the electron and the neutrino together into a single family, we might guess that there is a second-generation neutrino to go along with the muon and a third-generation neutrino to go along with the tau. This would be a good guess. All three generations interact by means of the same four exchange particles: the photon, W's, and Z's. The electroweak theory correctly predicts all of the observed interactions among all of these fundamental particles. Table 18.1 summarizes the theory.

Are there more generations of electroweak particles, heavier electrons accompanied by new kinds of neutrinos? In a surprising turn of events, astronomical observations indicate that there are only three generations. The argument comes out of a close connection between the large-scale universe and the microscopic world: Outer space and inner space are connected through a microscopic event that quickly became macroscopic. This event is the big bang.

After the first 4 minutes into the big bang, the universe was about 75% hydrogen and 25% helium (Section 16.3). These numbers were predicted by theoretical nuclear physics, and they agree with observations of the oldest material in the universe. The theoretically predicted helium fraction depends on the number of generations of electroweak particles: The predicted helium fraction grows larger if the number of generations grows larger. If there are three generations, this leads to a predicted helium fraction that agrees with the observed value; if there are four generations, this leads to a predicted helium fraction that is too high to agree with the observed helium fraction. So there can be only three generations.

The second and third generations play only a fleeting role in today's universe because they are unstable and vanish soon after they are created. However,

At first glance, all of this sounds like medieval mystics discussing the music of the spheres, angels on the head of a pin, or some similar early approach to cosmology. Is it just a mathematical game we are playing, is it just semantics, or is it reality?

LEON LEDERMAN AND DAVID SCHRAMM, IN *FROM QUARKS TO THE COSMOS*

They are all crazy in the same way.

RICHARD FEYNMAN, DESCRIBING THE SUBATOMIC PARTICLES

TABLE 18.1

Particles that interact by means of the electroweak force. The three generations interact among one another by means of the four exchange particles. The second and third generations mostly decayed during the early moments of the big bang and are practically nonexistent today because they are unstable.

Generation	Particle type	Mass (proton = 1)	Charge (proton = +1)
1	electron	0.0005	−1
1	electron neutrino	0?	0
2	muon (mu electron)	0.11	−1
2	muon neutrino	0?	0
3	tau (tau electron)	1.90	−1
3	tau neutrino	0?	0
	Exchange particles		
	photon	0	0
	W$^+$	86	+1
	W$^-$	86	−1
	Z	98	0

the two higher generations may have played a crucial role during the opening moments of the big bang. Without these phantomlike particles, the universe might be far different from what it is and might be incapable of sustaining life.

Unification is a recurring theme of science (Figure 18.11). For example, Copernicus unified Earth with the other planets; Newton unified Earth-based physics with physics throughout the heavens; and Maxwell found a field theory that unified electricity, magnetism, and light. By the end of the nineteenth century, scienⁱists believed that there were only two fundamental forces, electromagnetism and gravity. Einstein, after fashioning the new theory that explained gravity as a consequence of the geometry of space and time, spent much of his scientific career trying to unify electromagnetism with his gravitational theory, in the hope that a single "unified field theory" would show electricity and gravity to be different aspects of space–time geometry. He was not successful.

Lately we have sought unification at the microscopic level, based on quantum field theory. As we have seen, these efforts have achieved significant success by unifying quantum theory, special relativity, and the electric and weak forces. As we will see next, physicists today are trying to unify the electroweak with the

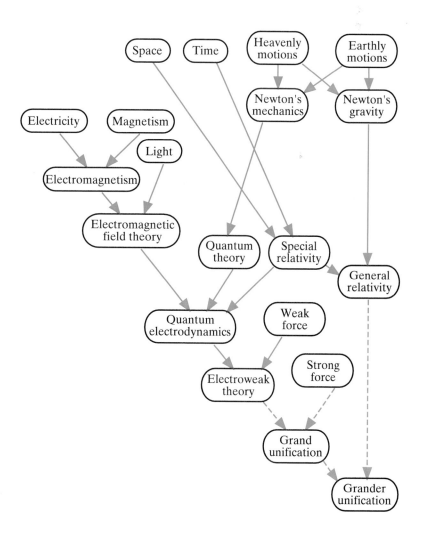

FIGURE 18.11
Some of the unifications in physics. The dashed lines represent unifications not yet established. Time runs from top to bottom.

strong force (Section 18.5) and to unify all of these with the gravitational force (Section 18.6).

18.5 GRAND UNIFICATION: QUARKS

As far as the measurements made so far can tell, all of the electroweak particles (Table 18.1) are **point particles**. That is, their force fields appear to be centered on a single point that itself takes up no volume. The electric charge of the electron, for example, appears to be concentrated at a single point.

But protons and neutrons are different. Experiments done in the 1950s showed that their electric and magnetic force centers are spread over a tiny volume about 10^{-15} meters across. Might they be composites that are made of still smaller particles?

There was a time early in the twentieth century when protons and electrons were thought to be the only subatomic particles. Then the discovery of both the neutron and the positron in 1932 initiated an era of particle discovery that, by 1960, had produced hundreds of new kinds of supposedly fundamental particles. The collection was frequently referred to as a "zoo." Fermi remarked that "if I could remember the names of all these particles I would have been a botanist." Surely the universe wasn't made of so many kinds of things.

Murray Gell-Mann (Figure 18.12) hoped to bring some order to this zoo. In about 1961 he found ways of grouping the known particles into families that corresponded to significant physical regularities among them. Gell-Mann's work was much like the work of the nineteenth-century chemists who found regularities in the chemical properties of the many known elements and grouped them accordingly into the pattern that we know today as the periodic table. It was only later that this periodic table found its natural explanation in a new model of the atom according to which the more than 100 elements are built of just three kinds of particles: electrons, protons, and neutrons. In a similar way, Gell-Mann's classification scheme led him to speculate on the existence of a few simpler entities, out of which protons, neutrons, and many other members of the particle zoo could be built. Gell-Mann originally made up a new word, "quork," for them, but changed the spelling to **quark** later when he ran across the line "Three quarks for Master Mark!" in James Joyce's novel, *Finnegan's Wake*.

That set experimentalists on a quark hunt. But despite strenuous searches among bubble-chamber tracks, nobody could come up with direct evidence for quarks, and so they were regarded merely as useful mathematical fictions.

FIGURE 18.12
Murray Gell-Mann (left) talking with Richard Feynman. Around 1960, Gell-Mann devised a classification scheme for the then-known sub-atomic particles. This scheme led to the prediction of new particles, and suggested the existence of a small number of simpler entities, called *quarks*, out of which protons, neutrons, and other particles could be built. Quarks were discovered experimentally in 1967.

HOW DO WE KNOW?

The discovery of quarks

When Richard Taylor, Jerome Friedman, and Henry Kendall (Figure 18.13), along with 12 coworkers, set out in 1967 to study the proton and the neutron, they weren't looking for quarks. Using the Stanford Linear Accelerator (SLAC, Figure 18.8(d)), they were following up on earlier experiments that had showed

(a)   **(b)**   **(c)**

FIGURE 18.13

(a) Richard Taylor, (b) Jerome Friedman, and (c) Henry Kendall. In much the same way that Rutherford probed the interior of the atoms by bombarding them with alpha particles (Figure 8.27), they probed the interior of protons and neutrons by bombarding them with high-energy electrons hurled by the Stanford Linear Accelerator. And just as the scattering of Rutherford's alpha particles revealed a small dense core within each atom—the nucleus—their experiment revealed that, within each proton and each neuton, lie three tiny hard points: quarks.

both protons and neutrons to be fuzzy balls about 10^{-15} meters across. Hoping to get a clearer picture of the structure of these fuzzballs, they hurled SLAC's high-energy electrons at protons and used huge detectors that they had built specifically to measure the energy and angular deviation of the electrons after they were deflected by the protons (Figure 18.14). At lower electron energies, their "scattered" electrons merely gave them a higher-resolution picture of the same old fuzzballs. But at energies so high that the electrons blew the protons and neutrons to bits, they found a surprise. Some of the electrons were deflected through very large angles, as though they were bouncing off hard little granules buried deep within the fuzzball. The experiment and its outcome paralleled Rutherford's discovery of a tiny hard nucleus deep within what had been supposed to be a fuzz-ball atom (Section 8.6). Only this time there appeared to be not one but three tiny force centers within the proton and within the neutron, a result that became clear only after 6 years of experimenting and sorting out data. As other labs confirmed these results, it gradually became clear that Taylor, Friedman, and Kendall had discovered Gell-Mann's quarks. The conclusion emerging from the experiments was that all protons and neutrons are made of two kinds of quarks, called the *u* and the *d quarks*. Protons are made of two u's and one d, and neutrons are made of one u and two d's.

Physicists have found a version of quantum field theory that describes the interactions between quarks and that has so far agreed with all experiments designed to test it. In this theory, the strong force acts directly between quarks, and the force acting between protons and neutrons is a consequence of the forces between their quarks.

FIGURE 18.14
The enormous electron detectors at SLAC. The electron beam enters from the left and collides with protons in a target. The deflected electrons are then analyzed by three detectors: the tank at the far left, the large detector in the foreground, and the other large detector in the background.

The force field that is quantized in this new theory is the strong force field, and the matter field that is quantized is the quark field. Think of the strong force field as analogous to the electric field, and the quark field as analogous to the electron psi field. The quanta of the quark field are quarks of two types, the u quark and d quark (and their antiparticles). They are the material particles of this theory, playing a role similar to that of the electron in quantum electrodynamics. The theory predicts that there are two stable configurations of u and d quarks, namely, u-u-d (the proton) and u-d-d (the neutron).

In addition to feeling and exerting the strong force, quarks must also experience the electric force, because we know that protons experience this force and quarks are supposed to explain protons.

■ **DIALOGUE 3**   Surprisingly, quarks turn out to be fractionally charged, the u possessing a positive charge of +2/3 of the electron's charge and the d possessing a negative charge of −1/3 of the electron's charge. In this case, what is the charge that this theory predicts for the proton? For the neutron?

The quanta of the strong force field are called **gluons** because they are the glue that binds quarks together, and on a larger scale they are the glue that binds the nucleus together. Think of them as the photons of the strong force. Like the photon, they have no mass and no charge. However, there is an important difference between the way that gluons operate in the strong force and the way that photons operate in the electric force. Gluons themselves directly feel the strong force, unlike photons, which are not electrically charged and do not directly feel the electric force. In quantum electrodynamics, "electric charge" can be thought of as "the ability to emit and absorb photons." In the same way, the property of feeling the strong force can be thought of as the ability to emit and absorb gluons. But gluons themselves feel the strong force, which means that gluons themselves can emit gluons, unlike photons, which cannot emit photons.

3. 2/3 + 2/3 − 1/3 = +1 electron charge, 2/3 − 1/3 − 1/3 = 0.

FIGURE 18.15
Here's why you can't separate quarks.

Suppose you start with the three quarks in a proton—

and begin pulling away one of the quarks. Gluons travel between the quarks.

As the quark is separated further, the gluons make more gluons, which makes the force between the quarks stronger.

Finally, all the gluon energy (strong field energy) creates a quark–antiquark pair; the new quark makes the proton whole again; and the new antiquark combines with the old quark to form an unstable quark–antiquark pair.

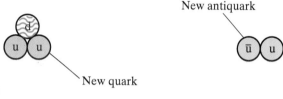

Proton

Gluons

More gluons

New antiquark

New quark

This ability of gluons to make more gluons explains one of the most curious features of quarks (Figure 18.15): The force between quarks grows stronger, not weaker, as they are separated, and as a result it is impossible to isolate single quarks. When a quark within a proton is pulled a short distance from its neighboring quarks, the gluons must fly farther away in order to reach from that quark to its neighbors. This gives these gluons more time to proliferate in flight, which makes more gluons, which makes the force larger as the distance becomes larger. As the quark is pulled farther away, energy quickly builds up in the strong force field, and this energy creates quark–antiquark pairs. After a brief reshuffling, a new quark is created in the proton from which the first quark had been removed! Furthermore, the removed quark and the new antiquark team up to form an unstable pair.

This theory provides a beautifully crazy explanation of why years of looking for isolated quarks in bubble chambers produced no results. Any attempt to pull a quark away from its neighbors just makes more nonisolated quarks. You can't find individual quarks.

As far as anybody can tell, quarks are truly point particles, like electrons and neutrinos. They seem to be good candidates, at last, for the truly fundamental "seeds" in nature's successive seeds within seeds (Figure 18.16).

Recall that there are three generations of electroweak particles (Table 18.1). In just the same way, observation reveals three generations of quarks. The second and third generations each consist of two quarks that are heavier and unstable (short-lived) copies of the u and d, just as the second and third generations of electroweak particles are heavier and unstable copies of the electron and neutrino.

Table 18.2 shows the entire setup for the strong force. The last quark to be discovered experimentally, the t quark, was tentatively confirmed by Fermilab (Figure 18.8(e)) in 1994. The t quark was the most difficult to discover because its mass turned out to be so much larger than the masses of the other five types

10^{-7} m

DNA

10^{-9} m

Molecule

10^{-10} m

Atom

10^{-14} m

Nucleus

10^{-15} m

Proton

Less than 10^{-18} m

? Quark

FIGURE 18.16
Nature's successive seeds within seeds, from DNA to quarks. Note the approximate size of each level.

TABLE 18.2

Particles that interact by means of the strong force.   The three generations of quarks interact by means of gluons. Only the first generation is active today, the others being unstable. The proton is made of u-u-d, and the neutron is made of u-d-d. The force that binds protons and neutrons into the nucleus is actually the strong force acting between their quarks.

Generation	Particle type	Mass (proton = 1)	Charge (proton = +1)
1	u-quark	0.004	+2/3
1	d-quark	0.007	−1/3
2	c-quark	1.6	+2/3
2	s-quark	0.16	−1/3
3	t-quark	185 (unconfirmed)	+2/3
3	b-quark	5.0	−1/3
	Exchange particles gluons	0	0

of quarks, which (because of $E = mc^2$) meant that much more energy was needed to create it. "Weighing" in at an estimated 185 proton masses, the t quark is about as massive as a gold atom! The resemblance between Table 18.1 and 18.2 is striking and points to a close connection between the electroweak and strong forces. Accordingly, physicists believe there should be a "grand unified theory" that would view the electroweak and strong forces as two facets of a single underlying force. So far, however, nobody has been able to formulate such a theory.

The particles and organization shown in Tables 18.1 and 18.2 is known as the **standard model**, a boring title for a theory with such fantastic predictions as antimatter, neutrinos, and quarks. To summarize:

THE STANDARD MODEL
Neglecting gravity, there are only two kinds of forces: the electroweak force and the strong force. Ordinary matter is made only of electrons, neutrinos, and u and d quarks. Electrons and neutrinos interact only through the electroweak force, whose exchange particles are photons, W's, and Z's. The u and d interact through the electroweak force and also the strong force whose exchange particles are gluons. In addition, there are two heavier unstable copies, or "generations," of the four basic particles. In all there are six kinds of electroweak particles and six kinds of quarks.

Think of quarks, electrons, and neutrinos as the bricks of the universe, and exchange particles (the photon and its partners) as the cement.

■ **DIALOGUE 4**   According to the standard model, which of the following are "composite" particles and which are "elementary"? Dust grain, electron, positron, hydrogen atom, photon, water molecule, neutrino, neutron, quark, muon, proton, antiproton.

---

4. Composite: dust grain, atom, molecule, neutron, proton, antiproton.

18.6 GRANDER UNIFICATION: GRAVITY AND THE INFLATIONARY UNIVERSE

As we have seen, two of the four fundamental microscopic forces have been unified into a single quantum field theory of the electroweak force. A similar quantum field theory has been worked out for the strong force, and the similarities between these two theories makes it plausible that a unified theory of the electroweak and the strong force will someday be invented.

That will still leave gravity out of the unified picture. One reason it has been so hard to work gravity into these theories is that so little is known about it at the microscopic level, because it is so weak at this level. For example, the gravitational attraction of a single proton for an electron is 10^{36} (a trillion trillion trillion) times smaller than the electric attraction between these two particles. Only if there are large concentrations of matter, as in a planet, are gravitational effects strong enough to be easily observed. In large concentrations of matter, the electric effects of the protons and electrons largely cancel out each other, while gravity adds up because it is always attractive, so that gravity dominates.

At the large-scale level, Einstein's general relativity appears to be the correct theory of gravity. It is a field theory whose field is the space–time curvature that is caused by masses (Section 11.6). The most reasonable way to combine this theory with the theories of the other forces would be to quantize the gravitational field to produce a quantum theory of gravity. But this turns out to be very difficult to do in any logically consistent way, because general relativity and quantum theory are such radically different theories.

One particle predicted by all the work on gravity is the **graviton**, the quantum of the gravitational field. Gravitons are similar to photons: They have zero mass, zero charge, and they move at lightspeed. From the quantum point of view, the gravitational forces between, say, Earth and the moon, occur by means of the exchange of gravitons between the two bodies. But gravitons have never been observed and perhaps never will be directly observed, because the gravitational force is so weak. For example, if a single proton absorbs a graviton, the proton should recoil, but this recoil is predicted to be so tiny that one cannot hope to observe it.

A marriage, or at least an affectionate relationship, has developed recently between high-energy microscopic physics and the astrophysics of distant realms of the universe. Inner and outer space are united by the fact that the big bang started microscopically. Physicists use high-energy accelerators to study events similar to those that occurred during the creation of the universe, and astronomers study evidence for a big bang that can be regarded as an extremely high-energy microscopic physics event. This union has stimulated some remarkable ideas.

In the early days of quantum theory, Max Planck pointed out that nature's fundamental numbers—lightspeed, Planck's constant, and the "gravitational constant" that specifies the strength of the gravitational force—can be combined in certain ways to obtain what appear to be "natural" units of length, time, and energy. In metric units, these natural **Planck scales** are

$$\text{Planck length} \approx 10^{-35} \text{ meters}$$

$$\text{Planck time} \approx 10^{-43} \text{ seconds}$$

$$\text{Planck energy} \approx 10^{-9} \text{ joules}$$

We now know that in the maelstrom of high density and temperature that existed in the early moments of the universe, only the most primordial objects could exist; any transient combinations, such as protons, let alone molecules, would decompose more quickly than a butterfly in the core of a volcano.

LEON LEDERMAN AND DAVID SCHRAMM, IN *FROM QUARKS TO THE COSMOS*

Where the telescope ends, the microscope begins. Which of the two has the grander view?

VICTOR HUGO

In the 1960s, John Wheeler (Figure 18.17) interpreted Planck's quantities in the light of general relativity and the uncertainty principle. We have seen (Section 18.3) that the uncertainty relations make the energy contained in small regions of space highly uncertain, so that the energy in a small region fluctuates randomly. These energy fluctuations are bigger and briefer in smaller regions of space. Wheeler found that in a sphere whose radius is the Planck length, and during time intervals whose duration is the Planck time, energy fluctuations as large as the Planck energy are likely to occur. The Planck energy is not large by macroscopic standards; in fact it is roughly the chemical energy in one automobile tankful of gasoline. But if this amount of energy is concentrated (because of a quantum fluctuation) in a sphere whose radius is only 10^{-35} meters, the result is remarkable: This energy is equivalent (because of $E = mc^2$) to a mass of about 20 billionths of a kilogram, and this much mass in such a tiny volume causes space–time to bend back upon itself and form a black hole that is cut off from the rest of the universe!

Wheeler pointed out that this phenomenon would break space and time into tiny pieces at this ultrasmall level, so that the Planck length and time are the smallest lengths and times that have any physical meaning at all. Viewed at this level, space and time are themselves broken up into a "quantum foam" pierced by tiny fluctuating black holes.

It is difficult to investigate such questions experimentally, because the energies going into individual microscopic events in today's large particle accelerators are far smaller than the Planck energy. However, the progress made with existing accelerators already points to a significant trend: The differences among the fundamental forces diminish as the energy rises. The theory of the electroweak force suggests, for example, that at higher energies the weak force increases in strength until it attains a strength that is roughly equal to the electric force. At still higher energies, the electroweak force attains a strength that is roughly equal to the strong force. And at even higher energies, namely the Planck energy, they all attain a strength that is roughly equal to that of gravity. This might seem surprising, because we pointed out earlier that the gravitational force is normally very weak. But at high energies, that is, at short separation distances (comparable to the Planck length) between particles, gravity becomes very strong.

All of this leads to a remarkable picture of the big bang, the ultimate high-energy microscopic event (Figure 18.18). This picture starts from the assumption that the big bang began as a microscopic quantum fluctuation. Such an event cannot have occurred in a region smaller in diameter than the Planck length or at a time earlier than the Planck time, because smaller volumes and times do not exist. At that time, the total energy (due to the quantum fluctuation) of the universe was only about 10^9 joules, the amount of energy in a tank of gasoline, and the universe measured only about 10^{-35} meters across! The Planck energy, when confined to a region this small, corresponds to an enormous temperature of about 10^{32} degrees. Then something called "inflation" occurred, which caused the universe to expand rapidly while it cooled.

Figure 18.18 traces the subsequent history of the universe. By now, the universe has cooled all the way down to about $-270°C$ (just 3 degrees above the lowest possible temperature, known as *absolute zero*), as astronomers have discovered by measuring the temperature of the cosmic background radiation that

Maybe the universe itself sprang into existence out of nothingness—a gigantic vacuum fluctuation which we know today as the big bang. Remarkably, the laws of modern physics allow for this possibility.

JOHN WHEELER

It is said that there's no such thing as a free lunch. But the universe is the ultimate free lunch.

ALAN GUTH, ORIGINATOR OF THE "INFLATION" IDEA THAT EXPLAINS HOW THE BIG BANG COULD HAVE CREATED OUR UNIVERSE OUT OF A VACUUM

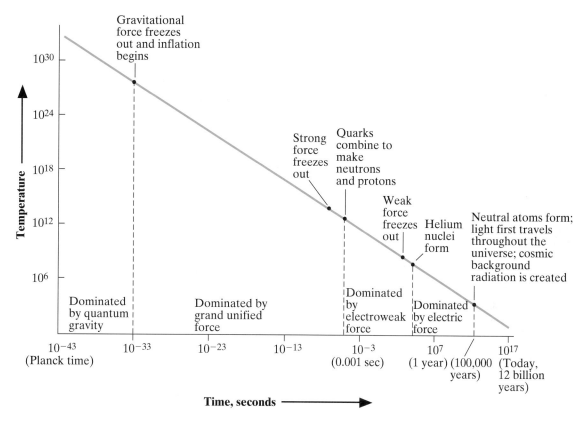

FIGURE 18.18

The thermal history of the universe according to current theories, starting from the earliest time that can physically exist. Each of the four fundamental forces had an era during which it dominated the early universe. Once neutral atoms formed, there was light throughout the universe. The cosmic background radiation originated at this time, and stars and galaxies formed after this time. Today, the universe has cooled to 3 K (3 degrees above absolute zero).

> Like a gleam in the darkness, we have appeared for an instant from the black nothingness of ever-unconscious matter, in order to make good the demands of Reason and create a life worthy of ourselves and of the Goal we only dimly perceive.
>
> ANDREI SAKHAROV, RUSSIAN PHYSICIST AND FATHER OF THE SOVIET FUSION BOMB

is left over from the big bang. This radiation was not emitted until 100,000 years after the big bang, because before that time the universe was so hot that electrons could not unite with nuclei to form electrically neutral atoms, and photons were unable to travel through the resulting electrically charged gas of electrons and bare nuclei. Measurements of the cosmic background radiation are seeing this universal first light (Figure 16.5).

As it cooled, the universe went through several abrupt transitions that were analogous to the transition that occurs when liquid water cools down to the freezing temperature and abruptly changes from liquid water into solid ice. In the initially very hot universe, all four fundamental forces were physically united as a single unified force. Physicists say that the four forces had the same *symmetries*, and thus did not exist as individual forces. As the universe cooled, the gravitational force "froze out" of the unified force; it lost the symmetry that had unified it with the other forces. This *symmetry breaking* is analogous to the loss of symmetry when water freezes: All directions are equivalent inside water, but ice crystals line up in specific directions. As the universe continued cooling, the

FIGURE 18.19
In 1979, high-energy particle physicist Alan Guth made a monumental discovery in the area of cosmology: the theory of cosmic inflation. This theory was developed from other theories of cosmology and high-energy physics, and agrees with those widely accepted theories. Furthermore, the theory has significant experimental support in satellite observations of the cosmic background radiation (see Figure 16.5) and in other cosmological observations. Thus, cosmic inflation is arguably the best scientific explanation to date of the creation of the universe and of the details of the first few moments of time.

There could be many, many other bubble universes existing out there, all out of communication with our universe. Thus, not only is the universe a very big place, but our universe might not be the only one.
LEON LEDERMAN AND DAVID SCHRAMM, IN *FROM QUARKS TO THE COSMOS*

Physicists are the Peter Pans of the human race. They never grow up, they keep their curiosity.
I. I. RABI, PHYSICIST

God used beautiful mathematics in creating the world.
PAUL DIRAC

strong force froze out and formed its own unique patterns, and finally the weak force did the same, leaving us with four distinct forces today.

The theory of **cosmic inflation**, invented in 1979 by physicist Alan Guth (Figure 18.19), provides our best current picture of the details of these early moments.* Cosmic inflation is a quantum field theory based on a new type of field called a *Higgs field*. According to the theory, a quantum fluctuation occurring in vacuum created, by chance, an energy of some billion joules within a volume about 10^{-35} meters across. At the colossal temperature associated with this much energy in such a tiny region, the symmetries of the Higgs fields unified all of nature's forces into a single force. Then, because of the particular properties of the Higgs fields, the universe went through a brief *inflationary phase* of very rapid expansion during which rapid cooling and an enormous volume increase occurred. This cooling broke the natural symmetries of the Higgs fields, "freezing out" the individual four forces as described above (Figure 18.18).

But where did all the mass and energy in the universe come from, if energy is conserved and if everything originated from only a billion joules—the amount of energy in a typical tank of gasoline? The answer is that the gravitational energy of compact matter is always negative (less than zero), because work must be done in order to pull matter apart into separated pieces. The inflation process didn't create any net energy, but instead created negative gravitational energy while creating positive kinetic energy. It's a little like a man who spends a lot of money by going into debt; he spends like a millionaire, but his net financial balance remains zero. Thus it is thought that the universe's net energy remains very close to zero, balanced between an enormous negative gravitational energy and a slightly more enormous but positive kinetic energy. As for the universe's mass—it exists because of the universe's energy coupled with $E = mc^2$; in other words, the universe has mass because it has energy and energy is equivalent to mass. Cosmic inflation is, as Alan Guth has said, "the ultimate free lunch": It created not only time and space, as described earlier in Section 11.7, but also the energy and mass that we observe around us.

As in any quantum field theory, the Higgs fields reveal themselves to observation as localized bundles of field energy, similar to the photons that are bundles of electromagnetic field energy. However, these predicted *Higgs particles* have not yet been observed because they are so energetic. Even the least energetic and therefore (because of $E = mc^2$) the least massive of the predicted Higgs particles is about 1000 times more massive than a proton, requiring more energy to create than our laboratories have so far been able to muster. Our most energetic particle accelerators are nearly up to this task today, so we might soon create, in the laboratory, these particles that facilitated the creation of our universe.

The theory of cosmic inflation predicts that, just as the direction in which ice crystals will freeze is an undetermined or random event, so the specific "direction" in which a Higgs field will "freeze" is, because of quantum uncertainties, also a random event. But in the case of the Higgs fields, different directions really mean different properties of the various fundamental forces as they freeze

---

*For an excellent recounting of this discovery and of the history of our understanding of the big bang, written in non-technical language for non-scientists, see Alan Guth's *The Inflationary Universe: The Quest for a New Theory of Cosmic Origins* (Reading, MA: Addison-Wesley Publishing Company, 1997).

As an extreme possibility, it is possible that there is only one theory . . . that is consistent with the existence of intelligent beings capable of wondering about the final theory. If this could be shown, then we would be as close as anyone could hope to a satisfactory explanation of why the world is the way it is.

STEVEN WEINBERG

out of the pre-existing symmetric unified force. In this process, basic properties of our universe such as the masses and charges of the fundamental particles would have been determined randomly. It seems possible that ours could be just one of many universes, each of them born in a new toss of the quantum dice and each of them characterized by different physical properties.

According to this view, in our universe the numbers happened to have just those lucky values that allowed us to evolve. In any other universe, in which these numbers were very different, life and intelligence might have been physically impossible. Life and intelligence—in other words, our own existence—might turn out to be the best explanation we have for these numbers having the values that they do. This idea—that our universe must be organized in the way that it is because any other organization would not allow intelligent beings to be here to ask the question in the first place—is called the **anthropic principle**.

And this outrageous but plausible connection between the big bang and our lives on Earth is a good place to end our excursion into concepts of physical reality and their connections to our lives.

SUMMARY OF IDEAS AND TERMS

Field view of reality   The view that the universe is made of fields, subject to the rules of relativity and quantum theory.

Field   A physical entity that is spread throughout a region of space. A **force field** exists wherever an object would feel a force.

Quantized field   A field that is treated as a real physical entity subject to the laws of quantum theory.

Field quanta   Discrete bundles of energy, predicted by quantum theory to be associated with any quantized field. Examples: Photons are quanta of the electromagnetic field, and electrons are quanta of the electron field (the psi field for electrons).

Quantum field theory   Everything is made of quantized fields that obey special relativity and quantum theory. All the particles of nature are field quanta. A field's intensity represents the probability for finding the particles that are the quanta of that field.

Quantum electrodynamics   The quantum field theory of electrons and photons. According to this theory, when we say that a particle is **electrically charged** we mean that it has the ability to emit and absorb photons. Particles exert electric forces on each other in tiny quantized increments, by **photon exchanges** in which one particle emits a photon that is then absorbed by the other particle.

Cloud chamber   A device that shows the path of a charged particle as a trail of droplets in water vapor. Its successor, the **bubble chamber**, shows a particle's path as a trail of bubbles in a liquid.

Antiparticle   The special relativity theory predicts that for every existing type of particle, there is an antiparticle carrying the opposite charge. Quantum uncertainties allow the **creation and annihilation of particle–antiparticle pairs.**

Positron   The electron's antiparticle. Identical to the electron except that it carries a positive charge.

Antimatter   is made of **antiprotons, antineutrons**, and positrons. Today's universe consists overwhelmingly of matter, not antimatter.

Particle accelerator   A device to accelerate microscopic particles to high energies.

Vacuum   A region that contains no matter. According to quantum field theory, fields exist even in vacuum. Since these fields are quantized, there is some probability that field quanta—photons, or particle-antiparticle pairs, will pop into and out of existence, even in vacuum. Furthermore, quantum uncertainties allow the energy present at any point in vacuum to undergo random **energy fluctuations** around its long-term average value.

Lamb shift   A small change in the energy levels of the hydrogen atom that is caused by vacuum energy fluctuations in the space surrounding the atom.

Muon, tau   The other two types of electrons that have been found. The only differences among the three are that the muon and tau are heavier than the electron and are unstable (they have short lifetimes).

Neutrino   A particle with no charge and possibly no mass. It exerts and feels only the weak force and gravity.

Electroweak force   The unified electromagnetic and weak forces. Electrons and neutrinos experience only the electroweak force. The **exchange particles**, or field quanta, for this force are the **photon, W^+, W^-**, and **Z**. W's are charged, and the photon and Z are not charged. There are three **generations** of electroweak particles: the electron and its neutrino, the muon and its neutrino, and the tau and its neutrino.

Point particle   A particle whose force field is centered on a single point and that itself takes up no volume. All of the fundamental particles described by quantum field theory appear to be point particles.

Strong force   One of the four fundamental forces. Its field quanta (or exchange particles) are **gluons**, and the particles that exert and feel this force are **quarks**. Two kinds of quarks, **u** and **d**, are important in ordinary matter. Protons are made of u-u-d, and neutrons are made of u-d-d. Quarks are only fractionally charged and are not found in isolation because any attempt to isolate them creates more quarks. There are two other (unstable) generations of quarks, so there are six kinds of quarks in all.

Standard model   The theory of the electroweak and strong force.

Graviton   The gravitational field quantum, predicted but not yet observed.

Planck scale   According to some theories of quantum gravity, the Planck length and time are the tiny ultimate units into which space–time itself is divided. Smaller dimensions have no physical meaning.

Big bang   The creation of our universe in a single event. The consensus today is that this event was initially microscopic and dominated by quantum gravity.

Cosmic inflation   A quantum theory of the big bang. It states that the big bang originated as a quantum energy fluctuation in the smallest existing volume of space. At the resulting high temperature, a quantum field called the **Higgs field** caused the universe to very rapidly expand or *inflate* while cooling. The cooling created our four known forces and the most basic properties of our universe. **Higgs particles**, the quanta of the Higgs field, are currently sought in high-energy accelerators.

Anthropic principle   Our universe must be organized in the way that it is because any other organization would not allow intelligent beings to be here to ask the question in the first place.

REVIEW QUESTIONS

QUANTIZED FIELDS

1. What two theories are combined to form quantum field theory?
2. What is a field? What is a quantized field?
3. Name the quanta of the electromagnetic field. Are electrons also "quanta"? Quanta of what?

QUANTUM ELECTRODYNAMICS AND ANTIMATTER

4. What role does the photon play in the electric force between two electrons?
5. Describe the events that are diagrammed in Figure 18.3(a) and (b).
6. What is a muon? A tau?
7. What is an antiparticle? Name two antiparticles.
8. What is antimatter?
9. Describe the creation of a particle–antiparticle pair.
10. Name and describe several devices that are used to observe the subatomic world.
11. Is empty space really empty? What happens there?

ELECTROWEAK UNIFICATION

12. Why is the neutrino so hard to detect? Which of the four fundamental forces does it experience?
13. Name the six particles that interact via the electroweak force.
14. Name the exchange particle for the electric force. Name the four exchange particles for the electroweak force.
15. The electroweak particles are laid out in "generations." Describe this pattern. How many generations are there?

GRAND UNIFICATION

16. Name the fundamental (not composite) particles that are responsible for the strong force.
17. How were quarks discovered?
18. Are protons fundamental particles? If they are composite particles, of what are they composites? What about electrons?
19. What force or forces do quarks exert on one another?
20. One property of quarks is that they exert and feel the strong force. List at least two other properties.
21. How many kinds of quarks are there? How many of these are found in ordinary matter?
22. Name the exchange particles that carry the strong force.
23. Why do we never observe an isolated quark?
24. Which four kinds of particles form the "bricks" of the universe, and which kinds form the "cement"?

GRANDER UNIFICATION

25. Which of the four fundamental forces can be felt over macroscopic distances?
26. What is a graviton? Has it been discovered experimentally? If so, how? If not, why not?
27. What is the significance of the Planck length and time?
28. What does it mean, in reference to Figure 18.18, to say that a fundamental force "freezes out"?
29. When did the cosmic background radiation originate?
30. In the theory of cosmic inflation, what is it that inflates? Name the field that causes this inflation.
31. Would it be consistent with presently understood science to assume that there are other universes with vastly different properties than ours?
32. What is the anthropic principle?

FOR DISCUSSION

1. Suppose that our universe had been created in such a way that life could not exist. One possibility, for instance, is that the universe could obey Newtonian physics even at the microscopic level. Chemical reactions would be impossible in such a universe because the existence of the chemical elements depends on the validity of quantum theory. Life would be impossible in a Newtonian universe. Since there is no possibility of any conscious being ever observing such a "dead" universe, would it make sense to say that such a universe *exists* or that it *could exist*?
2. In light of the anthropic principle, might there be a scientific sense in which the universe was made for humans, or at least for conscious beings?

EXERCISES

QUANTUM ELECTRODYNAMICS AND ANTIMATTER

1. In Figure 18.7(a) what is the evidence that each pair consists of two oppositely charged particles?
2. In Figure 18.7(a), which of the two pairs has the faster-moving particles? How do you know?
3. In Figure 18.7(a), why don't we see the tracks of the two photons that created the two pairs?
4. Each of the two photons created when an electron–positron pair annihilates has a frequency of about 10^{20} Hertz. To what region of the electromagnetic spectrum do these photons belong? If the electron and positron

were moving instead of being at rest, would it make this photon frequency higher or lower?

5. How would the photograph of Figure 18.5 be altered if the particle track had been made by an electron moving upward instead of a positron moving downward?

ELECTROWEAK UNIFICATION

6. Of the 10 electroweak particles (Table 18.1), which ones travel at (or near) lightspeed?

7. Of the 10 electroweak particles (Table 18.1), which ones can feel the electric force? Which can exchange photons?

8. Into which one of the boxes of Figure 18.11 should the discovery of beta decay be placed?

9. In what ways are the W and Z particles similar to photons? In what ways are they different?

GRAND UNIFICATION

10. In what ways are gluons similar to photons? In what ways are they different?

11. In what ways are quarks similar to electrons? In what ways are they different?

12. Give at least one specific reason (other than a general belief in unity) why scientists believe that there is probably a single theory that can unite the electroweak and the strong force into a single grand unified force.

13. According to Table 18.2, the total rest mass of the two u quarks and one d quark that make up a proton seems to be only 0.004 + 0.004 + 0.007 = 0.015 proton mass! Where do you suppose the remaining 0.985 proton mass comes from? In other words, where does some 98% of the rest mass of a proton come from?

GRANDER UNIFICATION

14. In what ways was the emergence of the four different fundamental forces during the big bang similar to the change of state of water from liquid to solid?

15. If the theory of cosmic inflation is correct, then what is undoubtedly nature's single most important example of quantum uncertainties?

16. According to the anthropic principle, is it possible that our universe could have been created in such a way that every isotope in the periodic table is unstable (radioactive) with half-lives of, say, less than 1 year?

17. Is it possible, according to the anthropic principle, that our universe could have been created in such a way that the proton is unstable, with a half-life of less than 1 year?

PROBLEMS

QUANTUM ELECTRODYNAMICS AND ANTIMATTER

1. Each of the two photons created when an electron–positron pair annihilates has a frequency of about 10^{20} Hertz. Find the energy of each photon (see Chapter 13).

2. A proton–antiproton pair, at rest, annihilate and create two photons. Using the information in the preceding problem, and the fact that a proton is 1800 times more massive than an electron, find the frequency of each photon.

3. A proton–antiproton pair, at rest, annihilate and create two photons. Find the energy of each photon from the fact that a proton's mass is 1.7×10^{-27} kg. Use this energy to find the frequency of each photon, and compare your answer with the preceding problem.

4. Making estimates. A large electric power plant generates 1000 MW of electricity. If the energy came from matter–antimatter annihilation, about what total mass of matter and of antimatter would be required each year, assuming that the electricity is generated at an energy efficiency of 50%?

5. Making estimates. Suppose 1 gram of matter is annihilated with 1 gram of antimatter. Show that the resulting energy could lift the entire U.S. population of about 270 million by about 1 km.

Epilogue: summing up

Tree Souls, by American sculptor Alison Saar, portrays humankind's dependence on the Earth for physical and spiritual survival.

I call Heaven and Earth to record this day against you, that I have set before you life and death, blessing and cursing:
Therefore choose life, that both thou and thy seed may live.

DEUTERONOMY

We have come some distance together, you and I, since we started this book. Now we are almost at journey's end. Let's step back and view the landscape through which we have passed. Although that landscape has ranged from the microscopic to the cosmological and from the ancient to the new, the varying details have been connected by four recurring themes: the social context of physics, energy, the significance of post-Newtonian physics, and how we know what we know in science.

Occasionally when I talk with groups about the *social context of physics*, I ask them to name any significant contemporary issues that come to mind. It doesn't take long to accumulate quite a list: overpopulation, urban decay, species extinctions, drug abuse, ozone depletion, AIDS, resource depletion, cults, deforestation, and so forth. After collecting 20 or 30 suggestions, we search for common themes among them. It soon becomes apparent that each one has a significant science and technology component.

Without science and technology, we would have other problems, such as early death by disease, but we would not have the particular problems we have today. For instance, because medical science has partly solved the problem of death by disease, we now have the problem of overpopulation. The death problem has been replaced, in a sense, by a birth problem. Because we have accepted science's help in solving the death side of the birth-and-death equation but have not simultaneously taken responsibility for the birth side, we have burdened the planet with more people than it can handle.

Science gives us great power, and we can use that power in either helpful or harmful ways. Without science and technology, we would not have the automobile or television, for example. When you turn on the switch of either device, you bring great power to bear on yourself, on others, and on Earth.

The problems of science and society come down to this: Humankind is not paying its dues for the fruits of the scientific age. We are quick to accept the speed of the automobile, the fun of television, and the cures of medicine, but we are slow to clean up our exhaust fumes, to maintain intelligent reading habits, or to control our birthrate. The ozone story is a good example: Humankind enjoyed its CFC-powered air conditioning and plastic foams for many decades before anybody took the trouble to consider the effects that all this might be having on the atmosphere, and even then it was 15 more years before we got serious about eliminating CFCs. Now stratospheric ozone is diminished, and it will not soon recover. Earth, and our own health, is paying the price. We dare not accept science's benefits without accepting its responsibilities.

Speaking as a science teacher who is doubtless prejudiced in the matter, my first suggestion is that all of us learn much more science. Scientists, who are usually only narrowly trained in their specialty, need broader scientific knowledge and improved communication skills. Nonscientists need to learn more about the physical universe and our planet. Humankind is using great power today, without knowledge. If we want our technology-based society to succeed, we had better try harder to understand what we are doing, for the use of power without knowledge is a prescription for disaster.

Energy unifies all physics, both Newtonian and post-Newtonian, and helps organize our thinking about any physical process. From automobiles to the big bang, the laws and concepts of energy can help clarify the essentials. On a practical level, our culture must get a grip on our use of energy if we are ever to fashion a prosperous future. Our most basic energy-related problem, global warming, casts an uncertain shadow over the twenty-first century. Although there are uncertainties in our understanding of this problem, enough is known to conclude that the probability of serious global harm is high. Given the inertia of the global climate, as well as the inertia of our energy economy, there are great risks in waiting to act.

We probably face the end of the fossil-fuel age. This is a challenging prospect but also an exciting one. Your help is needed.

On a deeper level, modern physics teaches us that the processes of the universe are not simply mechanical motions of matter but are more appropriately viewed as the organization and transformation of energy. Energy itself is the ability to do work and so bring about change, and it takes the form of motion and interactions. Ultimately, the universe appears to be made of fields, made of the "possibility of an interaction." The material objects of our everyday existence should not be regarded as made of permanent, solid parts but, rather, as made of vibrating fields of energy, bundles of possible interactions. The universe is not made of things; it is made of interaction and change.

The significance of post-Newtonian physics runs deep into the cultural roots of industrial civilization. Modern culture still assumes that the mechanical, Newtonian worldview is science's view of reality. It is a materialistic worldview that leaves little room for freedom, chance, or creativity. Many people would argue that it leaves little or no room for spiritual values.

But modern physics paints a quite non-Newtonian picture, a more abstract picture of fields and energy structured by relativity and quantum theory. Many nonmechanical forms of energy operate in this post-Newtonian universe, and reality emerges not as a predictable clockwork mechanism but as a dynamic and unpredictable network of energy. Material and nonmaterial particles pop unpredictably into and out of their fields. This is nothing like a clock. In many ways it is the opposite of a clock. Considering the hidden and often distant interactions that seem to be the essence of the physical world, many have suggested that if we are to use images at all (and physicists have used the clockwork image for many centuries now), the universe is more comparable to a live organism than it is to a clock.

It is by no means clear what worldview will emerge from all of this. I feel that the world is just beginning to absorb the impact of relativity and quantum theory. After all, more than a century elapsed after Copernicus's death in 1543 before Europe began to absorb the cultural impact of postmedieval science. So it is not surprising that the first post-Newtonian century will be the twenty-first, not the twentieth.

One practical example of the importance of forming a post-Newtonian worldview might be the ongoing reaction, in the United States at least, against the theory of biological evolution. The reaction comes from a perceived threat to religious beliefs. Religious fundamentalists typically view evolutionary theory as mechanical, deterministic, and materialistic, with no room for spiritual values. The real opposition is not so much to evolution itself as it is to evolution *as inter-*

Only by the fusion of science and the humanities can we hope to reach the wisdom appropriate to our day and generation.

I. I. RABI, PHYSICIST

preted through the concepts of Newtonian physics. A post-Newtonian culture might relieve these old religiously based anxieties about science.

Post-Newtonian physics suggests that we cannot control nature in the way that we can control a machine such as a clock, because we live in nature. Nature is a network, and we are part of that network. Efforts to subdue nature can react against us, to our own peril. Post-Newtonian physics suggests that our salvation lies not in imposing our own wills on nature but in aiding nature (and so ourselves) in nature's own pursuits.

All this is not to suggest that we should renounce the true benefits of modern science and technology, for if humankind and nature are one, then whatever truly benefits the one will benefit the other. But it is to suggest that we come more humbly to a natural and human world that surpasses our own understanding.

We have often inquired in this book: *How do we know?* The answer is surprisingly simple: We know by experience, as interpreted through intelligence. Science's method is really quite simple, but it is fundamental. It is to take nothing for granted, to form one's views on the basis of careful observations and hard honest thinking, and to be willing to modify those views in the light of new experience. In other words: *Trust the universe.*

The twentieth century has been torn by rigidly held and conflicting ideologies. The nationalistic, religious, and economic ideologies seem to come in every imaginable variety, many of them in utter contradiction to one another. Yet all those who believe them are absolutely convinced that they are right. The result has been war, prejudice, fanaticism, and other scourges. Science's view of this is that the danger lies not so much in the beliefs themselves as in their *absolute nature.* Even wrong or harmful beliefs can be corrected if one is willing to trust experience and to be intellectually honest. Even correct and healthy beliefs can become dangerous if accepted uncritically or absolutely.

In thinking about how we might do better in the twenty-first century than we have in the twentieth, we should perhaps ponder science's most basic value: *All ideas are subject to testing by experience and to challenge by critical rational thought.* It is a practical, simple, but demanding code. It can be uncomfortable, in fact painful, to honestly reevaluate one's beliefs in the light of experience. But it is a code that has worked surprisingly well for science. It might be science's most important benefit.

Science is much less a body of knowledge than it is a way of thinking, and that way of thinking, with its characteristic mix of rigorous skepticism and openness to new ideas, is desperately needed in every area of our lives—including social, economic, political, and religious arenas.

CARL SAGAN

For the belief in a single truth and in being the possessor thereof is the root cause of all evil in the world

MAX BORN, PHYSICIST

Answers to odd-numbered exercises and problems

CHAPTER 1

Exercises

1. Follow its position in the sky for a few weeks. If its position relative to the surrounding stars changes, it is a planet.

3. We are looking northward, and stars rise in the east (left-hand side of photo) and set in the west (right-hand side), so the stars in the photo are circling counterclockwise. If we were looking southward, toward the South Pole, the stars would be circling clockwise around a point in the southern sky.

5. Follow a planet every night until it noticeably brightens or dims—an indication that it is closer to or further from Earth.

7. This is possible, if Mars happens to lie a little to the east of the sun in the sky (in other words, close to a line joining Earth to sun in Figure 1.15) and if Venus is below the horizon. However, this combination of events isn't very likely.

9. Look near the rising or setting sun, just before it rises and just after it sets. If the sun is visible above the horizon, the dim light from Mercury will be obliterated by the light from the sun.

11. The observational imperfections in Ptolemy's and Copernicus's theories, that led Kepler to develop his theory, were first observed by Brahe. Since these imperfections had not yet been observed in Ptolemy's or Copernicus's times, Kepler's theory would have also agreed with the data known at those times.

13. The "eccentricity" of the elliptical orbits, in other words the displacement of the focus of the ellipse (the sun) away from the center of the ellipse, is similar to the displaced centers of Ptolemy and Copernicus.

15. Yes. The agreement of the data known during Copernicus's time with both the theory of Copernicus and the theory of Ptolemy is one example.

17. No it is not testable, because all meter sticks and other measuring devices would double too so there would be no observable change in anything. This is certainly not good science, and it is not even bad science. It is not science at all, because it doesn't refer to the observable universe.

19. No, some of the material (such as carbon dioxide—see Chapter 2) comes from the air. To test Aristotle's hypothesis, we could put a plant in a pot and carefully weigh it together with any water or fertilizer we put on it as it grows. If the plant's increase in weight is greater than the weight of the water plus fertilizer added to it, then the plant is getting some of its weight from someplace else (namely, the air).

21. This is not a testable hypothesis. So any ESP "theory" that includes this hypothesis should not be considered to be a science.

23. A photograph such as Figure 1.7. See Figure 1.8.

Problems

1. 100 billion times 100 billion = 10,000,000,000,000,000,000,000 (a one followed by 22 zeros).

3. 0.5 AU and 2.5 AU.

5. It exploded about 3500 years before A.D. 1054, or in about 2446 B.C.

CHAPTER 2

Exercises

1. No. General scientific principles are never certain. See Chapter 1.

3. 12 to 4, in other words 3 to 1.

5. $2 + 1 + 4 = 7$.

7. Helium is an element, carbon dioxide is a pure compound, polluted water is neither (it is a mixture), $C_6H_{12}O_6$ is a pure compound, gold is an element, steam is a pure compound (H_2O).

9. Molecule made of two or more atoms: Pure water (H_2O), atmospheric oxygen (O_2), H_2SO_4, carbon dioxide (CO_2), H_2. Single unattached atom: U, He, H.

11. The elements lying in the same column with chlorine in the periodic table: fluorine, bromine, iodine, astatine.

13. A piece of paper is too heavy to respond to be jostled noticeably (as observed by the unaided eye) by atoms.

15. The convict leaves many molecules from his or her body or clothing along the trail. When the dog sniffs these molecules into its nose, it is able to detect these molecules and identify them with the convict, just as your nose is able to identify the odor of violets.

17. Heating causes the air molecules to move faster. This causes them to hit the inner walls of the container harder, which increases the pressure.

19. The air outside the jar pushes downward on the lid more strongly than the air inside the jar pushes upward on the lid. So air pressure holds the lid on the jar.

21. With more air, the air molecules will hit the inside of the tire more often. With hotter air, the air molecules will hit the inside of the tire harder.

23. Proton, H_2, methane (CH_4), glucose molecule ($C_6H_{12}O_6$), hemoglobin molecule, dust particle.

25. Hydrogen and oxygen come in the 2-atom form H_2 and O_2. They combine to give water: $H_2 + O_2 ==> H_2O$.

27. Hydrocarbons are made of hydrogen (H) and carbon (C). When these burn in air, containing O_2, the H should

combine with O_2 to create H_2O, and the C should combine with O_2 to create CO_2.

29. From the atmosphere, which contains an abundance of N_2 and O_2. These elements don't combine at normal atmospheric temperatures, but at the high temperatures prevailing in automobile engines they do combine.

Problems

1. Assume that the coal is pure carbon (C). When the coal burns, each C atom attaches to two O atoms to make CO_2. If we assume, for simplicity, that C and O atoms have the same weight, then a CO_2 molecule would weigh three times as much as a single C atom. So the ton of coal makes three tons of carbon dioxide gas. The more precise answer, based on the weight ratio of 3 to 4 given in exercise 2, is that a ton of coal makes 11/3 (or 3.67) tons of carbon dioxide gas.

3. $10^9 = 1,000,000,000$
   $10^{-6} = 0.000\ 001$
   $3.6 \times 10^{13} = 36,000,000,000,000$
   $5.9 \times 10^{-8} = 0.000\ 000\ 059$

5. 3.84×10^5 km, 3.84×10^8 m, 3.84×10^{11} mm.

7. How many times does 10^{-10} m go into 0.1 mm? Since 0.1 mm $= 10^{-4}$ m, the answer is $10^{-4}/10^{-10} = 10^{-4+10} = 10^6$ atoms, or one million atoms thick.

9. An atom is about 10^{-10} m across (Section 2.3, also problem 7). Since 0.05 mm $= 5 \times 10^{-5}$ m, the number of atoms needed to stretch across a dust particle is $5 \times 10^{-5} / 10^{-10} = 5 \times 10^{-5+10} = 5 \times 10^5$ atoms, or 500,000 atoms (half a million).

CHAPTER 3

Exercises

1. Aristotle: This is the ball's natural motion. Galileo: Friction slowed it to a stop.

3. Nothing keeps them moving. According to the law of inertia, they keep moving simply because there is nothing to stop them.

5. 20 m/s. 20 m/s.

7. You would have to be able to verify that the speed of an object remains constant (assuming that there are no external influences on the object). So we would have to measure the speed, at several different times. Since speed is distance divided by time, we would have to measure distance and time. We would not need to measure weight.

9. No. You must turn several corners to get around the block.

11. They have different velocities, but the same speed.

13. Zero.

15. Although your foot is on the accelerator, you are not actually accelerating (in the physics sense of the word) when you are driving along a level highway at constant speed; also when you are climbing a straight (unchanging angle, or unchanging "slope") hill at constant speed. When you remove your foot from the accelerator, you slow down; this is an acceleration.

17. The upper ball is accelerated, while the lower ball is not. The lower ball catches up with the upper ball and passes

it at time "2," but then the upper ball speeds up and passes the lower ball at time "5." The two balls have the same speed during time "3" to time "4," since they both move the same distance during this time interval.

19. Yes, for example, a drag-racing car just as it starts up from rest. Yes, for instance, a comet that is speeding up very slightly as it approaches the sun, or Earth as it makes one circuit around the sun in one entire year.

21. The object moves with an unchanging velocity. Thus, distance changes; it gets larger and larger, at a steady rate. Speed does not change. Velocity does not change. Acceleration remains zero, so it does not change.

23. Distance changes; it gets larger and larger. Speed and velocity both change; they get larger and larger.

25. The object begins with a speed of 0 m/s and speeds up to a speed of 10 m/s by the end of the first second, so its *average* speed during the entire first second (from 0 to 1 s) is only 5 m/s.

27. By 10 m/s. By 10 m/s.

29. 1st second: About 10 m/s. 2nd second: About 20 m/s. 3rd second: About 30 m/s.

Problems

1. Upper ball: speed = d/t = 0.8 cm/0.2 s = 4 cm/s.
   Lower ball: speed = d/t = 1.2 cm/0.2 s = 6 cm/s.

3. 0.5 m = 500 mm. 500 mm / 2 mm/yr = 250 years.

5. The distance to the moon and back is 7.6×10^8 m. The speed is 300,000 km/s $= 3 \times 10^5$ km/s $= 3 \times 10^8$ m/s. Solving s = vt for t, we get t = s/v = (7.6×10^8 m) / (3×10^8 m) = 2.5 s.

7. total time = 1.0 hr + 3.0 hr + 0.5 hr = 4.5 hr.
   the total distance is 50 km (during first hr) + 250 km + 30 km = 330 km.
   overall av speed = 330 km / 4.5 hr = 73 km/hr.

9. acceleration = change in speed / time = $\dfrac{100 \text{ km/hr}}{10 \text{ s}}$ = 10 (km/hr)/s.
   acceleration = $\dfrac{400 \text{ km/hr}}{5 \text{ s}}$ = 80 (km/hr)/s.

11. The change in speed is 2.25 (km/hr)/s $\times$ 4 s = 9 km/hr. This must be added to the initial speed of 30 km/hr. Thus the final speed is 30 km/hr + 9 km/hr = 39 km/hr.

13. The answers to the preceding exercise would be the same on Earth. But the actual distance fallen would be different, because the acceleration due to gravity would be different.

15. s = (1/2)at^2 = (1/2) (9.8 m/s^2) (3 s)^2 = 44 m. v = at = (9.8 m/s^2) $\times$ (3 s) = 29 m/s.

17. Speed is proportional to time, so it is moving at 16 m/s after 2 s, and 24 m/s after 3 s. Distance is proportional to the square of the time so it has moved 4 times as far, or 16 m, after 2 s. It has moved 9 times as far, or 36 m, after 3 s.

CHAPTER 4

Exercises

1. A force must be exerted on you both when you speed up and when you slow down, in order to accelerate you. Newton's law of motion says so.

3. If the ball accelerates (speeds up, slows down, or changes direction), it must have a force on it.

5. Initially the book is at rest; then it quickly speeds up during the fraction of a second that the hammer is actually in contact with the book; after the hammer is no longer touching the book, the book gradually slows down to a stop. There is no net force on the book before the hammer hits it; then there is a large force in the forward direction while the hammer is in contact with the book; then there is a smaller force in the backward direction while the book is slowing down.

7. Since gravity is the only force acting on the ball, the net force on the ball is 8 newtons downward. Thus the ball's acceleration is also downward, opposite to the ball's upward velocity, because Newton's law of motion says that the acceleration is in the direction of the net force. Acceleration and direction are always in the same direction.

9. The object could have several forces on it, adding up to zero net force. For example, an object at rest on a table has two forces on it: weight acting downward, and perpendicular force by the table acting upward. The object has no net force on it.

11. No, your push cannot be the only force on the wall. Because the wall doesn't accelerate, the net force on the wall must be zero, and so there must be another force (provided by the concrete structure) pushing back in the other direction on the wall.

13. Each vehicle has zero net force on it. The truck has the larger perpendicular force on it.

15. Since 1 newton is about 1/4 pound, multiply your weight in pounds by 4 to get your approximate weight in newtons: 100 pounds is roughly 400 newtons, etc.

17. You would be better off having a hunk of gold whose weight is 1 N on the moon, because it would be a more massive hunk of gold (containing more gold) than one whose weight is 1 N on Earth.

19. The net force on the apple at rest is zero. The net force on the falling apple is 2 newtons downward. The net force on the apple moving upward is 2 newtons *downward* because the only force on the apple is gravity.

21. It would be easier to lift a rocket off the moon's surface, because the force of gravity on the rocket (i.e. the rocket's weight) is smaller.

23. Neglecting frictional effects, it would be just as hard to kick the rock on Earth; that is, it would be just as hard to accelerate the rock horizontally on Earth as on the moon, because the mass of the rock is the same on Earth as on the moon.

25. False. The boulder has a much larger acceleration than does Earth, because the boulder's mass is much smaller than Earth's mass.

27. The two vehicles exert equally strong forces on each other, and the two vehicles feel equally strong forces from the other vehicle. The car experiences the larger acceleration, because it has the smaller mass.

29. The string pulls upward on the apple, and Earth's gravitational force pulls downward. These do *not* form a force pair. The other members of the two force pairs are: (1) The apple pulling downward on the string. (2) The apple's gravitational force pulling upward on Earth.

31. Yes, you push on me too. The two forces are equally strong.

33. The diagram should show a large frictional force, acting backward, in addition to the forces of air resistance (backward), gravity (downward), and the road's perpendicular force (upward). The net force is backward. For a car coasting without braking, the forces are rolling resistance, air resistance, gravity, and the perpendicular force. The net force is strongest in the case of braking.

35. No, the car could be moving at constant velocity. However, it could be accelerating. The acceleration could be forward (if the car is speeding up) or backward (if the car is slowing down).

37. No a jet plane could not accelerate (except for the "natural" acceleration of 9.8 m/s^2 downward due to gravity). A rocket-driven airplane could accelerate.

Problems

1. a = F/M = 60 N / 3 kg = 20 m/s^2. After the kick, the acceleration must be zero (the law of inertia).

3. 600 N, acting upward, to balance the force of gravity. With no air, Ned's acceleration would be 9.8 m/s^2.

5. Solving a = F/M for F, we get F = Ma. Thus F = Ma = 0.5 kg $\times$ 50 m/s^2 = 25 N.

7. The net force is zero, because the car is not accelerated. The acceleration is zero. The drive force must be 500 newtons, to balance the resistive forces to make the net horizontal force zero.

9. The net force is now 500 − 100 = 400 newtons in the backward direction. The acceleration is: a = F/M = 400 newtons / 800 kg = 0.5 m/s^2 backward, i.e., it slows down.

11. The wall pushes with 50 N. a = F/M = 50 N / 40 kg = 1.25 m/s^2.

13. The truck exerts a 30,000-N force on the car, in the backward direction. The car's acceleration is a = F/M = 30,000 N / 1000 kg = 30 m/s^2. This will slow the car down. The truck's acceleration is a = F/M = 30,000 N / 60,000 kg = 0.5 m/s^2. This speeds the truck up, since the force is in the forward direction.

CHAPTER 5

Exercises

1. Iron. They fall equally fast.

3. The magnitude is your weight (in pounds, or in newtons), and the direction is downward.

5. No, you are not in orbit around Earth's center (although you are revolving around that center due to Earth's spinning motion). You are in orbit around the sun.

7. Yes. Yes.

9. It extends to infinity (but it is very small at great distances).

11. 8000 N, downward. 8000 N, upward.

13. A straight line. A straight line.

15. The law of force pairs tells us that these two forces are of equal strength.

17. Yes, this would decrease your weight.

19. The sun, Earth's moon, planets such as Mars and Venus.

21. Yes, the keys are falling, along with you. The keys have weight, although they are "apparently weightless."

23. You would weigh more in Los Angeles, because your distance from Earth's center would be smaller.

25. Yes, the moon is accelerated in the inward (downward) direction. The moon's velocity is sideways—along the moon's orbit.

27. The inward force on the Earth satellite would be larger (because Earth's mass is larger than the moon's mass), so the Earth satellite would have the larger acceleration (because of Newton's law of motion). In order to have this larger acceleration, the Earth satellite would have to be moving faster.

29. No. Earth does not have sufficient mass for it to collapse this far. Neither does the sun.

31. You would weigh more, because your distance from Earth's center would be less (the Earth's mass would be unchanged).

33. Special relativity and quantum theory.

Problems

1. $(1/4) \times 10^{-12}$ N. 4×10^{-12} N. $(1/9) \times 10^{-12}$ N. 16×10^{-12} N.

3. The force would be unchanged.

5. 100 times less than your present weight, i.e., your present weight divided by 100.

7. $F = 6.7 \times 10^{-11}\, m_1 \times m_2 / d^2 = 6.7 \times 10^{-11}$ (1 kg) $(7.4 \times 10^{22}$ kg$) / (1.7 \times 10^6$ m$)^2 = 1.7$ N. This is 17% (about one-sixth) of its weight on Earth (Problem 6).

9. On Earth, 1 kg weighs 9.8 N (Problem 6). At 200 km $(2 \times 10^5$ m) high, the distance from Earth's center is 6.6×10^6 m, so $F = 6.7 \times 10^{-11}\, m_1 \times m_2 / d^2 = 6.7 \times 10^{-11}$ (1 kg) $(6.0 \times 10^{24}$ kg$) / (6.6 \times 10^6$ m$)^2 = 9.2$ N. The % difference is $(9.8 - 9.2) / 9.8 = .06 = 6\%$.

11. The distance is d = 150×10^6 km = 150×10^9 m = 1.5×10^{11} m. Thus $F = 6.7 \times 10^{-11}\, m_1 \times m_2 / d^2 = 6.7 \times 10^{-11}$ $(2.0 \times 10^{30}$ kg$) (6.0 \times 10^{24}$ kg$) / (1.5 \times 10^{11}$ m$)^2 = 3.6 \times 10^{33}$ N. This is more than 10^{13} (10 trillion) times larger than the force by the moon on Earth (Problem 10).

13. A typical mass for a person is about 50 kg. A typical distance from you to a person next to you might be about 1 meter. So the gravitational force by one person on the other is about $(6.7 \times 10^{-11}) \times (50) \times (50) / (1)^2$ N, or $(6.7 \times 50 \times 50) \times 10^{-11}$ N, or about 1.7×10^{-7} N.

CHAPTER 6

Exercises

1. Yes, because Earth exerts a downward force on you as you walk downward.

3. Examples: Stretch a rubber band. Lift a rock. Throw a baseball.

5. Chemical, in the form of food. Chemical, in the form of fossil fuels.

7. Chemical, gravitational, elastic, chemical, chemical.

9. One example: A thrown baseball.

11. Thermal energy and radiant energy (because the explosion is hot, and bright). Chemical energy is consumed.

13. The most direct way would be to measure the mass and acceleration of the ball during the throw, and the distance the ball moves during the throw. Then force = mass × acceleration, and work = force × distance. An easier way would be to measure the ball's mass and the speed at which it is thrown. Then work = energy increase = $(1/2)ms^2$.

15. (a) 1000 km high. (b) Increased. (c`) At the lower altitude, 6000 km, even though the satellite's *energy* is larger at 12,000 km.

17. Examples: A hammer hitting a nail; a book slowing down as it slides along a table.

19. Chemical energy input, kinetic energy output. Chemical energy input (from the bicycler's body), kinetic energy output.

21. It decreases, because the air's energy increases. Its kinetic energy increases. Its thermal energy increases, because air resistance warms the rock.

23. The acid must increase its thermal energy, i.e., it must get warmer.

25. (a) Chemical energy (of human body) → kinetic energy (b) kinetic energy (of baseball) → thermal energy (of ball and of Jill's hand or baseball mitt).

27. Yes, it must, in both cases, because the energy must come from someplace.

29. It is moving slowest at 500 m, and fastest at 600 m.

31. Both processes impart the same energy, but the hammer blow has the greater power output because it occurs over a much shorter time than the arm-swing.

33. Watts.

Problems

1. Left hand does $2 \times 1.5 = 3$ joules of work, right hand does $4 \times 0.5 = 2$ joules of work. So the left hand does the most work, although the right hand exerts the largest force.

3. Tripled, because gravitational energy is proportional to height. Halved.

5. At twice the speed, you would slide 4 times as far. The reason: Your car has 4 times as much kinetic energy, so it will do 4 times as much work in coming to rest, so it must exert its sliding frictional force over 4 times as much distance (remember W = Fd) in coming to rest. Similarly, at half the speed you would have only 1/4 as much kinetic energy (because 1/2 squared is 1/4) and so you would slide only 1/4 as far.

7. (a) 160,000 J. (b) 160,000 J, 160,000 J. (c) One-quarter as much as it had at the high point, because it is one-quarter as high: 40,000 J. So it the remaining 120,000 J must be kinetic energy.

9. (a) $(1/2)\, ms^2 = 0.5 \times (1000$ kg$) \times (10$ m/s$)^2 = 50,000$ J. (b) At twice as fast, it would have four times as much kinetic energy, or 200,000 J.

11. Power = Work / time = 2 J / 0.1 s = 20 watt for the 2-joule work source. For the 1000-joule source, power = 1000 J / 3600 seconds = 0.27 watt, which is smaller than the power output of the 2-joule source. The 1000 J work source has the smaller power output, but the largest energy output.

13. 150/1000 = 15%.

15. To answer this, perform a calculation similar to problem 14 (above) for each appliance used in your home, getting the number of kW·h of electric energy consumed by each appliance during one month. Then add up the total consumption of all of the appliances.

CHAPTER 7

Exercises

1. Cold, below freezing. Cold, above freezing. A nice day. A hot day.
3. A Celsius degree.
5. Refrigerator; it moves thermal energy from inside the refrigerator to outside. It does not violate the second law, because the second law states that thermal energy cannot spontaneously flow from cold to hot; in a refrigerator, the flow is not spontaneous (it is assisted by the operation of the refrigerator, which draws the thermal energy out of the inside).
7. No, because the second law prohibits it. Yes; an example is dropping a book onto a table.
9. Hydroelectric power plant, bicycle.
11. The efficiency must be less than 100%.
13. Don't invest. Slick's scheme violates the second law of thermodynamics, because it purports to convert thermal energy entirely into the work needed to drive the automobile.
15. No. We can push the rock back to its starting condition. The precise process (without the push) cannot be reversed.
17. This does not violate the law of increasing entropy, because the egg is not an isolated system; it has outside help in the form of energy (thermal energy that is transferred into the egg).
19. Microscopic disorganization increases, so entropy increases.
21. You should get more useful work, because the efficiency would tend to be higher due to the higher input temperature.
23. Transportation.
25. Hydroelectric, because it is not a heat engine and so is not subject to the inefficiency implied by the second law of thermodynamics.
27. Electricity from a coal-fired generating plant, because such a plant is about 40% efficient while a car engine is only 10–15% efficient (Table 7.1).
29. June 29.
31. No, for example it stayed about level during 1930–35. About 150 BkW·h in 1935, about 350 BkW·h in 1945, 600 BkW·h in 1955, 1100 BkW·h in 1965, 2000 BkW·h in 1975. Each of these numbers is roughly twice the preceding number, so the growth is roughly exponential.
33. For all except geothermal, the original source is the sun. The source of geothermal energy is the heat within Earth. All are considered renewable except oil; strictly speaking, geothermal is not renewable—but this resource is so large that it is considered to be renewable.
35. Twice as many.

Problems

1. Energy input = 100 J + 400 J = 500 J. Efficiency = 100 J / 500 J = 20%.
3. Input temp = 500 + 273 = 773 K, efficiency = (input temp – exhaust temp) / input temp = 420 K / 773 K = 54%.
5. Assuming an efficiency of 13%, about 13 barrels go into getting the car down the road.

7. From Table 7.3, the car averages 20 mi/gal. Thus the amount of gasoline used is 2800 mi / 20 mi/gal = 140 gal. By air, at 50 mi/gal (Table 7.4), you use 56 gal. By bus, at 50 mi/gal, you also use 56 gal. By train, at 250 mi/gal, you use 11 gal.
9. According to Figure 7.19, the overall efficiency is 900 /2500, or 36%. Thus 36 tons go to electricity, while 64 tons are wasted as exhaust thermal energy.
11. The first plant would create half as much pollution.
13. 4000 BkW·h. Twice as many.
15. The doubling time is T = 70/P = 70/2 = 35 years.

CHAPTER 8

Exercises

1. No, because the water itself actually moves downhill.
3. It must have transformed into thermal energy.
5. No. The speed depends on the reaction time of the crowd.
7. One-half second later, all the ripples will have moved outward by 1/2 wavelength, so crests will be replaced by valleys and valleys will be replaced by crests. One second later, the surface will look just like it looks in the photo.
9. 3 cm. –3 cm (3 cm downward). 0 cm.
11. Because you cannot see in the dark, even though your eyes are open.
13. Firefly, heating element, flash-bulb, sun.
15. You would not hear the explosion, but you would see it.
17. The bag becomes charged, due to friction from the wool dress. This causes the sides to cling together electrically.
19. Charge must have flowed, or moved, from the metal sphere down to the leaves; this motion of charge is an electric current.
21. You will not be able to tell, because no matter which ends you put together, they will be attracted together.
23. The truck picks up charge from the road by friction (the tires rubbing against the road). It can be prevented by providing an easy pathway for charge back to the road—a chain hanging from the truck to the road, for example.
25. Roll a ball into the box to see whether the contents have high mass or low mass. Tie a string around it and pull it; the box's mass can be determined by measurement of the pulling force and the resulting acceleration. Fire bullets through the box from all directions, to see if they mostly pass straight through without hitting anything. Hit the box with a hammer to see if it causes anything inside to roll around.
27. 6, and 2. The ratio is about 12 to 4, or 3 to 1.
29. The ion will carry a positive charge, so it will be repelled by a positively-charged transparency, and attracted to a negatively-charged tissue. If only one electron were lost, the forces on the ion by the transparency and tissue would be only half as large, but the directions would still be the same.

Problems

1. Its frequency would halve, because it would take twice as long for each complete wavelength to pass. If you halved the wavelength, the frequency would double.

3. $s = f\lambda = 4$ Hz $\times$ 1.5 m = 6 m/s. At the lower frequency, the wavelength is doubled (since the frequency was halved): λ = 3 m.

5. Solving $s = f\lambda$ for λ, $\lambda = s/f = 2$ m/s / 3 Hz = 0.67 m. At a frequency of 4 Hz, λ = 2 m/s / 4 Hz = 0.5 m.

7. 10 m / 0.6×10^{-6} m = 1.7×10^7 = 17 million!

9. A 500-page stack of typing paper is about 5 cm thick, so the thickness of one sheet of paper is about 5 cm / 500 = 0.01 cm = 10^{-4} m. An atom is about 10^{-10} m across. So a piece of paper is about 10^{-4} / 10^{-10} = 10^6 (one million) atoms thick.

CHAPTER 9

Exercises

1. Yes, because electric currents create magnetic fields. You might be able to detect these fields with a small sensitive compass needle.

3. It will accelerate into motion. An electron would have a larger acceleration (because of its smaller mass) and it would move in a direction opposite to that of the proton (because the electron has a negative charge while the proton has a positive charge). In a gravitational field, the two would fall at the same rate.

5. At B there is an electric field (because of the charge at A), there is no electric force (because there is no charge at B), there is a magnetic field (because the charge at A is moving), there is no magnetic force (because there is no charge at B), and there is energy (because of the presence of an electromagnetic field).

7. It contains no matter. But it might contain energy, in the form of fields.

9. Each reply will be delayed by 20 minutes. The telescope would not speed things up.

11. Your ears would "hear" all the radio waves that your radio can receive, all at the same time.

13. Their wavelengths and frequencies are different. Light's wavelength is much shorter than radio's wavelength, and light's frequency is much higher.

15. Because ultraviolet radiation does not penetrate glass.

17. You would only see ultraviolet and visible light from unnatural sources, such as streetlights. But you would see infrared from all over Earth, with warmer places appearing brighter.

19. 92×10^6 Hertz. 300,000 km/s. According to Figure 9.7, one wavelength is on the order of one meter long (more precisely, about 3 meters long). Your ear is stimulated not by an electromagnetic wave, but by a sound wave.

21. So they will not react chemically inside your body. This inertness means that the gases do not react in the atmosphere, so the gases float up to the upper atmosphere with being broken down along the way.

23. Less than 1 ppb in 1950. It will reach the highest level around the year A.D. 2000. It will return to the 1975 level (about 2 ppb) around A.D. 2050. Without the Ozone Treaty, it would have been about 9 ppb in A.D. 2050.

25. Ozone is also continually created in the atmosphere, so that ozone is present in the atmosphere at a relatively constant level.

27. Electricity from solar power, electricity from nuclear power, ethanol from corn (harvesting the corn releases the carbon that was absorbed when the corn was grown, so there is no net carbon increase in the atmosphere).

29. The Ozone Treaty will end the use of CFCs, which not only destroy ozone but are also a greenhouse gas (see Figure 9.24).

31. Carbonated drinks. Raising the temperature causes the dissolved CO_2 gas to rise to the surface and move into the air. This would be a positive feedback on global warming.

Problems

1. (5000 km) / (300,000 km/s) = 0.017 seconds.

3. 3×10^5 km/s $\times$ 8.3 min $\times$ 60s/min = 1.5×10^8 km = 150 million km.

5. The distance light travels in 4 years is 300,000 km/s $\times$ 4 years $\times$ 365 days/yr $\times$ 24 hr/day $\times$ 60 min/hr $\times$ 60 s/min = 3.8×10^{13} km, or 38 trillion km.

7. Solving for λ, we get $\lambda = s/f = 3 \times 10^8$ m/s / 60 Hz = 5×10^6 m = 5000 km.

9. About 21% are O_2: $0.21 \times 2.7 \times 10^{19}$ = 5.7×10^{18} O_2 molecules. 350 out of every million ($350/10^6$) are CO_2: ($350/10^6$) $\times 2.7 \times 10^{19}$= 9.4×10^{15} = 9400 trillion CO_2 molecules. Less than 1 out of every million is an O_3 molecule: Less than ($1/10^6$) $\times 2.7 \times 10^{19}$ = 2.7×10^{13} = 27 trillion.

11. It reached 280 ppm about 140,000 years ago. The temperature at that time was about 1.0 degree (Celsius) above current levels.

13. 2.5% (in buildings) + 5.5% (in industry) = 8%. 1.5% + 3.5% + 9% = 14%.

CHAPTER 10

Exercises

1. No.

3. No.

5. According to Galilean relativity, she observes the light beam to move at 1.5c.

7. The ball would land on the dot.

9. No. The principle of relativity applies to *non-accelerated* reference frames; the bumps during the flight are accelerations.

11. If the plane is speeding up, you will feel pressed backward, and dropped objects will fall toward the rear of the drop point. If the plane is slowing down, you will feel pulled forward and objects will fall toward the front of the drop point. If the plane is turning to the right, you will feel pressed toward the left and objects will fall toward the left.

13. No. Only non-accelerated observers must necessarily measure the same speed for a lightbeam.

15. Yes. 300,000 km/s.

17. 300,000 km/s. 300,000 km/s.

19. The year (or Earth's orbital motion). The vibrations of an atom. The new moon (the moon's orbital motion around Earth). The day (or Earth's spinning motion). An observer moving past these clocks would observe all of them to go slow.

21. No. No. No.

23. 30,000 years. No, it would take more than 30,000 years as measured on Earth. Yes, a person could (theoretically)

travel there in as short an amount of their own time as desired; one second, for example. So a person could get there within their own lifetime. This is because of the relativity of time; the person would have to be moving very near lightspeed (relative to Earth), for nearly all of the trip.

25. No. The lifetime will be longer than 17 minutes, due to time dilation.

27. No, because you will age less than your father.

29. No. Time always goes forward, never backward.

Problems

1. Yes. 8 m/s southward.

3. 6 m/s northward. If Mort had thrown the ball southward, the ball's velocity relative to Velma would be 14 m/s southward.

5. 540 m/s as observed by the sheriff. 500 m/s as observed by the passenger. With the gun pointed toward the rear, the bullet moves at 460 m/s as observed by the sheriff, 500 m/s as observed by the passenger.

7. 300,000 km/s. 300,000 km/s. 300,000.04 km/s.

9. 299,970 km/s (300,000 − 30). 300,030 km/s.

11. Jet aircraft: 800 km/hr = 0.22 km/s. fraction = 0.22 km/s / 300,000 km/s = 7.3×10^{-7}. Bullet: 1 km/s / 300,000 km/s = 3.3×10^{-6}. Satellite: 8 km/s / 300,000 km/s = 2.7×10^{-5}. Time dilation would not be noticeable at these speeds, although it could be detected with sophisticated measurement techniques.

13. The speed is 0.1c, so Table 10.1 says that 1 of your seconds is observed on Earth as 1.005s. So when you return you will have aged by 24 hours while people on Earth have aged by 1.005 × 24 = 24.12 hours = 24 hours and 7.2 minutes. You have aged less than people on Earth, by 7.2 minutes.

CHAPTER 11

Exercises

1. According to Figure 11.4, the required speed is about 0.86c. She will observe the United States to be 0.86 × 5000 km = 4300 km wide.

3. According to Figure 11.4, a 10% length contraction (to 90% of the original length) occurs at a speed of about 0.4c.

5. Yes. Velma could be holding her meter stick perpendicular to the direction of motion.

7. Mort and Velma measure the same distance; Velma observes no length contraction because the distance is perpendicular to the direction of motion.

9. They both measure the same width, because the width is perpendicular to the direction of motion.

11. Galileo's relativity predicts 0.75c. Using Table 11.1, Einstein's prediction is 0.667c. Einstein is correct.

13. Using Figure 11.4, Mort measures the mass to be 1.6 × 10,000 kg = 16,000 kg, and the length to be 0.6 × 100 m = 60 m.

15. It is moving at about 0.87c, and is oriented crossways (perpendicularly) to its direction of motion.

17. They are normal; they have not changed at all.

19. It increases. This effect is so small that you wouldn't be able to detect it.

21. Yes; since the material inside the container loses energy, it also loses mass.

23. The second vat has more energy (because of the elastic energy in the second mouse trap), so it also has more mass (because of $E = mc^2$).

25. Twice as heavy as usual. Half as heavy as usual. Weightless.

27. Latitudinal lines are circles, but only the equator is a "straightest" line—the others are not. They do not meet.

29. The big bang! (The origin of the universe).

Problems

1. 90 J / c^2 = 90 J / 9 × 10^{16} m²/s² = 10^{-15} kg, or 0.000,000,000,000,001 kg.

3. mc^2 = (1 kg) × (9 × 10^{16} m²/s²) = 9 × 10^{16} J. This much energy goes into lifting. The energy needed to lift a weight through a height is weight × height. The weight of the U.S. population is about (250 × 10^6 people) × (600 N/person) = 1.5 × 10^{11} N. The height through which this mass could be lifted is 9 × 10^{16} J / 1.5 × 10^{11} kg = 6 × 10^5 m = 600 kilometers!

5. Since 1 watt = 1 joule/second, 1400 joules hit each square meter in each second. Earth's area facing the sun (the cross-sectional area facing the sun) is πR^2 = 3.14 × (6.4 × 10^6 m)² = 1.3 × 10^{14} m². The energy hitting Earth each second is thus 1400 J/m² × 1.3 × 10^{14} m² = 1.8 × 10^{17} J. The mass of this amount of energy is m = E/c^2 = 1.8 × 10^{17} J / (3 × 10^8 m/s)² = 2.0 kg.

CHAPTER 12

Exercises

1. No. Each neutron star and black hole must have gone through a supernova explosion, and this explosion would have destroyed any life that had developed previously. Similarly, each white dwarf must have gone through an unstable red giant phase that would destroy any pre-existing life. Our sun will go through such a red giant phase some 5 billion years in the future.

3. Simple sugars such as glucose $C_6H_{12}O_6$ (see Chapter 2).

5. (No unique answer).

7. No. It is impossible to prove a negative such as "technology does not exist elsewhere." It could exist without us detecting it.

9. The speeding up and slowing down phases require rocket power. During speeding up, the thrust is forward. During slowing down, the thrust is backward.

11. You would feel apparent gravity during the speeding up and slowing down phases. During speeding up, "downward" would be toward the rear. During slowing down, "downward" would be toward the front.

13. Any such planet should have a gravitational effect on the other planets, and this effect has not been observed.

15. There were no such reports. There were no stories reporting this fact. Generally, when far-fetched predictions such as this are made, there is little or no attempt to follow up on them.

Problems

1. 400 × 10^9 × 0.001 × 0.01 × 0.01 = 40,000 planets. Fraction = 40,000 / 400 billion = 10^{-7}, or 1 in 10 million. We would need to survey 10 million stars.

3. The ratio of the distances is the same as the ratio of 11 years to 8 minutes. Convert 11 years to minutes: 11 yrs × 365 days/yr × 24 hrs/day × 60 minutes/hr = 5,800,000 minutes (approximately). Dividing this by 8 minutes, we find that it is about 700,000 times farther to Tau Ceti than it is to the sun.

5. The round-trip distance is 22 light-years. She travels at half of lightspeed, so she takes twice as long as a light-beam would take to make the trip: 44 years (as measured on Earth). Because of time dilation, she measures a time that is less than 44 years. Figure 10.11 tells us that the time dilation factor is about 1.17. Thus the time as measured by Velma is 44 years / 1.17 = 38 years; Velma ages by only this amount.

7. Traveling at a speed of 0.8c (80% of lightspeed) will get you there in 5 years of Earth time.

CHAPTER 13

Exercises

1. Radiant to electromagnetic.

3. Green and violet will work, because green and violet photons have energies higher than the energy of yellow photons.

5. Ultraviolet radiation would work, but infrared would not (because UV photons have sufficient energy but IR photons do not).

7. No, the energies of the individual photons remain the same because the frequency remains the same. But the number of photons emitted in each second increases.

9. Infrared has greater energy. Radio has larger wavelength. Infrared has larger frequency. Both have the same rest-mass (zero). Both move at lightspeed.

11. Ultraviolet photons have more energy, enough to disrupt the chemistry of the cells in your skin.

13. Yes, it must emit photons because it emits electromagnetic radiation. It the comb is not charged, it emits no photons.

15. Similarities: both are small particles, both carry energy. Differences: photon is radiation while electron is matter, photon has no rest-mass while electron has rest-mass, photon moves at lightspeed while electron moves slower than lightspeed.

17. Very short.

19. A slow proton, because slower particles have larger wavelengths.

21. Yes. The wavelength of the protons is about 2000 times smaller than the wavelength of electrons (see Exercise 18), so the "interference lines" (the bright and dark lines on the screen) would be 2000 times smaller.

23. No, a coin-toss is "Newtonian"—the coin obeys Newtonian physics (to a very good approximation), so quantum uncertainties are insignificant. The uncertainties are due to the tosser's lack of detailed information and inability to carry out the calculations needed to predict the outcome.

25. We can predict the overall pattern of electron impacts on the screen. If a large number of electrons go through the experimental apparatus, we can predict what fraction of them will hit in each portion of the screen.

27. No, this would not be correct. Both electrons come through both slits; electrons coming through slit A or through slit B are inconsistent with the interference pattern.

Problems

1. Visible. Energy is proportional to frequency, so the ratio of the energies is the same as the ratio of the frequencies. Microwave frequencies are around 10^{11} Hz, while visible frequencies are around 10^{15} Hz (Figure 9.7). The ratio of these is 10^4, or 10,000. So a visible photon has about 10,000 times more energy than does a microwave photon.

3. $E = hf = (6.6 \times 10^{-34} \text{ J-s}) \times (10^{24} \text{ Hz}) = 6.6 \times 10^{-10}$ joules.

5. Visible light has a frequency of around 10^{15} Hz, so the energy of the 10 photons is about $10hf = 10 \times (6.6 \times 10^{-34} \text{ J-s}) \times (10^{15} \text{ Hz}) = 6.6 \times 10^{-18}$ joules.

7. This halves its wavelength.

9. The electron has the smaller mass, so it has longer wavelength. About 1800 times longer.

11. $s = 0.1c = 3 \times 10^7$ m/s. $\lambda = h/ms = (6.6 \times 10^{-34} \text{ J-s}) / (9.1 \times 10^{-31} \text{ kg} \times 3 \times 10^7 \text{ m/s}) = 2.4 \times 10^{-11}$ m.

13. $\lambda = h/ms = (6.6 \times 10^{-34} \text{ J-s}) / (38 \times 10^{-27} \text{ kg} \times 2 \text{ m/s}) = 8.7 \times 10^{-9}$ m $= 8.7 \times 10^{-6}$ mm (a few millionths of a mm).

CHAPTER 14

Exercises

1. Your radio detects radiation of specific frequencies (namely, the frequency of your radio dial). So you can use a radio to detect the radio-range frequencies that are present. Some of the differences from the spectroscope described in the text: The radio waves do not come from a glowing gas, no thin slit is used, no prism is used.

3. Yes (except for a possible difference in intensities). The sodium and mercury spectra would differ.

5. The wavelength could not be 2.1 m or 1.9 m, because a whole number of these wavelengths cannot be fit onto the string. But a wavelength of 0.5 m would fit; 4 of these wavelengths would just fit onto the string, so this is a possible standing wavelength.

7. Yes. An excited atom has more energy, so it has more mass.

9. 4 to 2. Longest wavelength is 4 to 3.

11. The 3 lines graphed come from the following 3 quantum jumps: N = 3 to 2, 4 to 2, and 5 to 2; they are the first three lines counting from the left in Figure 14.2. The other line (furthest to the right) comes from the N = 6 to 2 quantum jump.

13. Quantum uncertainties would be very large, and observable on the macroscopic level. Newtonian physics, including predictability, would not be even approximately correct.

15. Public opinion polls that alter public opinion are one example. Here are a few of the many other examples: A thermometer slightly alters the temperature of the object it is measuring, your own observation of your psychological responses can alter those responses, observation of another person might cause that person to become tense and thus alter that person's performance, the presence of a crowd at a sporting event can affect the outcome of the event. Observation of the moon is not an example.

17. It would collapse (or quantum-jump) into a much more spread-out wave, with a larger Δv and a smaller Δx.

19. Before. When the electron hits the screen, the spread out psi-wave collapses to a small point.

21. These correlations have nothing to do with quantum effects.

23. Non-Newtonian: Radioactive decay, the quantum states of an atom, the wave nature of electrons, the particle nature of radiation, quantum jumps, the line spectra of atoms. Newtonian: The path of a baseball, the forces on macroscopic objects, the motions of the planets, the recoil of a rifle, the swinging of a pendulum—macroscopic motions and macroscopic forces in general.

Problems

1. Solve $hf = E_2 - E_1$ (Section 14.2) for f: $f = (E_2 - E_1) / h = 16 \times 10^{-19}$ J / 6.6×10^{-34} J-s $= 2.4 \times 10^{15}$ Hz. This is a slightly higher frequency than visible frequencies, so it is in the UV region.

3. The energy difference $E_3 - E_1$ will just be the sum of the two preceding energy differences: 16×10^{-19} J + 3 × 10^{-19} J = 19×10^{-19} J. The frequency is $f = (E_3 - E_1) / h = 19 \times 10^{-19}$ J / 6.6×10^{-34} J-s $= 2.9 \times 10^{15}$ Hz. This is in the UV region.

5. Following the procedure of Problem 4, diameter = 10×10^{-7} m = 10^{-6} m = 0.001 mm. The giant atom's volume is larger by a factor of $10,000^3 = 10^{12}$, or 1 trillion. This is the number of ground-state hydrogen atoms that could fit inside the giant atom.

7. From $\Delta x \, \Delta s = h/m$, we get $\Delta x = h/m \, \Delta s = 6.6 \times 10^{-34}$ J-s / 9.1×10^{-31} kg × 1 m/s = 0.72×10^{-3} m = 0.72 mm.

9. $(\Delta x) \cdot (\Delta s) \approx h/m$, where $\Delta x \approx 10^{-10}$ meters and m ≈ 10^{-30} kg. So
$\Delta s \approx (6.6 \times 10^{-34}) / (10^{-30}) (10^{-10}) = 6.6 \times 10^6$ m/s.
As explained in the hint, this is also the approximate speed of the electron. The ratio of this speed to light-speed is $6.6 \times 10^6 / 3 \times 10^8 \approx 0.02$. So the electron moves at about 2% of lightspeed.

CHAPTER 15

Exercises

1. The strong force. If the electric force was stronger, the nucleus would fly apart.

3. $^{13}_{6}$C: 6 protons, 7 neutrons
$^{56}_{26}$Fe: 26 protons, 30 neutrons

5. Masses are in the ratio of 1 to 2 to 3. They all have the same amount of electric charge (the charge on 1 proton).

7. The fast-moving alpha and beta particles emitted during decay collide with, and thus warm, the atoms of the radioactive substance.

9. The light comes from atoms that were excited by the fast-moving alphas, betas, and gammas that came from radioactive decay.

11. No, a hydrogen atom doesn't contain enough nuclear particles to emit an alpha particle.

13. $^3_2$He, $^{218}_{84}$Po

15. 5 years is 5 half-lives, so 1/32 (about 0.03, or 3%) of it will remain.

17. The radioactive elements lighter than uranium are continually created by the decay of uranium and decay of the various particles that result from the decay of uranium.

19. The half-life of U is 0.7 billion years. So 4.5 billion years is 4.5 / 0.7 = 6.4 half-lives, or about 6 or 7 half-lives. So the amount of ^{235}U has dropped to between 1/64 and 1/128 of its original value. So there was between 64 and 128 (say around 100) times as much ^{235}U here on the newly-formed Earth as there is today.

21. Radon's half-life is 4 days. So 12 days is 3 half-lives. The amount remaining is 100 atoms × 1/8 = 12.5 atoms, in other words about 12 or 13 atoms. This is not a precise prediction, because of statistical uncertainties when dealing with such a small number of atoms.

23. Most stable: ^{238}U (longest half-life). Least stable: ^{214}Po (shortest half-life).

25. We cannot carbon-date rocks, because they contain no formerly-living material.

27. Yes, it would affect our estimate. It would cause us to estimate too young an age, because it would have caused a larger amount of ^{14}C to be deposited in the past.

29. This will cause future estimated ages to be younger than they really are. This will not affect the apparent age of an ax handle made in 1940 (before nuclear testing), but it will affect the apparent age of an ax handle made in 1990.

31. According to Table 15.4, the risk from the 4 cigarettes is about 3 times larger than the risk from the wine.

33. 30 / 3 = 10 times safer (that is, the motorcycle is 10 times more risky).

Problems

1. ^{131}I: 1/8 gram
^{222}Rn: 1/64 gram
^{234}Th: 1/2 gram

3. (1/16) × 100 = 6.25, so there will be approximately 6 or 7 left. This answer is approximate because of the uncertainties involved in radioactive decay.

5. According to Figure 15.10, the radioactivity drops to 20% of its original value after about 2.25 half-lives. Table 15.1 says that the half-life of C^{14} is about 6000 years, so the age of the ax handle is about 6000 × 2.25 = 13,500 years.

7. The halflife of ^{238}U is about the same as Earth's age (Tables 15.1 and 15.2), so about half of the original ^{238}U still remains.

9. ^{14}C halflife = 6000 years, so 50,000 years is a little more than 8 halflives, so the fraction remaining is only 1/256, or 0.4%. This dating procedure would be difficult to carry out, because the amount of ^{14}C-radioactivity would be so small that it would be hard to measure. In fact, 50,000 years is near the upper limit of ages that are measurable by ^{14}C radioactivity.

11. The risk has gone down, because the number of deaths has increased by less than 300% (less than 3 times) while the number of people has increased by 400%. More precisely, $315/(750 \times 10^6) = 0.44 \times 10^{-6}$, or 0.44 parts in a million risk in A.D. 1800.
$845/(3 \times 10^9) = 0.28 \times 10^{-6}$, or 0.28 parts in a million risk in the mid-twentieth century.

13. The number of days is 40 × 365 = 14,600 days, so the risk is $14,600 \times 10^{-6} = 0.014 = 1.4\%$. If you smoke 20 per day,

your risk is multiplied by 20/1.4, or about 14. So the over-all risk is now 20 × 1.4% = 28%. Note: We are assuming here that the risks simply add up, i.e. that previous smoking does not affect the risks involved in present smoking (so that the risk remains 10^{-6} for each 1.4 cigarettes throughout the smoker's life). This assumption is probably not quite correct, because the smoking itself must affect the health (and thus the risks) of the smoker.

CHAPTER 16

Exercises
1. The nucleus's mass is smaller than the sum of the masses of its protons and neutrons. This is true for any nucleus, as you can see from the nuclear energy curve.
3. Fission (note that this is not radioactive decay, because ^{16}O is not radioactive). $^{16}_{8}O ==> {}^{4}_{2}He + {}^{12}_{6}C$. According to the nuclear energy curve, the right side represents the most nuclear energy.
5. Use the nuclear energy curve. Carbon: fusion. Gold: fission. Iron: neither.
7. Nuclear reactor: measurable decrease (about 1%) in rest mass. TNT explosion: there is a decrease in rest mass, but it is so small that it cannot be measured. Fission bomb explosion: measurable decrease in rest mass. Throwing a baseball: Increase in rest mass, but it is so small that it cannot be measured.
9. It opposes nuclear fusion.
11. The sun's energy comes from nuclear fusion.
13. $^{24}_{12}Mg$. According to the nuclear energy curve, this released nuclear energy.
15. Gamma radiation is created in the center of the sun. The gamma radiation is all absorbed by the time this energy gets to the surface.
17. Strontium's atomic number is 38, and plutonium's is 94. 94 − 38 = 56, which is barium.
19. Similarities: Both convert microscopic forms of energy (chemical and nuclear) into thermal energy; rest-mass is lost in both cases; using the right materials, both can be made self-sustaining. Differences: Burning is a chemical process involving the orbital electrons, fission is a nuclear process; fission converts (or "releases") far more energy per reaction; fission releases neutrons while burning does not; fission is sustained by neutrons while burning is sustained simply by the thermal energy created.
21. The fission of ^{235}U emits enough neutrons to sustain the reaction, while the fission of ^{238}U does not.
23. A sphere, because it has a smaller surface area than does any other solid shape having the same volume (same amount of material). That is, the surface area of the sphere will be smaller than the surface area of the cylinder. So fewer neutrons will escape out through the surface.
25. Yes. All elements having atomic numbers greater than 83 are radioactive.
27. No. Neutrons are required to sustain the chain reaction.

Problems
1. 1% of 10 kg is 0.1 kg, or 100 grams. E = mc² = (0.1 kg) × $(3 \times 10^8 \text{ m/s})^2 = 9 \times 10^{15}$ joules.

3. The same amount of rest-mass would still have to vanish, because the same amount of electrical energy would still have to be created.
5. The weight of the U.S. population is 250 × 10^6 × 600 N = 1.5 × 10^{11} N. The energy available for lifting is 15 × 4.2 × 10^{12} J = 6.3 × 10^{13} J. Thus, from Grav E = weight × height (Chapter 6), we get height = Grav E / weight = 6.3 × 10^{13} J / 1.5 × 10^{11} N = 420 m, or 0.42 km. A 15 megaton bomb could do 1000 times as much lifting: 420,000 m, or 420 km!

CHAPTER 17

Exercises
1. Wood, trash, wind, hydroelectric.
3. Uranium, plutonium, and hydrogen for nuclear power. Geothermal energy. Also, conservation does not come from the sun—it comes from human ingenuity! (However, one might argue that the human energy comes from food which comes ultimately from the sun.)
5. For coal: 8 × 10^{18} J, 15 × 10^{18} J, 12 × 10^{18} J, 10 × 10^{18} J, 18 × 10^{18} J.
   For oil: 0, 2 × 10^{18} J, 7 × 10^{18} J, 22 × 10^{18} J, 38 × 10^{18} J.
7. From Figure 17.2: Wood in 1860, wood and coal were about even in 1880, coal in 1900, coal in 1920.
9. Thermal input: in the core of the reactor. Work output: at the turbine. Thermal output: At the condenser.
11. Nuclear power plants do not have a stack, because they do not operate by combustion of fuel.
13. We do not want the radioactive water that goes through the reactor to move outside of the containment dome. If there were only one loop, either radioactive water would have to move outside the dome to reach the turbine, or the turbine and condenser would have to be moved inside the containment dome (which would pose great dangers for the workers who must service the turbine and condenser).
15. No. Even if neutrons hit a rich uranium deposit, there is insufficient ^{235}U (as compared with ^{238}U) in natural uranium to sustain a chain reaction.
17. Very large coal supply in U.S., no proliferation concerns, few terrorism concerns, less expensive to build plant, a large accident is less likely.
19. Gas is a more efficient fuel—more joules of energy can be gotten per ton of carbon.
21. A breeder reactor makes Pu fuel out of the ^{238}U, while the material consumed for energy is ^{235}U. A reactor that consumes more ^{238}U than ^{235}U will create more fuel than it consumes. Two separate reactions occur here: Fission of ^{235}U, and creation of Pu from ^{238}U. Both reactions obey the law of conservation of energy.
23. But creating plutonium in nuclear power reactors, by processing plutonium (separating plutonium from the other materials in the used fuel rods), and by training nuclear scientists and technologists.
25. Hydroelectric: large land areas flooded, rivers disrupted, possible dam failures (Johnstown flood of 1889 killed 2200). Biomass: can contribute to global warming, crops could instead be used for food, energy and chemicals used to create crops. Geothermal: hot water and steam

are a finite resource, hot rock geothermal is speculative and in the distant future at best. Wind: land use, materials use, wind is an intermittent (not continuous) source. Solar-thermal, photovoltaic cells, and active solar heating: land use, materials use, the sun is an intermittent source. Passive solar: The sun is an intermittent source.

27. Yes. Examples: electric trains that get energy from hydroelectric plants, electric cars and other small electric vehicles, sailboats, canoes and barges and other boats that float downstream, methanol- and ethanol-powered vehicles.

29. Geothermal: thermal to electric. Wind: kinetic to electric. Solar-thermal: radiant to thermal to electric. Photovoltaic: radiant to electric.

31. Wind: 1985. Solar thermal: 1992. Photovoltaic: 1996.

33. Yes. From Figure 17.19, Japan and Britain use less than half the U.S. energy use per person.

Problems

1. Approximate percentages for coal: 80%, 70%, 50%, 20%, 20%.

3. Energy use roughly doubled during 1880–1900, and doubled again during 1900–1920, so the doubling time was 20 years. From Section 7.8, $P = 70/T = 70/20 = 3.5$, so the annual percentage increase was 3.5%.

5. $30/85 = 35\%$.

7. Approximately: 3 exajoules (EJ), 10 EJ, 25 EJ, 40 EJ, 40 EJ. Approximate percentages: 4%, 10%, 25%, 30%, 28%.

9. It is the same for both plants. Efficiency = useful output / total input, from which total input = useful output / efficiency. So, for 1J of electrical energy, the total energy input = 1J / 0.4 = 2.5J. Thus the exhausted thermal energy (i.e. the thermal pollution) is 2.5J – 1J = 1.5J.

11. PV: 2006. Solar thermal: Sometime after 2010. If the price of coal were 8¢/kWh, then PV becomes (or rather, would have become!) competitive in 1998 while solar thermal would have become competitive in 1996.

13. To provide 400×10^9 watts at 20% efficiency, 2000×10^9 watts (2×10^{12} watts) must be collected. This amount of solar power falls on an area of 2×10^{12} watts / (200 watts/m^2) $= 10^{10}$ m^2. But 1 km^2 = 1000m × 1000m $= 10^6$ m^2, so the land required is 10^4 km^2, or 10,000 km^2. Complications: The sun is an intermittent source, so an energy-storage method would have to be found. Energy-storage would make the process less efficient. It would be difficult to distribute the energy from Arizona to the rest of the country; solar power plants would need to be located in many less sunny regions in order to solve the distribution problem. There would be land-use problems and materials problems (supply, disposal, resource use, pollution).

CHAPTER 18

Exercises

1. Both the upper pair of tracks and the lower pair of tracks curve in opposite directions, indicating the forces on the pair of charged particles are in opposite directions, so the particles must have opposite charges.

3. Photons are not charged, and only charged particles make tracks in bubble-chambers.

5. The portion of the track below the barrier would have been straighter than the portion above the barrier.

7. The charged particles: electron, muon, tau, W$^+$ and W$^-$. These are also the particles that can exchange photons, because microscopically "exchanging photons" is what we *mean* by "feeling the electric force."

9. Similarities: the W and Z are exchange particles; the Z is uncharged. Differences: the W and Z have rest-mass; the W is charged; the W and Z move at less than lightspeed.

11. Similarities: quarks are material particles (they have rest-mass); quarks move at less than lightspeed; quarks are charged (i.e. they feel the electric force). Differences: Quarks feel the strong force while electrons do not; quarks are fractionally charged (+2/3 or –1/3 of proton charge).

13. It comes from the energy in the strong force fields that bind the quarks together.

15. According to the theory of cosmic inflation, the big bang was a microscopic event, and would have exhibited extreme quantum effects such as quantum uncertainties. These uncertainties would surely be nature's most important quantum uncertainties.

17. No. If the proton were unstable, we would not be here.

Problems

1. From Chapter 13, E = hf =$(6.6 \times 10^{-34}$ J-s) × $(10^{20}$ Hz) = 6.6×10^{-14} J.

3. Each photon's energy is mc^2, where m represents the mass of either a proton or antiproton (they have the same mass). Thus E = mc^2 = $(1.7 \times 10^{-27}$ kg) × $(3 \times 10^8$ m/s)^2 = 1.5×10^{-10} J. f = E/h = 1.5×10^{-10} J/6.6×10^{-34} J-s = 2.3×10^{23} Hz.

5. 2 gm, or 0.002 kg, of rest-mass is annihilated. The energy of this mass is E = mc^2 = $(0.002) \times (3 \times 10^8)^2$ = 1.8×10^{14} J. Assume that the average weight of a U.S. citizen is 600 N (about 130 pounds). Then the weight of the U.S. population is $270 \times 10^6 \times 600$N = 1.6×10^{11} N. Since GravE = weight × height: height = energy / weight = 1.8×10^{14} J / 1.6×10^{11} N = 1.1×10^3 m, or about one kilometer!

Glossary

A-bomb See *fission bomb.*

acceleration An accelerated object is one whose velocity is changing. Quantitatively, the acceleration is the change in velocity during a time-interval, divided by the duration of that time-interval. It can be measured in (km/hr)/s, or in (m/s)/s = m/s^2.

acceleration due to gravity The acceleration of any freely falling object. On Earth it is about 10 m/s^2 or more precisely 9.8 m/s^2.

action-reaction cycle Mutually reinforcing cycle of increased armaments in arms races involving two or more hostile nations.

active solar energy A renewable energy resource in which a solar-heated liquid or gas is pumped elsewhere for space or water heating.

ages See *geological ages.*

air A mixture of several chemical compounds: N_2 (about 80%), O_2 (about 20%), argon (about 1%), and a smattering of trace gases.

air resistance The resistance to the motion of an object through the air due to the object's collisions with numerous air molecules.

alpha decay See *radioactive decay.*

alpha particle A helium nucleus (two protons attached to two neutrons) that breaks off of a larger nucleus during radioactive decay. See *radioactive decay.*

alpha radiation Steams of alpha particles that move outward from a radioactive source. See also *alpha particle.*

AM radio waves See *electronic waves* and *electromagnetic spectrum.*

amplitude The half-width of vibration of each part of a wave's medium.

analysis Breaking a question into parts.

annihilation See *matter-antimatter annihilation.*

anthropic principle The idea that our universe must be organized in the way that it is because any other organization would not allow intelligent beings to be here to ask the question in the first place.

antimatter Matter that is made of *antiprotons, antineutrons,* and *positrons.* These three types of particles are the antiparticles of protons, neutrons, and electrons, respectively. Today's universe consists overwhelmingly of matter, not antimatter.

antiparticle Special relativity predicts that, for every existing type of particle, there is an antiparticle carrying the opposite charge. Quantum uncertainties allow the creation and annihilation of particle-antiparticle pairs.

Aristarchus's theory of the universe Greek sun-centered theory of the layout of the universe. Rejected because it seemed to conflict with everyday observations.

Aristotelian elements Earth, water, air, fire, and ether.

Aristotelian physics Aristotle's theory of physics. This theory was accepted from the time of ancient Greece until about the 17th century. According to Aristotelian physics, the fundamental elements of the universe are earth, water, air, fire, and ether; falling is one form of natural motion, and horizontal motion is always unnatural or violent. See also *Aristotelian physics, difficulties with.*

Aristotelian physics, difficulties with Contrary to Aristotelian predictions, heavy and light objects often fall at the same speed, and horizontally-moving objects would move forever if there were no external forces.

artificial radiation Radiation from human-made sources. The main artificial source is medical and dental x-rays and other medical technology.

astrology The belief that events on Earth are influenced by the positions and motions of the planets. Scientists overwhelmingly reject astrology as pseudoscientific and false.

astronomy The scientific study of the stars and other objects in space.

atom The smallest particle of an element. Also see *atom, models of.*

atom, models of The *Greek atom* is a tiny indestructible object, like a small and rigid pea. The *planetary model of the atom* is made of parts, including a tiny central nucleus containing protons and neutrons, and one or more electrons orbiting far outside the nucleus. The *quantum theory of the atom* is based on the post-Newtonian quantum theory.

atomic bomb See *fission bomb.*

atomic materialism See *materialism.*

atomic number The number of protons in an atom, also the number of electrons in a neutral atom. An atom's atomic number determines its chemical properties and thus the *element* to which it belongs. The atomic number of a chemical element is the number of protons in the nucleus of an atom of that element.

atomic theory of matter All matter is made of tiny particles, too small to be seen.

atomism The notion that nature can be reduced to the motion of tiny material particles. See also *materialism.*

average speed An object's average speed, during a time-interval, is the distance traveled during that time-interval, divided by the duration of that time-interval. s = d/t.

Bell's interconnectedness principle Entangled particles cooperate in a way that can only be explained by the existence of real non-local (acting instantaneously across a distance) connections between them. Entanglement of this sort is predicted by quantum theory and has been confirmed by experiments.

beta decay See *radioactive decay.*

beta particle An electron that, during radioactive decay, is created in the nucleus and then emitted from the nucleus. See *radioactive decay.*

beta radiation Streams of beta particles that move outward from a radioactive source. See also *beta particle.*

big bang A single event some 12 billion years ago that gave birth to our universe. The big bang created time, space, energy, and matter, and sent them hurtling outward. The universe has been expanding ever since. The consensus today is that this event was initially microscopic, and dominated by quantum gravity. See also *cosmic inflation*.

biological effects of radiation The main types of damage caused by ionizing radiation are *radiation sickness* (damage to the red blood-forming cells of the bone marrow) which can appear quickly following exposure, and two longer-term effects: *mutation* (an inheritable alteration of the genetic material in a sperm or egg cell) and *cancer*.

biologically convergent property A biological feature that, because of its survival value, tends to occur again and again during biological evolution.

biology and the second law The entropy of a growing plant decreases, at the expense of a far greater entropy increase in the absorbed and re-radiated solar energy that passes through the plant. A similar situation exists for all biological processes, including the evolution of species. Growth and biological evolution are organized by solar radiation and are in full accordance with the second law.

biomass energy A renewable energy resource. The chemical energy of organic substances.

birth of stars See *star birth*.

black hole Any object that has gravitationally collapsed into a single point. Nothing can escape from its vicinity. When very massive stars (greater than 30 solar masses before collapse) run out of fuel, they explode and the remaining remnant collapses to become a black hole. Giant black holes having millions of solar masses are thought to exist at the centers of many galaxies.

breeder reactor A reactor that creates more than one ^{239}Pu nucleus (from ^{238}U) for each ^{235}U nucleus it fissions, and so creates more fuel than it uses.

bubble chamber See *cloud chamber*.

burning or *combustion* A chemical reaction that creates heat by combining oxygen from air with a fuel such as carbon or hydrogen. See also *chemical reaction*.

c The symbol customarily used for lightspeed, 300,000 km/s.

cancer See *biological effects of radiation*.

carbon dating See *radioactive dating*.

CFC See *chlorofluorocarbons*.

chain reaction A series of neutron-induced fission reactions that proceeds from one nucleus to the next by means of the neutrons that are released during each fission reaction.

charge See *electrically charged object*.

chemical compound A pure substance that can be chemically decomposed. Its smallest particle is a *molecule*, two or more atoms connected into a single unit. A compound's molecules are all identical.

chemical decomposition Any process that changes a single substance into two or more other substances.

chemical element A substance that cannot be chemically decomposed. 112 are known. They are tabulated in the *periodic table*, which numbers them with their *atomic number*. An *atom* is the smallest particle of an element.

chemical energy One of the physical forms of energy; energy due to molecular structure. See also *energy*.

chemical origin of life The hypothesis that life originated on Earth by chemical processes, beginning with the action of various energy sources on the early Earth atmosphere to create the organic chemicals that are the building blocks of life.

chemical reaction A rearrangement of atoms into new molecular forms.

chemistry The study of the properties and transformations of substances (chemical compounds). See *chemical compound*.

Chernobyl Site of history's worst nuclear power accident, in 1986. Tens were killed from short-term effects, and thousands are expected to die of long-term cancers. The fuel melted down. The reactor suffered a "slow nuclear explosion," comparable to a chemical bomb. A large amount of radioactivity escaped.

chlorofluorocarbons Also called *CFCs*. Chemicals whose molecules are made of chlorine, fluorine, and carbon. Manufactured as coolants, spray propellants, foaming agents, and solvents. They destroy stratospheric ozone.

closed universe See *geometry of the universe*.

cloud chamber A device that shows the path of a charged particle as a trail of droplets in water vapor. Its successor, the *bubble chamber*, shows a particle's path as a trail of bubbles in a liquid.

cogeneration The simultaneous production of both electricity and useful thermal energy.

combustion See *burning*.

compound See *chemical compound*.

condenser See *steam-electric power plant*.

conservation of energy See *law of conservation of energy*.

conservation of matter The idea that the total amount of matter involved in any chemical reaction, or in any other physical process, is the same after the reaction as it was before. This is also called conservation of mass. Although it is a useful theory, especially in chemistry, experiments have proved that it is only a useful approximation in chemical reactions, and entirely wrong in many other situations.

constructive interference See *wave interference*.

continuous spectrum A spectrum that extends over a continuous range of values. See *spectrum*.

cooling tower See *steam-electric power plant*.

Copenhagen interpretation The consensus interpretation of quantum theory. Uncertainties, complementarity, and nonlocality are inherent in nature; observations help create the observed properties; microscopic reality is *contextual*, that is, it is created by the full experimental environment of the phenomenon.

Copernican revolution Rejection of the idea that humankind is central to the universe.

Copernican viewpoint The view that Earth is not a unique place in the universe, that the same principles of nature apply everywhere. Since intelligent life occurred here, this view argues that it should have occurred elsewhere.

Copernicus's theory Sun-centered theory of the layout of the universe, similar to Aristarchus's. The planets, including Earth, circle the sun, and Earth spins on its axis. Accepted today, with many modifications. See also *Kepler's theory of the universe*.

cosmic inflation A quantum theory of how and why the big bang occurred. It states that the big bang originated as a quantum energy fluctuation in the smallest existing volume of space. At the resulting high temperature, a quantum field called the *Higgs field* caused the universe to very rapidly expand or "inflate" while cooling. The cooling created our four known forces and the most basic properties of our universe. *Higgs particles*, the quanta of the Higgs field, are currently sought in high-energy accelerators. See also *big bang.*

cosmic rays High-energy particles that travel through outer space and occasionally impact Earth.

creationism The belief that the Bible's Old Testament can be read literally, as scientific and historical truth, and that Earth and the biological organisms, including humans, were created separately, just a few thousand years ago. Scientists overwhelmingly reject creationism as pseudoscientific and false.

critical mass The minimum amount of fissionable material that will sustain a chain reaction.

curved space According to Einstein's general theory of relativity, gravity bends lightbeams. It follows that gravity bends space-time. In Einstein's theory, gravity is this curvature of space-time caused by masses. See *four-dimensional space-time.*

dark matter Astronomical material that is not visible. Observation of gravitational effects in our own and other galaxies shows that much of the universe's mass is invisible. Dark matter has been discovered in the outer halo of our galaxy in the form of non-glowing star-like or planet-like objects called *MACHOS* (massive compact halo objects). Other forms of dark matter might not be made of the ordinary matter (protons, neutrons, and electrons) that we are familiar with on Earth. Candidates for such *exotic dark matter* include neutrinos with mass, and entirely new forms of matter. Widely accepted principles of the big bang and nuclear physics imply that the ordinary matter in the universe can constitute only about 10% of the mass that would be needed to cause the universe to be either flat or closed. Thus, if cosmic inflation's prediction of a flat universe is correct, some 90% of the universe must be made of exotic dark matter.

daughter nucleus The nucleus that remains after a radioactive decay has occurred.

decay curve A graph of the amount of a radioactive material remaining, versus time.

destructive interference See *wave interference.*

diffusion The spreading out of a collection of particles due to their random jostling by each other or by other neighboring particles.

double-slit experiment When photons or material particles such as electrons are sent through two narrow slits, the distribution of many impacts on a screen forms a wave-interference pattern. This shows that each individual particle also behaves like an extended wave.

double-slit experiment with electrons See *double-slit experiment.*

double-slit experiment with light The interference of light coming from two synchronized sources is observed on a screen, demonstrating that light is a wave. See also *wave interference.*

doubling time See *exponential growth.*

drive force The force, exerted on a self-propelled (self-energized) object, that accelerates it or keeps it moving against resistive forces. Most drive forces are due to friction, but unlike sliding friction they act in the forward direction.

dualism Descartes' idea that there are two realities, physical and spiritual. In the physical realm, the real or *primary qualities* are objective, impersonal phenomena such as the motion of atoms. Human sense impressions are considered to be *secondary qualities*, caused by the primary qualities.

E = mc² See *mass-energy equivalence.*

efficiency Useful output divided by total input. See *energy efficiency,* or *transportation efficiency.*

Einstein's relativity Einstein's two theories of time and space. See *special theory of relativity* and *general theory of relativity.*

elastic energy One of the physical forms of energy; energy due to the ability of a deformed system to snap back. See also *energy.*

electric current A flow or motion of electrically charged particles. Electric currents in wires are due to electrons moving along the wire.

electric discharge A flow of electric current through a gas. Sparks are electric discharges.

electric field See *electromagnetic field.*

electric force See *electric force law.*

electric force law Electrically charged objects exert forces on each other at a distance. Like charges repel each other, and unlike charges attract each other.

electric generator See *steam-electric power plant.*

electrically charged object Any object that can exert or feel the electric force. There are two types of charge, *positive* and *negative.* Like charges repel each other, and unlike charges attract each other.

electromagnetic energy One of the physical forms of energy; energy due to electromagnetic forces. See also *energy.*

electromagnetic field Exists wherever an object could feel an electromagnetic force. Every charged object creates an electromagnetic field throughout the space around it. A stationary charged object creates only an *electric field*, while a moving charged object creates in addition a *magnetic field*. These two kinds of fields, taken together, are called an electromagnetic field. See also *force field.*

electromagnetic force The total (electric and magnetic) force between charges. See also *electric force law* and *magnetic force law.*

electromagnetic radiation Electromagnetic waves, or photons, that move outward ("radially") from a region of vibrating charged objects. See also *electromagnetic wave* and *photon.*

electromagnetic spectrum The complete range of electromagnetic radiations or electromagnetic waves. Divided into the following six regions, listed from longest to shortest wavelength (lowest to highest frequency): *electronic or radio waves, infrared, visible light, ultraviolet, X-rays, gamma rays.* See also the listing for each of these regions.

electromagnetic wave A disturbance (or wave) in an electromagnetic field, that moves through the field at 300,000 km/s (lightspeed). Every vibrating charged object creates such an electromagnetic wave. Also called *electromagnetic radiation.* There are six major types: *electronic or radio waves, infrared, visible light, ultraviolet, X-rays, gamma rays* (see these entries also).

electromagnetic wave theory of light Every vibrating charged object creates a disturbance in its own electromagnetic field, that spreads outward through the field at 300,000 km/s. Light is just such an *electromagnetic wave.*

electron A small negatively-charged sub-atomic particle, about 2000 times lower in mass than a proton. One or more electrons orbit the nucleus of each atom. Most electrical currents are due to the flow of electrons, often within wires. See also *planetary model of the atom.*

electron interference experiment See *double-slit experiment.*

electron microscope Uses electron probability waves to form images of microscopic objects.

electron wave interference The wave interference patterns that are observed in experiments involving electrons. Such patterns occur for example when electrons pass through a thin metal foil, and when electrons pass through a pair of thin slits.

electronic waves, or *radio waves.* Electromagnetic radiation that is created by humans in such forms as AM and FM radio, TV, radar, and microwave.

electroweak force The unified electromagnetic and weak forces, which constitute two of the four fundamental forces. Electrons and neutrinos experience only the electroweak force. The *exchange particles* for this force are the *photon, W+, W−, and Z.* Ws are charged, and the photon and Z are uncharged. There are three *generations* of electroweak particles: the electron and its neutrino, the muon and its neutrino, and the tau and its neutrino.

element See *chemical element.*

emission of radiation This occurs when an excited atom quantum-jumps into a lower energy level, emitting a photon. Since the photon's frequency is determined by the atom's energy levels, an atom's spectrum can be predicted from Schroedinger's equation.

energy The ability to do work. Quantitatively, the amount of energy in any system is the amount of work the system can do. Units: joule (J), Calorie, kilowatt-hour (kW·h).

energy conservation Measures to reduce energy consumption, including both efficiency measures to save energy while providing the same services, and changes in the quality of services.

energy efficiency The energy efficiency of any process is the useful energy produced by the process, divided by the total energy input to the process.

energy flow diagram A diagram showing the energy transformations occurring during some process. Often, the various forms of energy are pictured as though energy were a fluid flowing through pipes.

energy fluctuation The random energy variations that occur at every point within a quantum field. See also *vacuum.*

energy level The precise, predictable energy an atom has when it is in a particular quantum state. See *quantum state.* An *energy-level diagram* shows the collection of possible energy levels for an atom.

energy level diagram A diagram showing the collection of energy levels for an atom. See *energy level.*

energy quality Thermal energy is considered to be of lower quality than other energy forms, because of the second law's restrictions on converting it to other forms of energy.

energy resource A natural resource containing useful energy. The major U.S. resources today are: fossil (coal, oil, natural gas), nuclear (uranium), hydroelectric, and wood. A resource

is *renewable* if it is continuously replaced or replaced within a human lifetime; otherwise it is *nonrenewable.*

enrichment See *isotope separation.*

entanglement Two particles are entangled when they interact and then separate in such a way that they form a single inseparable psi field.

entropy The amount of microscopic disorganization a system possesses. See also *second law of thermodynamics, law of increasing entropy,* and *microscopic interpretation of the second law.*

ether The gas-like material substance that nineteenth-century scientists believed filled the universe. Ether's mechanical motions were thought to be responsible for light waves and for the transmission of forces such as the electromagnetic force that act at a distance. Einstein's work led to the rejection of the ether theory, and the acceptance of the electromagnetic field as a nonmaterial, but nonetheless physically real, medium.

ether theory The idea that a gas-like, non-atomic, material medium, the ether, fills the entire universe, and that light-waves are waves traveling through this medium. This theory was rejected after 1905 because it contradicted Einstein's theory.

excited states Any quantum state of an atom whose energy is higher than the atom's lowest possible energy; that is, any quantum state other than the ground state. A gas can be excited by heating, and by *electric discharge,* in other words an electric current flowing through it.

exhaust See *heat engine.*

exhaust temperature See *heat engine.*

exotic dark matter See *dark matter.*

expansion of the universe The general theory of relativity predicts that the universe's three-dimensional space must either expand or contract. Observations show that expansion is, in fact, occurring, as evidenced by the fact that distant galaxies are all moving away from each other.

experiment See *observation.*

exponential growth Growth by a fixed percentage in each unit of time. It has a fixed *doubling time,* related to its fixed percentage *growth rate* by $T \approx 70/P$.

external combustion engine See *steam-electric power plant.*

feedback effect Any effect that further influences the phenomenon that caused it. *Negative feedback* diminishes its cause, while *positive feedback* enhances its cause.

field A physical entity that is spread throughout a region of space. A *force field* exists wherever an object would feel a force.

field quanta Discrete bundles of energy, predicted by quantum theory to be associated with any quantized field. Examples: photons are quanta of the electromagnetic field, and electrons are quanta of the electron field (the psi field for electrons).

field view of reality The view that physical reality is a set of fields in motion, and that material objects arise from the energy of these fields.

first law of thermodynamics An alternative way (not stated or used in this book) of stating the law of conservation of energy. See *law of conservation of energy.*

fission See *nuclear fission.*

fission bomb Also called *atomic bomb* or *A-bomb*. A bomb that gets its energy from a fission chain reaction in ^{235}U or Pu. In the ^{235}U bomb, the design can be as simple as bringing two sub-critical masses together to form a critical mass. If the bomb contains Pu, a subcritical mass is made critical by squeezing it to high density. See also *critical mass* and *nuclear fission*.

fission fragment One of the two pieces that results from the fissioning of a nucleus.

flat universe See *geometry of the universe*.

fluorescent bulb A type of electric light bulb that is filled with a gas such as mercury vapor. An electric current flows through the gas, exciting the gas and causing it to emit ultraviolet radiation, which is absorbed by the phosphor coating, which emits light. New *high-frequency ballasts* that cause the current to oscillate at a much higher-than-normal frequency have recently been developed for increased efficiency.

FM radio waves See *electronic waves* and *electromagnetic spectrum*.

force A body exerts a force on another body whenever the first body causes the second body to accelerate. A force is an action by one body on another; it is not a thing or a property of a body. Every force is similar to a push or a pull.

force diagram A diagram that shows all of the individual forces acting on an object. Each force is shown as an arrow pointing in the direction of that force.

force field The effect that the source of a force has on the surrounding space. A force field exists wherever an object would feel a force. See also these specific examples: *gravitational field* and *electromagnetic field*.

force of gravity See *gravitational force*.

force pair The two forces that two bodies exert on each other, as predicted by the law of force pairs. See also *law of force pairs*.

fossil fuel Remains of ancient life, used as an energy resource: coal, oil, natural gas.

four fundamental forces These are the gravitational force, electromagnetic force, strong force, and weak force. Every known force is explainable in terms of these four forces acting at the microscopic level.

four-dimensional space-time The intertwined combination of space and time described by Einstein's relativity. Gravity affects space and time in a manner best-described as the *bending of space-time*.

free fall Falling that is influenced only by gravity, and not by air resistance or other influences.

free fall, quantitative description of For an object that starts from rest and then falls freely: the object's speed is proportional to the time of fall ($s \propto t$), and the distance fallen is proportional to the square of the time ($d \propto t^2$). These proportionalities are correct for any motion that starts from rest and maintains an unchanging acceleration in a straight line.

friction The resistive force exerted by a surface on a sliding object, caused by the roughness of the two surfaces.

fundamental forces See *four fundamental forces*.

fusion See *nuclear fusion*.

fusion bomb, or hydrogen bomb (H-bomb). A bomb that gets its energy from the fusion of hydrogen, triggered by a fission bomb.

fusion reactor A nuclear reactor that gets its thermal energy from fusion rather than fission. Now under development, it might be commercially viable by the middle of the 21st century.

galaxy A large aggregation of stars. Most galaxies, such as our own *Milky Way*, have a disk-like, pancake shape, and revolve about their centers.

Galilean relativity The intuitive theory of relativity, in which time and space are absolute (in other words, different observers measure the same time intervals and the same distances), and light has different speeds relative to different reference frames.

Galileo's principle of falling Neglecting air resistance, any two objects dropped together will fall together, regardless of their weights.

gamma particle A photon of gamma radiation. A gamma particle is usually emitted during each radioactive decay process. See *radioactive decay*.

gamma radiation See *gamma ray*.

gamma ray Also called *gamma radiation*. Electromagnetic radiation that is created in atomic nuclei. Gamma rays, which have very high energy, are one type of *ionizing radiation*, because they can ionize biological molecules. See *electromagnetic spectrum* and *radioactive decay*.

gas See *states of matter*.

gas pressure The outward push caused by gas molecules hitting the walls of the gas's container.

gaseous diffusion One method of isotope separation in uranium and other materials, used during the Manhattan Project. It is based on the fact that gaseous molecules containing ^{235}U diffuse faster (because they are lighter) than gaseous molecules containing ^{238}U.

general relativity See *general theory of relativity*.

general theory of relativity Einstein's theory based on the principle of general relativity. Describes the observations of accelerated observers, while the special theory of relativity applies to non-accelerated observers. See *special theory of relativity* and *principle of general relativity*.

geological ages The ages of important events and periods in Earth's history. Some approximate ages, as determined by radioactive and other methods, are: creation of Earth = 4.6 billion years; origin of life = 3 billion years; origin of humans (hominids) = 4 million years.

geometry of the universe According to the general theory of relativity, the large-scale structure of the three-dimensional universe must have one of three possible shapes: A *closed universe* bends back on itself to form a three-dimensional space that is analogous to the two-dimensional surface of a sphere; it has a finite total volume, and will eventually stop expanding and collapse back down to zero size. An *open universe* is analogous to a two-dimensional saddle-shaped surface; it has an infinite total volume and will continue expanding forever. A *flat universe* has no overall large-scale curvature, and is analogous to a flat two-dimensional surface; it has infinite total volume and will continue expanding forever but with an expansion speed that approaches zero at long times.

geothermal energy The thermal energy of hot underground steam, water, or rock. Usually considered to be a renewable energy resource.

global warming Additional greenhouse-effect warming of Earth, caused by fossil-fuel use, deforestation, and other human activities. See *greenhouse effect*.

gluon See *strong force*.

gravitational collapse Matter falling together due to gravity.

gravitational energy One of the physical forms of energy; energy due to gravitational forces. Grav E = wt × ht. See also *energy*.

gravitational field Exists wherever an object could feel a gravitational force. Every material object creates a gravitational field throughout the space around it. See also *force field*.

gravitational force The attraction that every object has for every other object, due to the object's mass. An important example is the downward force by Earth on objects in Earth's vicinity, a force commonly known as the *weight of the object*.

graviton The gravitational field quantum, predicted but not yet observed due to the smallness of its effect.

gravity See *gravitational force, curved space,* and *law of gravity*.

gravity in general relativity See *curved space*.

Greek atom See *atom, models of*.

Greek model of the atom This model pictures the atom as a tiny indestructable object, like a small and rigid pea. See *atom, models of*.

greenhouse effect The warming created by Earth's surrounding blanket of atmospheric gases, caused by the absorption and re-radiation of infrared radiation by some atmospheric gases. See *greenhouse gas*.

greenhouse gas The atmospheric gases, mostly water vapor and carbon dioxide, that cause the greenhouse effect.

ground state An atom's quantum state of lowest possible energy. See also *excited states*.

growth rate See *exponential growth*.

H-bomb See *fusion bomb*.

half-life The half-life of a radioactive isotope is the time during which half of a macroscopic amount of the isotope will decay.

heat engine Any cyclic device that uses thermal energy to do work. Its *energy efficiency*, the fraction of its input thermal energy that is converted to work, will be higher if the *input temperature* (the temperature at which the energy enters the engine) is higher and the *exhaust temperature* (the temperature at which the unused energy is exhausted from the engine) is lower. The portion of the input energy that is not converted to work is call the *exhaust*. See also *law of heat engines*.

heating The spontaneous flow of thermal energy from a higher temperature object to a lower temperature object. See also *law of heating*.

Heisenberg's uncertainty principle. Every material particle has an inherent *uncertainty in position* (Δx) and *uncertainty in velocity* (Δv). Although either one of these uncertainties can take on any value, the two uncertainties are related through the fact that their product must approximately equal Planck's constant divided by the particle's mass.

Higgs field See *cosmic inflation*.

Higgs particle See *cosmic inflation*.

high-level nuclear waste Used reactor fuel rods, containing highly radioactive fission products.

Hiroshima A Japanese city, the first city to be fission-bombed. The second was Nagasaki. Both bombings occurred in August of 1945, near the end of World War II.

hydroelectric energy The gravitational energy of raised water. A renewable energy resource.

hydrogen bomb See *fusion bomb*.

hypothesis An educated guess; a tentative theory.

inertia A body's ability to stay at rest or to maintain an unchanging velocity. Quantitatively, a body's inertia is its degree of resistance to acceleration, also called the *mass* of the body. See also *law of inertia*.

inertial mass or *mass*. The amount of inertia possessed by an object, measured in kilograms.

inflation See *cosmic inflation*.

infrared radiation Electromagnetic radiation created by the thermal motion of molecules. See *electromagnetic spectrum*.

input temperature See *heat engine*.

instantaneous speed Average speed during a time-interval that is so short that the speed hardly changes. Speedometers measure this. Measured in meters per second (m/s).

interference See *wave interference*.

internal combustion engine A heat engine in which burning occurs within the hot gases that push directly on mechanical parts, such as a piston, to provide useful work. Typically, its energy efficiency is low, largely because of the second law.

ion Any atom having an excess or deficiency of electrons. See also *planetary model of the atom*.

ionizing radiation Any atomic or sub-atomic emissions that are capable of ionizing biological molecules. See also the specific types: *alpha rays, beta rays, gamma rays,* and *x-rays*. See also *natural radiation* and *artificial radiation*.

irreversibility The second law of thermodynamics says that most processes are *irreversibile*, i.e. that physical systems proceed spontaneously toward states of higher entropy, and will not proceed spontaneously in the reverse direction. This principle is responsible for the difference between forward and backward in time.

irreversible See *irreversibility*.

isotope A particular type of nucleus, having a particular number of protons and a particular number of neutrons. Specified by its *atomic number* (number of protons) and *mass number* (total number of protons and neutrons). A symbol like $^{14}_{6}C$ represents the isotope having atomic number 6 and mass number 14.

isotope separation In uranium, this is any process that increases the percentage of ^{235}U relative to ^{238}U. Also called *enrichment* of uranium. One method is based on *gaseous diffusion*. Weapons-grade uranium is *highly enriched* to about 90% U^{235}.

Kepler's theory The planets orbit the sun in ellipses with the sun at one focus. Agrees with Tycho Brahe's observations. Similar to Copernicus's theory, but with elliptical orbits rather than circular orbits.

kilo- (k) Prefix meaning *thousand*.

kilogram (kg) The unit of mass. The mass (or inertia) of the object known as the *standard kilogram*. Any object that has the same inertia as the standard kilogram has a mass of 1 kilogram.

kilometer (km) 1000 meters. See *meter.*

kiloton The amount of energy that would be released in the explosion of one thousand tons of TNT.

kinetic energy One of the physical forms of energy; energy due to motion. Kin E = $(1/2)ms^2$. See also *energy.*

Lamb shift A small change in the energy levels of the hydrogen atom that is caused by vacuum energy fluctuations in the space surrounding the atom.

law A single theoretical idea, often within a larger theory. Synonymous with *principle.* See *theory.*

law of conservation of energy The total energy of all the participants in any process must be unchanged throughout that process. There are no known exceptions. This law is equivalent to the first law of thermodynamics.

law of force pairs Forces always come in pairs: Whenever one body exerts a force on a second body, the second exerts a force on the first. The two forces are equal in strength but opposite in direction.

law of gravity See *Newton's law of gravity.*

law of heat engines Any cyclic process that uses thermal energy to do work must have a thermal energy exhaust. See also *second law of thermodynamics.*

law of heating Thermal energy flows spontaneously from higher to lower temperatures. See also *second law of thermodynamics.*

law of increasing entropy The total entropy of all the participants in any physical process must either increase or remain unchanged. See also *second law of thermodynamics.*

law of inertia A body that is subject to no external forces will maintain an unchanging velocity. More briefly: all bodies have inertia.

law of motion See *Newton's law of motion.*

length contraction See *relativity of space.*

light clock A clock whose time-keeping is based on the motion of a light beam.

light-year The distance that light travels in one year.

lightspeed The speed at which light moves: In metric units, it is 300,000 km/s, or 3×10^8 m/s. It is customarily represented by the symbol *c.*

limitations of Newtonian physics See *Newtonian physics, limitations of.*

line spectrum A spectrum that has only specific allowed numerical values; its name stems from the fact that when a line spectrum is viewed with a spectroscope, the spectrum appears as a series of lines on the viewing screen. See *spectrum.*

linear growth Straight-line growth; it increases by the same amount (rather than the same percentage, as in exponential growth) in each unit of time.

liquid state of matter See *states of matter.*

MACHOS See *dark matter.*

macroscopic Large enough to be seen with the unaided eye, as opposed to *microscopic,* or too small to be seen with the unaided eye.

magnetic field See *electromagnetic field.*

magnetic force See *magnetic force law.*

magnetic force law Charged objects that are moving exert and feel an additional force, called the *magnetic force,* beyond the electric force that exists when they are at rest.

magnetic poles The ends of a permanent magnet. There are two types, *north* and *south.* Like poles attract, and unlike poles repel.

Mandel's experiment. Demonstrates that a detector can cause quantum jumps even if the detector is remote from the particle being detected.

Manhattan Project The U.S. project, during World War II, to build fission bombs.

mass A body's mass is its amount of inertia. We find a body's mass by comparing its inertia with the inertia of a standard kilogram. Except at high speeds where relativistic effects become important, a body's mass is also approximately its quantity of matter. See also *inertia.*

mass-energy equivalence, or $E = mc^2$ All mass has energy, and all energy has mass. A system with m units of mass has mc^2 units of energy. A system with E units of energy has E/c^2 units of mass.

mass number The total number of protons and neutrons in a nucleus. It gets its name from the fact that the mass of any nucleus is proportional to its mass number.

materialism The philosophy that only matter is real, and that everything is determined by its impersonal workings. The early Greek atomists were *atomic materialists,* materialists who also believed that all matter is made of atoms. Newtonian physics is compatible with this philosophy.

matter Anything that has rest-mass. Matter is distinct from *electromagnetic radiation* or simply *radiation,* which has no rest-mass. Matter moves at less than lightspeed, while radiation always moves at lightspeed.

matter-antimatter annihilation The transformation of matter into high-energy radiation. This occurs when a sub-atomic particle such as an electron is brought close to its antiparticle (such as an anti-electron). See *antimatter.*

measurement See *observation.*

mechanical universe The philosophical view, accompanying Newtonian physics, that the impersonal, clocklike workings of material atoms, as prescribed by natural laws, determine everything else. Thus the universe's future behavior is entirely determined by its state at any single time, and there is no free will.

medium The substance that does the waving in any wave motion.

mega- (M) Prefix meaning *million.*

megaton The amount of energy that would be released in the explosion of one million tons of TNT.

melt-down Melting together of the fuel in a nuclear reactor accident. See *Three Mile Island* and *Chernobyl.*

metabolic rate The rate, usually measured in Cal/s, at which an animal transforms its bodily chemical energy into thermal energy and other forms of energy.

meter (m) The basic metric unit of distance. About 39 inches.

metric ton, or **tonne** 1000 kilograms, about 2200 pounds.

Michelson–Morley experiment An 1887 experiment to compare the speeds of lightbeams moving in different directions. The ether theory predicted that the lightbeams should have slightly different speeds, but the experiment detected no such difference.

micro- (μ) Prefix meaning *one-millionth.*

microscopic Too small to be seen with the unaided eye, as opposed to *macroscopic,* or large enough to be seen with the unaided eye.

microscopic interpretation of the second law Entropy increase results from the fact that microscopic disorganization is overwhelmingly likely to increase.

microscopic interpretation of warmth Temperature is associated with disorganized, or random, microscopic motions that are not visible macroscopically.

microwaves See *electronic waves* and *electromagnetic spectrum.*

Milky Way Our own galaxy. See *galaxy.*

milli- (m) Prefix meaning *one-thousandth.*

millimeter (mm) 1/1000th meter. See *meter.*

millirem 1/1000th (one-thousandth) of a rem. See *rem.*

model A theory that can be visualized. See *theory.*

models of the atom See *atom, models of.*

molecule The smallest particle of a chemical compound. See also *chemical compound.*

moon See *solar system.*

multiple transparent spheres theory An extension of the Pythagorean theory to account for "wandering" of the planets. Each planet has several transparent Earth-centered spheres, with the planet riding on the outermost sphere. See *Pythagorean theory of the universe.*

muon The muon and the tau are the other two types of electrons that have been found. The only differences between the three types of electrons are that the muon and tau are heavier than the electron, and they are unstable (they have short lifetimes).

mutation See *biological effects of radiation.*

natural motion Unassisted motion. According to Newtonian (and current) physics, any motion at constant velocity is natural motion, and falling is not natural motion.

natural radiation Ionizing radiation from natural sources. These sources include *radon gas,* cosmic rays, the ground, and internal consumption. See also *artificial radiation* and *radon gas.*

negative charge See *electrically charged object.*

net force The total, overall, force on an object. The net force due to two forces acting in the same direction is the sum of the two. The net force due to two forces acting in opposite directions is the difference between the two, and acts in the direction of the stronger force.

neutrino A particle with no charge and possibly no mass. It exerts and feels only the weak force, and gravity.

neutron An uncharged, or electrically neutral, sub-atomic particle, having about the same mass as a proton. The nuclei of most atoms contain one or more neutrons. See also *planetary model of the atom.*

neutron star A compact star a few kilometers in diameter, resembling a giant nucleus made of neutrons. It spins rapidly, sending out radio beeps and light flashes. When massive stars (10 to 20 solar masses prior to collapse) run out of fusion fuel, they explode and the remaining remnant collapses to become a neutron star.

newton (N) The unit of force. The amount of force that can accelerate a 1-kg mass by 1 m/s^2.

Newtonian physics The ideas about motion, force, and gravity developed by Isaac Newton and others around A.D. 1700.

Newtonian physics, limitations of Newtonian physics gives incorrect predictions for (1) fast-moving objects, (2) strong gravitational forces or large distances, and (3) small objects. Special relativity, general relativity, and quantum theory do give correct predictions for each of these three classes of phenomena, respectively.

Newtonian physics and democracy All humans are fundamentally equal, because all are ultimately governed by the same universal natural laws. Newtonian attitudes toward natural law pervade the strivings toward democracy of the past several centuries, and are particularly evident in the documents establishing the United States.

Newtonian worldview The philosophical notions associated with Newtonian physics, especially the mechanical universe, and the democratic ideals implied by universal natural laws. See also *Newtonian physics and democracy* and *mechanical universe.*

Newton's law of gravity Between any two objects there is an attractive force proportional to the product of the two objects' masses and proportional to the inverse of the square of the distance between them: $F \propto M_1 \times M_2 / d^2$.

Newton's law of motion An object's acceleration is proportional to the net force exerted on it by its surroundings, and proportional to the inverse of its mass. The direction of the acceleration is the same as the direction of the net force. $a \propto F/M$. In the appropriate units (m/s^2, N, kg), $a = F/M$.

nonlocal effect. An instantaneous connection between two particles that are separated in space.

nonrenewable resource A natural resource that cannot be replaced within a human lifetime. Its use usually follows a curve that begins exponentially, levels off, and declines.

nuclear energy One of the physical forms of energy; the energy resulting from the structure of a material's nuclei. See also *energy.*

nuclear energy curve A graph that shows the energies of the different nuclei, versus their mass number. The graph shows a lowest point at mass number 52 (iron), indicating that nuclear energy can be released by the fusion of lighter nuclei, and also by the fission of heavier nuclei.

nuclear fission A nuclear reaction in which a large single nucleus splits into two roughly equal smaller nuclei. Nuclear energy is released whenever a heavy nucleus is fissioned into two nuclei that are both heavier than iron.

nuclear fusion A nuclear reaction in which two nuclei combine to form a single larger nucleus. Nuclear energy is released (converted to other forms) whenever two light nuclei are fused to create a nucleus that is lighter than iron.

nuclear power reactor A device in which chain-reacting nuclei transform nuclear energy into thermal energy for electric power.

nuclear reaction Any process that alters the structure of a nucleus. The main types are *radioactive decay, nuclear fission, nuclear fusion* (see these entries also).

nuclear reactor A device that controllably transforms nuclear energy into other energy forms.

nuclear waste See *high-level nuclear waste.*

nuclear weapons proliferation The spread of nuclear weapons to more and more nations.

nucleus The tiny central core of each atom, containing protons and neutrons. Nearly all the mass of an atom resides in its nucleus. See also *planetary model of the atom.*

objectivity An experiment is objective if its outcome is uninfluenced by humans. The Newtonian worldview assumes that perfect objectivity is possible, at least in principle.

observation The fact-gathering process. A *measurement* is a quantitative observation. An *experiment* is a controlled observation.

open universe See *geometry of the universe.*

ozone The O_3 molecule. A dilute layer of ozone gas fills the upper atmosphere or *stratosphere*. Ozone absorbs, and shields biological life from, most ultraviolet radiation.

Ozone Treaty An international treaty, signed by most nations, that calls for a nearly total phase-out of CFCs and most other ozone-destroying chemicals by 2000.

particle accelerator A device to accelerate microscopic particles to high energies.

particle theory of radiation Electromagnetic radiation is created by vibrating charged particles whose energy is quantized. A *photon* is created when a charged particle executes an instantaneous jump from one allowed energy to another lower energy. Each photon has energy hf, where f is the radiation's frequency and h is *Planck's constant*, 6.6×10^{-34} joule-seconds.

passive solar energy A renewable energy resource in which solar radiation, natural air flows, energy storage, and insulation provide direct solar heating and cooling.

per In each.

periodic table A table showing all of the chemical elements in numerical order according to their atomic number, and in which elements having similar chemical properties are listed beneath each other in columns. See also *chemical element.*

perpendicular force The force, perpendicular to a solid surface, that is exerted by any solid surface on any object touching it.

photoelectric effect Light and other radiation shining onto a metal surface can eject electrons from the surface, thereby providing evidence that radiation is made of photons.

photon A particle of radiation. A photon moves at lightspeed, has zero rest-mass, and carries a *quantum* (a specific allowed amount) of energy. See also *particle theory of radiation.*

photosynthesis A chemical reaction in plants that combines atmospheric carbon dioxide and water to form carbohydrate molecules such as glucose, along with oxygen. See also *chemical reaction.*

photovoltaic electricity A renewable energy resource. Generated from solar radiation using the photoelectric effect. Two thin layers of silicon, constructed somewhat differently, cause electrons to move from one layer to the other when illuminated

with light. The device acts like a battery. Silicon is a typical *semiconductor*—a material having properties lying midway between insulators and conductors. Normally, it is difficult to make electrons flow through a semiconductor. But if some of a semiconductor's electrons are given a small amount of energy, these electrons can flow easily, as in a conductor.

physics The branch of science that studies the most general principles underlying the natural world.

Planck scale According to some theories of quantum gravity, the Planck length and time are the tiny ultimate units into which space-time itself is divided. Smaller dimensions have no physical meaning.

Planck's constant A fundamental number that appears in several of the basic principles of quantum theory. In metric units, Planck's constant is 6.6×10^{-34} joule-seconds. It is customarily represented by the symbol h.

planet To the Greeks, these were objects that looked like stars but that wandered, out of step with the stars. Today, we view planets as objects similar to Earth that orbit the sun in nearly circular orbits.

planetary model of the atom Atoms are made of negative *electrons*, positive *protons*, and uncharged *neutrons*. From one to a few hundred protons and neutrons form a tiny central *nucleus*, which the electrons orbit. See also *electron, proton, neutron, nucleus* and *atom, models of.*

point particle A particle whose force field is centered on a single point and that itself takes up no volume. All of the fundamental particles described by quantum field theory appear to be point particles.

position-entanglement experiment Demonstrates non-local connections between two entangled particles that each go through a double-slit apparatus. The statistics of the impact points of the two particles are instantaneously correlated across a distance.

positive charge See *electrically charged object.*

positron The electron's antiparticle. Identical to the electron, except that it carries a positive charge.

power When a force does work, the power developed by the force is the rate at which the force does work: power = work/time. Units: J/s, watt (W = J/s), horsepower.

powers of 10 10 raised to some positive or negative power. Powers of ten are used to express very large or very small numbers.

primary qualities See *dualism.*

principle A single theoretical idea, often within a larger theory. Synonymous with *law*. See *theory.*

principle of falling See *Galileo's principle of falling.*

principle of general relativity No experiment performed within a sealed room can tell you whether your room is accelerating in the absence of gravity, or is at rest in the presence of gravity. In other words, every accelerated observer experiences the same laws of nature.

principle of mass-energy equivalence See *mass-energy equivalence.*

principle of relativity Every nonaccelerated observer observes the same laws of nature. "Unless you look outside, you can't tell how fast you're moving."

principle of the constancy of lightspeed Light (and other electromagnetic radiation) has the same speed for all nonaccelerated observers, regardless of the motion of the light source or of the observer.

probability The probability of an event is that event's fractional number of occurrences in a long series of trials. Probabilities are useful for describing uncertain events.

probability wave A probability pattern that moves through space in a wave-like manner. According to quantum theory, every microscopic particle is represented by a probability wave.

projectile motion or *particle motion* An actual transfer of matter. The motion of a material object. See also *wave motion*.

proportional One quantity is proportional to a second quantity if the first is always multiplied by whatever the second is multiplied by. One quantity is *proportional to the square of* a second quantity if the first is always multiplied by the square of whatever the second is multiplied by. The symbol ∝ means *is proportional to*.

proportional to the inverse A quantity is proportional to the inverse of another quantity if the first is proportional to (equal to some number times) 1 divided by the second quantity.

protocells Non-living precursors of single-celled organisms.

proton A positively-charged sub-atomic particle, about 2000 times more massive than an electron. The nucleus of each atom contains one or more protons. See also *planetary model of the atom*.

pseudoscience The dogmatic belief in an appealing idea that purports to be scientific but that is not supported by scientific methods.

psi field The psi field for a particle is roughly the same as the particle's probability wave. The psi field is spread out in space, and represents the possible places where the particle might be found upon observation. More specifically, the square of the psi field's value at any particular point is the probability for finding the particle at that point. Thus, the particle is more likely to be found where the psi field has a larger magnitude, and will not be found where the psi field is zero. Quantum theory predicts a particle's psi field.

Ptolemy's theory Greek Earth-centered theory of the layout of the universe, in which the planets move in circles within circles, or loop-the-loops. A good theory that survived for 1500 years.

Pythagorean theory of the universe The earliest Greek theory of the layout of the heavens. In this theory, the stars, sun, moon, and each of five planets, circle Earth on uniformly spinning transparent Earth-centered spheres to which each is attached.

Pythagoreans The earliest scientists. A society of ancient Greeks who believed that the fundamental principles of the universe are precise mathematical ideas.

quantitative risk estimates The quantitative evaluation of human-made and natural risks, especially for the purpose of comparing different risks.

quantized A phenomenon is quantized if its range of possible values is limited to specific allowed quantities, rather than covering an entire continuous range.

quantized field A field that is treated as a real physical entity that is subject to the laws of quantum theory.

quantum A specific allowed amount of any physical property such as energy.

quantum electrodynamics The quantum field theory of the electromagnetic field. See also *quantum field theory*. According to quantum electrodynamics, when we say that a particle is electrically charged we mean that it has the ability to emit and absorb photons. Particles exert electric forces on each other in tiny quantized increments, by photon exchanges in which one particle emits a photon that is then absorbed by the other particle.

quantum field theory According to this theory, everything is made of quantized fields that obey special relativity and quantum theory. All the particles of nature are field quanta. A field's intensity represents the probability for finding the particle that is the quantum of that field.

quantum jump An instantaneous change of an entire psi field. When an atom quantum-jumps, the psi field for its electron changes to a new quantum state.

quantum model of the atom This theory of the atom represents an atom's electrons as standing waves surrounding the nucleus. The possible wave patterns, or psi field patterns, are predicted by Schroedinger's equation. This model agrees with all experiments performed to date.

quantum numbers The numbering system used to label an atom's quantum states. The states of hydrogen are labeled by three quantum numbers, usually called N, L, and M.

quantum state The quantum states of the hydrogen atom are the various possible psi fields for its electron. Each quantum state is a standing wave pattern that obeys Schroedinger's equation, and represents a possible probability pattern for the atom's electron. The electrons in more complex atoms are also described by quantum states.

quantum theory The currently-accepted physical theory of the microscopic world. Its central notion is that nature is discontinuous or *quantized* at the microscopic level, in contrast to the Newtonian assumption that nature is smooth or continuous at all levels. According to this theory, both matter and radiation occur as particles whose behavior is unpredictable but whose probabilities form predictable spatially extended psi fields, or wave patterns.

quantum theory of the atom See *atom, models of*.

quantum uncertainties Experiments show that identical physical situations often lead to different outcomes. Thus, there are irreducible uncertainties in nature itself.

quark The underlying particles from which protons and neutrons are made. See *strong force*.

radar See *electronic waves* and *electromagnetic spectrum*.

radiant energy One of the physical forms of energy. Energy carried by a lightbeam or by other electromagnetic radiations: radio, infrared, ultraviolet, x-ray, gamma-ray. See also *electromagnetic radiation* and *energy*.

radiation Any phenomenon that moves outward (or "radially") from a central source. Examples include *alpha radiation*, *beta radiation*, and *electromagnetic radiation* (see these entries also). See also *matter*.

radiant energy The energy of an electromagnetic wave.

radiation sickness See *biological effects of radiation*.

radio A device designed to receive radio or electronic waves. See *electronic waves* and *electromagnetic spectrum*.

radioactive dating Determining the ages of old objects by using radioactive methods. In *carbon dating*, radioactive ^{14}C in a dead organism is measured as a fraction of the total carbon in order to determine how long the ^{14}C has been decaying and thus when the organism died. There are many other forms of radioactive dating, based on other radioactive isotopes.

radioactive decay The spontaneous emission, by a nucleus, of an *alpha particle* (a helium nucleus that breaks off of a larger nucleus) or a *beta particle* (an electron created in the nucleus). These two processes are called *alpha decay* and *beta decay*, respectively. At the time of decay, a *gamma particle*, or high-energy gamma-ray photon, is also emitted. The streams of particles that are emitted by a macroscopic sample of radioactive material are called *alpha rays, beta rays,* and *gamma rays.*

radioactive decay sequence A sequence of radioactive decays that occurs when a radioactive isotope decays to become another isotope that is also radioactive.

radioactive fallout Dust that falls to the ground carrying radioactive isotopes from a nuclear explosion or nuclear accident.

radioactive isotope An isotope that is radioactive. An isotope that is not radioactive is said to be *stable.*

radioactive nucleus A nucleus that can undergo radioactive decay. Also called an *unstable nucleus.*

radioactivity The phenomenon of radioactive decay. See *radioactive decay.*

radio telescope Receivers that detect radio waves from space.

radio waves See *electronic waves.*

radon gas The radioactive gas emitted by Earth's rocks and soil. For most people, radon is the main source of ionizing radiation in the environment. It often collects inside buildings.

realm of possibilities The range of possible positions and speeds that a microscopic particle can have at any particular time, as determined by the particle's psi field. This range cannot be smaller than is permitted by the uncertainty principle.

reference frame The laboratory or other surroundings within which an observer makes measurements. Measurements made in a particular reference frame are said to be *relative to* that frame.

relative motion Two objects are in relative motion whenever they have different velocities.

relativity of mass An object's inertia (in other words, its mass) increases with its speed, so its mass is different for different observers.

relativity of space or *length contraction* Moving objects are contracted along their direction of motion, so an object's length is different for different observers.

relativity of time or *time dilation* The elapsed time (the number of seconds) between two particular events, such as two ticks on a particular clock or the birth and death of a person, is different for two observers who are in relative motion. The elapsed time between two ticks is longer for observers for whom the ticking clock is moving (that is, who are in motion relative to the clock) than it is for observers for whom the clock is at rest: moving clocks go slow.

relativity of velocities Both Galileo's and Einstein's theories predict that velocities are relative, but velocities "add" differently in Einstein's theory.

rem The unit of biological damage caused by ionizing radiation.

renewable resource A natural resource that is continually replaced or replaced within a human lifetime. Its use typically begins exponentially, then levels off at some sustainable level (provided that it is not over-consumed).

resistive force Any force that acts on a moving body in a direction opposite to the body's motion. *Friction* is the resistive force by a surface on an object sliding across that surface. *Air resistance* is the resistive force that air molecules exert on an object moving through the air. *Rolling resistance* is the resistive force by a surface on a rolling object.

resource See *natural resource.*

respiration A chemical reaction in animals that combines oxygen and carbon-based molecules such as glucose to generate useful energy along with carbon dioxide and water. See also *chemical reaction.*

rest-mass The mass of an object at rest. Rest-mass represents quantity of matter.

retrograde motion A temporary change in the direction that a planet moves relative to the stars, as seen from Earth.

risk estimates See *quantitative risk estimates.*

rocket drive A particular type of drive force. When material is thrown away from, or ejected from, any vehicle, the material exerts a reaction force back on the vehicle because of the law of force pairs. When this reaction force is used as a drive force, it is called rocket drive.

rolling resistance The resistive force by a surface on a rolling object.

satellite A body in orbit around a larger astronomical body. Inertia keeps satellites moving, while the gravitational force exerted by the central body holds satellites into their orbits.

Schroedinger's equation Predicts the psi field for material particles. See also *psi field.*

science The observation and theoretical understanding of the natural world. See also *scientific method.*

scientific method The dynamic interplay between experience (in the form of experiments and observations) and thought (in the form of creatively constructed theories and hypotheses that correlate exeriences). See also *science.*

second law of thermodynamics This law describes the general tendency of non-thermal forms of energy to end up as thermal energy. This law can be stated in three logically equivalent forms: *law of heating, law of heat engines,* and *law of increasing entropy* (see these entries also).

secondary qualities See *dualism.*

semiconductor See *photovoltaic electricity.*

SETI The search for extra-terrestrial intelligence. Usually conducted with *radio telescopes*: receivers that detect radio waves from space.

shape of the universe See *geometry of the universe.*

short-lifetime hypothesis Technological civilizations do not survive for long because overindulgence destroys their capacity to continue being technological. This is a possible answer to Fermi's question "Why haven't aliens visited Earth by now?"

solar radiation Electromagnetic radiation from the sun. Concentrated mainly in the *infrared, visible, and ultraviolet.* See *infrared radiation, visible light, ultraviolet radiation.*

solar system The sun and the objects that orbit the sun. It includes the nine *planets*, and the *moons* that orbit many of the planets.

solar-thermal electricity A renewable energy resource. Generated from thermal energy created by the sun.

solid state of matter See *states of matter.*

space In astronomy, *space* means the universe outside of Earth, Earth's atmosphere, and other astonomical bodies; space begins about 100 km above Earth's surface. In mathematics, a *space* is a region without edges; a space may have any number of dimensions, and may be flat or curved, infinite or finite.

special theory of relativity Einstein's theory based on the principle of relativity and the principle of the constancy of lightspeed. This theory applies only to non-accelerated observers. In this theory, time and space are not absolute, and light has the same speed in all non-accelerated reference frames.

spectroscope A device that measures the *spectrum,* or set of frequencies, emitted by a source of electromagnetic radiation. Spectroscopes operate by spreading a spectrum out on a viewing screen.

spectrum The range of possible values. A *continuous spectrum* is one that extends over a continuous range of values. A *line spectrum* is one that has only specific allowed numerical values; its name stems from the fact that when a line spectrum is viewed with a spectroscope, the spectrum appears as a series of lines on the viewing screen. Also see *electromagnetic spectrum,* to which *spectrum* usually refers.

speed See *instantaneous speed* and *average speed.*

stable isotope An isotope that is not radioactive; a nucleus that will not spontaneously fall apart.

stable nucleus A nucleus that is not radioactive; a nucleus that will not spontaneously fall apart.

standard model The theory of the electroweak and strong force. See *electroweak force* and *strong force.*

standing wave A wave in which the medium vibrates in a wave pattern, but the pattern doesn't move.

star birth Stars form from collapsing gas clouds. Gravitational collapse heats the gas, which initiates nuclear fusion in the center, which stops the collapse.

states of matter Nearly every substance can exist in any of the three states known as *solid, liquid,* and *gas.* A solid's molecules are locked closely together in a regular pattern, a liquid's molecules are close together but not fixed in position, and a gas's molecules are far apart and move around rapidly.

statistical correlation Two uncertain events are statistically correlated if the outcome of one event affects the probabilities of the other.

steam-electric power plant Uses thermal energy from an external source such as burning coal, to turn water into steam that pushes on a *steam turbine* that provides work to generate electricity. The device that converts the rotational motion of the turbine into electricity is called an *electric generator.* At the back end of the turbine, the *condenser* converts the steam back to water, using a cooling lake or river or an evaporative *cooling tower.* Since the steam is heated by fuel that burns outside of the boiler, the plant is an *external combustion engine.* Typically, its energy efficiency is less than 50%, mostly because of the second law.

steam turbine See *steam-electric power plant.*

strong force One of the four fundamental forces. The force that holds the nucleus together. It acts between nuclear particles (protons, neutrons), is strongly attractive at separations of around 10^{-15} m, and negligible at larger distances. Its field quanta (or exchange particles) are *gluons*, and the particles that exert and feel this force are *quarks*. Two kinds of quarks, u and d, are important in ordinary matter. Protons are made of u-u-d, and neutrons are made of u-d-d. Quarks are fractionally charged, and are not found in isolation because any attempt to isolate them creates more quarks. There are two other (unstable) generations of quarks, so there are six kinds of quarks in all.

supernova explosion The explosion of a giant star. Supernova explosions spread the chemical elements into space, and so are the source of the elements heavier than helium in our solar system. Furthermore, the elements heavier than iron are created by fusion during supernova explosions, using the energy of the explosion to provide the energy to fuse these heavy elements. See also *neutron star* and *black hole.*

tau The muon and the tau are the other two types of electrons that have been found. The only differences between the three types of electrons are that the muon and tau are heavier than the electron, and they are unstable (they have short lifetimes).

technological imperative The tendency to build whatever technology is possible. See also *technological momentum.*

technological life Any form of life having a sufficiently well-developed technology to be capable of sending out radio signals.

technological momentum The tendency to continue a technological project once it is started. See also *technological imperative.*

technology assessment The process of clarifying the scientific and technological issues, and objectively listing the costs and benefits of the available options, as a prelude to making a technology-related decision.

television See *electronic waves* and *electromagnetic spectrum.*

temperature A quantitative measure of warmth. The units normally used to measure temperature are called *degrees Celsius.* (°C)*;* the freezing point of water is 0°C, and the boiling point is 100°C Any device that measures temperature is called a *thermometer.*

the second law and biology See *biology and the second law.*

theory An idea, or several related ideas, that explains or unifies a group of observations. A *model* is a theory that can be visualized. A *principle* or *law* is a single idea, often within a larger theory.

theory of relativity Any theory that provides answers to questions about observers who are in relative motion. See *Galilean relativity* and *Einstein's relativity.*

thermal energy One of the physical forms of energy; energy due to temperature. Equivalently, thermal energy is microscopic energy, the kinetic (and other) energy of molecules that cannot be observed macroscopically. See also *energy.*

thermodynamics The study of the general properties of energy. Thermal energy plays a central role in understanding these properties. See also *law of conservation of energy* and *second law of thermodynamics.*

thermometer Any device that measures temperature. See also *temperature.*

thermonuclear reaction A fusion reaction that creates the thermal energy needed to sustain itself.

Three Mile Island Site of the most significant U.S. nuclear power plant accident. Little radioactivity escaped even though the fuel suffered a *melt-down* in which the fuel melted together into a solid radioactive mass that slumped downward inside the reactor.

three states of matter See *states of matter.*

time Time is defined by clocks, in other words by the operations we go through to measure time. The lightclock, based on the motion of lightbeams, is one simple instrument to define time.

time dilation See *relativity of time.*

time travel An observer who accelerates to a high speed and then returns to the initial reference frame experiences a shorter elapsed time than an observer who remains in the initial frame. Thus objects in the initial reference frame have aged more than the traveler. This effect makes it possible to travel to stars that are many light-years distant, in only a few years' travel time. It also makes travel to the future possible, by going on a fast trip and returning.

trace gases Gases that form only a minute fraction of the atmosphere, far less than one percent in all. Includes ozone, carbon dioxide, water vapor, and many other chemical compounds.

transportation efficiency Useful output (such as distance traveled, passengers moved, freight moved, or mass moved) per unit of fuel input.

turbine See *steam-electric power plant.*

Tycho Brahe's observations Tycho Brahe gathered highly accurate data on planetary positions. This data disproved both Ptolemy's and Copernicus's theories.

UFO Unidentified object in the sky. Two popular beliefs about UFOs are: (1) UFOs are visitations by aliens today; and (2) aliens visited Earth within the past few thousand years. There is no evidence to support either belief; scientists overwhelmingly reject them as pseudoscientific and false.

UV See *ultraviolet radiation.*

ultraviolet radiation Electromagnetic radiation that is created by higher-energy motions of electrons in atoms. Has sufficient energy to damage biological molecules, but not enough to ionize them. Can cause mutations and cancers. See *electromagnetic spectrum.*

uncertainty in position (Δx) Refers to a particle whose position is uncertain. The quantitative extent of the spatial range within which such a particle's position might lie.

uncertainty in speed (Δs) Refers to a particle whose speed is uncertain. The quantitative extent of the range within which such a particle's speed might lie.

uniform circular motion Motion in a circle at an unchanging or *uniform* speed.

unstable nucleus A radioactive nucleus; a nucleus that will eventually and spontaneously fall apart.

vacuum A region that contains no atoms or other material particles. According to quantum field theory, fields exist even in vacuum. Since these fields are quantized, there is some probability that field quanta such as photons, or particle-antiparticle pairs, will pop into and out of existence, even in vacuum. Furthermore, quantum uncertainties allow the energy present at any point in vacuum to undergo random *energy fluctuations* around its long-term average value.

velocity The combined instantaneous speed and instantaneous direction of motion.

visible light Electromagnetic radiation that is detectable by the human eye. Created by lower-energy motions of electrons in atoms. Colors are due to different wavelengths, ranging from red (longest) to violet (shortest). See *electromagnetic spectrum.*

wave frequency The number of vibrations that any part of the medium completes in each second. Also the number of complete wavelengths sent out in each second. Higher frequency waves have shorter wavelengths and (assuming equal amplitudes) higher energies. Unit: vib/s = hertz (Hz).

wave interference The effects that occur when two waves of the same type are present at the same time and place. Interference is *constructive* or *destructive* depending on whether the two waves reinforce or cancel.

wavelength The distance from any point to the next similar point along a continuous wave.

wave motion A disturbance that travels through a *medium* and that transfers energy without transferring matter. See also *projectile motion.*

wave packet. A psi field moving through space and spread out over only a limited uncertainty in position.

wave-particle duality of radiation and matter Both radiation and matter have both wave-like (that is, field-like) and particle-like aspects.

wavespeed The speed at which a disturbance (a wave) moves through a medium.

wave theory of matter Every material particle is accompanied by a wave having wavelength h/mv, where m is the particle's mass and v is its speed.

weight The weight of an object means the net gravitational force exerted on it by all other objects. On Earth, an object's weight means the gravitational pull exerted on it by Earth.

weightlessness Bodies are weightless only when they are far from all other bodies. Bodies in orbit around Earth are not weightless, but they seem weightless because they are falling freely around Earth.

white dwarf A planet-size, compact star. When stars of about the sun's mass run out of fuel, they flare up and then collapse to become white dwarfs.

wind energy A renewable energy resource. The kinetic energy of moving air.

work Object A does work on object B if A exerts a force on B while B moves. Unit: newton-meter (N·m) = joule (J). Quantitatively, the work done on an object by a force equals the force (in N) multiplied by the distance (in m) through which the object moves: Work = Fd

work-energy principle Work is an energy transfer. The amount of energy transferred when one system does work on another system is equal to the amount of work done.

X-rays Electromagnetic radiation that is created by the highest-energy motions of electrons in atoms. X-rays are one type of *ionizing radiation*, because they can ionize biological molecules. See *electromagnetic spectrum.*

Photo credits

All cartoons © Sidney Harris, used with permission.

Figure	Photographer/Source
PO1	NASA Headquarters
1.1	UPI/Corbis-Bettmann
1.2a	NASA/Johnson Space Center
1.2b	European Southern Observatory
1.2c	NASA Headquarters
1.4	National Optical Astronomy Observatories
1.7	NASA Headquarters
1.12	Courtesy of New York Public Library Still Pictures Collection
1.14	American Institute of Physics/Emilio Segre Visual Archives
1.18	Courtesy of New York Public Library Still Pictures Collection
1.19	American Institute of Physics/Emilio Segre Visual Archives
1.20	Corbis-Bettmann/From *The World of Copernicus*, p. 135
1.21	American Institute of Physics/Emilio Segre Visual Archives
1.27	NASA Headquarters
1.28	NASA Headquarters
1.29	M. Seldner, B. L. Siebers, E. J. Groth, and P. J. E. Peebles, *Astronomical Journal* 82, 249, 1977/Princeton University Department of Physics
2.8	Geoscience
2.10	National Oceanic and Atmospheric Administration
2.12	University of Chicago Archives
2.14	Courtesy of International Business Machines Corporation. Unauthorized use not permitted.
2.15	Courtesy of International Business Machines Corporation. Unauthorized use not permitted.
2.17	Corbis-Bettmann
PO2	Gregory G. Dimijian/Photo Researchers, Inc.
3.1	Louvre, Dept. des Antiquites Grecques/Romaines, Paris, France. © Photograph by Erich Lessing/Art Resource, NY.
3.4	Editorial Photocolor Archives
3.6	From *PSSC Physics* by Haber-Schaim, Dodge, Gardner, Shore. Dubuque, Iowa: Kendall/Hunt Publishing Co., © 1991.
3.12	From *PSSC Physics* by Haber-Schaim, Dodge, Gardner, Shore. Dubuque, Iowa: Kendall/Hunt Publishing Co., © 1991.
4.1	Newton's Mathematical Principles. Courtesy AIP Emilio Segre Visual Archives
4.6	National Institute of Standards and Technology
5.9a	NASA Headquarters
5.9b & c	NASA/Johnson Space Center
5.12	NASA Headquarters
5.14	© W. K. Hartman/Planetary Science Institute
5.15	National Optical Astronomy Observatories
5.16	Julian Baum/New Scientist/Science Photo Library/Photo Researchers, Inc.
PO3	Martin Dohm/Science Photo Library/Photo Researchers, Inc.
6.4	*Physics Today* Collection, AIP Emilio Segre Visual Archives
7.13	© 1988 General Motors Corp. Used with permission of GM Media Archives.
7.14	Toyota Motor Corporation Services
7.15	Honda of America Mfg., Inc.
7.16	Honda of America Mfg., Inc.
7.18	© 1988 General Motors Corp. Used with permission of GM Media Archives.
7.21	Grapes-Michaud/Photo Researchers, Inc.
8.1	Martin Dohm/Science Photo Library/Photo Researchers, Inc.
8.2	From *PSSC Physics* by Haber-Schaim, Dodge, Gardner, Shore. Dubuque, Iowa: Kendall/Hunt Publishing Co., © 1991.
8.12	From *PSSC Physics* by Haber-Schaim, Dodge, Gardner, Shore. Dubuque, owa: Kendall/Hunt Publishing Co., © 1991.
8.13	From *PSSC Physics* by Haber-Schaim, Dodge, Gardner, Shore. Dubuque, Iowa: Kendall/Hunt Publishing Co., © 1991.
8.17	Art Hobson
8.20	From *PSSC Physics* by Haber-Schaim, Dodge, Gardner, Shore. Dubuque, Iowa: Kendall/Hunt Publishing Co., © 1991.
8.21	From *PSSC Physics* by Haber-Schaim, Dodge, Gardner, Shore. Dubuque, Iowa: Kendall/Hunt Publishing Co., © 1991.
8.23	Gordon R. Gore
8.25	American Institute of Physics/Emilio Segre Visual Archives
8.26	C. E. Wynn-Williams/ American Institute of Physics/Emilio Segre Visual Archives
9.4	American Institute of Physics/Emilio Segre Visual Archives
9.13	The Nobel Foundation
9.14	Susan Solomon
9.21	Graph provided by David Parker and Matthew O'Donnell of the Hadley Centre for Climate Prediction and Research at the U.K. Meteorological Office.
PO4	University of Pennsylvania, Van Pelt Library
10.1	UPI/Corbis-Bettmann
10.2	University of New Hampshire Lotte Jacobi Archives

11.11	UPI/M. Manni/Corbis-Bettmann
12.1	NASA Headquarters
12.2	Douglas Faulkner/Photo Researchers, Inc.
12.3	Sidney Fox/Science VU/Visuals Unlimited
12.4	Michael Abbey/Science Source/Photo Researchers, Inc.
12.5	NASA Headquarters
12.6	NASA Goddard Space Flight Center
12.7	NASA Headquarters
13.3	American Institute of Physics/Emilio Segre Visual Archives
13.7	Dr. Albert Rose/Plenum Press with permission
13.8	AIP Meggers Gallery of Nobel Laureates
13.14	Dr. Claus Johnson
13.15	Akira Tonomura/Hitachi, Ltd.
13.18	AIP Emilio Segre Visual Archives. Gift of Jost Lemmerich.
13.20	Corbis-Bettmann
14.2	Sargent-Welch Scientific Company
14.5	Educational Development Center, Inc.
14.13	American Institute of Physics/Emilio Segre Visual Archives
14.18a	Mark Edwards/Still Pictures
14.18b	Courtesy of CERN
14.18c	Courtesy of Lawrence Berkeley Laboratory
14.18d	Courtesy of Alain Aspect
14.24	AIP Emilio Segre Visual Archives. Margrethe Bohr Collection
14.26	AIP Emilio Segre Visual Archives. Margrethe Bohr Collection
PO5	Cern/Science Photo Library/Photo Researchers, Inc.
15.3	AIP Emilio Segre Visual Archives. W. F. Meggers Collection
15.13	Masao Tsuzuki and Gon'ichi Kimura/AP/Wide World Photos
16.5	NASA Goddard Institute for Space Studies
16.6	NOAO/Science Photo Library/Photo Researchers, Inc.
16.7	Société Francaise de Physique, Paris. Courtesy AIP Emilio Segre Visual Archives.
16.8	Photograph by Lotte Meitner-Graf, courtesy of Egon Weiss. Leo Szilard Biography
16.9	AIP Emilio Segre Visual Archives. Goudsmit Collection.
16.10	Herzfeld Collection/American Institute of Physics/Emilio Segre Visual Archives
16.16	Chicago Historical Society
16.17	Martin Marietta Energy Systems, Inc.
16.18a	Courtesy of Lawrence Berkeley National Laboratory
16.18b	Photo by Bainbridge/Bainbridge Collection. Courtesy AIP Emilio Segre Visual Archives.
16.21	National Archives at College Park (NARA)
16.22	National Archives at College Park (NARA)
17.7	Catherine Pouedras/Science Photo Library/Photo Researchers, Inc.
17.8	Rollin R. Geppert
17.11	National Renewable Energy Lab
17.12	Solar Energy Research Institute
17.13	National Renewable Energy Lab
17.15	National Renewable Energy Lab
17.20	Osram Sylvania, Inc.
18.1a	California Institute of Technology, Palomar Observatory
18.1b	Courtesy of Julian Schwinger
18.4	Max Planck Institute. Courtesy AIP Emilio Segre Visual Archives
18.5	Courtesy of Lawrence Berkeley National Laboratory
18.7a	Courtesy of Lawrence Berkeley National Laboratory
18.7b	Fermilab Visual Media Services
18.8a	University of California at Berkeley Lawrence Radiation Laboratory
18.8b	Lawrence Radiation Laboratory. Courtesy AIP Emilio Segre Visual Archives
18.8c	Courtesy of Lawrence Berkeley National Laboratory
18.8d	Stanford Linear Accelerator Center/U.S. Department of Energy
18.8e	Fermi National Accelerator Laboratory/Fermilab Visual Media Services
18.9	International Centre for Theoretical Physics
18.10	American Institute of Physics/Emilio Segre Visual Archives
18.12	AIP Emilio Segre Visual Archives, Marshak Collection
18.13a	Henry Kendall
18.13b	Donna Coveney/Massachusetts Institute of Technology
18.13c	American Institute of Physics/Meggers Gallery of Nobel Laureates
18.14	Stanford Linear Accelerator Center/U.S. Department of Energy
18.17	Photo by Ulli Steltzer. Courtesy AIP Emilio Segre Visual Archives
18.19	Donna Coveney/Massachusetts Institute of Technology
Epilogue	Alison Saar, "Tree Souls" 1994, wood and copper. Location: Denver Art Museum Sculpture Garden. Photo by William O'Connor. Denver Art Museum Exhibition: Landsacpe as Metaphor: Visions of America in the Late Twentieth Century.

Index

QUICK REFERENCE LIST: HOW THINGS WORK